CIRCUIT
THEORY
with computer methods

CIRCUIT
THEORY
with computer methods

OMAR WING

Professor of Electrical Engineering
Columbia University

HOLT, RINEHART AND WINSTON, INC.

NEW YORK   CHICAGO   SAN FRANCISCO   ATLANTA
DALLAS   MONTREAL   TORONTO   LONDON   SYDNEY

To Camella

PREFACE

This book is about electrical circuits. It tells what they are, how to describe and analyze them, and what we can do with them. The style of presentation is informal. When mathematical rigor is called for, the presentation is unhurried. Systematic procedures for solving problems are emphasized. This algorithmic approach leads naturally to computer solutions. However, "pencil-and-paper" techniques are given equal prominence, especially in a first course on circuits, for the reason that students should first learn to solve problems by simple reasoning rather than by the unreasoned use of the computer.

The components of this book are a presentation of basic theory, a logical development of analysis techniques, and a repertoire of computer programs. Basic theory covers those properties of circuits that are immediately deducible from first principles. These properties form the formal mathematical basis for writing equations describing the behavior of circuits. Analysis techniques include the formal solutions of circuits in the frequency and time domains as well as solutions "by inspection," which are useful in practice. The computer programs range from one for finding a tree in a graph to one for the frequency analysis of a general linear circuit. They are written in FORTRAN IV and are completely documented.

Other computer programs will be written by the student as exercises. These programs are simple and limited in application, since it is felt that the student should devote his time to learning about circuits rather than to the debugging of his programs.

To motivate the student and to show him what circuits can do, examples of circuits for use in communication systems and digital computers are given throughout the book. Thus, as early as in Chapter 2, the idea that a circuit can be used to extract a wanted signal in the presence of unwanted noise is presented. In Chapter 5, it is shown how diodes can be connected together to perform simple logic functions. In Chapters 4 and 8, circuits that are used in smoothing or delaying a signal are illustrated.

To encourage students "to do their own thing" and to get some appreciation of realistic circuits, a number of experiments are suggested at the end of each chapter. The purpose and general guidelines of each experiment are given, the details being left to the instructor.

Problems are also given at the end of each chapter and are of three types: (1) exercises to give experience and confidence in analyzing circuits, (2) simple design problems to provide a taste of engineering practice, and (3) "fun" problems, which may be difficult, but are interesting. Those in the last group are mostly for recreation, but some are designed to tempt the student to study circuit theory further.

There are ten chapters. Chapter 1 presents the Kirchhoff voltage and current laws. The all-important question of the maximal number of independent voltage and current equations is resolved here by using the concept of trees, chords, cut

v

sets, and fundamental loops. Multiterminal elements as well as two-terminal elements are treated. The physical basis of the voltage and current laws as far as circuits are concerned are discussed with care, with particular attention as to what is observable (conduction currents, potential differences) and what is not (displacement currents, magnetic flux). A computer program to find a tree and its associated chords is given.

Electrical elements are introduced in Chapter 2. The distinction between a physical element and a mathematical element is clearly pointed out. Examples are presented to illustrate that circuits can be used as signal processors such as filters and amplifiers. Linear and nonlinear elements are included as are dependent sources. Circuit analysis based on the direct use of trees, chords, fundamental cut sets, and loops is developed here.

In Chapter 3, the principle of superposition is introduced. Thévenin and Norton equivalent circuits are presented, as are the series-parallel reduction and the star-mesh transformation. Emphasis in this chapter is on analysis by repeated application of the principle of superposition and the reduction techniques of equivalent circuits. The ladder network is introduced, and its input resistance is related to a continued-fraction expansion. A computer program to evaluate a continued fraction and to calculate the voltages and currents on a ladder is given.

First- and second-order circuits are introduced and analyzed in detail in Chapter 4. Formulation of the differential equations is in terms of the state variables. The complete solution is carefully developed and is shown to be the sum of a homogeneous solution and a convolution integral of this solution with the excitation, the latter obtained by the method of variational parameters. Examples of such practical circuits as the sawtooth generator, the compensated attenuator, and a smoothing filter are given. The fourth-order Runge-Kutta algorithm for the numerical solution of differential equations is presented, together with the computer program to implement the algorithm for both the first-order and second-order cases. Error analysis of numerical integration of differential equations is discussed briefly. A computer routine to plot one array of numbers versus another array, on a computer printout, is included.

Chapter 5 deals with simple, nonlinear circuits. Since there are no general analysis techniques for them, case studies of simple, practical, nonlinear circuits are given. Diode circuits are presented as rectifiers, limiters, and AND-OR gates. The Newton-Raphson algorithm to solve nonlinear resistor circuits by iteration is described. Nonlinear first-order circuits are studied, and their solution by numerical integration is illustrated.

Chapter 6 has to do with the formal methods of writing circuit equations. The general case of a resistor circuit containing both voltage and current sources is treated first using the concept of a proper tree. The formulation leads to tree and chord analysis (hybrid), and the solution algorithm has been implemented in a computer program named RNET. Loop, mesh, and node equations follow. In circuits containing inductors and capacitors, these equations are shown to be "unnatural" for solution in the time domain in that they do not describe the state of the circuits.

The sinusoidal steady state is introduced in Chapter 7, and reasons for treating it are carefully given. In particular, it is shown that in the case of a periodic signal, it can be represented as a sum of sine and cosine components on a discrete frequency spectrum by means of the Fourier series. A nonperiodic signal can be similarly resolved into sine and cosine components on a continuous spectrum. The use of exponential excitation and complex numbers to simplify the algebraic manipulations of circuit equations is introduced. Examples of low-pass filters, band-pass filters, and impedance bridges are used to illustrate the loop, mesh, and node equations.

In Chapter 8, the frequency characteristics of circuits are examined in detail. The concept of poles and zeros of network functions is introduced, and it is shown that they are related to the natural frequencies of a circuit. Frequency response of such practical circuits as filters, lumped delay lines, and amplifiers is presented. The ladder analysis program of Chapter 3 is modified to calculate the voltages and currents on a ladder in the sinusoidal steady state. By modifying the program RNET, the program DZNET is produced. DZNET computes the frequency response of a linear circuit containing resistors, inductors, capacitors, mutually coupled coils, and dependent sources.

Chapter 9 deals with the analysis of circuits in the time domain. The methods of Chapter 4 are extended to the nth-order case. State equations are formulated systematically for any linear circuits, including those with capacitor loops, inductor cut sets, or dependent sources. Their complete solution is given in terms of eigenvalues, eigenvectors, and the characteristic matrix. Examples of a complete solution in a second-, third-, and fourth-order circuit are given. Analysis of circuits with repeated eigenvalues is included. Numerical integration of state equations by the Runge–Kutta algorithm is presented. Lastly, a novel numerical solution scheme without the use of state equations is described. The scheme is simple to understand and is based on the use of implicit integration of the terminal equations of the inductors and capacitors in a circuit. It has the distinctive advantage in that the complicated topological procedure for setting up the state equations is completely avoided.

Finally, Chapter 10 treats the topic of multiterminal circuits, with emphasis placed on analysis by the use of indefinite admittance matrices. Hybrid and $ABCD$ parameters are introduced, and the utility of the latter in the analysis of filter structures is illustrated by a computer program to calculate the frequency response of a fifth-order elliptic low-pass filter. Other practical examples of multiterminal circuits, such as the operational amplifier and the RC bridged-T oscillator, are also given.

It is seen that the book thus contains all of the essential topics in circuit analysis. It can be said that a student who studies it from cover to cover can be expected to acquire the skill necessary to analyze any circuit in the frequency and time domains. He should also gain some appreciation as to what circuits can do and, lastly, he will acquire experience in using the computer to solve circuit problems.

This book is intended to be a first course on circuit analysis given at the sophomore or junior level. Its contents are organized so as to suit the various course structures at different universities and colleges. Difficult topics are marked by an asterisk after the section heading and can be omitted on the first reading. Some possible arrangements of course outline are given below.

Format I: A two-semester, three-hour per week course set at a leisurely pace. **Outline:** First semester – fundamentals (Chapters 1, 2, 3, and 6) and the sinusoidal steady state (Chapter 7). Second semester – time-domain analysis (Chap-

ters 4 and 9), nonlinear circuits (Chapter 5), frequency-domain analysis (Chapter 8), and multiterminal circuits (Chapter 10).
Prerequisites: First semester – none; second semester – differential equations and some linear algebra.

Format II: A one-semester, four-hour per week course set at a brisk pace.
Outline: Fundamentals, sinusoidal steady state, frequency-domain analysis, time-domain analysis, and nonlinear circuits.
Prerequisites: Differential equations and some linear algebra.

Format III: A one-semester, three-hour per week course, briskly paced.
Outline: Fundamentals, sinusoidal steady state, time-domain analysis, and frequency-domain analysis.
Prerequisites: Differential equations and some linear algebra.

Format IV: A two-quarter, four-hour per week course set at medium speed.
Outline and Prerequisites: Same as I.

Format V: A two-quarter, three-hour per week course, briskly paced.
Outline: Same as II.
Prerequisites: First quarter – none; second quarter – differential equations and some linear algebra.

Solutions to problems are given in the Instructor's Manual, available from the publisher to teachers who have adopted the book for class use. The computer programs listed in this book are also available on magnetic tape from the same source.

In preparation of this book, I was given much encouragement by my colleagues at Columbia University, among them, Professor Jacob Millman and Professor Ralph J. Schwarz. I am indebted to Professor Basil R. Myers of Notre Dame University for his complete and careful reading of the manuscript. His constructive criticisms, together with those of Professor Robert W. Newcomb of the University of Maryland, had resulted in many significant improvements in the style of presentation and choice of topics. I am grateful to my former students who accepted the early drafts of the manuscript as their text, and who pointed out many of the errors. For their assistance in preparing the problem solutions and computer programs, my thanks go to many of my former teaching assistants and doctoral students. For a first-rate typing job well-done, I am much obliged to the secretarial staff, in particular the late Mrs. Jean Chapman, of the Department of Electrical Engineering and Computer Science. Finally, I wish to thank my children, David and Jeannette, for their rough drafts of some of the figures, and my wife, for her constant encouragement, without whose support this book would not have been possible.

Pomona, New York   OMAR WING
February 1971

CONTENTS

ix

3

6

7

8

9

10

introduction

Electrical engineering is concerned with the design, manufacture, and sales of systems for the generation, transmission, processing, control, and distribution of energy and information in electrical form. It is a broad discipline and is ever-expanding, as it transforms the new discoveries in practically all branches of science, physics, mathematics, chemistry, and metallurgy into useful products and systems. It is not possible to describe all that an electrical engineer does. Perhaps an appreciation of the range of his activities can be gained by taking a brief look at the electrical and electronic industry, a synopsis of which is given in Table 1.

TABLE 1 A synopsis of the electrical and electronic industry.

Communications	Computer
Telephone, telegraph, telemetry; deep space, underwater, satellite; computer data, information retrieval; radio, television broadcast; radio astronomy	Data processing, scientific computation, process control, flight guidance control, traffic control; medical diagnosis; typesetting, editing; simulation
Power Systems	**Instrumentation**
Consumer power, industrial power, portable power; generation, transmission, distribution; energy conversion devices and systems; electroautomotive systems; magnetohydrodynamics, solar power, nuclear power, geothermal power	Physical, chemical, biological, physiological, mechanical, thermal, geological, oceanographical, optical, and environmental measurements; meters, oscilloscopes, bridges, spectrum analyzers; pollution control, weather forecasting
Defense	**Entertainment**
Missile guidance, flight control, radar, sonar, submarine defense, missile defense, satellite surveillance, fire power control, "secured" intelligence, invulnerable communication system	AM, FM broadcast, television, phonograph, tape recorder, home video recording, educational television, cable television, computer-aided instruction
Components	**Devices**
Amplifiers, filters, equalizers, delay lines, wave-shaping networks, integrated circuits, logic circuits, analog-digital converters, magnetic core memory, tape and disk memory; antennas, waveguides, coaxial cables; motors, generators; loudspeakers, microphones; lasers and masers	Transistors, diodes, resistors, capacitors, inductors, transformers, vacuum tubes, magnetic cores, crystals, semiconductor material, relays, switches, circuit breakers, photocells, lamps, connectors, sockets

From the table, one can discern that the principal and creative aspect of electrical engineering is the design of systems. But in order to design, one must understand the properties of the components that make up a system. He must be well versed in the mathematical description of the components and the analytical techniques necessary to predict and evaluate the performance of the system. In short, he must be able to analyze his design.

A *circuit* is an interconnected collection of elements or devices: resistors, capacitors, inductors, transistors, diodes, transformers, and so forth, and circuits are the principal components of an electrical or electronic system. The radio receiver is a simple system consisting of circuits which are the RF (radio-frequency) amplifier, local oscillator, mixer, IF (intermediate-frequency) amplifiers, detector, audio amplifiers, and power supply. The digital computer is a complex system, which consists of thousands of circuits performing the functions of addition and multiplication of numbers represented by voltage or current pulses. The radar system is another complex system. It sends a signal out to the target, receives its echo, and by examining the nature of the returned signal determines the position, velocity, acceleration, and the probable course of the target. All these functions are performed by circuits. The telephone system is another example of a system which is made up of a large number of circuits. The circuits perform the function of transmitting a signal from one point to another, possibly thousands of miles away, with a high degree of fidelity. The electric power distribution system of a local utility company can be regarded as a large circuit. It consists of the generators, the power lines, and the appliances in the homes of the customers, often millions in number.

Circuit design is the art of creating a configuration of elements of the proper kinds to meet some performance specifications. We can create circuits to perform arithmetic operations of time functions (signals), to differentiate and to integrate a time function, and to solve algebraic and differential equations. We can create circuits to extract a wanted signal among a multitude of unwanted signals. We can create circuits to detect and analyze brain waves, seismic waves, and radio waves from the stars. We can also create circuits to measure the stress and strain on structures, to count atomic particles, to guide a rocket to the moon, to control traffic, and to simulate a mechanical, a fluid flow, or a thermal system. Indeed, the list is endless. In Figures 1 through 9, on the following pages, examples of circuits are shown.

That circuits are so versatile is due in part to the availability of the large variety of elements having wide ranges of values (zero to $10^{\pm 12}$), to the unlimited number of configurations that can be created, to the small physical size (as small as 10^{-6} in.), to the extremely fast response time (as fast as the speed of light) of many elements, and in no small measure to the fact that circuits are amenable to almost exact mathematical modeling. The last point is important, for it means that the behavior of circuits can be predicted mathematically with precision. It also means that circuits can be conveniently classified according to their peculiar mathematical properties and that their theoretical limitations and characteristics can be deduced and studied. We are further aided by the fact that the physical laws governing the behavior of circuits are few in number and simple in principle. The consequence is that the mathematical description of circuits is, as we shall see, also simple.

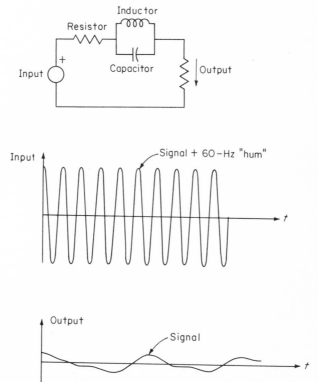

Figure 1 Illustration of a circuit, which is used to remove an unwanted interference. A weak signal (seismic wave, electrocardiograph, and so on) is buried in a strong 60-Hz "hum." The composite waveform is applied to the circuit shown. The interference is removed by the circuit, and the signal is recovered at the output. Computer solution was used to generate the input and output waveforms.

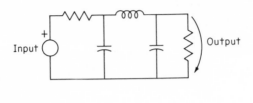

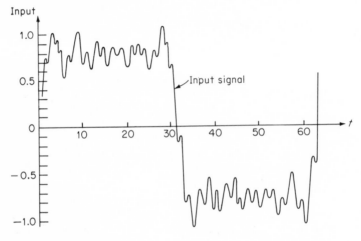

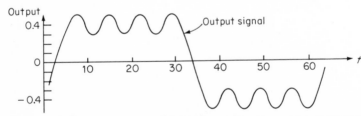

Figure 2 Illustration of a circuit, which is used to remove high-frequency noise. The input signal (submarine echo, radar pulse, computer data, and so on) is superimposed on high-frequency noise. The input is applied to the low-pass filter shown. The high-frequency noise is removed by the filter. The analysis was done on a computer.

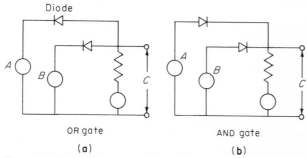

Diode

OR gate

AND gate

(a)

(b)

Figure 3 **(a)** The output C is low, if input A or input B is low, or if both A and B are low. Otherwise the output is high. The circuit realizes the logical OR function. **(b)** The output C is low if and only if both A and B are low. Otherwise the output is high. The circuit realizes the logical AND function.

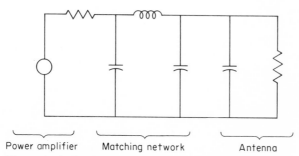

Power amplifier   Matching network   Antenna

Figure 4 A circuit as a matching network. The power delivered to the antenna is substantially increased if a matching network is inserted between it and the power amplifier in a broadcast or communication system.

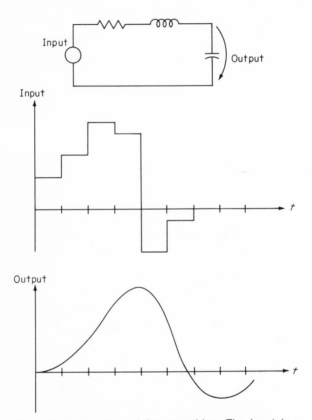

Figure 5 A circuit used for smoothing. The input is a staircase voltage, which may represent a voice signal in a digital communication system. The voltage is smoothed by the circuit, and the output is essentially the original voice. The solution was obtained by numerical integration of a system of differential equations on the computer.

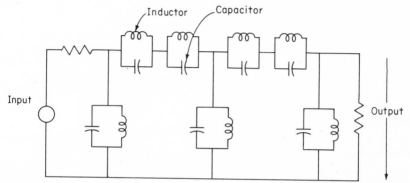

Figure 6 A realistic band-pass filter used in a communication system. The input is a signal having frequency components over a wide spectrum. It is "filtered" by the circuit, and only those components over a desired, narrow range are selected. Analysis of the circuit is greatly facilitated by the use of computer.

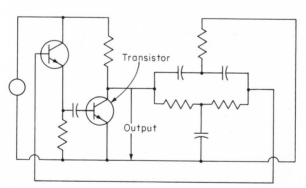

Figure 7 A transistor oscillator. The circuit generates a steady, sinusoidal voltage at the output.

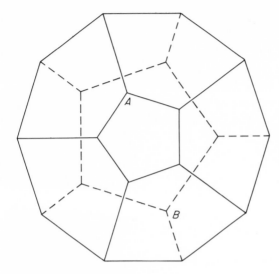

Figure 8 Each side of the pentagonal dodecahedron is a 1-Ω resistor. Can you show that the resistance between the two opposite vertices A and B is $\frac{7}{6}\,\Omega$?

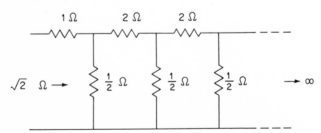

Figure 9 A rational approximation of an irrational number by means of continued-fraction expansion, which can be theoretically realized by a circuit of infinite length as shown.

This book is concerned with the analysis of circuits. Specifically, we shall be dealing with the problem of calculating the response of a circuit to which an excitation is applied. The excitation (input) is an electrical signal, which may represent energy or information as a function of time. The response (output) is another signal having some desired properties that the input does not possess. We shall develop the mathematical techniques and algorithms by which this response can be found by hand and by computer. Essential to this development is the formulation of equations which describe the behavior of the circuit. We shall also deduce properties of circuits and, by means of examples, show what circuits can do. Analysis is the foundation of design, since it is through analysis that design procedures are derived.

1
fundamental laws

1.1 Objective

To analyze a circuit, one must have a mathematical description of the circuit in terms of a set of variables. Preferably, the variables should correspond to physically observable quantities, and the description should be an expression of some physical laws. In this chapter, we shall introduce these variables and laws. The laws are simple, but in spite of their simplicity, some very profound and interesting properties can be said about them. These properties follow from some careful observations of the laws and are the foundation of circuit analysis.

1.2 Voltage and Current

We begin by specifying the two electrical quantities in terms of which the behavior of a circuit is described. They are *voltage* and *current*.

Figure 1.1 shows a circuit in which the elements are represented by rectangular boxes, and the conductors that connect them are represented by lines. The points (black dots) at which the conductors are joined electrically are called the nodes of the circuit. The nature of the elements is unimportant at the moment, but it is understood that we have no physical access to the interior of the elements. Furthermore, the conductors are assumed to be perfect so that their length or physical shape is unimportant. Figure 1.1 can be redrawn as shown in Figure 1.2(a) and (b), both being "topologically" and electrically equivalent to Figure 1.1. Another point to note is that in Figure 1.1, there are three two-terminal elements: A, B, and C, one three-terminal element: D, and one four-terminal element: E.

We speak of the voltage, measured in volts (abbreviated as V), between two nodes in a circuit as the electric potential that exists between these points. There may or may not be an element connected directly between these nodes. We shall not be concerned with the physical mechanism which gives rise to the potential. It may come about by virtue of the presence of electrical charges or by virtue of a changing magnetic field. The charges and the flux produced by the magnetic field shall be taken to

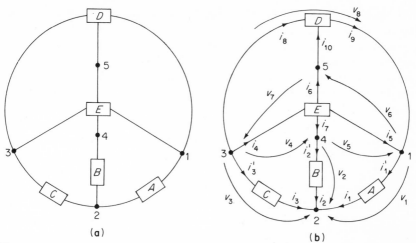

Figure 1.1 **(a)** A circuit consisting of two-, three-, and four-terminal elements; **(b)** assignment of voltages and currents.

be *internal* to an element and are not observable as far as the circuit description is concerned. Thus, in Figure 1.1 we speak of the voltage between nodes 1 and 2, between nodes 3 and 1, and so on. These voltages, in general, will be functions of time (t) and can be denoted as follows.

$v_1(t)$ = voltage between nodes 1 and 2, taken in the direction from 1 to 2;

$v_8(t)$ = voltage between nodes 3 and 1, taken in the direction from 3 to 1;

$v_4(t)$ = voltage between nodes 3 and 4, and so on.

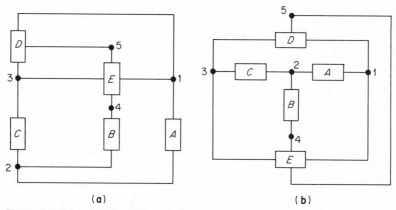

Figure 1.2 The circuit of Figure 1.1(a), redrawn in two different ways as shown in **(a)** and **(b)** here.

The directions assigned to the voltages are completely arbitrary, as they serve only as references by which the voltages are measured.

We speak of the current, measured in amperes (abbreviated as A), in a conductor as the *conduction* current that exists there. Again we shall not be concerned with the physical mechanism which gives rise to the conduction current. It may come about by virtue of the presence of a displacement current (changing electric field), the presence of a magnetic field, or by the flow of charged particles as in the case of a vacuum tube or transistor. The flux, the changing electric field, and the flow of charged particles are all regarded as internal to the elements and are not observable.

In Figure 1.1, we may refer to the current that exists at one of the terminals of element A as $i_1(t)$, that at one of the terminals of E as $i_4(t)$, and so on. The directions assigned to the currents are again completely arbitrary.

1.3 Kirchhoff Voltage and Current Laws

There are two fundamental laws which govern the voltages and currents in a circuit. They are the *Kirchhoff Voltage Law* and the *Kirchhoff Current Law*. Regardless of the composition of the circuit, the voltages and the currents of the circuit are constrained to satisfy these two laws at all times.

The Kirchhoff Voltage Law (abbreviated as KVL) states that *the sum of the voltages around a closed path in a circuit is zero*. The closed path (loop) may be taken over a set of two-terminal elements or over a set of nodes between which multiterminal elements are connected. For example, in Figure 1.1 we can write, among others, the following equations:

$$v_5(t) + v_1(t) - v_2(t) = 0 \qquad (1.1)$$
$$v_4(t) + v_2(t) - v_3(t) = 0 \qquad (1.2)$$
$$v_5(t) + v_6(t) + v_7(t) + v_4(t) = 0 \qquad (1.3)$$
$$v_4(t) + v_2(t) - v_1(t) - v_8(t) = 0 \qquad (1.4)$$
$$v_3(t) - v_2(t) + v_5(t) + v_6(t) + v_7(t) = 0 \qquad (1.5)$$
$$v_7(t) + v_4(t) + v_2(t) - v_1(t) + v_6(t) = 0 \qquad (1.6)$$

For every loop in the circuit, we can write a KVL equation. The set of equations is, in fact, a set of linear, homogeneous, algebraic equations in the variables $v_1, \ldots, v_8$. We shall see later that it has nonzero solutions.

The Kirchhoff Current Law (abbreviated as KCL) states that *the sum of the currents entering an element is zero*. We shall regard a node as a special case of an element. In a two-terminal element, the current entering into it is the same as the current leaving it. Thus in Figure 1.1, we have $i_1'(t) = i_1(t)$, $i_2(t) = i_2'(t) = i_2(t)$, and so on. In addition, we can write the following KCL equations (the functional dependence on t being implicit):

$$i_5 + i_9 - i_1 = 0 \qquad (1.7)$$
$$i_1 + i_2 + i_3 = 0 \qquad (1.8)$$
$$-i_3 - i_4 - i_8 = 0 \qquad (1.9)$$
$$i_7 - i_2 = 0 \qquad (1.10)$$
$$i_6 - i_{10} = 0 \qquad (1.11)$$
$$i_8 + i_{10} - i_9 = 0 \qquad (1.12)$$
$$i_4 - i_5 - i_6 - i_7 = 0 \qquad (1.13)$$

This is a set of linear, homogeneous, algebraic equations in the variables $i_1, \ldots, i_{10}$. We shall see later that it has nontrivial solutions.

These two laws, which are extraordinarily simple, govern the behavior of all circuits. From the point of view of the mathematical theory of circuits, they constitute the basic axioms from which theorems concerning circuits are deduced. From the physical point of view, the equations based on the two laws are the equilibrium equations of the physical system, which is the circuit. At all times, the voltages and currents in a circuit must be such that the KVL equations are satisfied for every closed loop and the KCL equations are satisfied at every node and for every element.

The actual values of the voltages and currents in a circuit will depend on the types of elements that are present. It is possible, for example in Figure 1.1, that the currents and voltages take on the values given below:

$$
\begin{array}{ll}
v_1 = 3 \text{ V} & v_5 = -2 \text{ V} \\
v_2 = 1 \text{ V} & v_6 = 2 \text{ V} \\
v_3 = -1 \text{ V} & v_7 = 2 \text{ V} \\
v_4 = -2 \text{ V} & v_8 = -4 \text{ V} \\
\\
i_1 = 1 \text{ A} & i_6 = 1 \text{ A} \\
i_2 = 0 \text{ A} & i_7 = 0 \text{ A} \\
i_3 = -1 \text{ A} & i_8 = 2 \text{ A} \\
i_4 = -1 \text{ A} & i_9 = 3 \text{ A} \\
i_5 = -2 \text{ A} & i_{10} = 1 \text{ A} \qquad (1.14)
\end{array}
$$

Another set of voltages and currents may be:

$$
\begin{array}{ll}
v_1 = \cos 2t & v_5 = \sqrt{2} \cos(2t - 2.36) \\
v_2 = \sin 2t & v_6 = 2 \cos 2t \\
v_3 = 2 \cos 2t & v_7 = -3 \cos 2t \\
v_4 = \sqrt{5} \cos(2t + 0.465) & v_8 = \cos 2t \\
\\
i_1 = \sin 2t & i_5 = \sqrt{5} \cos(2t - 2.68) \\
i_2 = -\sqrt{2} \cos(2t - 0.785) & i_6 = 2\sqrt{2} \cos(2t + 0.785) \\
i_3 = \cos 2t & i_7 = -\sqrt{2} \cos(2t - 0.785) \\
i_4 = -\sqrt{5} \cos(2t - 1.105) & i_8 = 2 \sin 2t \qquad (1.15)
\end{array}
$$

We can verify, with these values, that all the KVL and KCL equations are satisfied.

1.4 Observations on the KVL Equations – Linear Dependency

First, let us start with a digression. A set of mathematical objects, $\phi_1, \phi_2,..., \phi_n$ (vectors, polynomials, linear forms, equations), in m variables, $x_1, x_2,..., x_m$, is said to be *linearly dependent* if it is possible to find a set of numbers, $C_1, C_2,..., C_n$, not all zero, such that the linear combination

$$C_1\phi_1 + C_2\phi_2 + \cdots + C_n\phi_n$$

is zero identically for all values of the variables x_i. Otherwise, the set is said to be *linearly independent*. For example, the equations

$$\phi_1 \triangleq 5x_1 - 2x_2 + x_3 = 2$$
$$\phi_2 \triangleq x_1 + 7x_2 + x_3 = 1$$
$$\phi_3 \triangleq 13x_1 - 20x_2 + x_3 = 4$$

are linearly dependent because

$$3\phi_1 - 2\phi_2 - \phi_3 \equiv 0$$

The result is that we can express one of the equations, say ϕ_3, as a linear combination of the others; for example, $\phi_3 = -2\phi_2 + 3\phi_1$. On the other hand, the following equations are linearly independent:

$$\phi_1 \triangleq 5x_1 - 2x_2 + x_3 = 2$$
$$\phi_2 \triangleq x_1 + 7x_2 + x_3 = 1$$
$$\phi_3 \triangleq x_1 - x_2 + x_3 = 3$$

The linear combination $C_1\phi_1 + C_2\phi_2 + C_3\phi_3$ gives

$$(5C_1 + C_2 + C_3)x_1 + (-2C_1 + 7C_2 - C_3)x_2$$
$$+ (C_1 + C_2 + C_3)x_3 = (2C_1 + C_2 + 3C_3)$$

To make this combination identically zero, we must have the following set of simultaneous equations in the unknowns C_1, C_2, and C_3:

$$5C_1 + C_2 + C_3 = 0$$
$$-2C_1 + 7C_2 - C_3 = 0$$
$$C_1 + C_2 + C_3 = 0$$
$$2C_1 + C_2 + 3C_3 = 0$$

We can show by systematic elimination of variables that the only solution is $C_1 = C_2 = C_3 = 0$, and the equations in the x's are linearly independent. Now we reach the end of digression.

It is of interest to note at this time that not all of the KVL equations that can be written for a circuit are linearly independent. One or more of them can be expressed as linear combinations of the others. By inspection of Equations (1.1) to (1.6), we see that Equation (1.5) can be expressed as the difference of Equations (1.2) and (1.3), and that Equation (1.6) as the difference of Equations (1.1) and (1.3). Equations (1.5) and (1.6) are, therefore, not necessary to the description of the circuit. The questions are: (1) What is the maximum number of independent KVL equations? (2) How should the independent KVL equations be chosen? These questions will be examined and answered precisely in Section 1.9.

Another not so obvious observation is that there is a set of voltages whose specification determines all the other voltages in the circuit. For example, in Figure 1.1, if v_1, v_2, v_3, and v_6 are known, the remaining voltages can be found from KVL as follows:

$$v_4 = v_3 - v_2 \qquad\qquad (1.16)$$
$$v_5 = v_2 - v_1 \qquad\qquad (1.17)$$
$$v_7 = v_1 - v_6 - v_3 \qquad\qquad (1.18)$$
$$v_8 = v_3 - v_1 \qquad\qquad (1.19)$$

The significance of this observation is that a subset of the set of all KVL equations is sufficient to describe the circuit, insofar as the voltages of the circuit are concerned.

1.5 Observations on the KCL Equations – A Theorem

Inspection of Equations (1.7) to (1.13) reveals that the sum of all of the equations is identically zero in the variables i_1, i_2,..., i_{10}. Therefore, any one equation can be expressed as a linear combination of the remainder. This is a general result. A KCL equation can be written for every node or element in the circuit. In the set of all KCL equations, every current variable appears exactly twice, once with a coefficient +1 and once with −1. See Equations (1.7) to (1.13). This is so because every current must enter some node or element and leave another. The result is that the sum of all KCL equations is an equation with zero coefficients.

The following is a consequence of this observation:

Theorem 1.1 Let a circuit be partitioned into two disjoint parts. Let the currents in the conductors that connect the two parts be i_1, i_2,..., i_m. Then

$$i_1 + i_2 + \cdots + i_m = 0 \qquad\qquad (1.20)$$

Proof With reference to Figure 1.3, consider the set of KCL equations written for all the nodes and elements contained in part 1. Each of those currents, which

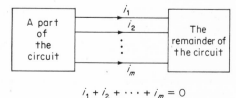

Figure 1.3 A generalization of the Kirchhoff Current Law.

are not any of i_1, i_2,..., i_m, appears exactly twice in the set of equations, once with a coefficient $+1$ and once with -1. On the other hand, each of the currents i_1, i_2,..., i_m appears exactly once in the set of equations, all having the same sign. Now if we add the set of equations written for all the nodes and elements contained in part 1, we obtain the equation of the theorem.[1]   ■

The theorem can be verified for the circuit of Figure 1.1. For example, using the values of currents given by Equation (1.14), we have

$$i_8 + i_6 + i_5 - i_1 = 2 + 1 - 2 - 1 = 0 \qquad (1.21)$$
$$i_9 - i_6 + i_4 + i_3 = 3 - 1 - 1 - 1 = 0 \qquad (1.22)$$

Another observation is that the KCL equations can be regarded as special cases of Theorem 1.1. One of the two parts of the circuit separated by the set of conductors can be a single node. Therefore, Theorem 1.1 sometimes is called the *generalized* KCL.

A less obvious observation is that there is a set of currents whose specification determines all the other currents in the circuit. For example, in Figure 1.1, if i_1, i_3, i_8, and i_9 are known, the other currents can be found as follows:

$$i_2 = -i_1 - i_3$$
$$i_4 = -i_3 - i_8$$
$$i_5 = i_1 - i_9$$
$$i_6 = -i_8 + i_9 \qquad (1.23)$$

These relations can be verified, using the values of currents given by Equation (1.14). More will be said about them in Section 1.11.

1.6   Another Observation — Computational Point of View

We have seen, without much mathematical exactitude at this time, that not all of the KVL equations and the KCL equations are required in the description of a circuit. Some of the equations are obtainable from the others by linear combination and can be regarded as "surplus." These surplus equations can be put into good use.

In the numerical solution of equations, there will always be round-off errors and sometimes cumulative errors. The voltage and currents computed numerically from the circuit equations will be approximate. The goodness of approximation can be determined by using the surplus equations to check if the voltages and currents, in fact, satisfy all the KVL and KCL equations, respectively. This checking scheme is particularly important in the solution of large circuits and should be incorporated in every computer program for circuit analysis.

[1] The symbol ■ denotes the end of a Proof, Solution, Remark, or of other statements where the ending is not clear.

1.7 Elements of Graph Theory

We now return to the question of the maximum number of independent KVL equations. To answer it we will need certain elementary concepts from the theory of graphs. These concepts are intuitively obvious and will now be introduced one by one in an informal way.

Graph A graph is a collection of points or nodes joined by line segments called *edges*. The shape and length of the line segments are unimportant. Figure 1.4(a) shows a graph of five nodes and seven edges. Figure 1.4(b) shows a graph of ten nodes and eleven edges. We say that a graph is *connected* if it is possible to reach every node from every other node via some path formed of edges. For example, Figure 1.4(a) is a connected graph, whereas Figure 1.4(b) is not. We shall be dealing only with connected graphs in this book.

Figure 1.4 (a) A graph of 5 nodes and 7 edges. **(b)** A graph of 10 nodes and 11 edges.

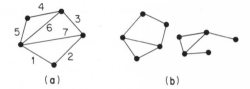

(a)　　　　　　(b)

Tree A tree of a connected graph of n nodes is a connected subgraph of the graph that contains all n nodes but no loops. In Figure 1.4(a) the subgraph consisting of the edges 1, 7, 6, 5 is a tree, as is the subgraph consisting of edges 1, 3, 4, 6. In general, a graph has many trees. Some of the trees of Figure 1.4(a) are shown in Figure 1.5, the others having been omitted for brevity. From the definition of a tree, we can deduce the following remark.

Remark 1.1 A tree in a connected graph of n nodes has $n-1$ edges.

Proof We can prove this by induction. For a graph of two nodes, the formula is obviously true. Assume it is true for all graphs with k or less nodes. Now consider a graph with $k+1$ nodes. Construct a tree. By definition, it has $k+1$ nodes but no loops. Remove an edge from the tree. The resulting subgraph has two separate parts, one having n_1 nodes and the other n_2 nodes, with $n_1 + n_2 = k+1$. Since $n_1 \leqslant k$ and $n_2 \leqslant k$, the part with n_1 nodes is a tree with $n_1 - 1$ edges, and the part with n_2

Figure 1.5 Some of the trees of the graph of Figure 1.4(a).

nodes is a tree with $n_2 - 1$ edges. If the removed edge is restored, we obtain the original tree and it has $(n_1 - 1) + (n_2 - 1) + 1 = n_1 + n_2 - 1 = (k + 1) - 1$ edges, and the remark is proved. ∎

We shall call the edges of a tree the *tree edges*. Those edges that are not tree edges shall be called the *chords* associated with the given tree. Figure 1.6 shows a graph in which the chords 1, 4, 7, corresponding to

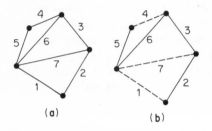

Figure 1.6 **(a)** A graph. **(b)** A tree (solid lines) of the graph and the chords associated with the tree (dotted lines).

the tree 5, 6, 3, 2, are shown in dotted lines. The chords corresponding to a different tree, say 1, 7, 6, 5, would be the edges 2, 3, 4.

If there are b edges in a graph, the number of chords is $b - n + 1$.

Fundamental loop A fundamental loop is a loop which consists of exactly one chord and just sufficient tree edges to form a loop. In Figure 1.6, the fundamental loop associated with the chord numbered 1 is $\{1,2,3,6\}$; with 7, $\{7,3,6\}$; with 4, $\{4,5,6\}$. There are as many fundamental loops as there are chords. Also there is one, and only one, fundamental loop associated with a chord. (Can you prove this statement?)

Cut set A cut set is a minimal set of edges, whose removal separates the graph into two parts. In Figure 1.6(a), $\{1,7,3\}$ is a cut set, as are $\{4,5\}$, $\{5,6,7,2\}$, and $\{2,3,7\}$. Theorem 1.1 can now be stated as: "The sum of the currents in a cut set is zero."

Fundamental cut set A fundamental cut set with respect to a tree is a set of edges which consists of exactly one tree edge and just sufficient chords to form a cut set. In Figure 1.6(b), $\{4,5\}$ is the fundamental cut set associated with the tree edge 5; $\{1,2\}$ is that associated with 2; $\{1,7,3\}$ with 3; $\{1,7,6,4\}$ with 6. Note that the fundamental cut set associated with a given tree edge is unique. (Can you prove this statement?)

There are other notions in graph theory, but what we have enumerated above is sufficient for the purpose at hand.

1.8 Voltage Graph of a Circuit

A circuit may contain two-, three-, or more-terminal elements. These elements are connected to one another at nodes. Between every pair of nodes we can, if we care to, define a voltage. The voltage relations can best be described by what is called the *voltage graph* of a circuit. The voltage graph is constructed as follows. For every node in the given circuit, there corresponds a node in the voltage graph. For every voltage variable defined in the circuit, there corresponds an edge in the voltage graph. For example, the voltage graph of the circuit of Figure 1.7(a) is shown in Figure 1.7(b). Note that the edges of the voltage graph are

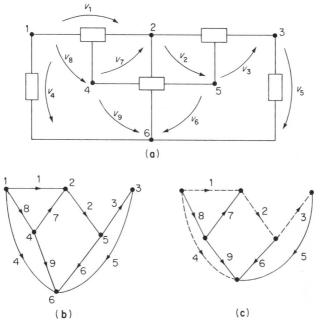

Figure 1.7 **(a)** A circuit with voltage variables assigned as shown; **(b)** the voltage graph; **(c)** its chosen tree and the associated chords.

assigned the same orientations as those of the corresponding voltage variables in the circuit. Note, too, that the voltage graph is not unique for a given circuit, but depends on a labeling scheme.

The voltage graph is just a contrivance which reveals, in pictorial fashion, the voltage relations of a circuit. It does not have physical significance, but it does describe completely the circuit insofar as the voltages are concerned, if a sufficient number of voltages is defined. By inspection of Figure 1.7(b), we can write for the circuit of Figure 1.7(a) the following KVL equations in the form of a system of linear, homogeneous, algebraic equations.

$$\phi_1 \triangleq v_1 \qquad\qquad\qquad\qquad -v_7 - v_8 \qquad = 0$$
$$\phi_2 \triangleq \qquad v_2 \qquad\qquad +v_6 + v_7 \quad -v_9 = 0$$
$$\phi_3 \triangleq \qquad\qquad v_3 \quad +v_5 - v_6 \qquad\qquad = 0$$

$$\phi_4 \triangleq \qquad\qquad\qquad v_4 \qquad\qquad\qquad - v_8 - v_9 = 0$$
$$\phi_5 \triangleq v_1 + v_2 \qquad - v_4 \qquad + v_6 \qquad\qquad\qquad = 0$$
$$\phi_6 \triangleq \qquad v_2 + v_3 \qquad + v_5 \qquad + v_7 \qquad - v_9 = 0$$
$$\phi_7 \triangleq v_1 + v_2 + v_3 - v_4 + v_5 \qquad\qquad\qquad = 0$$
$$\phi_8 \triangleq v_1 + v_2 \qquad\qquad + v_6 \qquad - v_8 - v_9 = 0$$
$$\phi_9 \triangleq \qquad v_2 + v_3 - v_4 + v_5 \qquad + v_7 + v_8 \qquad = 0$$
$$\phi_{10} \triangleq v_1 \qquad\qquad - v_4 \qquad\qquad - v_7 \qquad + v_9 = 0$$
$$\phi_{11} \triangleq \qquad v_2 \qquad - v_4 \qquad + v_6 + v_7 + v_8 \qquad = 0 \qquad (1.24)$$

We shall refer to these equations in later discussion.

1.9   Independent KVL Equations

We are now in a position to deduce the maximum number of independent KVL equations. Suppose we pick a tree in the voltage graph of a circuit. We note that we can assign arbitrary values to the voltages of the tree edges. This is so because a tree does not contain any loops, and the tree voltages are not constrained by any equations. Now if a chord is added to the tree, it forms a (unique) fundamental loop. By means of KVL, we can express the chord voltage as a linear combination of the tree voltages uniquely. For example, suppose edges 5, 6, 7, 8, 9 of Figure 1.7(b) are selected as tree edges as shown in solid lines in Figure 1.7(c). The chord voltages v_1, v_2, v_3, and v_4 can be expressed in terms of the tree voltages as follows:

$$v_1 = v_8 + v_7$$
$$v_2 = v_9 - v_7 - v_6$$
$$v_3 = v_6 - v_5$$
$$v_4 = v_8 + v_9 \qquad (1.25)$$

These observations lead to the following remark:

Remark 1.2   Let there be b voltage variables in a circuit of n nodes. Let a tree be chosen in its voltage graph. Then the $b - n + 1$ chord voltages are uniquely determined from the $n - 1$ tree voltages, which may be arbitrarily specified.   ■

A corollary of this remark is that we must define at least $n - 1$ voltage variables in a circuit in order to describe it insofar as the voltages are concerned. These $n - 1$ variables must be chosen so that their corresponding branches form a tree in the voltage graph. Another way of stating the remark is that all of the voltages of a circuit can be uniquely expressed as linear combinations of any $n - 1$ voltages that correspond to a tree. To illustrate this, we rewrite the set of Equations (1.25) and append the equations for the tree voltages and obtain the following:

$$v_1 = 0v_5 + 0v_6 + v_7 + v_8 + 0v_9$$
$$v_2 = 0v_5 - v_6 - v_7 + 0v_8 + v_9$$
$$v_3 = -v_5 + v_6 + 0v_7 + 0v_8 + 0v_9$$

$$v_4 = 0v_5 + 0v_6 + 0v_7 + v_8 + v_9$$
$$v_5 = v_5$$
$$v_6 = v_6$$
$$v_7 = v_7$$
$$v_8 = v_8$$
$$v_9 = v_9 \qquad (1.26)$$

This set of equations is the complete solution to the system of Equations (1.24) in the sense that arbitrary values can be assigned to the tree voltages $v_5, \ldots, v_9$, and the chord voltages will be such that *all* the KVL equations are satisfied. For example, suppose we let the tree voltages have the following values:

$$v_5 = 1 \text{ V}, \qquad v_6 = 2 \text{ V}, \qquad v_7 = 3 \text{ V}, \qquad v_8 = 4 \text{ V}, \qquad v_9 = 5 \text{ V}$$

The chord voltages are determined from Equations (1.25) and are found to be

$$v_1 = 7 \text{ V}, \qquad v_2 = 0 \text{ V}, \qquad v_3 = 1 \text{ V}, \qquad v_4 = 9 \text{ V}$$

Now the fifth equation of (1.24) is

$$v_1 + v_2 - v_4 + v_6 = 7 + 0 - 9 + 2 = 0$$

The sixth equation is

$$v_2 + v_3 + v_5 + v_7 - v_9 = 0 + 1 + 1 + 3 - 5 = 0$$

The other equations can be similarly verified.

Now we note that the chord voltages were determined from the four KVL equations, which are the first four equations of (1.24). It appears that if the voltages of a circuit are such that they satisfy the KVL equations written for the fundamental loops, all the other KVL equations are automatically satisfied. This conjecture is true. We must now find out what is so peculiar about these four equations and must prove the conjecture. But first, we need a digression.

We recall certain properties of systems of linear, homogeneous, algebraic equations. Suppose we have a set of three such equations in five unknowns given below:

$$x_1 - x_2 + 2x_3 - x_4 + 2x_5 = 0$$
$$-x_1 + x_2 - x_3 + x_4 + 3x_5 = 0$$
$$-2x_1 - 3x_2 + x_3 + x_4 - x_5 = 0 \qquad (1.27)$$

The set of equations has a nontrivial solution of maximal order 3 if, and only if, the equations are linearly independent. In our case, they are, and we can solve for three (any three) of the unknowns in terms of the remaining two. For example, we can write

$$x_1 - x_2 + 2x_3 = x_4 - 2x_5$$

$$-x_1 + x_2 - x_3 = -x_4 - 3x_5$$
$$-2x_1 - 3x_2 + x_3 = -x_4 + x_5$$

We can assign arbitrary values to x_4 and x_5, and x_1, x_2, and x_3 are then uniquely determined. We find the solution to Equations (1.27), which is in fact the complete solution, to be:

$$x_1 = \frac{4}{5}x_4 + \frac{18}{5}x_5$$

$$x_2 = -\frac{1}{5}x_4 - \frac{22}{5}x_5$$

$$x_3 = 0x_4 - \frac{25}{5}x_5 \tag{1.28}$$

If the equations are not linearly independent, we can take a maximum number of independent equations and solve for the unknowns as before. For example, the following set of equations is linear dependent. The fourth equation is expressible as the sum of the first two minus the third, the fifth as the difference of the first two, and the sixth as the sum of the first and the third.

$$x_1 - x_2 + 2x_3 - x_4 + 2x_5 = 0$$
$$-x_1 + x_2 - x_3 + x_4 + 3x_5 = 0$$
$$-2x_1 - 3x_2 + x_3 + x_4 - x_5 = 0$$
$$2x_1 + 3x_2 + 0x_3 - x_4 + 6x_5 = 0$$
$$2x_1 - 2x_2 + 3x_3 - 2x_4 - x_5 = 0$$
$$-x_1 - 4x_2 + 3x_3 + 0x_4 + x_5 = 0 \tag{1.29}$$

The maximum number of independent equations is three. We can choose the first three (or any three independent ones) and solve for x_1, x_2, and x_3 in terms of x_4 and x_5 as before. The other equations are automatically satisfied, as can be checked using the solution given in Equations (1.28).

In general, given a system of m linear, homogeneous, algebraic equations in b unknowns, if the maximum number of independent equations is r, we can solve for r variables uniquely in terms of the remaining $b - r$ variables. Conversely, if we can express r variables uniquely in terms of the remaining $b - r$ variables, then the maximum number of independent equations is necessarily r. To see this, we note first of all that the maximum number certainly cannot be less than r, for if it were, we would have a set of r dependent equations yielding a unique solution in r unknowns. This is impossible. Now the maximum number cannot be greater than r either, for suppose it were, we could assign arbitrary constant values to the $b - r$ variables, and we would have a system of independent equations with more equations than unknowns. But such a system does not have a nontrivial solution. We can summarize these observations in a remark which is well known in the theory of linear equations.

Remark 1.3 Given a system of m linear, homogeneous equations in b unknowns, we can obtain a unique solution in r variables in terms of the remaining $b - r$ variables if, and only if, the maximum number of independent equations is r. ∎

This remark is the basis of the following important theorem in circuit theory:

> *Theorem 1.2* The maximum number of independent KVL equations in a circuit is $b - n + 1$, where b is the number of voltage variables and n the number of nodes of the circuit.

Proof The set of KVL equations is a system of linear, homogeneous equations in the unknowns which are the voltage variables of the circuit. We have already established in Remark 1.2 that for a given tree, the chord voltages can be expressed uniquely in terms of the tree voltages. Since a tree always exists, we can always find a set of $n - 1$ voltages in terms of which the remaining can be uniquely expressed. By Remark 1.3, the maximum number of independent KVL equations must be r such that $b - r = n - 1$, or $r = b - n + 1$. ∎

The interpretation of this theorem is that of all the KVL equations that can be written for a circuit, at most $b - n + 1$ of them are linearly independent. The rest can be obtained by taking linear combinations of any $b - n + 1$ independent ones. We shall see an illustration of this fact shortly.

It remains to find a systematic way of selecting $b - n + 1$ independent KVL equations. The following theorem shows a way:

> *Theorem 1.3* The set of KVL equations written for the fundamental loops associated with a tree is a maximal set of independent KVL equations.

Proof In the set of KVL equations specified in the theorem, we have that each of the chord voltage variables appears exactly once. Let $k = b - n + 1$. Let v_1, $v_2,..., v_k$ be the chord voltages. Let ϕ_1, $\phi_2,..., \phi_k$ be the KVL equations written for the fundamental loops associated with the chords. v_1 appears in ϕ_1 and in no other equations; v_2 appears in ϕ_2 and in no others, and so on. In the linear combination $C_1\phi_1 + \cdots + C_k\phi_k$, we would have

$$C_1v_1 + C_2v_2 + \cdots + C_kv_k + (\cdot)\, v_{k+1} + \cdots + (\cdot)\, v_b$$

where $(\cdot)$ denotes some linear expression of the C's. For this combination to be identically zero, we must have that $C_1 = C_2 = \cdots = C_k = 0$. Hence the equations are linearly independent. Since there are $b - n + 1$ fundamental loops, the set is a maximal set. ∎

As an example, in Figure 1.7(c), the chords are 1, 2, 3, 4. The set of KVL equations written for the fundamental loops are the first four equations in (1.24). They are clearly independent. Moreover, the remaining KVL equations can be expressed as linear combinations of the four as follows.

$$\phi_5 = \phi_1 + \phi_2 - \phi_4$$
$$\phi_6 = \phi_2 + \phi_3$$
$$\phi_7 = \phi_1 + \phi_2 + \phi_3 - \phi_4$$
$$\phi_8 = \phi_1 + \phi_2$$
$$\phi_9 = \phi_2 + \phi_3 - \phi_4$$
$$\phi_{10} = \phi_1 - \phi_4$$
$$\phi_{11} = \phi_2 - \phi_4 \tag{1.30}$$

where ϕ_1 denotes the first equation, ϕ_2 the second, and so on.

We must now note that not every set of $b - n + 1$ KVL equations is linearly independent. For example, the following set, extracted from Equations (1.24), is not independent, even though every voltage variable is included in the set.

$$\phi_5 \triangleq v_1 + v_2 + v_6 - v_4 = 0$$
$$\phi_6 \triangleq v_2 + v_3 + v_5 - v_9 + v_7 = 0$$
$$\phi_1 \triangleq v_1 - v_7 - v_8 = 0$$
$$\phi_{11} \triangleq v_2 + v_6 - v_4 + v_8 + v_7 = 0 \tag{1.31}$$

To summarize what appears to be a long, long section, we can now state that of all the KVL equations that we can write for a circuit of b voltage variables and n nodes, at most $b - n + 1$ of them are independent. If we choose the set written for the fundamental loops, we would have a maximal set of independent equations. All of the other KVL equations can be generated by taking linear combinations of the fundamental loop equations. The other equations are, therefore, not needed in the description of the circuit, insofar as the voltages are concerned. Moreover, once the tree voltages are known, all other voltages are uniquely determined, and all the KVL equations are satisfied automatically.

Lastly, we can make a stronger statement, namely, that any set of $b - n + 1$ *independent* KVL equations is sufficient to describe the circuit voltages. This follows from the fact that given a maximal set of independent equations, all the other equations, in particular those for the fundamental loops, can be generated by linear combinations. See Problem 1.17.

We now turn our attention to similar questions concerning the current equations.

1.10 Current Graph

We introduce the notion of the current graph of a circuit. It is a graph in which for every node or element in the circuit, there corresponds a node in the current graph, and for every current variable in the circuit, there corresponds an edge in the current graph. For example, the current graph of Figure 1.8(a) is shown in Figure 1.8(b). The edges of the current graph are given the same orientations as those of the corresponding current variables in the circuit.

The current graph, like the voltage graph, is a contrivance which is

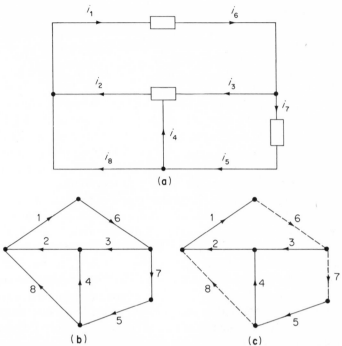

Figure 1.8 **(a)** A graph with current variables assigned as shown;
(b) the current graph; **(c)** a chosen tree and its associated chords.

used to describe the relations between the current variables in a circuit.
For each node in the graph, we can write a KCL equation, and for each
cut set in the graph, we can write an equation as stated in Theorem 1.1,
namely, such that the sum of the currents in a cut set is zero. We shall
call both types of equations the current equations. For example, in Figure
1.8, we have the following:

$$
\begin{aligned}
\phi_A &\triangleq & i_1 & & & -i_6 & & &= 0 \\
\phi_B &\triangleq & & & -i_3 & +i_6 -i_7 & &= 0 \\
\phi_C &\triangleq & & -i_2 + i_3 + i_4 & & & &= 0 \\
\phi_D &\triangleq & & -i_4 + i_5 & & & -i_8 &= 0 \\
\phi_E &\triangleq & & -i_5 & & +i_7 & &= 0 \\
\phi_F &\triangleq & -i_1 + i_2 & & & & +i_8 &= 0 \\
\phi_G &\triangleq & i_2 & & & -i_6 & +i_8 &= 0 \\
\phi_H &\triangleq & & +i_4 & & & -i_7 + i_8 &= 0 \\
\phi_I &\triangleq & & i_3 & +i_5 - i_6 & & &= 0 \\
\phi_J &\triangleq & -i_1 & +i_3 & +i_5 & & &= 0 \\
\phi_K &\triangleq & -i_1 & +i_3 & & +i_7 & &= 0 \\
\phi_L &\triangleq & i_1 - i_2 & +i_4 - i_5 & & & &= 0 \\
\phi_M &\triangleq & i_1 - i_2 & +i_4 & & -i_7 & &= 0 \\
\phi_N &\triangleq & & i_3 + i_4 & -i_6 & +i_8 &= 0 & & \quad (1.32)
\end{aligned}
$$

The first six equations are KCL equations, and the remainder are cut set equations. They constitute a system of linear, homogeneous, algebraic equations. We shall show later that the system has nontrivial solutions.

1.11 Independent Current Equations

As in the case of the KVL equations, not all of the current equations are linearly independent. We have already noted that one of the KCL equations can be expressed as a linear combination of the rest. We shall show in this section that the maximum number of independent current equations is $n - 1$, where n is the number of nodes in the current graph and that any $n - 1$ KCL equations are linearly independent.

Suppose we pick a tree in the current graph of a circuit. Figure 1.8(c) shows a tree of Figure 1.8(b). The chords are shown in dotted lines. Associated with each tree edge is a unique fundamental cut set. The $n - 1$ fundamental cut sets of Figure 1.8(c) are $\{1,6\}$, $\{2,6,8\}$, $\{3,6,7\}$, $\{4,7,8\}$, and $\{5,7\}$. Now each fundamental cut set contains precisely one tree edge. If we write the current equations for the fundamental cut sets and solve for the current variables of the tree edges, we would have the tree currents uniquely expressed in terms of the chord currents. Moreover, we can assign arbitrary values to the chord currents. To see this, we note the following remark:

Remark 1.4 The set of chords does not contain a cut set as subset.

Proof Assume the contrary. Suppose there is a subset of chords that forms a cut set. The removal of the cut set from the graph will leave a remainder with two separate parts. Since chords are edges of a graph that are not tree edges, this would mean that the tree associated with the given chords is contained in the two separate parts. This is absurd, and hence the remark is proved. As a result, the chord currents do not have to satisfy any current equation and may, therefore, assume any value. With this remark and the observations on the tree currents, we deduce the following remark:

Remark 1.5 The $n - 1$ tree currents in the current graph can be expressed uniquely as linear combinations of the $b - n + 1$ chord currents, which may assume arbitrary values, where n is the number of nodes, and b is the number of edges in the current graph. ∎

As an example, with reference to Figure 1.8(c), we can write the following current equations for the fundamental cut sets:

$$
\begin{aligned}
\phi_1 &\triangleq i_1 & & & & & -i_6 & & = 0 \\
\phi_2 &\triangleq & i_2 & & & & -i_6 & +i_8 & = 0 \\
\phi_3 &\triangleq & & i_3 & & & -i_6 + i_7 & & = 0 \\
\phi_4 &\triangleq & & & i_4 & & -i_7 + i_8 & & = 0 \\
\phi_5 &\triangleq & & & & i_5 & -i_7 & & = 0
\end{aligned}
\qquad (1.33)
$$

Note that the equations are independent. Solving for the tree currents, we have

$$i_1 = i_6$$
$$i_2 = i_6 - i_8$$
$$i_3 = i_6 - i_7$$
$$i_4 = i_7 - i_8$$
$$i_5 = i_7 \tag{1.34}$$

The chord currents may assume any values, and the tree currents are determined.

A corollary of Remark 1.5 is that we must define at least $b - n + 1$ current variables in order to describe the circuit, insofar as the currents are concerned. These $b - n + 1$ variables are those of the chords of some tree of the current graph.

Following an argument entirely similar to that used in deducing the maximum number of independent voltage equations, we obtain the two important theorems given below:

Theorem 1.4 The maximum number of independent current equations is $n - 1$, where n is the total number of nodes and elements in a circuit.

An immediate corollary of this theorem is that the maximum number of independent KCL equations is also $n - 1$.

Theorem 1.5 The set of current equations written for the $n - 1$ fundamental cut sets is a maximum set of linearly independent equations.

The significance of these theorems is that Equations (1.34) are the complete solutions to Equations (1.32) and that all the current equations, including the KCL equations, can be obtained by taking linear combinations of the equations written for the fundamental cut sets. For example, the KCL equations of (1.32) can be obtained from the fundamental cut set Equations (1.33) as follows:

$$\phi_A = \phi_1$$
$$\phi_B = \phi_3$$
$$\phi_C = \phi_3 + \phi_4 - \phi_2$$
$$\phi_D = \phi_5 - \phi_4$$
$$\phi_E = \phi_5$$
$$\phi_F = \phi_2 - \phi_1 \tag{1.35}$$

In addition, the other current equations can also be obtained from $\phi_1, \ldots, \phi_5$. For example, the last equation of (1.32) is $\phi_3 + \phi_4$, the next to the last is $\phi_1 - \phi_2 + \phi_4$, and so on.

We now verify the fact that arbitrary values can be assigned to the chord currents. The tree currents, which are uniquely determined from the chord currents, will be such that all the current equations, in particular the KCL equations, are satisfied. We make reference to Equations (1.33), which are the current equations for the fundamental cut sets defined by the tree 1, 2, 3, 4, 5 of Figure 1.8(c). Suppose we let the chord currents i_6, i_7, and i_8 have the following values:

$$i_6 = 1 \text{ A}, \qquad i_7 = 2 \text{ A}, \qquad i_8 = 3 \text{ A}$$

The tree currents are found to be

$$i_1 = 1 \text{ A}, \qquad i_2 = -2 \text{ A}, \qquad i_3 = -1 \text{ A}, \qquad i_4 = -1 \text{ A}, \qquad i_5 = 2 \text{ A}$$

The last equation of (1.32) is

$$i_3 + i_4 - i_6 + i_8 = -1 - 1 - 1 + 3 = 0$$

The next to the last equation is

$$i_1 - i_2 + i_4 - i_7 = 1 + 2 - 1 - 2 = 0$$

The other equations can be similarly verified.

We have one more important theorem to consider. It is stated as follows.

Theorem 1.6 Any $n - 1$ KCL equations written for the nodes of a current graph form a set of linearly independent equations.

Proof Consider the set of all n KCL equations. We recall that in the set, every current variable appears exactly twice, once with a coefficient $+1$ and once -1. The consequence is that there is a linear relation among the equations, namely, that the sum of all KCL equations is identically zero. This relation allows us to write the following:

$$Q_1 = -Q_2 - \cdots - Q_n \tag{1.36}$$

in which Q_1 denotes the first KCL equation, Q_2 the second, and so on. We wish to show that this is the only linear relation. There can be no others. To do this, we assume the contrary. Suppose there is another linear relation. Then it would be possible to find constants C_1, C_2,..., C_n, not all zero, such that

$$C_1 Q_1 + C_2 Q_2 + \cdots + C_n Q_n = 0$$

Suppose

$$C_1 \neq 0, \qquad C_2 \neq 0,..., C_k \neq 0, \qquad \text{and } C_{k+1} = \cdots = C_n = 0$$

We would have

$$C_1 Q_1 + C_2 Q_2 + \cdots + C_k Q_k = 0 \tag{1.37}$$

Note that in Equation (1.37), some of the KCL equations have been omitted. Since the graph is connected, there will be current variables in Equation (1.37) that appear only once. As a result, Equation (1.37) cannot be an identity, and we conclude that there does not exist another linear relation among the n KCL equations. This is to say that in the set of n KCL equations, Equation (1.36) is the only linear relation. Therefore, the $n - 1$ KCL equations Q_2, $Q_3,..., Q_n$ must be linearly independent. Since the numbering of the equations is arbitrary, any $n - 1$ KCL equations form a set of linearly independent equations. ∎

The significance of this theorem is that since the maximum number of independent current equations is $n - 1$, any set of $n - 1$ KCL equations is a maximal set. The consequence is that as in the case of the set of current equations written for the fundamental cut sets, all other current equations can be generated by the set of $n - 1$ KCL equations by linear combinations. For example, we can express the fundamental cut set current equations (1.33) in terms of the KCL equations $\phi_A,..., \phi_E$, in Equations (1.32) as follows:

$$\phi_1 = \phi_A$$
$$\phi_2 = \phi_B - \phi_C - \phi_D + \phi_E$$
$$\phi_3 = \phi_B$$
$$\phi_4 = \phi_E - \phi_D$$
$$\phi_5 = \phi_E \qquad\qquad (1.38)$$

In addition, the other current equations can also be obtained from $\phi_A,..., \phi_E$. For example, the last equation of (1.32) is $-\phi_B - \phi_D - \phi_E$, the next to last is $\phi_A + \phi_B + \phi_C$, and so on.

1.12 Further Remarks on KVL and KCL Equations

We have taken a great deal of pain to find the maximum number of independent KVL and KCL equations and to give a systematic way of selecting a maximal set of independent voltage and current equations. The labor is well rewarded because the results that we have just obtained are fundamental in circuit theory. No matter how complicated a circuit may be, no matter what types of elements it contains, no matter how many terminals the elements may have, these results always hold true. They are directly deducible from the two physical laws. As long as these two laws are valid, so are Theorems 1.1 to 1.6.

Furthermore, as we shall see later, the results of the last four sections form the basis on which we shall develop techniques to formulate equations to describe the behavior of circuits in a systematic manner. The formulation of equations is the first step in circuit analysis.

1.13 Circuits Containing Two-Terminal Elements Only

If a circuit contains two-terminal elements only, the voltage graph and the current graph are particularly simple. If we assign a voltage to each and every element, the two graphs can be made topologically equivalent in the sense that if the current graph is simplified, it will coincide with the voltage graph.

For example, Figure 1.9(a) shows a circuit containing two-terminal elements only. We define a voltage variable across each element. The voltage graph is shown in Figure 1.9(b). The current graph is shown in Figure 1.9(c). Now by KCL we know $i_1 = i_1'$, $i_2 = i_2'$, and so on, so that we can disregard the nodes which represent the elements. The graph is then simplified into one shown in Figure 1.9(d). If we assign directions to the currents in the same way as the voltages, the simplified current graph coincides with the voltage graph.

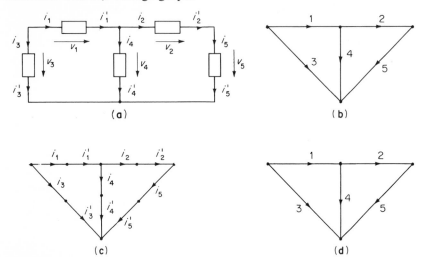

Figure 1.9 **(a)** A circuit with two-terminal elements only; **(b)** its voltage graph; **(c)** its current graph; **(d)** the reduced current graph.

If there are b elements in the circuit, there will be b voltage variables and b current variables. If there are n nodes in the circuit, there will be n nodes in the simplified current graph. The maximum number of independent KVL and KCL equations will now be simply $b - n + 1$ and $n - 1$, respectively.

1.14 Dualism

If we review the last few sections, we find that there was certain dualism about circuits. It seems that whatever we can say about the set of voltages in a circuit, we can say the same about the currents if certain key words are changed. For example, we write KVL equations for a loop and KCL equations for a node. We have tree voltages on one hand and chord currents on the other. We have fundamental loops on one and fundamental cut sets on the other. There are $n - 1$ tree edges and $b - n + 1$ chords, and there are at most $b - n + 1$ independent voltage equations and $n - 1$ independent current equations. We shall see more of these dual relations throughout the book.

1.15 Physical Origin of the Voltage and Current Laws

The voltage and current laws are regarded as the fundamental laws of circuits. We accept them as axioms. However, one might ask if there are physical laws which are more fundamental than these from which the two laws are derived. With the assumption that in a circuit, the interior of an element is not accessible and that the electric field, magnetic field, charges, and flux exist only in the interior of an element, we can derive the two laws from Maxwell's equations, which govern all electromagnetic phenomena. The voltage law follows from the fact that the line integral of the electric field that exists in the elements along a closed path, which does not enclose any changing magnetic field, is zero. For such a field, we can define the voltage as the line integral of the electric field that exists between the terminals of an element. The closed line integral of the electric field becomes the sum of the voltages along the closed path, and we have the voltage law.

The current law follows from the fact that the surface integral of the current density on the surface of a volume enclosing the exterior of an element is the negative rate of change of the total charges enclosed by the volume. Since the only currents that are observable are the conduction currents and since the charges and displacement currents are internal to the element, the integral is zero, and the surface integral becomes the sum of the conduction currents. Thus we obtain the current law.

1.16 Lumped Circuits

The elements and conductors in a circuit have finite physical dimensions. The electrical quantities, the voltage and current, are functions of time. If these quantities vary rapidly, say, at a rate equivalent to a change of sign once every billionth of a second, the wave phenomenon sets in. That is to say, the electromagnetic waves generated by the rapidly changing voltages and currents have wavelengths comparable to the physical dimensions of the elements and conductors. The result is that the voltages and currents become not only functions of time, but also functions of the spatial variables. Moreover, there will be electric and magnetic fields created external to the elements, and the voltage and current laws as stated are no longer valid.

A circuit in which the voltage and current are functions of time and position is called a *distributed-parameter* circuit. One in which the voltage and current are functions of time only is called a *lumped-parameter* circuit, or simply lumped circuit. We shall be concerned only with lumped circuits in this book.

Theoretically, any time-varying field produces an electromagnetic wave. Whether a circuit should be considered as lumped or distributed depends entirely on the relative order of magnitude of the wavelength and the physical dimension of the elements. For example, a circuit that is excited by a sinusoidal source of 60 cycles per second (hertz — abbreviated as Hz) and has physical dimensions of the order of one meter is a lumped circuit because the wavelength is six orders of magnitude larger. But at 10^8 Hz (FM broadcast frequency), it is a distributed circuit. On the other hand, the electric power lines that span distances of 1000 miles constitute a distributed-parameter circuit even at 60 Hz.

1.17 Summary

1. The voltage and current that exist in a circuit must obey the Kirchhoff Voltage Law and the Kirchhoff Current Law at all times.

2. Of all the KVL equations that we can write, at most $b - n + 1$ of them are linearly independent, where b is the number of voltage variables and n is the number of nodes of the circuit.

3. Of all the current equations that we can write, at most $n - 1$ of them are independent, where n is the total number of nodes and elements in the circuit.

4. The set of KVL equations written for the fundamental loops associated with a chosen tree is a maximal set of independent equations.

5. The set of current equations written for the fundamental cut sets associated with a chosen tree is a maximal set of independent equations.

6. Any set of $n - 1$ KCL equations is a maximal set of independent equations.

7. Every KVL equation can be expressed as a linear combination of the KVL equations written for the fundamental loops.

8. Every current equation can be expressed as a linear combination of $n - 1$ KCL equations or the current equations written for the fundamental cut sets.

SUGGESTED EXPERIMENTS

1. Obtain ten resistors of various resistances. (The precise values are unimportant.) Connect them into a circuit of arbitrary topology. Excite the circuit by a battery. Measure the voltage and current. Verify KVL and KCL.

2. Obtain three batteries and connect them into a tree. Connect resistors to the tree to form a circuit of arbitrary topology. Verify that the voltages across the resistors are as expected.

3. In Experiment 1, replace one or two resistors by capacitors. Replace the battery by an audio generator (a voltage source that produces a sinusoidal wave at audio frequencies). Carefully observe the waveforms of the currents and voltages and verify KVL and KCL.

COMPUTER PROGRAMMING

1. How would you describe a circuit to a computer? In the case of a circuit containing two-terminal elements only, we can number the elements and the nodes of the circuit and say that element 1 is connected between nodes 4 and 5, element 2 is connected between nodes 5 and 3, and so on. The connections can thus be described by a set of triplets x,y, and z, which is to mean that element x is connected between nodes y and z. The topology of the circuit can be entered as input data to the computer in the form of a set of triplets. How would you describe the topology if the circuit contains multiterminal elements?

2. The concept of a tree of a connected graph is fundamental in the description of a circuit. Suppose the topology of the graph is described by means of a list of branches in terms of triplets. How would you find a tree from the input data? The following is one of the many algorithms for selecting a tree.

    Let there be n nodes and b branches. Let the list of branches be $B(1), B(2),..., B(b)$.

    a. Start with $B(1)$. Let its nodes be n_1 and n_2.
    b. Assign n_2 to the variable $NHEAD$.
    c. Construct a node list X. Assign n_1 to $X(1)$ and n_2 to $X(2)$.
    d. Construct a tree list T. Assign $B(1)$ to $T(1)$.
    e. If the length of X is equal to n, write out the list T and stop. Otherwise, go on.
    f. Search the list B for the next branch which has a node equal to $NHEAD$. Assign its other node to the variable $NTAIL$. If none is found, go to i.
    g. If $NTAIL$ is on X, discard the branch and go to f. If not, assign $NTAIL$ to the end of the list X. Assign the branch name to the end of the list T.
    h. Assign $NTAIL$ to $NHEAD$ and go to e.
    i. Assign the element of X, which precedes the current element of X to $NHEAD$ and go to f.

    The list T contains the tree branches. Write a computer program to implement the algorithm. See the Appendix at the end of this chapter.

3. Write a program to identify the fundamental loops associated with a given tree. Use the tree-finding program of the Appendix to produce the given tree.

4. Write a program to identify the fundamental cut sets associated with a given tree.

PROBLEMS

1.1 Deduce the rule which states that the voltage appearing across each of any number of two-terminal elements connected in parallel is the same. See the following figure.

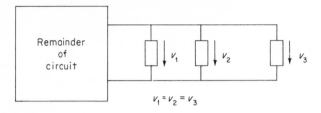

1.2 Deduce the rule which states that the current in each of any number of two-terminal elements connected in series is the same. See the following figure.

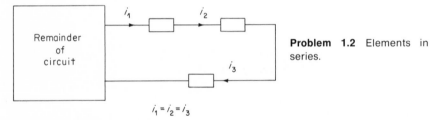

Problem 1.2 Elements in series.

1.3 In the circuit shown, the voltage variables are as defined.
   a. Write the set of all KVL equations.

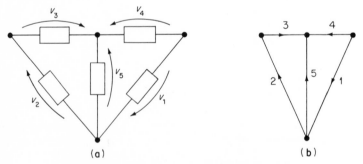

Problem 1.3 **(a)** A circuit; **(b)** its voltage graph.

   b. Pick the tree 3, 4, 5. Write the KVL equations for the fundamental loops.
   c. Express the KVL equations as linear combinations of the fundamental loop equations.

     d. Let $v_3 = 1$ V, $v_4 = 2$ V, and $v_5 = 3$ V. Find the other voltages. Verify that all the KVL equations are satisfied.

1.4 Suppose in Problem 1.3, we pick a tree 1, 2, 3. Let $v_1 = -1$ V, $v_2 = 2$ V, and $v_3 = 1$ V. Find the other voltages. Verify that all the KVL equations are satisfied.

1.5 In the following circuit, the voltages are defined as shown on the voltage graph.

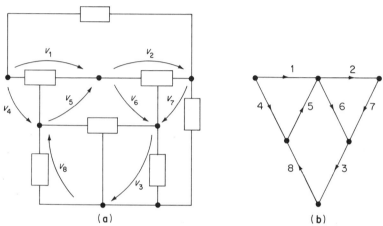

Problem 1.5 **(a)** A circuit; **(b)** its voltage graph.

     a. Write the set of all KVL equations.

     b. Pick the tree 4, 5, 6, 7, 8. Write the KVL equations for the fundamental loops. Express all KVL equations as linear combinations of the fundamental loop equations.

     c. Let $v_4 = 4$ V, $v_5 = 5$ V, $v_6 = 6$ V, $v_7 = 7$ V, and $v_8 = 8$ V. Verify that all the KVL equations are satisfied.

1.6 Suppose in Problem 1.5 we let $v_1 = 1$ V, $v_2 = 2$ V, $v_3 = 3$ V, $v_4 = 6$ V, and $v_5 = -5$ V. Can you find the other voltages? Why?

1.7 In the circuit shown, the currents are as defined.

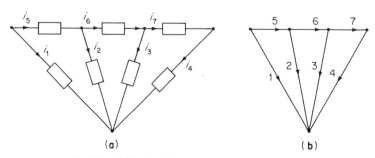

Problem 1.7 **(a)** A circuit; **(b)** its current graph.

   a. Select 1, 2, 3, 4 as a tree. Write the set of current equations for the set of fundamental cut sets.

   b. Express the other current equations in terms of the fundamental cut set equations.

   c. Suppose $i_5 = 1$ A, $i_6 = 2$ A, and $i_7 = 3$ A. Find the other currents. Verify that all the current equations are satisfied.

1.8 Suppose in Problem 1.7, we let $i_2 = A$ A, $i_6 = B$ A, and $i_3 = C$ A. Find the other currents. Verify that all the KCL equations are satisfied.

1.9 In the circuit shown, the currents are as defined.

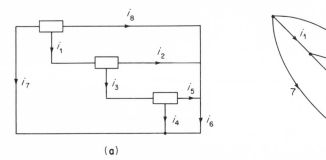

(a)   (b)

Problem 1.9 **(a)** A circuit; **(b)** its current graph.

   a. Select 1, 2, 3, 4 as a tree. Write the set of current equations for the set of fundamental cut sets.

   b. Express the KCL equations in terms of the fundamental cut set equations.

   c. Let $i_5 = 1$ A, $i_6 = 2$ A, $i_7 = 3$ A, and $i_8 = 4$ A. Find the tree currents. Verify that all KCL equations are satisfied.

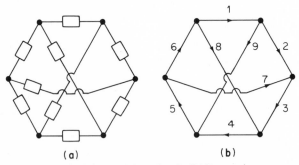

(a)   (b)

Problem 1.10 **(a)** A circuit; **(b)** its graph.

1.10 In the above circuit, it is known that
$$v_1(t) = \sin 3t$$
$$v_2(t) = \cos t + 5 \sin 3t$$

$$v_9(t) = -\sin t$$
$$v_4(t) = \sin t + \cos t$$
$$v_5(t) = \cos 3t$$

Find the remaining voltages at $t = 0$, $t = \pi/4$, $t = \pi/2$, and $t = \pi$.

1.11 In the circuit of Problem 1.10, suppose we write the following KVL equations. Are they sufficient to describe the behavior of the circuit, insofar as the voltages are concerned? Why?

$$v_1 + v_9 + v_5 + v_6 = 0$$
$$v_1 + v_2 + v_3 - v_8 = 0$$
$$v_1 + v_9 - v_4 - v_8 = 0$$
$$v_1 + v_9 + v_5 + v_7 + v_3 - v_8 = 0$$

1.12 In the following set of two equations written for the circuit of Problem 1.10:

$$v_1 + v_9 + v_5 + v_7 + v_3 - v_8 = 0$$
$$v_1 + v_2 + v_3 + v_4 + v_5 + v_6 = 0$$

every voltage variable of the circuit is included in the set. However, the set is not sufficient to describe the behavior of the circuit, insofar as the voltages are concerned. Explain.

1.13 What is the answer to Problem 1.11 if the equations are

$$v_1 + v_9 + v_5 + v_7 + v_3 - v_8 = 0$$
$$v_1 + v_2 + v_3 + v_4 + v_5 + v_6 = 0$$
$$v_1 + v_2 - v_7 + v_6 = 0$$
$$v_2 - v_7 + v_6 + v_8 + v_4 - v_9 = 0$$

1.14 The following current equations are written for the circuit shown below. Are they sufficient to describe the circuit, insofar as the currents are concerned? Why?

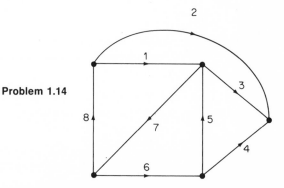

Problem 1.14

$$i_1 + i_2 - i_8 = 0$$
$$i_8 - i_7 + i_6 = 0$$
$$i_4 + i_5 - i_6 = 0$$
$$i_2 + i_3 + i_4 = 0$$

1.15 What is the answer to Problem 1.14 if the equations are

$$i_1 + i_2 - i_7 + i_6 = 0$$
$$i_2 + i_3 - i_5 + i_6 = 0$$
$$i_1 - i_7 + i_5 - i_3 = 0$$
$$i_8 - i_7 + i_5 + i_4 = 0$$

1.16 In the circuit of Problem 1.14, suppose the following three current equations are written:

$$i_2 + i_3 - i_5 + i_7 - i_8 = 0$$
$$i_8 - i_1 - i_2 = 0$$
$$i_6 - i_5 - i_4 = 0$$

The equations include all the current variables, but they are not sufficient to describe the circuit, insofar as the currents are concerned. Explain.

1.17 Show that the following set of KVL equations is a maximal set of independent equations for the circuit shown. It will be noted that it is not possible to find a tree such that these equations are the

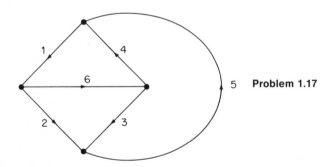

5   **Problem 1.17**

KVL equations for the fundamental loops. Show that it is possible to generate all the KVL equations of the circuit by taking linear combinations of these equations.

$$v_1 + v_6 + v_3 + v_5 = 0$$
$$v_1 + v_2 - v_3 + v_4 = 0$$
$$v_2 + v_5 - v_4 - v_6 = 0$$

1.18 Indicate which of the following statements are true or false. Justify your answers. (The statements apply to a circuit of two-terminal elements which has n nodes and b branches.)

a. Any $b - n + 1$ KVL equations are independent.

b. Any k KVL equations are independent, $k < n - 1$.

c. Any $n - 1$ current equations are independent.

d. Any k current equations are independent, $k < n - 1$.

e. Any $b - n + 1$ independent KVL equations are sufficient to describe the circuit, insofar as the voltages are concerned.

f. Any $n - 1$ independent current equations are sufficient to describe the circuit, insofar as the currents are concerned.

1.19 Indicate which of the following statements are true or false. Justify your answers.

a. If a connected graph has but one tree, it has no loops.

b. If a connected graph has but one loop, it has only one tree.

c. The number of fundamental cut sets is always less than the number of fundamental loops.

d. The number of fundamental loops is always less than the number of fundamental cut sets.

e. The number of branches that are common to a fundamental loop and a fundamental cut set is either zero or two.

APPENDIX

A Program to Find a Tree and Its Associated Chords

```
      INTEGER X(100),T(100)
      INTEGER CHRD(100)
      DIMENSION N(100),M(100)
C     WE FIRST FIND THE TREE BRANCHES
C     NODE IS THE NUMBER OF NODES AND NBRN IS THE NUMBER OF BRANCHES
      READ(5,80)NODE,NBRN
   80 FORMAT(2I5)
      WRITE(6,84)NODE,NBRN
   84 FORMAT(1H ,'NUMBER OF NODES =',I5,15X,'NUMBER OF BRANCHES =',I5)
      READ(5,80)(N(I),M(I),I=1,NBRN)
      WRITE(6,83)
   83 FORMAT(1H,'BRANCH NO.',10X,'NODE',10X,'NODE')
      WRITE(6,82)(I,N(I),M(I),I=1,NBRN)
   82 FORMAT(1H ,3X,I5,10X,I5,11X,I5)
C     LTREE IS THE LENGTH OF THE TREE BEING CONSTRUCTED
      LTREE=1
      T(1)=1
      X(1)=N(1)
      X(2)=M(1)
      LX=2
      ISKIP=1
    1 IF(LTREE .EQ. NODE-1)GO TO 1000
      INDEX=LX
C     NHEAD IS THE NODE TO WHICH NEW BRANCH OF THE TREE IS ADDED
      NHEAD=X(INDEX)
C     FIND THIS NEW BRANCH
   24 DO 10 I=1,NBRN
      IF(I.EQ.ISKIP) GO TO 10
      IF(NHEAD.EQ.N(I)) GO TO 20
      IF(NHEAD.EQ.M(I))GO TO 21
      GO TO 10
C     NTAIL IS THE OTHER NODE OF THE NEW BRANCH
```

```
      20 NTAIL=M(I)
         GO TO 22
      21 NTAIL=N(I)
C        CHECK IF THE OTHER NODE (NTAIL) IS ALREADY ON THE TREE
      22 DO 40 J=1,LX
         IF(J.EQ.INDEX)GO TO 40
         IF(NTAIL.EQ.X(J)) GO TO 10
      40 CONTINUE
         LX=LX+1
         X(LX)=NTAIL
         LTREE=LTREE+1
         T(LTREE)=I
         ISKIP=I
         GO TO 1
      10 CONTINUE
         INDEX=INDEX-1
         ISKIP=T(LTREE-1)
         NHEAD=X(INDEX)
         GO TO 24
    1000 WRITE(6,91)
      91 FORMAT(1H ,'TREE BRANCHES ARE')
         WRITE(6,90)(T(I),I=1,LTREE)
      90 FORMAT(1H ,I10)
C        WE NOW FIND THE CHORDS
         LCHRD=0
         DO 30 I=1,NBRN
         DO 50 J=1,LTREE
         IF(I.EQ.T(J))GO TO 30
      50 CONTINUE
         LCHRD=LCHRD+1
         CHRD(LCHRD)=I
      30 CONTINUE
         WRITE(6,95)
      95 FORMAT(1H ,3X,'CHORDS ARE')
         WRITE(6,90)(CHRD(I),I=1,LCHRD)
         STOP
         END
```

NUMBER OF NODES =	10	NUMBER OF BRANCHES =	13
BRANCH NO.	NODE	NODE	
1	1	2	
2	2	3	
3	4	3	
4	4	5	
5	6	5	
6	6	7	
7	7	2	
8	4	7	
9	1	8	
10	8	9	
11	9	2	
12	9	10	
13	3	10	

```
TREE BRANCHES ARE
         1
         2
         3
         4
         5
         6
        13
        12
        10
   CHORDS ARE
         7
         8
         9
        11
```

2

electrical elements

2.1 Mathematical Elements

An electrical element is any object that supports the flow of an electric current and maintains a voltage across its terminals. Practically any substance will react in some way under the influence of electricity. The variety of electrical elements is, therefore, very large.

We shall describe the behavior of an electrical element, that is, its reaction to an electrical excitation, by means of a mathematical model. The model is essentially a functional equation in the variables, which are the current and the voltage that exist on the electrical element. As it turns out, the mathematical model itself has components, which we shall call the *mathematical elements*. In spite of the large variety of electrical elements that exist, the number of different types of mathematical elements that are required to describe things in the real world is very small.

Mathematical elements are abstractions. Given an electrical element, we cannot always identify the part that corresponds to one mathematical element in the model and another part to another mathematical element. It is by a judicious choice of the kinds of mathematical elements and the configuration in which they are put together that we construct the mathematical model which approximates the behavior of an electrical element. The more complicated and sophisticated the model, the better the approximation. The model that we finally arrive at is always a compromise between simplicity and accuracy.

As an example, a solenoid is an electrical element, which, when a current passes through it, produces a magnetic field. If we increase the current, we notice that the temperature of the solenoid begins to rise. The magnetic field is produced by the "inductive" effect of the solenoid and the heat by the "resistive" effect. We can describe these effects by a mathematical model which consists of a mathematical inductor (defined later) and a mathematical resistor. The former accounts for the inductive effect, or the presence of the magnetic field, and the latter the heat. But we note that we cannot point to the solenoid and identify the mathematical elements separately as physical entities. Now if we pass a current through the solenoid which changes rapidly with time, we will observe that in addition to the magnetic field, an electric field is produced. Our model must now be modified to include a mathematical capacitor to account for the presence of the electric field. Again, the mathematical capacitor does not have a physical counterpart.

The construction of mathematical models to represent electrical elements requires an understanding of the physical mechanisms that give rise to the observable effects produced by the external excitations, and it also requires skill, experience, and at times intuition. The complexity of the model will depend on the desired degree of accuracy and, to a large extent, on the nature of the excitations. As we gain experience in analyzing circuits and in understanding their properties, we shall be able to construct mathematical models.

We shall now introduce the mathematical elements one by one.

2.2 Voltage Source

A voltage source is a mathematical element with two terminals and has the property that the voltage across the terminals is independent of the current that passes through the element. The voltage may be a constant, or it may be a function of time. Figure 2.1 shows the schematic repre-

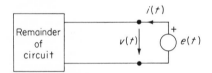

Figure 2.1 Schematic representation of a voltage source. Note the direction of $v(t)$ in relation to the symbol +.

sentation of a voltage source. The rectangular box contains the rest of the circuit and is unimportant for the discussion. If we denote the value of the source by $e(t)$ and the voltage across its terminals by $v(t)$, then the mathematical description of the voltage source is

$$v(t) = e(t) \qquad \text{independent of } i(t) \qquad (2.1)$$

Regardless of what the current $i(t)$ may be, the voltage across the terminals remains to be $e(t)$.

The mathematical voltage source does not exist physically. However, such electrical elements as the battery, the power generator, can be represented approximately by a voltage source. In addition, practically any source of electricity, be it generated by a rotating dynamo, a microphone under pressure, a thermocouple being heated, a photocell under illumination, or a dipole antenna in the presence of an electromagnetic field, can be represented by a mathematical model which will include the voltage source as one of its elements.

In Equation (2.1), if we have the special situation that $e(t) \equiv 0$, then of course $v(t) = 0$ independent of the current $i(t)$. A two-terminal element whose voltage is identically zero is known as a short circuit. This element can be realized in practice by a short length of "ideal" conductor. Figure 2.2 shows the schematic presentation of a short circuit.

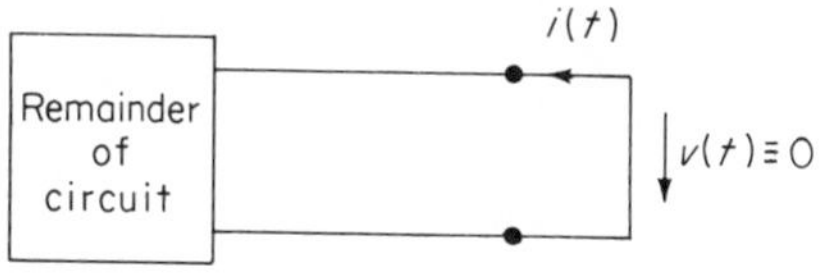

Figure 2.2 A short circuit.

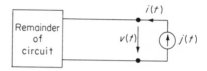

2.3 Current Source

A current source is a mathematical element with two terminals and has the property that the current passing through the element is independent of the voltage across the terminals. The current may be a constant or a function of time. Figure 2.3 shows the schematic representation of the

Figure 2.3 Schematic representation of a current source. Note the direction of $i(t)$ in relation to the arrowhead of $j(t)$.

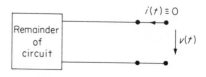

current source. The rectangular box contains the rest of the circuit and is unimportant. If we denote the value of the source by $j(t)$ and the voltage and current by $v(t)$ and $i(t)$, respectively, then the mathematical description of the current source is given by

$$i(t) = j(t) \qquad \text{independent of } v(t) \qquad (2.2)$$

Regardless of what the voltage may be, the current $i(t)$ remains to be $j(t)$. Any source of electricity — dynamo, thermocouple, photocell, or antenna — can be represented by a mathematical model which will include the current source as one of its elements.

In Equation (2.2), if $j \equiv 0$, the current $i(t) = 0$ regardless of what the voltage may be. A two-terminal element whose current is identically zero is known as an *open circuit*. This element can be realized in practice by cutting a length of conductor into two. The two ends constitute the element. Figure 2.4 shows the schematic representation of an open circuit.

Figure 2.4 An open circuit.

Example 2.1   In the circuit of Figure 2.5, the voltage and current sources are as specified. Find the voltages and currents in all the elements.

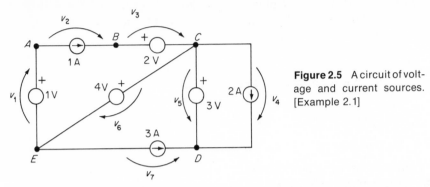

Figure 2.5 A circuit of voltage and current sources. [Example 2.1]

Solution   Since the circuit contains only two-terminal elements, we choose a voltage graph to coincide with its current graph. The voltage graph is shown in Figure 2.6, in which the numbering and orientations of the voltages and currents

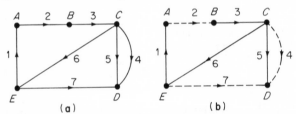

(a)   **(b)**

Figure 2.6   (a) The graph. **(b)** The chosen tree (solid lines). [Example 2.1]

are as shown. Since the voltages of the voltage sources can be arbitrarily specified, we assign them to be those of the tree edges. In our case, the tree is 1,3,5,6 with chords 2, 4, and 7. The fundamental loops are $\{2,3,6,1\}$, $\{4,5\}$, and $\{7,5,6\}$. From the KVL equations written for the fundamental loops, we have the chord voltages given by

$$
\begin{aligned}
v_2 &= -v_1 - v_3 - v_6 \\
v_4 &= v_5 \\
v_7 &= -v_6 + v_5
\end{aligned}
\tag{2.3}
$$

From the definition of a voltage source, we have

$$
\begin{aligned}
v_1 &= -1 \text{ V} \\
v_3 &= 2 \text{ V} \\
v_5 &= 3 \text{ V} \\
v_6 &= 4 \text{ V}
\end{aligned}
\tag{2.4}
$$

Therefore, the voltages across the current sources are given by

$$
\begin{aligned}
v_2 &= 1 - 2 - 4 = -5 \text{ V} \\
v_4 &= 3 \text{ V} \\
v_7 &= -4 + 3 = -1 \text{ V}
\end{aligned}
\tag{2.5}
$$

We can verify that all the KVL equations are satisfied.

The fundamental cut sets are $\{1,2\}$, $\{3,2\}$, $\{5,4,7\}$, and $\{6,2,7\}$. From the current equations written for the fundamental cut sets, we find the tree currents to be

$$i_1 = i_2$$
$$i_3 = i_2$$
$$i_5 = -i_4 - i_7$$
$$i_6 = i_2 + i_7 \tag{2.6}$$

From the definition of a current source, we have

$$i_2 = 1 \text{ A}$$
$$i_4 = 2 \text{ A}$$
$$i_7 = 3 \text{ A} \tag{2.7}$$

hence the tree currents are found to be

$$i_1 = 1 \text{ A}$$
$$i_3 = 1 \text{ A}$$
$$i_5 = -5 \text{ A}$$
$$i_6 = 4 \text{ A} \tag{2.8}$$

We can verify that all the current equations are satisfied. ∎

The solution of this example follows a systematic procedure. First we select a tree and place all the voltage sources in the tree. Next we identify the chords. We then express the chord voltages in terms of the tree voltages. Application of the definition of the elements gives us the chord voltages. We identify the fundamental cut sets next, and we express the tree currents in terms of the chord currents. Applying the definition of the elements, we obtain the values of the tree currents. This procedure turns out to be universally applicable to all circuits, and we shall use it often and shall formalize it in Chapter 6.

2.4 Resistor

A resistor is a mathematical element with two terminals and has the property that the voltage across its terminals is proportional to the current that passes through it. The proportionality constant is known as the *resistance* of the resistor, measured in ohms (abbreviated as Ω). Figure 2.7 shows the schematic representation of a resistor. The rectangular box contains the rest of the circuit and is unimportant to the discussion.

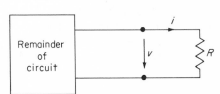

Figure 2.7 Schematic representation of a resistor. Note the directions of v and i.

The mathematical equation that describes the resistor is

$$v(t) = Ri(t) \qquad (2.9)$$

where $v(t)$ is the voltage, $i(t)$ the current, and R the resistance. We can also describe the resistor in an inverse manner, namely,

$$i(t) = Gv(t) \qquad (2.10)$$

The current through the resistor is proportional to the voltage. The proportionality constant, G, is the *conductance* of the resistor, measured in mhos (abbreviated as ℧). Clearly, we have

$$G = \frac{1}{R} \qquad (2.11)$$

A physical resistor is made of a short length of resistive material. Carbon film and nichrome wire (65 percent Ni, 12 percent Cr, 23 percent Fe) are the most common substances used. Figure 2.8 shows two pos-

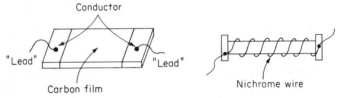

Figure 2.8   Two realizations of a resistor.

sible realizations of a physical resistor. The resistor is the electrical element which can be most closely approximated by a mathematical model. For a large class of excitations, the physical resistor behaves very much like a mathematical resistor.

Example 2.2   Find the voltages and currents in each of the elements of the circuit of Figure 2.9(a).

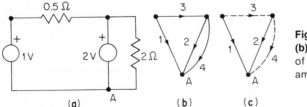

Figure 2.9   (a) A circuit; **(b)** its graph; **(c)** selection of tree and chords. [Example 2.2]

Solution   The voltage graph and the current graph coincide, and the graph is shown in Figure 2.9(b). We place the voltage sources in a tree. The tree edges are 1, 2 and the chords are 3, 4. Expressing the chord voltages in terms of the tree voltages by means of the KVL equations for the fundamental loops, we have

$$v_3 = v_1 - v_2 = -1 \text{ V}$$
$$v_4 = v_2 = 2 \text{ V} \qquad (2.12)$$

Expressing the tree currents in terms of the chord currents by means of the current equations for the fundamental cut sets, we have

$$i_1 = -i_3$$
$$i_2 = i_3 - i_4 \tag{2.13}$$

To find the currents i_3 and i_4, we apply the definition of a resistor to the two resistors of the circuit, namely,

$$i_3 = 2v_3 = -2 \text{ A}$$
$$i_4 = \tfrac{1}{2}v_4 = 1 \text{ A} \tag{2.14}$$

Substituting these values into the equations for i_1 and i_2, we obtain

$$i_1 = 2 \text{ A}$$
$$i_2 = -3 \text{ A} \tag{2.15}$$

We have now found all the voltages and currents. To check our answers, we write a KVL equation around the loop 1, 3, 4 and get

$$-v_1 + v_3 + v_4 = -1 - 1 + 2 = 0 \text{ V} \tag{2.16}$$

We write a KCL equation at node A and get

$$i_1 + i_2 + i_4 = 2 - 3 + 1 = 0 \text{ A} \tag{2.17}$$

We see that our answers satisfy all the KVL and KCL equations of the circuit.

Example 2.3 Find the voltages and currents in each of the elements in the circuit of Figure 2.10(a).

Figure 2.10 (a) A circuit; **(b)** its graph; **(c)** selection of tree and chords. [Example 2.3]

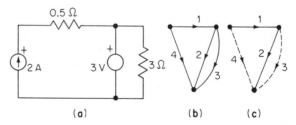

(a)   (b)   (c)

Solution The graph of the circuit is shown in Figure 2.10(b). Since the current of a current source can be arbitrarily specified, we reserve the current source to be a chord. Place the voltage source in a tree. We need to complete the tree by including the $\frac{1}{2}$-Ω resistor. The tree and the chords are shown in Figure 2.10(c). We express the chord voltages in terms of the tree voltages by means of the KVL equations for the fundamental loops and get

$$v_4 = v_1 + v_2 = v_1 + 3$$
$$v_3 = v_2 = 3 \text{ V} \tag{2.18}$$

Next we express the tree currents in terms of the chord currents by means of the current equations written for the fundamental cut sets, and we get

$$i_1 = -i_4 = 2 \text{ A}$$
$$i_2 = -i_4 - i_3 = 2 - i_3 \tag{2.19}$$

To find v_1 and i_3, we apply the definitions to the resistors:

$$v_1 = \tfrac{1}{2} i_1 = 1 \text{ V}$$
$$i_3 = \tfrac{1}{3} v_3 = 1 \text{ A} \tag{2.20}$$

Having these, we find v_4 and i_2 to be

$$v_4 = v_1 + 3 = 4 \text{ V}$$
$$i_2 = 2 - i_3 = 1 \text{ A} \tag{2.21}$$

As a check we note that $v_1 + v_3 - v_4 = 1 + 3 - 4 = 0$ V.

2.5 Linear Elements

The mathematical equation that defines a resistor is a linear relation between the voltage and the current. In Equation (2.9) we may regard the current as an excitation which, when applied to the resistor, produces a response $v(t)$. Or in Equation (2.10) we may regard the voltage as an excitation, which produces a response $i(t)$. In either case, the response is simply proportional to the excitation. Regarding the current as the excitation for the moment, we note that if $i(t)$ is multiplied by a constant, then the response is multiplied by the same constant. Let the new excitation be $i'(t)$ given by

$$i'(t) = Ki(t) \tag{2.22}$$

The new response is

$$v'(t) = Ri'(t) = RKi(t) = Kv(t) \tag{2.23}$$

as asserted.

Now suppose the new excitation is a sum of two currents $i_1(t)$ and $i_2(t)$, namely,

$$i'(t) = i_1(t) + i_2(t) \tag{2.24}$$

The new response will be

$$v'(t) = Ri'(t) = Ri_1(t) + Ri_2(t) \tag{2.25}$$

We see that it is the sum of two responses, one being the response due to $i_1(t)$ alone and the other due to $i_2(t)$ alone. The resistor is seen to have two properties: (1) The response due to an excitation multiplied by a constant is the response obtained by first applying the excitation alone and then by multiplying the result by the same constant. (2) The response due to the sum of two excitations is the sum of the responses due to the excitations applied separately. Property (1) is known as the *homogeneity* property, and Property (2) the *additivity* property. An element that obeys these two properties is called a *linear element*. The resistor is a linear element.

2.6 Capacitor

A capacitor is a two-terminal mathematical element which has the property that the current passing through the element is proportional to the rate of change of the voltage across the terminals. The proportionality constant is the capacitance of the capacitor, measured in farads (abbreviated as F). The mathematical description is

$$i(t) = C \frac{dv}{dt} \tag{2.26}$$

where $i(t)$ is the current, $v(t)$ the voltage, and C the capacitance. The schematic representation of a capacitor is shown in Figure 2.11. Alternately, we can also write

$$v(t) = \frac{1}{C} \int_a^t i(u)\ du + v(a) \tag{2.27}$$

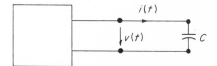

Figure 2.11 Schematic representation of a capacitor.

where $v(a)$ is a constant, which is the value of the voltage at the time instant $t = a$. It follows from Equation (2.27) that as long as the current is finite, the voltage at the instant $t = a$ remains to be $v(a)$ no matter what $i(t)$ may be. Even if the current is discontinuous at $t = a$, the voltage remains unchanged at this instant. This observation is commonly stated as: *"The voltage across a capacitor cannot change instantaneously."*

If we measure time from 0, then from Equation (2.27), we have

$$v(t) = \frac{1}{C} \int_0^t i(u)\ du + v(0) \tag{2.28}$$

and $v(0)$ is known as the initial voltage. If the initial voltage is not zero, we say that the capacitor is initially "charged." If $v(0)$ is zero, the capacitor is initially uncharged.

If we describe the capacitor by the differential equation (2.26) and regard the current as the response and the voltage the excitation, we see that the capacitor is a linear element.

To show this, we let $v'(t) = Kv(t)$. The new response will be

$$i'(t) = C \frac{d(Kv)}{dt} = K\left(C \frac{dv}{dt}\right) = Ki(t) \tag{2.29}$$

If we let $v'(t) = v_1(t) + v_2(t)$, then the new response will be

$$i'(t) = C \frac{d(v_1 + v_2)}{dt} = C \frac{dv_1}{dt} + C \frac{dv_2}{dt}$$
$$= i_1(t) + i_2(t) \tag{2.30}$$

where $i_1(t)$ is the current due to the excitation $v_1(t)$, and $i_2(t)$ is that due to $v_2(t)$.

Alternately, from Equation (2.28), we note that the capacitor voltage $v(t)$ consists of two parts, one due to the current $i(t)$ and the other an initial voltage $v(0)$. We may, therefore, regard the current as one excitation and the initial voltage as another. In other words, we now have a set of excitations $i(t)$, $v(0)$. If we replace the excitations by $Ki(t)$ and $Kv(0)$, the response $v(t)$ becomes $Kv(t)$. If we replace $i(t)$ by $i_1(t) + i_2(t)$, we would have

$$v'(t) = \frac{1}{C} \int_0^t i_1(u) \, du + \frac{1}{C} \int_0^t i_2(u) \, du + v(0)$$
$$= v_1(t) + v_2(t) + v(0)$$

Again, the response is the sum of responses due to excitations applied one at a time.

Example 2.4   Find the current in each of the elements of the circuit shown in Figure 2.12. Assume the capacitor is initially uncharged.

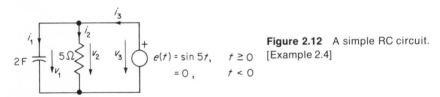

Figure 2.12 A simple RC circuit. [Example 2.4]

Solution   Since the three elements are connected in parallel, we have $v_1 = v_3$, and we do not need to use trees and chords. From the definition of the source, we have

$$v_3 = e(t) = \sin 5t \qquad t \ge 0$$
$$= 0 \qquad\qquad t < 0 \tag{2.31}$$

From the definition of the capacitor,

$$i_1(t) = C \frac{dv_1}{dt} = 2 \frac{dv_3}{dt} = 10 \cos 5t \qquad t \ge 0$$
$$= 0 \qquad\qquad\qquad\qquad t < 0 \tag{2.32}$$

From the definition of the resistor,

$$i_2(t) = \tfrac{1}{5} v_2(t) = \tfrac{1}{5} v_3(t) = \tfrac{1}{5} \sin 5t \qquad t \ge 0$$
$$= 0 \qquad\qquad\qquad\qquad t < 0 \tag{2.33}$$

Finally, from KCL we get

$$i_3(t) = i_1(t) + i_2(t) = 10 \cos 5t + \tfrac{1}{5} \sin 5t \qquad t \geq 0$$
$$= 0 \qquad\qquad\qquad\qquad\qquad t < 0 \qquad (2.34)$$

Example 2.5   Find all the voltages in the circuit of Figure 2.13. The initial voltage across the capacitor is -0.5 V. The current source $j(t)$ is as shown in Figure 2.14(a).

Figure 2.13 A series RC circuit. [Example 2.5]

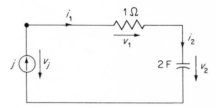

Solution   Since the elements are connected in series, we have that $i_1(t) = i_2(t) = j(t)$

$$v_2(t) = \frac{1}{C} \int_0^t j(u) \ du - 0.5 \qquad (2.35)$$

The integral between 0 and 1 is a straight line of slope 2. The value of the integral

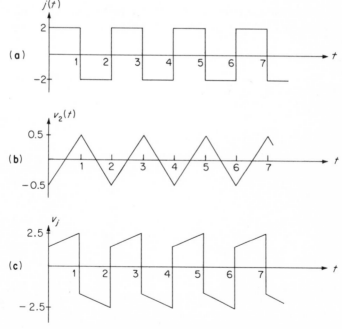

Figure 2.14   (a) Waveform of current source *j(t);* **(b)** voltage across capacitor *v₂(t);* **(c)** voltage across current source *vⱼ(t).* [Example 2.5]

at $t = 1$ is 2, and $v_2(1)$ is, from Equation (2.35), $\frac{1}{2}(2) - 0.5 = 0.5$ V. The integral from 1 to 2 is a straight line of slope -2. The voltage at $t = 2$ is

$$v_2(2) = \frac{1}{2} \int_1^2 j(u) \ du + v_2(1) = -1 + 0.5 = -0.5 \text{ V} \qquad (2.36)$$

Thus the resultant waveform of $v_2(t)$ is as given in Figure 2.14(b). Now the voltage across the resistor, by definition, is

$$v_1 = Ri_1 = 1j(t) \qquad (2.37)$$

It is the same as j. The voltage across the current source, by KVL is

$$v_j = v_1 + v_2 \qquad (2.38)$$

It is the sum of the two waveforms of Figures 2.14(a) and 2.14(b), and it is given as shown in 2.14(c). ∎

We see from this example that a capacitor has the ability to integrate a waveform. It is this property that makes the capacitor a useful element in circuits which are used to find the average value of a waveform at any given instant of time, in circuits to find the velocity from the acceleration of a moving object, in circuits to find the number of events over a period of time, and in circuits to "smooth" out the noise present in a signal. See Problems 2.6 and 2.7.

A physical capacitor is made of two conducting plates or films separated by some insulating material such as air, paper, oil, ceramic, poly-

Figure 2.15 A mathematical model of a realistic capacitor.

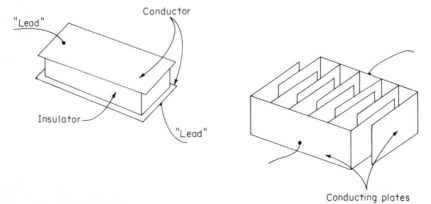

Figure 2.16 Two possible realizations of a capacitor.

styrene, and so on. The mathematical capacitor approximates the physical capacitor very well, except in cases where the insulator is not perfect. In such cases, there is heat loss in the insulator, and the mathematical model of the physical capacitor consists of a mathematical capacitor in a parallel with a resistor as shown in Figure 2.15. Figure 2.16 shows two possible realizations of a physical capacitor.

2.7 Inductor

An inductor is a mathematical element with two terminals and has the property that the voltage across the terminals is proportional to the rate of change of the current passing through it. The proportionality constant is known as the inductance, measured in henrys (abbreviated as H). The schematic representation is shown in Figure 2.17. The mathematical description is given by

$$v(t) = L\frac{di}{dt} \tag{2.39}$$

where $v(t)$ is the voltage, $i(t)$ the current, and L the inductance. Equivalently, we can describe the inductor in the integral form:

$$i(t) = \frac{1}{L}\int_a^t v(u)\ du + i(a) \tag{2.40}$$

Figure 2.17 Schematic representation of an inductor.

where $i(a)$ is the value of the current at the time instant $t = a$. If $a = 0$, we call $i(0)$ the initial current in the inductor. It follows from Equation (2.40) that no matter what the voltage may be at $t = a$ (it may be discontinuous), as long as it is finite, the current $i(t)$ will remain to be $i(a)$ at this instant. This observation is commonly stated as: *"The current in an inductor cannot change instantaneously."* Like the capacitor, the inductor is a linear element.

Example 2.6 The electron beam in a TV tube is deflected from left to right by means of a magnetic field produced by the "horizontal deflection coil." The beam traverses across the face of the tube at a constant rate and returns to the extreme left quickly. This process is repeated 525 times per picture frame, and it takes 1/30 second to cover a frame. By convention, the time it takes to return the beam should be about 16 percent of the time it takes to complete one cycle of the process. The current, which is necessary to produce the magnetic field, has the waveform shown in Figure 2.18(a). Because of the unavoidable resistance in the coil, its mathematical model consists of an inductor in series with a resistor

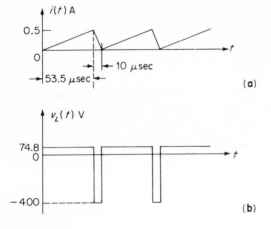

(a)

(b)

Figure 2.18 Waveforms of a TV deflection coil. [Example 2.6]

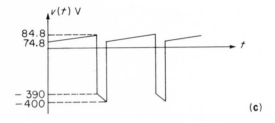

(c)

as shown in Figure 2.19. Find the voltage of the source necessary to produce the current.

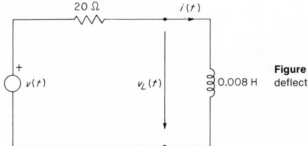

Figure 2.19 Circuit of a TV deflection coil. [Example 2.6]

Solution   The time it takes to complete one cycle is $(1/30)(1/525) = 1/15750$ second or about 63.5 microseconds (μsec). The return trace takes about $0.16 \times 63.5 = 10$ μsec. From the definition of an inductor, the voltage $v_L(t)$ is given by

$$v_L(t) = 0.008 \frac{di}{dt}$$

$$= 0.008 \frac{0.5}{53.5} 10^6 = 74.8 \text{ V} \qquad 0 \le t \le 53.5 \text{ } \mu\text{sec}$$

$$= -0.008 \frac{0.5}{10} 10^6 = -400 \text{ V} \qquad 53.5 \le t \le 63.5 \text{ } \mu\text{sec}$$

Figure 2.18(b) shows the waveform of $v_L(t)$. From the definition of a resistor, the voltage $v_R(t)$ is

$$v_R(t) = 20i(t)$$

At $t = 53.5$ μsec, $v_R = 10$ V. The voltage of the source, by KVL, is

$$v(t) = v_R(t) + v_L(t)$$

and it is given as shown in Figure 2.18(c).

Example 2.7 In the circuit shown in Figure 2.20, it is desired that the voltage

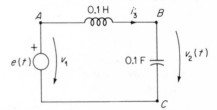

Figure 2.20 A simple LC circuit. [Example 2.7]

across the capacitor $v_2(t)$ be sin 10t. Find the voltage of the source which produces the desired $v_2(t)$.

Solution The circuit contains two-terminal elements only. The voltage graph and the current graph coincide. Place the voltage source and the capacitor in a tree. We have, from Figure 2.21, that the chord voltage v_3 is expressible in terms of the tree voltages:

$$v_3(t) = v_1 - v_2 \qquad (2.41)$$

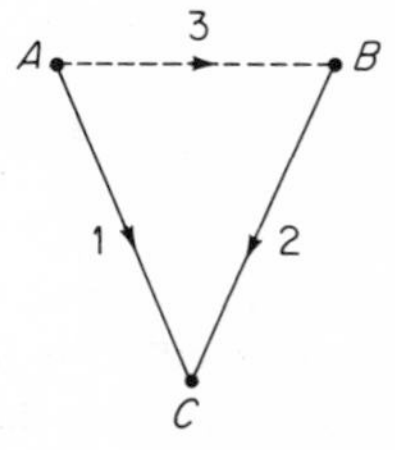

Figure 2.21 Tree of graph of Example 2.7.

Expressing the tree currents in terms of the chord currents, we have

$$i_1 = -i_3 \qquad (2.42)$$
$$i_2 = i_3 \qquad (2.43)$$

From the definitions of the elements, we have

$$v_3 = L \frac{di_3}{dt} = 0.1 \frac{di_3}{dt} \qquad (2.44)$$

$$i_2 = C \frac{dv_2}{dt} = 0.1 \frac{dv_2}{dt} \qquad (2.45)$$

Substituting Equation (2.44) into (2.41) and (2.45) into (2.43), we obtain a set of two simultaneous differential equations in the unknowns v_2 and i_3:

$$0.1 \frac{di_3}{dt} = v_1 - v_2$$

$$0.1 \frac{dv_2}{dt} = i_3 \qquad (2.46)$$

Replacing v_2 by its desired value, we get

$$i_3 = \cos 10t$$

Putting it into Equation (2.46), we obtain

$$-\sin 10t = v_1 - \sin 10t$$

so that v_1 is identically zero. This example shows that it is possible to have a non-zero voltage and current in a circuit in the absence of a source. The voltage and current exist by virtue of an initial charge in the capacitor or an initial current in the inductor. ∎

A physical inductor is made of a length of conductor wound around some material such as air, iron, or ferrite (a ferromagnetic material such as $Ca(FeO_2)_2$). Figure 2.22 shows two possible realizations of an inductor. Because conductors always have ohmic loss (resistance), the mathematical model of a physical inductor must always include a resistor and

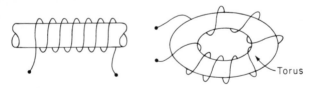

Figure 2.22 Two possible realizations of an inductor.

a mathematical inductor. Moreover, if the length of conductor is long compared with the wavelength of the excitation, the voltage will be distributed along the conductor, and there will be a potential difference between adjacent turns of the coil. The result is that a displacement current exists across the turns, and this effect must be accounted for by including a capacitor in the mathematical model.

2.8 Coupled Inductors

A system of coupled inductors is a mathematical element with two pairs of terminals at which the voltages and currents obey the following relations:

$$v_1(t) = L_1 \frac{di_1}{dt} + M \frac{di_2}{dt}$$

$$v_2(t) = M \frac{di_1}{dt} + L_2 \frac{di_2}{dt} \qquad (2.47)$$

where v_1 is the voltage across one of the pairs of terminals and v_2 the voltage across the other, i_1 the current in the same pair and i_2 the current in the other. Figure 2.23 shows the schematic representation of such a

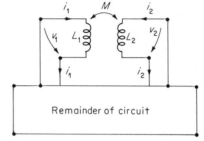

Figure 2.23 Schematic representation of two coupled inductors.

system of coupled coils. The coefficients L_1, L_2, and M are known as the self-inductance of inductor "1," the self-inductance of inductor "2," and the mutual inductance of the coupled inductors, respectively, all in henrys. Unlike any of the elements that have been introduced so far, the coupled inductors have four terminals, or actually two pairs of terminals. By convention, the direction of the voltages and currents on a set of coupled inductors will always be defined as shown in Figure 2.23.

Physically, the coupled inductors can be realized by placing two inductors in proximity to each other so that the magnetic flux of one couples or links the other. Figure 2.24 shows two possible realizations.

The direction of the flux produced by a coil depends on the direction

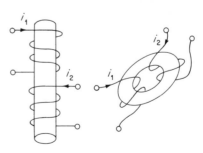

Figure 2.24 Two possible realizations of coupled inductors.

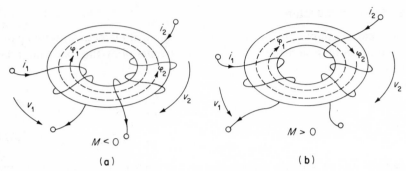

Figure 2.25 Sign convention of mutual inductance. (a) $M < 0$ if ϕ_1 and ϕ_2 are in opposite direction; (b) $M > 0$ if ϕ_1 and ϕ_2 are in the same direction.

of the current and obeys the "right-hand screw" rule as shown in Figure 2.25(a). If the flux of coil 1 links coil 2 in such a way as to induce a voltage (electromotive force) in coil 2 in a direction that tends to increase $v_2(t)$ (as defined), the mutual inductance is positive. Simply stated, with the adopted convention of the voltage and current directions, M is positive if the flux ϕ_1 and the flux ϕ_2 are in the same direction, and is negative if they oppose each other. See Figure 2.25.

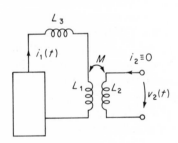

Figure 2.26 Generation of high voltage to accelerate an electron beam. [Example 2.8]

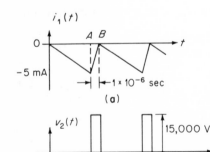

Figure 2.27 Waveforms of circuit of Example 2.8.

Example 2.8 The high voltage, which is required to accelerate the electron beam in a television tube, is generated by the much simplified circuit shown in Figure 2.26. The current $i_1(t)$ produces a magnetic field in the horizontal deflecting coil L_3 and has a waveform shown in Figure 2.27(a). The high voltage is taken across L_2, which is coupled to the inductor L_1. Find the value of the mutual inductance, which is required to produce a voltage $v_2(t)$ having a peak value of 15,000 V.

Solution Since the second inductor is connected to an "open circuit," i_2 is identically zero. From the definition of a system of inductors, we have

$$v_2(t) = M \frac{di_1}{dt}$$

The slope of i_1 between A and B of Figure 2.27(a) is $0.005/10^{-6}$ or 5×10^3 A/sec. The mutual inductance required is then

$$M = \frac{15,000}{5 \times 10^3} = 3 \text{ H}$$

The waveform of $v_2(t)$ is shown in Figure 2.27(b). In practice, the voltage $v_2(t)$ is further processed to produce a constant value slightly less than 15,000 V.

Example 2.9 A much simplified version of the "input" circuit of a radio receiver is shown in Figure 2.28. The current source represents the signal which is impressed on the receiving antenna. The voltage $v_2(t)$ is applied to an amplifier

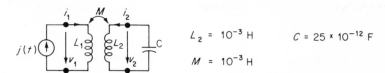

$$L_2 = 10^{-3} \text{ H} \qquad C = 25 \times 10^{-12} \text{ F}$$
$$M = 10^{-3} \text{ H}$$

Figure 2.28 Input circuit of a radio receiver. [Example 2.9]

(a vacuum tube or a transistor), which is not shown. It is found that $v_2(t)$ is the sum of two sine waves

$$v_2(t) = \sin(2\pi \times 10^6 t) + 10^{-2} \sin(2\pi \times 10^3 t)$$

Find the signal $j(t)$.

Solution From the definition of coupled inductors, we have

$$v_1 = L_1 \frac{di_1}{dt} + M \frac{di_2}{dt} \tag{2.48}$$

$$v_2 = M \frac{di_1}{dt} + L_2 \frac{di_2}{dt} \tag{2.49}$$

From the definition of a capacitor, we have

$$i_2 = -C \frac{dv_2}{dt}$$

or

$$\frac{di_2}{dt} = -C\frac{d^2v_2}{dt^2}$$

Substituting the above into Equation (2.49) and collecting terms, we have

$$L_2C\frac{d^2v_2}{dt^2} + v_2 = M\frac{di_1}{dt}$$

Substituting the given v_2 and element values into the above, we get

$$\frac{di_1}{dt} = 13\sin(2\pi \times 10^6 t) + 10\sin(2\pi \times 10^3 t)$$

Integrating, we get

$$i_1 = -(2.07 \times 10^{-6})\cos(2\pi \times 10^6 t) - (1.59 \times 10^{-3})\cos(2\pi \times 10^3 t) + K$$

where K is a constant of integration whose value depends on the initial value of i_1. We see that the signal $j = i_1$ also has two sinusoidal components. We may regard the component that has a frequency of 10^6 Hz as the wanted signal and the other component as the unwanted signal. The wanted signal has an amplitude which is about 1000 times smaller than the unwanted signal. After passing through the circuit, the signal appears as v_2, which again has two sinusoidal components. But note now that the wanted signal has an amplitude which is 100 times larger than the unwanted signal. This example illustrates that circuits can be designed to extract a wanted signal in the presence of a strong unwanted signal.

2.9 Ideal Transformer

An ideal transformer is a mathematical element which has two pairs of terminals and has the property that the voltages and currents obey the following relations:

$$v_1 = \frac{n_1}{n_2}v_2 \qquad (2.50)$$

$$i_1 = \frac{n_2}{n_1}i_2 \qquad (2.51)$$

where n_1 and n_2 are constants, and n_1/n_2 is known as the *turns ratio* of the transformer. The voltages and currents are as defined in Figure 2.29, which also shows the schematic representation of the ideal transformer.

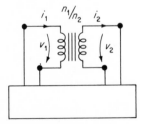

Figure 2.29 Schematic of an ideal transformer.

The ideal transformer does not exist physically. However, in the mathematical model of a physical transformer, the ideal transformer is one of the elements. A physical transformer is somewhat like that shown in Figure 2.30. L_1 and L_2 are the "primary" and "secondary" inductances.

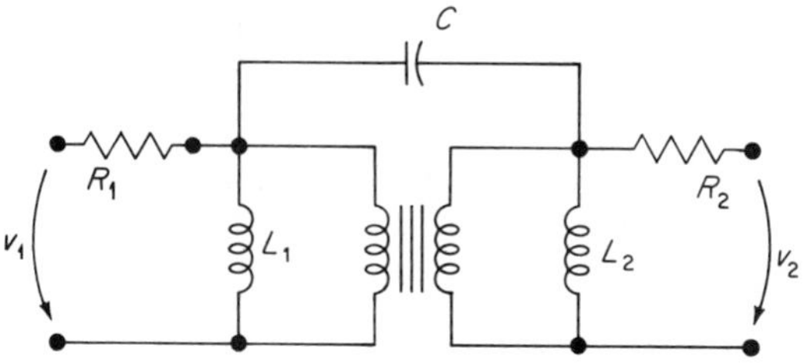

Figure 2.30 Circuit model of a realistic transformer.

The resistors account for the losses in the windings, and the capacitors for the interwinding capacitances. The transformer action is accounted for by the ideal transformer.

Example 2.10 Find all the voltages and currents in the circuit shown in Figure 2.31.

Figure 2.31 The resistance seen by the source is $(\frac{1}{100})$ *R*. [Example 2.10]

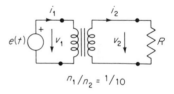

$$n_1/n_2 = {}^1/10$$

Solution From the definitions of the elements, we have

$$v_1 = e$$

$$v_1 = \frac{n_1}{n_2} v_2$$

$$i_1 = \frac{n_2}{n_1} i_2$$

$$v_2 = Ri_2$$

Making the obvious substitutions, we get

$$i_1 = \frac{n_2}{n_1} \frac{v_2}{R} = \left(\frac{n_2}{n_1}\right)^2 \frac{v_1}{R} = \left(\frac{n_2}{n_1}\right)^2 \frac{e}{R}$$

$$i_2 = \frac{n_2}{n_1} \frac{e}{R}$$

$$v_2 = \frac{n_2}{n_1} e$$

Suppose $R = 1$, $n_1/n_2 = 0.1$, and $e = 1$ V. Then

$$i_1 = \frac{1}{100} \text{ A}$$

$$i_2 = 10 \text{ A}$$

$$v_1 = 1 \text{ V}$$

$$v_2 = \frac{1}{10} \text{ V}$$

The observation to make is that the secondary voltage is "stepped" down by a ratio of 10 to 1, and the secondary current is "stepped" up by the same ratio. On the other hand, the ratio of v_1 and i_1 is

$$\frac{v_1}{i_1} = \left(\frac{n_2}{n_1}\right)^2 R = \frac{1}{100} \ \Omega$$

As far as the voltage source is concerned, it sees a resistance of $1/100$ Ω in the sense that what is connected across the source has the property of a resistor of $1/100$ Ω.

Example 2.11 Find the voltages and currents that exist for each of the elements in the circuit of Figure 2.32. The source current is $j(t) = \sin 2t$. The initial voltage on the capacitor is zero and the turns ratio is 2.

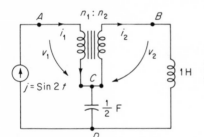

Figure 2.32 Analysis of a circuit with a three-terminal element. [Example 2.11]

Solution Assigning voltages as shown, we construct the voltage graph of Figure 2.33(a). Pick 1, 3, 4 as a tree. Expressing the chord voltages in terms of the tree voltages, we have

$$v_2 = v_4 - v_3$$
$$v_5 = v_1 + v_3$$

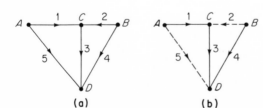

Figure 2.33 (a) Voltage graph of Example 2.11; **(b)** a tree and its chords.

Assigning a node to the transformer (which is an element), we construct the current graph as shown in Figure 2.34. Pick 1, 3, 4, 6 as a tree. We express the tree currents in terms of the chord currents and get

$$i_1 = -i_5$$
$$i_3 = -i_2 - i_5$$
$$i_4 = i_2$$
$$i_6 = -i_2 - i_5$$

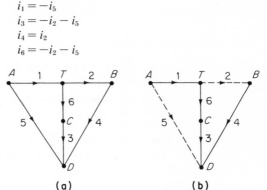

Figure 2.34 (a) Current graph of Example 2.11; **(b)** a tree and its chords. Node *T* corresponds to the element that is the transformer.

From the definitions of the elements, we have

$$v_4 = L\frac{di_4}{dt} = \frac{di_4}{dt}$$
$$i_3 = C\frac{dv_3}{dt} = \frac{1}{2}\frac{dv_3}{dt}$$
$$i_2 = \frac{n_1}{n_2}i_1 = 2i_1$$
$$v_1 = \frac{n_1}{n_2}v_2 = 2v_2$$
$$i_5 = -j = -\sin 2t$$

These equations describe the circuit completely. Combining the equations, we obtain

$$\frac{1}{2}\frac{dv_3}{dt} = -i_2 - i_5 = -2i_1 + j = -2j + j = -\sin 2t$$

Integrating, we get

$$v_3 = \cos 2t$$

From the equation for the inductor, we have

$$v_4 = \frac{di_4}{dt} = \frac{di_2}{dt} = 2\frac{di_1}{dt} = 2\frac{dj}{dt} = 4\cos 2t$$

The remaining unknowns are found from the voltage and current equations as follows:

$$v_2 = v_4 - v_3 = 3\cos 2t$$
$$v_1 = 2v_2 = 6\cos 2t$$
$$v_5 = v_1 + v_3 = 7\cos 2t$$

$$i_1 = \sin 2t$$
$$i_2 = 2 \sin 2t$$
$$i_3 = -\sin 2t$$
$$i_4 = 2 \sin 2t$$
$$i_6 = -\sin 2t$$

2.10 Dependent Sources

Dependent sources, also known as controlled sources, are mathematical sources whose values are not arbitrarily specified but dependent upon, or controlled by, the voltages or currents that exist at some other part of the circuit. There are four types.

Voltage-dependent voltage source   The schematic representation is shown in Figure 2.35. The value of the source, v, is proportional to a voltage that

Figure 2.35   Voltage-dependent voltage source.

exists at some other part of the circuit. The voltage-dependent voltage source does not have a physical counterpart. It is a mathematical element which is usually included in the mathematical model of a vacuum tube. For example, under certain conditions, the mathematical model of a triode has the configuration shown in Figure 2.36, where k is a constant

Figure 2.36   Circuit model of a triode.

peculiar to the type of vacuum tube, and the resistor accounts for the fact that the plate-to-cathode voltage v decreases as the current increases. Note that the triode itself is a three-terminal device or electrical element, but its mathematical model is a circuit consisting of two-terminal elements only.

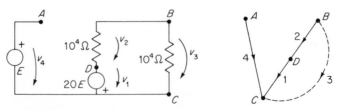

Figure 2.37   An equivalent circuit of an amplifier.   [Example 2.12]

Example 2.12 A vacuum tube amplifier commonly used in radio receivers and transmitters is shown, after much simplification, in Figure 2.37. Calculate v_3/E.

Solution The graph of the circuit is shown in Figure 2.37 next to the circuit diagram. Choose 1, 2, 4 as the tree. Expressing the chord voltage in terms of the tree voltage, we have

$$v_3 = v_2 + v_1$$

Expressing the tree currents in terms of the chord currents, we have

$$i_1 = -i_3$$
$$i_2 = -i_3$$

From the definitions of the elements, we relate the tree voltages to the tree currents and the chord current to the chord voltage and get

$$v_2 = 10,000i_2$$
$$v_1 = -20E$$
$$i_3 = 10^{-4}v_3$$

Substituting the above into the voltage equation, we get

$$v_3 = 10^4 i_2 - 20E = -10^4 i_3 - 20E = -v_3 - 20E$$

Solving for v_3, we get

$$\frac{v_3}{E} = -10$$

If E represents an input signal (speech, radio signal, computer data), the output voltage v_3 is 10 times the signal. For this reason, the circuit is called an amplifier. The ratio v_3/E is called the *gain*.

Voltage-dependent current source The schematic representation is shown in Figure 2.38. The value of the current source is proportional to

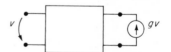

Figure 2.38 Voltage-dependent current source.

a voltage that exists elsewhere in the circuit. This source appears in the mathematical model of a transistor.

Example 2.13 Figure 2.39 shows a simplified version of a transistor amplifier. Find the gain of the amplifier v_5/E.

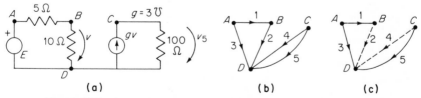

Figure 2.39 An equivalent circuit of an amplifier. [Example 2.13]

Solution The graph of the circuit is shown in Figure 2.39(b), and the tree is chosen as shown in Figure 2.39(c). Expressing the chord voltages in terms of the tree voltages, we have

$$v_2 = v_3 - v_1$$
$$v_4 = v_5$$

Expressing the tree currents in terms of the chord currents, we get

$$i_1 = i_2$$
$$i_3 = -i_2$$
$$i_5 = -i_4$$

Relating the chord voltages to the chord currents and the tree voltages to the tree currents by means of the definitions, we have

$$i_4 = -gv_2 = -3v_2$$
$$v_1 = 5i_1$$
$$v_3 = E$$
$$i_2 = 0.1v_2$$
$$v_5 = 100i_5$$

Using the equation for v_2, we have

$$v_2 = -5i_1 + E = -5i_2 + E = -0.5v_2 + E$$

or

$$v_2 = \tfrac{2}{3} E$$

Using the equation for v_5, we have

$$v_5 = 100i_5 = -100i_4 = 300v_2 = 200E$$

The gain of the amplifier, v_5/E, is 200.

Current-dependent voltage source The schematic representation is shown in Figure 2.40. The value of the voltage source is proportional to a current that exists elsewhere in the circuit. This source appears in the mathematical model of a transistor.

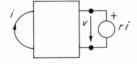

Figure 2.40 Current-dependent voltage source.

Example 2.14 A hypothetical transistor amplifier is shown in Figure 2.41(a). Find the gain of the amplifier, v_5/E.

Solution The graph and the chosen tree are shown, respectively, in Figures 2.41(b) and (c). Expressing the chord voltages in terms of the tree voltages, we have

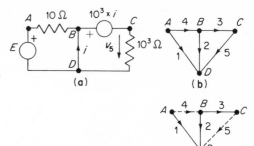

Figure 2.41 A hypothetical transistor amplifier. [Example 2.14]

$$v_4 = v_1 - v_2$$
$$v_5 = -v_3 + v_2$$

Expressing the tree currents in terms of the chord currents, we have

$$i_1 = -i_4$$
$$i_2 = i_4 - i_5$$
$$i_3 = i_5$$

From the definitions of the elements, we have

$$v_1 = E$$
$$v_2 \equiv 0 \qquad \text{(a short circuit)}$$
$$v_3 = 10^3 i = -10^3 i_2$$
$$i_4 = 0.1 v_4$$
$$i_5 = 10^{-3} v_5$$

Substituting the above into the voltage equations, we have

$$v_4 = E$$
$$v_5 = -v_3 = 10^3 i_2 = 10^3 i_4 - 10^3 i_5 = 10^2 v_4 - v_5$$

Solving for v_5, we get

$$v_5 = 50E$$

The gain of the circuit is 50.

Current-dependent current source The schematic representation is shown in Figure 2.42. The value of the current source is proportional to

Figure 2.42 Current-dependent current source.

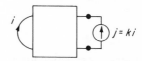

a current that exists elsewhere in the circuit. This source is found also in the mathematical model of a transistor.

Example 2.15   The circuit shown in Figure 2.43(a) has been proposed for use as a differentiator. Show that the output voltage v_3 is proportional to the derivative of the input voltage $e(t)$.

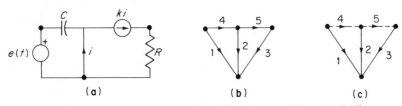

Figure 2.43   A circuit model of a differentiator.   [Example 2.15]

Solution   The graph and the chosen tree are shown in Figure 2.43(b) and (c). Expressing the chord voltages in terms of the tree voltages, we have

$$v_4 = v_1 - v_2$$
$$v_5 = v_2 - v_3$$

Expressing the tree currents in terms of the chord currents, we have

$$i_1 = -i_4$$
$$i_2 = i_4 - i_5$$
$$i_3 = i_5$$

Using the definitions of the elements, we have

$$v_1 = e(t)$$
$$v_2 = 0 \quad \text{(a short circuit)}$$
$$v_3 = Ri_3$$
$$i_4 = C \frac{dv_4}{dt}$$
$$i_5 = ki$$
$$i = -i_2$$

From the equation for i_2, we have

$$i_2 = C \frac{dv_4}{dt} - ki = C \frac{dv_4}{dt} + ki_2 = C \frac{de}{dt} + ki_2$$

Solving for i_2, we get

$$i_2 = \frac{1}{1-k} C \frac{de}{dt}$$

Using the equation for v_3, we get

$$v_3 = Rki = -Rki_2 = -\frac{RCk}{1-k} \frac{de}{dt}$$

which is what was required to be demonstrated.

2.11 A Pause and a Few Observations

It would be desirable to pause for a moment and to review what we have presented so far. We have introduced a number of mathematical elements, that are approximations of the physical elements or are the components of the mathematical model of a physical element. By putting together a collection of these mathematical elements, we were able to create circuits to do simple things. For example, we created a circuit to extract a signal in the presence of an unwanted signal. We created a circuit whose output voltage is proportional to the derivative of the input voltage. The number of different circuits that we can create is in fact unlimited.

The three basic elements of a circuit are the resistor, the capacitor, and the inductor. Note that the voltage and the current on each are simply related. In the case of a resistor, the voltage is simply proportional to the current. In the case of the capacitor, the current is proportional to the derivative of the voltage, and in the case of the inductor, the voltage is proportional to the derivative of the current. In the absence of capacitors and inductors, a circuit would be *static* in the sense that the currents and voltages are identical in form, but differ only in their relative magnitude. The equations that describe the behavior of the circuit are *algebraic* equations. If capacitors and inductors are present, the circuit has *dynamic* behavior in the sense that the description of the circuit includes the manner as to how the currents and voltages change with time. In other words, the equations describing the behavior of the circuit are *differential* equations.

To find the solution of a circuit, namely, to calculate the voltages and currents that exist in a circuit, a systematic procedure is followed. First we construct the graph or graphs of the circuit. Pick a tree and express the chord voltages in terms of the tree voltages by means of the KVL equations written for the fundamental loops. The tree currents are then expressed in terms of the chord currents by means of the current equations written for the fundamental cut sets. We use the definitions of the elements to relate the tree voltages to the tree currents and the chord currents to the chord voltages. Substituting the latter into the voltage and current equations, we find all of the voltages and currents. Note that the KVL equations for the fundamental loops and the current equations for the fundamental cut sets are used. The voltages and currents thus calculated satisfy all the KVL and all the current equations of the circuit. The set of voltages and currents that satisfy all the voltage and current equations and the definitions of the elements is called the *solution* of the circuit. Except in highly unrealistic cases, the solution is unique.

The systematic procedure outlined above will be followed throughout the book. Later on, we shall see how we may, under certain conditions, combine the three sets of equations into one and how we may, for simple cases, derive the final set of equations directly from the circuit by inspection. Almost all computer programs for circuit analysis are based on the systematic procedure of picking a tree, expressing the chord voltages in terms of the tree voltages, expressing the tree currents in terms of the chord currents, and relating the voltages and currents by definitions. More will be said about this later.

There are a number of interesting and important questions we can pose now. We have been concerned with analysis. Given a circuit, we now have a systematic procedure to find the voltages and currents. We have seen examples of useful circuits. A fundamental question is: Given a circuit containing an arbitrary number of certain types of elements, say resistors and capacitors only, what is the class of time functions to which the voltages and currents belong for a given class of time functions of the sources (excitations)? What is the answer to the same question if the circuit contains resistors, inductors, and capacitors? Dependent sources? These are important questions, for the answers tell us what we can and cannot do with various classes of circuits. These questions are the subjects of theoretical investigation considered in advanced courses on circuit theory. We shall point out some simple cases from time to time in this book.

2.12 Multiterminal Elements

There are physical elements which have more than two terminals. The transistor is an example of a three-terminal element. It can be characterized by a mathematical model consisting of two-terminal elements as we did earlier. Or it can be characterized by a set of equations on the voltages and currents on the external terminals.

As an example, many three-terminal resistive elements, including the transistor, can be characterized by the following equations relating the voltages and currents at the terminals as defined in Figure 2.44.

$$v_1 = r_{11}i_1 + r_{12}i_2$$
$$v_2 = r_{21}i_1 + r_{22}i_2 \qquad (2.52)$$

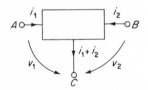

Figure 2.44 A three-terminal element.

The coefficients r_{11}, r_{12}, r_{21}, r_{22} have dimensions of resistance. Note that it is unnecessary to define the voltage from A to B since it is always $v_1 - v_2$. Note also that the equations that define the element are linear equations. The voltages may be regarded as the responses and the currents the excitations.

To analyze circuits containing three-terminal elements, we proceed as before except that now the voltage graph and the current graph do not coincide.

Example 2.16 A three-terminal resistive element is characterized by the equations

$$v_1 = 10i_1 + 10i_2$$
$$v_2 = 2000i_1 + 1000i_2$$

The element is embedded in the circuit of Figure 2.45. Find the value of R such that v_3/E is greater than unity.

Figure 2.45 An amplifier circuit consisting of a three-terminal element embedded in a resistor circuit. [Example 2.16]

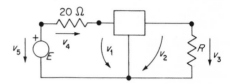

Solution Define the voltages as shown in Figure 2.45. The voltage graph is shown in Figure 2.46(a). Choose a tree as shown. Expressing the chord voltages in terms of the tree voltages, we have

$$v_4 = v_5 - v_1$$
$$v_3 = v_2$$

Figure 2.46 The voltage graph of Example 2.16 and its tree.

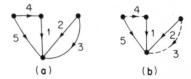

(a)   (b)

The current graph is shown in Figure 2.47(a) and the chosen tree in Figure 2.47(b). Expressing the tree currents in terms of the chord currents, we have

$$i_4 = i_1$$
$$i_5 = -i_1$$
$$i_6 = i_1 + i_2$$
$$i_3 = -i_2$$

Figure 2.47 The current graph of Example 2.16 and its tree.

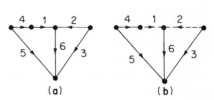

(a)   (b)

From the definitions of the elements, we have

$$v_4 = R_4 i_4 = 20 i_4$$
$$v_3 = R i_3$$
$$v_1 = 10 i_1 + 10 i_2$$
$$v_2 = 2000 i_1 + 1000 i_2$$

Using the equations for v_3 and v_4 and making substitutions, we get two equations in two unknowns:

$$30 i_1 + 10 i_2 = E$$
$$2000 i_1 + (1000 + R) i_2 = 0$$

Solving for i_2, we get

$$i_2 = \frac{-2000E}{10,000 + 30R}$$

and

$$\frac{v_3}{E} = \frac{2000R}{10,000 + 30R}$$

The ratio will be greater than unity if

$$R > \frac{10,000}{1970} > 5$$

For example, if $R = 1000 \ \Omega$,

$$\frac{v_3}{E} = 50$$

The circuit can be used as an amplifier. For $E = 1$ V, the other voltages and currents are found to be

$$v_2 = 50 \ \text{V}$$
$$i_1 = \tfrac{1}{20} \ \text{A}$$
$$v_4 = 1 \ \text{V}$$
$$v_1 = 0 \ \text{V}$$

2.13   Time-Varying Elements

The elements introduced so far are all time invariant. The element values — the resistance, capacitance, and inductance — are constants independent of time. It is easy to construct a physical element whose value varies with time. For example, if we vary the separation of the conducting plates of a capacitor periodically, we shall have a time-varying capacitor whose capacitance is a periodic function of time. To describe such an element, we must use electrical quantities that are more fundamental than the voltage or the current. In this case, we relate the charge $q(t)$, measured in coulombs, that is stored in the capacitor, to the voltage across it by

$$q(t) = C(t)v(t) \qquad (2.53)$$

where $C(t)$ is the time-varying capacitance. The conduction current $i(t)$, observable at the terminal, is the rate of change of the charge

$$i(t) = \frac{dq}{dt} \qquad (2.54)$$

The voltage-current relation of a time-varying capacitor is, therefore, given by

$$i(t) = C(t)\frac{dv}{dt} + v\frac{dC}{dt} \qquad (2.55)$$

In the case of a time-varying inductor, the quantity that should be used is the flux $\phi(t)$, measured in Webers, that is generated by the current $i(t)$ in the inductor. The flux is related to the current by

$$\phi(t) = L(t)i(t) \qquad (2.56)$$

The voltage (electromotive force) that is produced by the flux is the rate of change of the flux. The voltage-current relation of a time-varying inductor is, therefore, given by

$$v(t) = L(t)\frac{di}{dt} + i\frac{dL}{dt} \qquad (2.57)$$

Note that the voltage-current equations (2.55) and (2.57) are both linear. The time-varying capacitor and inductor, as defined here, are both linear elements. Note also the dualism of the two equations.

In the case of a time-varying resistor, we simply have

$$v(t) = R(t)i(t) \qquad (2.58)$$

where $R(t)$ is the time-varying resistance. Note that the time-invariant elements are special cases of time-varying elements. Unless otherwise stated, we shall be dealing exclusively with time-invariant elements throughout the book.

2.14 Nonlinear Elements

All of the elements that have been defined so far are linear elements in that the voltage-current relations which characterize them are linear equations. There are elements in the real world that cannot be approximated by mathematical models consisting of only linear elements. Many of the elements used in a digital computer, for example, are nonlinear.

In fact, it is the nonlinear property of these elements that makes them useful devices in the construction of the computer. We now present a few simple nonlinear elements and note their properties.

2.15 Nonlinear Resistor

A nonlinear resistor is an element whose voltage and current are related by a nonlinear equation. If the current can be expressed as an explicit function of the voltage, we call such an element a *voltage-controlled* nonlinear resistor. If the converse is true, the element is called a *current-controlled* nonlinear resistor. For the former, the mathematical expression can be stated as

$$i = f(v) \tag{2.59}$$

and for the latter

$$v = h(i) \tag{2.60}$$

where f and h are some nonlinear functions of their argument. Note that if f is a single-valued, nondecreasing function, it has an inverse which is precisely the function h. But unfortunately we do not always have a function f that possesses an inverse. We shall see examples of such functions in Chapter 5.

An example of a voltage-controlled nonlinear resistor is a p–n junction made of semiconductor material. The voltage-current relation is given by

$$i = I_s \left[\exp\left(\frac{qv}{kT}\right) - 1 \right] \tag{2.61}$$

where I_s is a constant peculiar to the type of material, q the electronic charge (1.602×10^{-19} coulomb), k the Boltzman constant (1.380×10^{23} joule/K), and T the temperature in absolute degree. The schematic representation of this device and its current-voltage characteristic are shown in Figure 2.48. Clearly, this element is nonlinear, for if we apply the homogeneity and the additivity tests, both would fail. (Please try.)

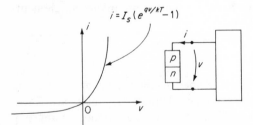

Figure 2.48 The i–v characteristic of a junction diode.

A linear resistor can be regarded as a special case of a nonlinear resistor. Using the current-voltage characteristic, we would have, for the case of a linear resistor, a straight line of slope equal to the conductance of the resistor, as shown in Figure 2.49.

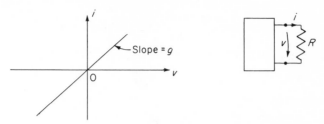

Figure 2.49 The *i–v* characteristic of a linear resistor.

Example 2.17 A diode is a nonlinear element with an approximate current-voltage (frequently abbreviated as *i–v*) characteristic shown in Figure 2.50. If the voltage across the diode is positive, the current is proportional to the voltage.

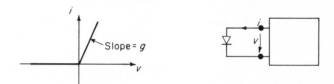

Figure 2.50 The *i–v* characteristic of a hypothetical diode.

If the voltage is negative, the current is zero. This element is useful in a circuit where we wish to cut off, for example, the negative part of a waveform. Consider the circuit shown in Figure 2.51. Find the voltages and currents everywhere.

Figure 2.51 A diode embedded in a circuit. [Example 2.17]

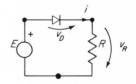

Solution The elements are in series, and the current everywhere is the same. From the voltage equation we have

$$v_D = E(t) - v_R$$

From the definition of the diode, we have

$$i = gv_D, \quad v_D \geq 0$$
$$i = 0, \quad v_D < 0$$

Assume for the moment v_D is positive. We would have

$$i = gE(t) - gRi$$

or

$$i = \frac{g}{1 + gR} E(t),$$

$$v_D = \frac{E(t)}{(1 + gR)} \geq 0, \quad \text{if } E(t) \geq 0$$

The assumption is correct if $E(t)$ is positive or zero. Now assume $v_D < 0$. Then

$$i = 0$$

and

$$v_D = E(t)$$

The voltage across the diode is the same as $E(t)$ whenever $E(t)$ is negative. Figure 2.52 shows the various waveforms for the case in which $E(t) = \sin t$.

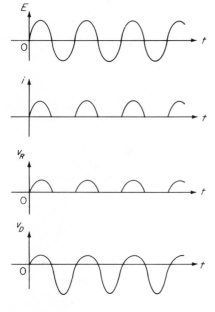

Figure 2.52 Waveforms of Example 2.17; output voltage is zero when the input voltage is negative.

2.16 Nonlinear Capacitor

A nonlinear capacitor is one whose stored charge is a nonlinear function of the voltage across it. An example is the semiconductor junction capacitor, which under certain conditions, has the following charge-voltage relation:

$$q = Kv^{1/2} \qquad (2.62)$$

where q is the charge, v the voltage, and K a constant peculiar to the type of material used to make the junction. The current $i(t)$ is related to the voltage by

$$i(t) = \frac{dq}{dt} = \left[\left(\frac{K}{2} \right) v^{-1/2} \right] \frac{dv}{dt} \qquad (2.63)$$

so that we may define a nonlinear capacitance C as a function of the voltage v as

$$C(v) = \left(\frac{K}{2} \right) v^{-1/2} \qquad (2.64)$$

Example 2.18 Suppose the voltage across the nonlinear capacitor in the circuit shown in Figure 2.53 is $V(1 + a \sin t)$, where $a \ll 1$. Find the current $i(t)$.

Figure 2.53 A nonlinear capacitor used to generate the second harmonic of $v(t)$. [Example 2.18]

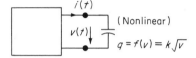

Solution Using Equation (2.63), we have

$$i(t) = \frac{K}{2} \frac{V\, a \cos t}{\sqrt{V(1 + a \sin t)}}$$

$$\cong \frac{K}{2}\, a\sqrt{V} \cos t \left(1 - \frac{1}{2}\, a \sin t \right)$$

$$\cong \frac{K}{2}\, a\sqrt{V} \left(\cos t - \frac{1}{4}\, a \sin 2t \right)$$

It is seen that an excitation consisting of a sinusoidal component superimposed upon a constant produces a response which consists of two sinusoidal components, one having the same frequency as the excitation and one having twice the frequency. Can you think of any use for such a capacitor?

2.17 Nonlinear Inductor

A nonlinear inductor (Figure 2.54) is one in which the flux ϕ is a nonlinear function of the current $i(t)$. The voltage across the inductor is

$$v(t) = \frac{d\phi}{dt} \qquad (2.65)$$

Figure 2.54 A nonlinear inductor.

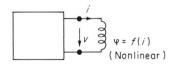

An example of a nonlinear inductor is the ferrite core used as a memory element in a digital computer. The flux as a function of the current is somewhat like that shown in Figure 2.55. If the current i is positive and is increasing, the flux increases linearly from A to B. From thereon, any further increase in the current does not produce any change in the flux.

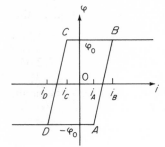

Figure 2.55 ϕ–i characteristic of a magnetic core.

(The core is said to be saturated.) In fact, as long as the current is greater than i_c, the flux will remain at the value ϕ_0. When the current decreases to a value smaller than i_c, the flux decreases along the path CD. At D, no change in the flux will result for any value of the current as long as it is less than i_A.

Example 2.19   Suppose the ferrite core is subjected to a current having the waveform shown in Figure 2.56. Find the voltage across the terminals.

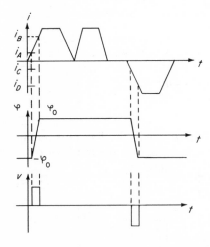

Figure 2.56   Waveforms of i, ϕ, and v of a magnetic core.   [Example 2.19]

Solution   The waveforms are as shown in Figure 2.56 and are self-explanatory. Note that the flux remains at ϕ_0 until a negative pulse of current of sufficient amplitude is applied, and it will remain at $-\phi_0$ until a sufficiently large positive current pulse is applied. Can you think of any applications for this element?

2.18 Energy and Power

A capacitor stores electric charges. The work that is required to assemble the collection of charges is precisely the electric energy that is stored in the capacitor. The incremental increase of the energy dw is related to the incremental increase in the charge dq by

$$dw = v \, dq \qquad (2.66)$$

where v is the voltage across the capacitor. The total energy stored in the capacitor containing a total amount of charge Q is

$$w = \int_0^Q v \, dq \qquad (2.67)$$

In the case of a linear capacitor, the charge is proportional to the voltage

$$q = Cv \qquad (2.68)$$

or the charge is a linear function of the voltage as shown in Figure 2.57. From Equation (2.67), the energy is simply the area of the shaded triangle and is given by

$$w = \tfrac{1}{2} CV^2 \qquad (2.69)$$

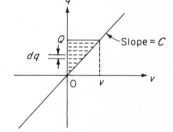

Figure 2.57 The q–v characteristic of a linear capacitor.

The unit of w is joule (abbreviated as J). In the case of a nonlinear capacitor, the charge is a function of the voltage, somewhat as shown in Figure 2.58. The energy is the area of the shaded region in the figure.

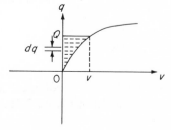

Figure 2.58 The q–v characteristic of a nonlinear capacitor.

In general, both voltage and charge are functions of time. The rate of change of the energy (rate of doing work) is the power, denoted by $p(t)$, given by

$$p(t) = \frac{dw}{dt} \tag{2.70}$$

The unit of power is watt (abbreviated as W). From Equation (2.67), we have the power taken by the capacitor given by

$$p(t) = \frac{dw}{dt} = v\frac{dq}{dt} = vi \tag{2.71}$$

where i is the current on the terminal of the capacitor.

When a current passes through an inductor, it sets up a magnetic field. The work required to produce the field is the magnetic energy stored in the inductor. An incremental increase in the energy is proportional to an incremental increase in the flux. The relation is given by

$$dw = i\,d\phi \tag{2.72}$$

where i is the current. The energy stored in the inductor which produces an amount of flux Φ is

$$w = \int_0^\Phi i\,d\phi \tag{2.73}$$

In the case of a linear inductor, we have

$$\phi = Li \tag{2.74}$$

where L is the inductance. This linear relation is shown in Figure 2.59. The shaded area of the triangle is the energy stored in the inductor and is given by

$$w = \tfrac{1}{2} Li^2 \tag{2.75}$$

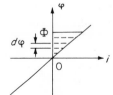

Figure 2.59 The ϕ–i characteristic of a linear inductor.

In the case of a nonlinear inductor, the flux-current characteristic may be as shown in Figure 2.60. The area of the shaded region is the energy stored.

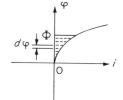

Figure 2.60 The φ–i characteristic of a nonlinear inductor.

From Equation (2.73), the power taken by the inductor to store an amount of energy w is

$$p(t) = \frac{dw}{dt} = i\frac{d\phi}{dt} = vi \qquad (2.76)$$

where v is the voltage across the inductor.

A resistor is an element which does not store any energy, but dissipates it in the form of heat. The rate at which the heat is generated is the power and is given by the relation

$$p(t) = Ri^2(t) = vi \qquad (2.77)$$

where i is the current, v the voltage, and R the resistance.

We see that the power is given by the product of the voltage and the current at the terminals. We may now generalize the concept to say that the power taken by a circuit N as shown in Figure 2.61 is

$$p = vi \qquad (2.78)$$

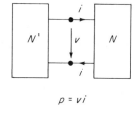

Figure 2.61 Definition of power in a two-terminal element.

$$p = vi$$

where the directions of the voltage and current are as shown in the figure. Still more generally, we can speak of the power taken by a multiterminal

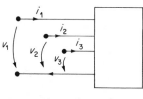

Figure 2.62 Definition of power in a multiterminal element.

$$p = v_1 i_1 + v_2 i_2 + v_3 i_3$$

element or circuit to be the sum of the products of the voltages and currents at all the terminals as shown in Figure 2.62.

$$p = v_1 i_1 + v_2 i_2 + \cdots + v_n i_n \qquad (2.79)$$

Example 2.20   A realistic voltage source always has some resistance associated with it. Suppose the source is connected to a resistor as shown in Figure 2.63.

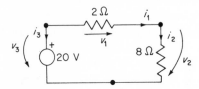

Figure 2.63   Calculation of power in a resistor circuit.   [Example 2.20]

Find the power taken by the resistor and the power wasted in the internal resistance.

Solution   Defining the voltages and currents as shown, we have

$$v_1 + \quad v_2 = 20 \text{ V}$$
$$2i_1 + \quad 8i_2 = 20$$
$$i_1 = i_2$$

so that

$$i_1 = 2A = -i_3$$

The power taken by the 8-Ω resistor is

$$p_8 = 8i_2^2 = 32 \text{ W}$$

The power wasted in the internal resistance is

$$p_2 = 2i_1^2 = 8 \text{ W}$$

The power delivered by the voltage source is

$$p_E = -20i_3 = 40 \text{ W}$$

which is seen to be the sum of the powers dissipated in the resistors. Is this always true? The answer is to be found in what follows.

2.19   Tellegen's Theorem

We now wish to prove an important theorem in circuit theory.

Theorem 2.1   Consider a circuit of b two-terminal elements and n nodes. Let $v_1, v_2,\ldots, v_b$ be the voltages that satisfy the KVL equations. Let $i_1, i_2,\ldots, i_b$ be the currents that satisfy the current equations (KCL and cut set). Then

$$\sum_{k=1}^{b} v_k i_k = 0 \qquad (2.80)$$

Moreover, let $v_1', v_2',\ldots, v_b'$ be another set of voltages that satisfy the KVL equations and let $i'_1, i'_2, \ldots, i'_b$ be another set of currents that satisfy the current equations on the same circuit. Then

$$\sum_{k=1}^{b} v'_k i_k = 0 \qquad (2.81)$$

$$\sum_{k=1}^{b} v_k i'_k = 0 \qquad (2.82)$$

Before we prove the theorem, it is worthwhile to consider its significance. First, the theorem is true regardless of the type of elements that we have. The elements may be linear or nonlinear, time invariant or time varying. The circuit may contain independent or dependent sources. Second, each term in the sum (2.80) represents the power taken by the element. The statement of the theorem is that the sum of the powers taken by all the elements of a circuit is zero.

Proof The voltages satisfy the KVL equations, and the currents the KCL or the cut set equations. By the observations of Chapter 1, we know that if we pick a tree, the chord voltages can be uniquely expressed in terms of the tree voltages, and the tree currents can be uniquely expressed in terms of the chord currents. Now consider the sum of (2.80). Break it up into two summations, one being summed over all the tree branches and the other over the chords. Denote the tree voltages and the tree currents as $v_{t1}, v_{t2}, \ldots, v_{tp}$ and $i_{t1}, i_{t2}, \ldots, i_{tp}$, respectively, where $p = n - 1$. Denote the chord voltages and chord currents as $v_{c1}, v_{c2}, \ldots, v_{cm}$ and $i_{c1}, i_{c2}, \ldots, i_{cm}$, respectively, where $m = b - n + 1$. The sum becomes

$$\sum_{k=1}^{b} v_k i_k = \sum_{k=1}^{p} v_{tk} i_{tk} + \sum_{k=1}^{m} v_{ck} i_{ck}$$

Now let us focus on a typical fundamental cut set as shown in Figure 2.64. The

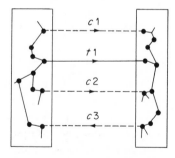

Figure 2.64 A typical fundamental cut set in the proof of Tellegen's theorem.

cut set contains one, and only one, tree branch, say $t1$, the rest of the branches being chords, say $c1$, $c2$, and $c3$. The current equation for this fundamental cut set yields

$$i_{t1} = -i_{c1} - i_{c2} + i_{c3}$$

and

$$v_{t1} i_{t1} = -v_{t1} i_{c1} - v_{t1} i_{c2} + v_{t1} i_{c3}$$

which is one of the p terms in the first summation. Now, consider chord $c1$. It defines a fundamental loop. Moreover, the fundamental loop must contain the tree branch $t1$, otherwise $t1$ would not be in the fundamental cut set. In the KVL equation for the fundamental loop, we have

$$v_{c1} = v_{t1} + \text{(other tree voltages)}$$

so that

$$v_{c1}i_{c1} = v_{t1}i_{c1} + \text{(other tree voltages)} \, i_{c1}$$

Similarly, from the fundamental loops defined by $c2$ and $c3$, we have

$$v_{c2}i_{c2} = v_{t1}i_{c2} + \text{(other tree voltages)} \, i_{c2}$$
$$v_{c3}i_{c3} = -v_{t1}i_{c3} + \text{(other tree voltages)} \, i_{c3}$$

The three terms $v_{c1}i_{c1}$, $v_{c2}i_{c2}$, and $v_{c3}i_{c3}$ appear in the second summation. But note that no other terms in the second summation, when each is expanded in terms of the tree voltages, contain the tree voltage v_{t1}, for if there were, the chord corresponding to the fundamental loop containing v_{t1} would have been included in the fundamental cut set associated with $t1$. In other words, for every term $v_{tk}i_{tk}$ in the first summation, when it is expanded in terms of the chord currents, there are precisely the same terms in the second summation, but with opposite signs. If we repeat the argument for all the fundamental cut sets, we have the assertion of the theorem.

Since the only requirements on the voltages and the currents are that they obey the KVL, KCL, and cut set equations, any set of voltages and currents will satisfy the assertion (2.80) as long as they satisfy the KVL, KCL, and cut set equations. Therefore,

$$\sum_{k=1}^{b} v_k i_k' = 0$$

and

$$\sum_{k=1}^{b} v_k' i_k = 0$$

which is what was required to be demonstrated.

In addition to being of fundamental importance, this theorem offers another artifice by which the accuracy of the numerical results of a computer solution can be checked. After all the voltages and currents have been computed, we compute the sum of the products of the corresponding terms. The sum should be zero for all values of time.

Example 2.21 Applying Tellegen's theorem to Example 2.20, we have

$$v_1 i_1 + v_2 i_2 + v_3 i_3 = (16)(2) + (4)(2) - (20)(2) = 0$$

The v–i products are the powers taken by the elements. The observation made at the end of the example is, therefore, always true.

Now, suppose we change the source to 60 V. We then have

$$v_3' = 60 \text{ V}$$
$$v_1' = 12 \text{ V}$$
$$v_2' = 48 \text{ V}$$
$$i_1' = i_2' = -i_3' = 6 \text{ A}$$

The sum of products of the voltages and currents is

$$v_1'i_1' + v_2'i_2' + v_3'i_3' = (12)(6) + (48)(6) - (60)(6) = 0$$

Moreover, we have

$$v_1 i_1' + v_2 i_2' + v_3 i_3' = (16)(6) + (4)(6) - (20)(6) = 0$$
$$v_1' i_1 + v_2' i_2 + v_3' i_3 = (12)(2) + (48)(2) - (60)(2) = 0$$

and Tellegen's theorem is verified.

Example 2.22 Verify Tellegen's theorem in the circuit of Figure 2.65, in which the following set of voltages and currents exists.

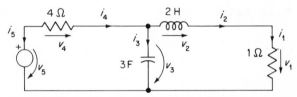

Figure 2.65 Verification of Tellegen's theorem in a circuit in which the voltages and currents are functions of time. [Example 2.22]

$v_1 = \sin 2t$	$i_1 = \sin 2t$
$v_2 = 4\cos 2t$	$i_2 = \sin 2t$
$v_3 = 4\cos 2t + \sin 2t$	$i_3 = -24\sin 2t + 6\cos 2t$
$v_4 = -92\sin 2t + 24\cos 2t$	$i_4 = -23\sin 2t + 6\cos 2t$
$v_5 = -91\sin 2t + 28\cos 2t$	$i_5 = 23\sin 2t - 6\cos 2t$

Coefficients on:

Terms	$\sin^2 2t$	$\cos^2 2t$	$\sin 2t \cos 2t$
$v_1 i_1$	1	0	0
$v_2 i_2$	0	0	4
$v_3 i_3$	-24	24	-90
$v_4 i_4$	2116	144	-1104
$v_5 i_5$	-2093	-168	1190
Sum	0	0	0

Another set of voltages and currents may be the following:

$v_1' = e^{-t}$	$i_1' = e^{-t}$
$v_2' = -2e^{-t}$	$i_2' = e^{-t}$
$v_3' = -e^{-t}$	$i_3' = 3e^{-t}$
$v_4' = 16e^{-t}$	$i_4' = 4e^{-t}$
$v_5' = 15e^{-t}$	$i_5' = -4e^{-t}$

Forming the products $v_k i'_k$ $(k = 1,\ldots, 5)$, we have

Coefficients on:

Terms	$e^{-t} \sin 2t$	$e^{-t} \cos 2t$
$v_1 i'_1$	1	0
$v_2 i'_2$	0	4
$v_3 i'_3$	3	12
$v_4 i'_4$	-368	96
$v_5 i'_5$	364	-112
Sum	0	0

Thus, Tellegen's theorem is verified.

2.20 Summary

The elements presented in this chapter are the constituents of circuits. The elements are mathematical abstractions. However, by putting together elements of the proper types and values, we construct mathematical models which approximate the electric elements in the real world.

Elements are characterized by the voltage-current relations that exist at the external terminals of the elements. Linear elements obey the homogeneity and the additivity properties.

Tellegen's theorem holds for any circuit.

SUGGESTED EXPERIMENTS

1. Design an experiment using realistic elements to demonstrate the smoothing property of the circuit of Problem 2.6.
2. Construct a circuit similar to that of Problem 2.7 to demonstrate the filtering property of the circuit.
3. Construct a circuit to demonstrate that a very high voltage can be produced by interrupting the current in an inductor.
4. Design an experiment to demonstrate that a high current pulse can be created in a capacitor by interrupting the voltage across it.
5. Construct a circuit similar to that of Example 2.17 and verify the waveforms of the voltages and currents.
6. Construct a circuit similar to that of Example 2.18. Observe the waveforms of the currents and voltages as the amplitude is varied. Apply the current to a spectrum analyzer and find all the significant sinusoidal components.
7. Display the nonlinear property of a ferrite core on an oscilloscope. Demonstrate the saturation property using the circuit of Example 2.19 and verify all waveforms.
8. Construct a simple resistor circuit containing two voltage sources. Measure all the voltages and currents. Verify Tellegen's theorem

that the sum of the power is zero. Now change the value of one of the sources. Measure all the voltages and currents. Verify the second part of Tellegen's theorem.

PROBLEMS

2.1 Find the voltages across and the currents in each of the elements of the circuit shown.

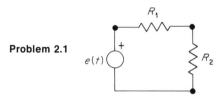

Problem 2.1

2.2 Find the currents in the resistors and voltage sources. Verify that all KVL and KCL equations are satisfied.

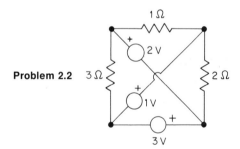

Problem 2.2

2.3 Find the voltages across the resistors and current sources. Verify that all KVL and KCL equations are satisfied.

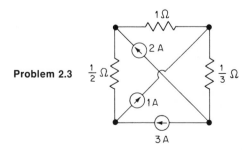

Problem 2.3

2.4   It is desired to produce a voltage $v(t)$ having a waveform shown using the circuit below. Find the source voltage $e(t)$ required.

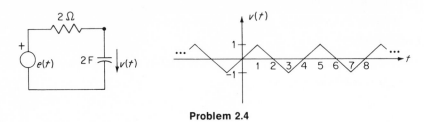

Problem 2.4

2.5   It is desired to produce a current $i(t)$ having a waveform shown using the circuit below. Find the source current $j(t)$ required.

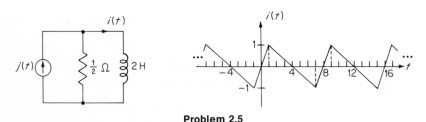

Problem 2.5

2.6   The voltage $v(t)$ consists of two sinusoidal components: $v(t) = 10^{-2} \cos(2\pi \times 10^3 t) + \cos(2\pi \times 10 t)$. The ratio of their amplitudes is 100. Find the ratio of the amplitudes of the two components of the input voltage $e(t)$.

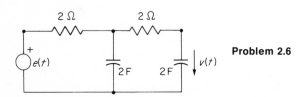

Problem 2.6

2.7   Repeat Problem 2.6 for the following circuit. If we wish to filter out the high-frequency component of the signal $e(t)$, should we use this circuit or the one in Problem 2.6?

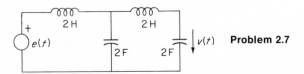

Problem 2.7

2.8 The following circuit has been proposed for use as a counter. The events are represented as short rectangular pulses of height 0.1 V and duration 1 μsec. How many pulses have been counted if the voltage $v(t)$ is 100 V? Sketch $v(t)$.

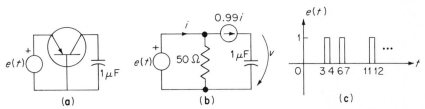

(a)   (b)   (c)

Problem 2.8 A circuit to count pulses. **(a)** The circuit; **(b)** the circuit model; **(c)** a series of pulses.

2.9 A simplified integrator circuit is shown below. Show that the voltage $v(t)$ is proportional to the integral of the input voltage $e(t)$.

Problem 2.9 A simplified integrator circuit.

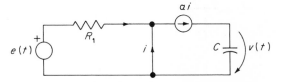

2.10 If the capacitor in the circuit of Problem 2.9 is replaced by an inductor, what possible uses would you suggest for the circuit?

2.11 The output voltage $v(t)$ of the following circuit is shown. Find the input voltage $e(t)$. Can you suggest some uses for this circuit?

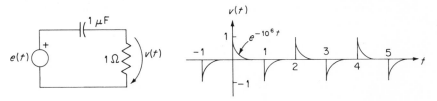

Problem 2.11 A peaking circuit.

2.12 In the circuit of Problem 2.11, if the voltage $v(t)$ is $\sin 10^3 t + 0.01 \sin 10t$, what is $e(t)$? Suggest some uses of the circuit.

2.13 Find the current and voltage that exist in each element in the circuit shown. $e(t) = \cos t$, $j(t) = \sin t$.

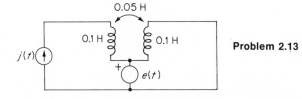

Problem 2.13

2.14 Find v/e in the following circuit.

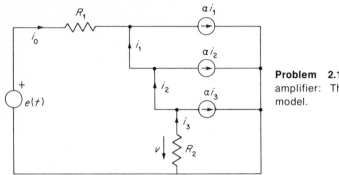

Problem 2.14 A transistor amplifier: The mathematical model.

2.15 Find v/e in the following circuit. Suggest some uses of the circuit, assuming $\alpha \simeq 1$.

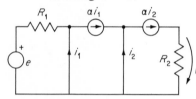

Problem 2.15 A transistor amplifier: The mathematical model.

2.16 a. Find the voltages and currents in all the elements. It is known that $v(t) = \sin 3t$.

   b. Find the energy stored in the capacitor and in the inductor. What is the average value of the energy in each element?

   c. Find the power taken by each element. Verify Tellegen's theorem.

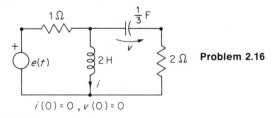

Problem 2.16

$i(0) = 0, v(0) = 0$

2.17 a. Find the voltage and current in each element.

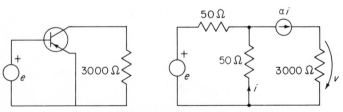

Problem 2.17 A simple transistor amplifier.

b. Calculate the power taken by each element. Verify Tellegen's theorem.

2.18 A three-terminal resistive element is characterized by the following equations relating the voltages and currents at the terminals:

$$v_1 = 10i_1 + 50i_2$$
$$v_2 = 100i_1 + 1000i_2$$

The element is embedded in a circuit shown. Find v_3/E. Also find the power taken by each element.

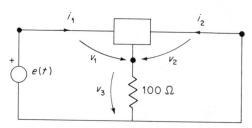

Problem 2.18 A three-terminal element in a resistor circuit.

2.19 A three-terminal element is characterized by the following v–i characteristic:

$$v_1 = 1i_1 + 2i_2$$
$$v_2 = 3i_1 + 4i_2$$

Two such elements are connected as shown. Find v/E.

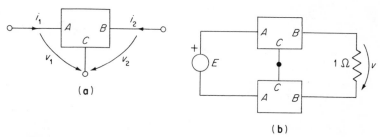

(a)

(b)

Problem 2.19 (a) A three-terminal element; (b) used twice in a circuit.

2.20 Sketch the voltages and currents in the following circuit. Verify Tellegen's theorem.

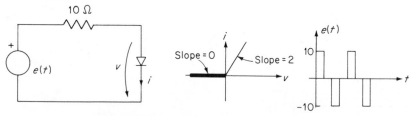

Problem 2.20 A diode circuit.

2.21   A nonlinear resistor has an *i–v* characteristic given by

$$i = v + v^3 + v^5$$

Find all the sinusoidal components of *i* if $v(t) = \sin \omega t$.

2.22   The charge-voltage relation of a nonlinear capacitor is as shown. Find the current in the capacitor if $v(t)$ is as shown. Sketch the energy stored in the capacitor as a function of time.

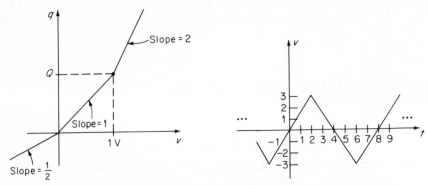

Problem 2.22   A nonlinear capacitor.

2.23   A nonlinear element (NL) has an *i–v* characteristic as shown. Find and sketch the voltages and currents everywhere in the circuit below.

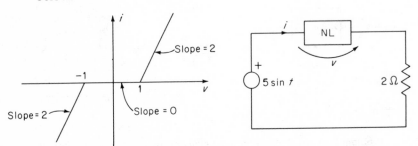

Problem 2.23   A nonlinear resistor circuit.

2.24   A nonlinear inductor has a ϕ–*i* characteristic as shown. Find and

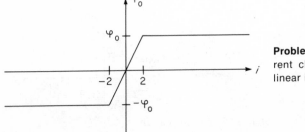

Problem 2.24   The flux-current characteristic of a nonlinear inductor.

sketch the voltage and flux in the inductor if the current is $i(t) = 10$ sin t. Find the energy stored in the inductor also.

2.25 A magnetic core has a ϕ–i characteristic as shown. If the current is as shown, sketch the flux and voltage.

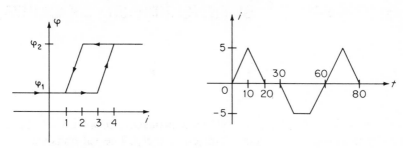

Problem 2.25 Calculation of voltage and flux in a magnetic core.

2.26 Find the voltages and currents in all the elements. Verify Tellegen's theorem.

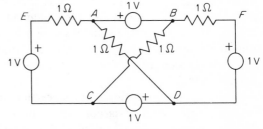

Problem 2.26 Verification of Tellegen's theorem.

2.27 An imperfect voltage source having an internal resistance of R_s is connected to a resistor of R Ω. What is the value of R such that the power taken by R is maximum? What is the power taken by the internal resistor under this condition?

2.28 If an ideal transformer is inserted between the source and the resistor of Problem 2.27, for a fixed value of R, what must be the turns ratio of the transformer in order that the maximum power is delivered to R?

3

analysis of simple linear circuits

In Chapter 2, we presented a systematic procedure to write a set of equations that describes a given circuit completely. The set consists of the KVL equations written for the fundamental loops, the KCL or current equations written for the fundamental cut sets, and the equations that define the elements of the circuit. While this procedure is universally applicable to any circuit, large or small, linear or nonlinear, it is nevertheless unnecessarily cumbersome when it is applied to simple circuits consisting of, say, one or two loops and a few elements. It seems that we should be able to develop analysis schemes which would allow us to find the solution of a simple circuit by *inspection,* and that we should reserve the more powerful procedure for large circuits or for computer programs which cannot "see."

These simple analysis schemes are all deducible from the most important principle of linear circuits, namely, the principle of superposition. In this chapter, we will derive this principle, which states that the response and excitation of a linear circuit are linearly related; that is, they are governed by the homogeneity and additivity properties. As a result, the response to a collection of excitations can be calculated by adding the responses to the excitations applied one at a time. Also, when it is convenient to do so, an excitation may be decomposed into a sum of excitations which, when applied one at a time, are more amenable to analysis than the excitation applied in whole. This is particularly true when the excitation is a signal that is a complicated function of time. In that case, we may wish to express it as a linear combination of elementary functions, for example. This and other far-reaching consequences of the principle of superposition will appear in later chapters.

One important consequence which will be taken up in this chapter is the idea of "equivalent circuits." This idea is usually applied to simplify the analysis of circuits. But more importantly, it enables us to replace a complicated circuit, or a circuit whose internal structure is not known, by one that consists of a few elements. The elements are determined from external measurements made on the circuit. It is this idea that we use to describe, or to regard, an antenna, microphone, loudspeaker, thermocouple, photoelectric cell, or the human heart as an electric circuit.

3.1 Principle of Superposition

We saw in Chapter 2 that for a single linear element, the response and the excitation satisfy the homogeneity and additivity properties. We now wish to extend this concept to a circuit containing linear elements only. Such a circuit will henceforth be called a *linear circuit*.

In a circuit, the excitations are the independent sources, the initial capacitor voltages, and the initial inductor currents. The response is the collection of voltages and currents on the elements. What can we say about the response of a linear circuit?

Consider the circuit shown in Figure 3.1. Let the voltages be $v_1, v_2,...,$ v_b and the currents be $i_1, i_2,..., i_b$. For convenience, let $e(t)$ and $j(t)$ be

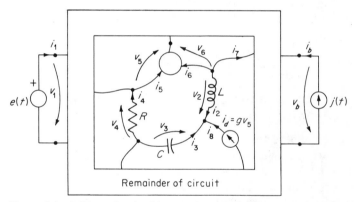

Figure 3.1 A linear circuit with a portion of its interior exposed (proof of the superposition principle).

the independent sources. The voltages and currents satisfy the KVL equations, KCL equations, and definitions of the elements. Typically, with reference to Figure 3.1, we can write the following equations:

$$\text{KVL:} \quad v_5 - v_6 + v_2 - v_3 + v_4 = 0 \tag{3.1}$$
$$\text{KCL:} \quad -i_6 - i_2 - i_7 = 0 \tag{3.2}$$
$$i_2 + i_3 + i_8 = 0 \tag{3.3}$$

Definitions of the elements:

$$v_2 = L\frac{di_2}{dt} \quad \text{or} \quad i_2 = \frac{1}{L}\int_0^t v_2(x)\ dx + i_2(0) \tag{3.4}$$

$$i_3 = C\frac{dv_3}{dt} \quad \text{or} \quad v_3 = \frac{1}{C}\int_0^t i_3(x)\ dx + v_3(0) \tag{3.5}$$

$$v_4 = Ri_4 \tag{3.6}$$

$$\left.\begin{array}{l} v_5 = r_{11}i_5 + r_{12}i_6 \\ v_6 = r_{21}i_5 + r_{22}i_6 \end{array}\right\} \quad \text{(3-terminal element)} \qquad (3.7)$$

$$i_8 = gv_5 \qquad \text{(dependent source)} \qquad\qquad (3.8)$$

$$v_1 = e(t) \qquad \text{(independent source)} \qquad (3.9)$$

$$i_b = j(t) \qquad \text{(independent source)} \qquad (3.10)$$

Now, suppose we multiply each of the equations by a constant K. We would get, among others,

KVL: $Kv_5 - Kv_6 + Kv_2 - Kv_3 + Kv_4 = 0$

KCL: $\qquad\qquad\qquad -Ki_6 - Ki_2 - Ki_7 = 0$

Definitions: $Kv_2 = KL\dfrac{di_2}{dt} = L\dfrac{d(Ki_2)}{dt} \qquad$ or

$$Ki_2 = \frac{K}{L}\int_0^t v_2(x)\,dx + Ki_2(0) = \frac{1}{L}\int_0^t (Kv_2)\,dx + Ki_2(0)$$

$$Ki_3 = KC\frac{dv_3}{dt} = C\frac{d(Kv_3)}{dt} \qquad \text{or}$$

$$Kv_3 = \frac{1}{C}\int_0^t (Ki_3)\,dx + Kv_3(0)$$

$$Kv_4 = KRi_4 = R(Ki_4)$$

$$Ki_8 = Kgv_5 = g(Kv_5)$$

$$Kv_1 = Ke(t)$$

$$Ki_b = Kj(t)$$

From the above, it is clear that the set of voltages $\{Kv_1, Kv_2,..., Kv_b\}$ and the set of currents $\{Ki_1, Ki_2,..., Ki_b\}$ satisfy all of the KVL, KCL, and the element equations. The excitations now are $Ke(t)$, $Kj(t)$, $Ki_2(0)$, and $Kv_3(0)$. The responses are Kv_2, $Kv_3,...,$ Ki_2, $Ki_3,...,$ and so on. These observations lead to the following remark:

Remark 3.1 (Homogeneity Property) In a linear circuit, let $\{v(t)\}$ be the set of voltages and $\{i(t)\}$ the set of currents resulting from the collection of excitations consisting of the set of independent sources $\{e(t), j(t)\}$, the set of initial capacitor voltages $\{u(0)\}$, and the set of initial inductor currents $\{l(0)\}$. The voltages and currents resulting from the excitations $\{Ke(t), Kj(t)\}, \{Ku(0)\}$, and $\{Kl(0)\}$ are $\{Kv(t)\}$ and $\{Ki(t)\}$, respectively. ∎

With reference to Figure 3.1 again, let v_1', $v_2',...,$ v_b' be the voltages and i_1', $i_2',...,$ i_b' be the currents resulting from the excitations $e'(t), j'(t), i_2'(0)$, and $v_3'(0)$. Let v_1'', $v_2'',...,$ v_b'' be the voltages and i_1'', $i_2'',...,$ i_b'' be the currents resulting from the excitations $e''(t), j''(t), i_2''(0)$, and $v_3''(0)$. We now wish to show that $v_1' + v_1''$, $v_2' + v_2'',...,$ $v_b' + v_b''$ are the voltages and $i_1' + i_1''$, $i_2' + i_2'',...,$ $i_b' + i_b''$ are the currents resulting from the excitations $e'(t) + e''(t), j'(t) + j''(t), i_2'(0) + i_2''(0)$, and $v_3'(0) + v_3''(0)$.

The voltages v_1', $v_2',...,$ v_b' and the currents i_1', $i_2',...,$ i_b' satisfy KCL, KVL, and the element equations. Therefore, we have, among others,

$$v_5' - v_6' + v_2' - v_3' + v_4' = 0$$
$$-i_6' - i_2' - i_7' = 0$$

$$v_2' = L\frac{di_2'}{dt} \quad \text{or} \quad i_2' = \frac{1}{L}\int_0^t v_2' \, dx + i_2'(0)$$

$$i_3' = C\frac{dv_3'}{dt} \quad \text{or} \quad v_3' = \frac{1}{C}\int_0^t i_3' \, dx + v_3'(0)$$

$$v_4' = Ri_4'$$
$$i_8' = gv_5'$$
$$v_1' = e'(t)$$
$$i_b' = j'(t)$$

Similarly, we have

$$v_5'' - v_6'' + v_2'' - v_3'' + v_4'' = 0$$
$$-i_6'' - i_2'' - i_7'' = 0$$

$$v_2'' = L\frac{di_2''}{dt} \quad \text{or} \quad i_2'' = \frac{1}{L}\int_0^t v_2'' \, dx + i_2''(0)$$

$$\vdots$$

$$v_1'' = e''(t)$$
$$i_b'' = j''(t)$$

We now add the corresponding equations in the two sets of equations and obtain

$$(v_5' + v_5'') - (v_6' + v_6'') + (v_2' + v_2'') - (v_3' + v_3'') + (v_4' + v_4'') = 0$$
$$- (i_6' + i_6'') - (i_2' + i_2'') - (i_7' + i_7'') = 0$$

$$(v_2' + v_2'') = L\frac{d}{dt}(i_2' + i_2'')$$

or

$$(i_2' + i_2'') = \frac{1}{L}\int_0^t (v_2' + v_2'') \, dx + [i_2'(0) + i_2''(0)]$$

$$\vdots$$

$$(v_1' + v_1'') = [e'(t) + e''(t)]$$
$$(i_b' + i_b'') = [j'(t) + j''(t)]$$

It is seen that the voltages $v_1' + v_1''$, $v_2' + v_2''$,..., $v_b' + v_b''$ and the currents $i_1' + i_1''$, $i_2' + i_2''$,..., $i_b' + i_b''$ satisfy the KCL, KVL, and the element equations. The excitations are now $e'(t) + e''(t)$, $j'(t) + j''(t)$, $i_2'(0) + i_2''(0)$, and $v_3'(0) + v_3''(0)$. These observations lead to the following remark:

Remark 3.2 (Additivity Property) In a linear circuit, let $\{v'\}$ be the set of voltages and $\{i'\}$ be the set of currents resulting from the excitations consisting of independent sources $\{e'(t), j'(t)\}$, initial capacitor voltages $\{u'(0)\}$, and initial inductor currents $\{l'(0)\}$. Let $\{v''\}$ and $\{i''\}$ be the voltages and currents resulting from another set of excitations $\{e'', j''(t)\}$, $\{u''(0)\}$, and $\{l''(0)\}$. The voltages and currents resulting from the excitations $\{e' + e'', j' + j''\}$, $\{u'(0) + u''(0)$, and $\{l'(0) + l''(0)\}$ are $\{v' + v''\}$ and $\{i' + i''\}$, respectively.

Remarks 3.1 and 3.2 together constitute what is known as the *principle of superposition* and are summarized in Figure 3.2.

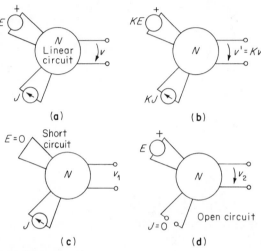

(a)   (b)

(c)   (d)

Figure 3.2 Summary of the principle of superposition. **(a)** A circuit with excitations E and J and response v; **(b)** the response is multiplied by K if the excitations are each multiplied by K; **(c)** the response due to J alone; **(d)** the response due to E alone. The response v is the sum of the responses v_1 and v_2.

If we let all but one of the excitations, sources, and initial conditions be zero in turn, we obtain the following useful rule:

In a linear circuit the response due to a collection of excitations applied to the circuit collectively is the sum of the responses due to the excitations applied individually.

It should be noted that it is the linearity of the elements that is really the basis of the principle of superposition. KVL and KCL hold for any circuits, but if the elements are nonlinear, the principle of superposition will not hold.

3.2 Two Useful Formulas

Many simple circuits take the form shown in Figure 3.3. The voltages can be found as follows:

Figure 3.3   Voltage divider circuit.

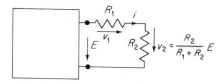

$$E = v_1 + v_2 = R_1 i + R_2 i$$
$$i = \frac{E}{R_1 + R_2} \tag{3.11}$$

and we have

$$v_1 = \frac{R_1}{R_1 + R_2} E \tag{3.12}$$

$$v_2 = \frac{R_2}{R_1 + R_2} E \tag{3.13}$$

Equations (3.12) and (3.13) are known as the *voltage divider formulas.*
They are useful in the analysis and simplification of circuits. Another

Figure 3.4   Current divider circuit.

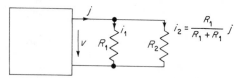

simple circuit is that shown in Figure 3.4. The currents can be found as
follows:

$$j = i_1 + i_2 = g_1 v + g_2 v = (g_1 + g_2)v$$
$$v = \frac{j}{g_1 + g_2}$$

$$i_1 = \frac{g_1}{g_1 + g_2} j = \frac{R_2}{R_1 + R_2} j \tag{3.14}$$

$$i_2 = \frac{g_2}{g_1 + g_2} j = \frac{R_1}{R_1 + R_2} j \tag{3.15}$$

Equations (3.14) and (3.15) are known as the *current divider formulas.*

Example 3.1   Find the voltages and currents in the circuit of Figure 3.5.

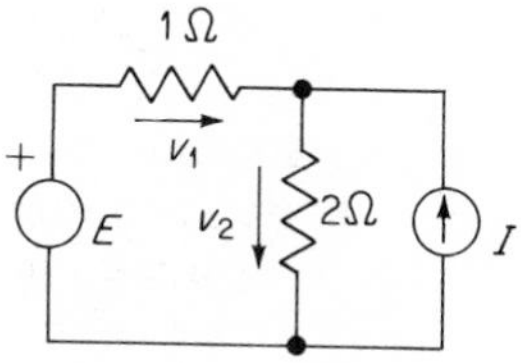

Figure 3.5   Illustration of superposition.   [Example 3.1]

Solution Since there are two sources, we may apply the principle of superposition. Applying the voltage source alone, we have the circuit of Figure 3.6(a).

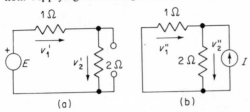

(a)   (b)

Figure 3.6 Application of the principle of superposition to the circuit of Figure 3.5. **(a)** Circuit with excitation E only; **(b)** circuit with excitation I only.

Applying the voltage divider formula, we get

$$v_1' = \frac{1}{1+2} E = \frac{1}{3} E$$

$$v_2' = \frac{2}{3} E$$

Applying the current source alone, we have the circuit shown in Figure 3.6(b). Using the current divider formula, we get

$$v_1'' = (-1) \frac{2}{1+2} I = \frac{-2}{3} I$$

$$v_2'' = (2) \frac{1}{1+2} I = \frac{2}{3} I$$

The voltages in the original circuit are

$$v_1 = v_1' + v_1'' = \tfrac{1}{3} E - \tfrac{2}{3} I$$
$$v_2 = v_2' + v_2'' = \tfrac{2}{3} E + \tfrac{2}{3} I$$

The voltages are linear combinations of the excitations as expected. As a check, we note that the KVL equation around the loop is

$$E - v_1 - v_2 = E - \tfrac{1}{3} E - \tfrac{2}{3} E = 0$$

Example 3.2 Find the voltages v_{AC}, v_{BD}, v_{AD}, and v_{BC} in the circuit of Figure 3.7.

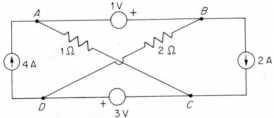

Figure 3.7 Illustration of superposition. [Example 3.2]

Solution Break up the problem into four smaller problems as shown in Figures 3.8(a), (b), (c), and (d). The circuits are obtained by setting all but one of the sources to zero in turn. In Figure 3.8(a), applying the voltage divider formula, we have

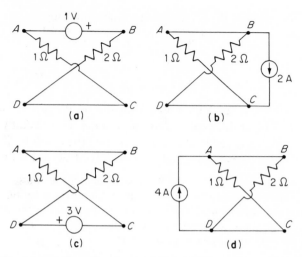

Figure 3.8 By superposition, the circuit of Figure 3.7 can be decomposed into four simpler circuits.

$$v_{AC} = \tfrac{1}{3}\,(-1) = -\tfrac{1}{3}\ \text{V}$$
$$v_{BD} = \tfrac{2}{3}\ \text{V}$$
$$v_{AD} = v_{AC} = -\tfrac{1}{3}\ \text{V}$$
$$v_{BC} = v_{BD} = \tfrac{2}{3}\ \text{V}$$

In Figure 3.8(b), applying the current divider formula, we get

$$v_{AC} = (1)(\tfrac{2}{3})(-2) = -\tfrac{4}{3}\ \text{V}$$
$$v_{BD} = (2)(\tfrac{1}{3})(-2) = -\tfrac{4}{3}\ \text{V}$$
$$v_{AD} = v_{AC} = -\tfrac{4}{3}\ \text{V}$$
$$v_{BC} = v_{BD} = -\tfrac{4}{3}\ \text{V}$$

In Figure 3.8(c), we have

$$v_{AC} = (\tfrac{1}{3})3 = 1\ \text{V}$$
$$v_{BD} = -(\tfrac{2}{3})3 = -2\ \text{V}$$
$$v_{AD} = v_{BD} = -2\ \text{V}$$
$$v_{BC} = v_{AC} = 1\ \text{V}$$

In Figure 3.8(d), we have

$$v_{AC} = (1)(\tfrac{2}{3})(4) = \tfrac{8}{3}\ \text{V}$$
$$v_{BD} = (2)(\tfrac{1}{3})(4) = \tfrac{8}{3}\ \text{V}$$
$$v_{AD} = v_{AC} = \tfrac{8}{3}\ \text{V}$$
$$v_{BC} = v_{BD} = \tfrac{8}{3}\ \text{V}$$

The voltages in the original circuit are

$$v_{AC} = -\tfrac{1}{3} - \tfrac{4}{3} + 1 + \tfrac{8}{3} = 2\ \text{V}$$
$$v_{BD} = \tfrac{2}{3} - \tfrac{4}{3} - 2 + \tfrac{8}{3} = 0\ \text{V}$$
$$v_{AD} = -\tfrac{1}{3} - \tfrac{4}{3} - 2 + \tfrac{8}{3} = -1\ \text{V}$$
$$v_{BC} = \tfrac{2}{3} - \tfrac{4}{3} + 1 + \tfrac{8}{3} = 3\ \text{V}$$

As a check, we note that the KVL equation around the loop $ABCD$ yields

$$v_{AB} + v_{BC} + v_{CD} + v_{DA} = -1 + 3 - 3 + 1 = 0$$

Also,

$$v_{AC} + v_{CD} + v_{DB} + v_{BA} = 2 - 3 + 0 + 1 = 0$$ ∎

The preceding two examples illustrate the utility of the principle of superposition. The problem of finding the solution of a circuit with a multitude of excitations is reduced to one of finding the solution of a number of simpler problems, each of which can be solved by inspection.

Example 3.3   The circuit of Figure 3.9 is a simplified model of a so-called three-phase power system. The three voltage sources are created by three specially placed windings on a generator. The system supplies power to three equal loads represented by the three R-Ω resistors. The resistor R_g represents the "return" wire or "ground." Find the power wasted in R_g.

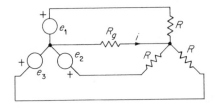

Figure 3.9 Illustration of superposition. [Example 3.3]

$e_1(t) = \sin t$

$e_2(t) = \sin(t + 2\pi/3)$

$e_3(t) = \sin(t - 2\pi/3)$

Solution   Reducing two of the three sources to zero in turn, we have the three circuits shown in Figure 3.10. By symmetry and homogeneity, we have

$$i_1' = Ce_1$$
$$i'' = Ce_2$$
$$i''' = Ce_3$$

where C is a constant. By additivity, we have

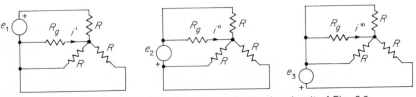

Figure 3.10   Application of superposition to the circuit of Fig. 3.9.

$$i = i' + i'' + i''' = C(e_1 + e_2 + e_3)$$
$$= C\left[\sin t + \sin\left(t + \frac{2\pi}{3}\right) + \sin\left(t - \frac{2\pi}{3}\right)\right]$$
$$= 0$$

so that the power in R_g is zero, and there is no power wasted, no matter how poor a conductor the return wire may be.

3.3 Equivalent Circuits*

We speak of the idea of one circuit being *equivalent* to another to mean that the voltage-current relations at the external terminals of one circuit are indistinguishable from the voltage-current relations at the external terminals of the other. Figure 3.11 shows two circuits, C1 and C2. The

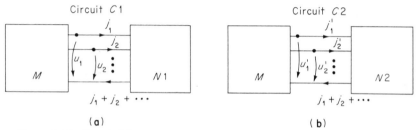

Figure 3.11   Circuit **(a)** is equivalent to circuit **(b)** if $u_i = u_i'$ and $j_i = j_i'$ for all i.

subcircuit $N1$ is connected to the subcircuit M in C1, and the subcircuit $N2$ is connected to a subcircuit which is identical to M. $N1$ and $N2$ may be structurally different from each other. One may contain more nodes or branches, for example. Let $u_1, u_2,..., u_n$ and $j_1, j_2,..., j_n$ be the voltages and currents at the external terminals of $N1$ as shown in Figure 3.11. Let $u_1', u_2',..., u_n'$ and $j_1', j_2',..., j_n'$ be the voltages and currents, respectively, at the external terminals of $N2$. $N1$ and $N2$ are said to be equivalent if, and only if, the following is true:

$$\{u\} = \{u'\} \leftrightarrow \{j\} = \{j'\} \tag{3.16}$$

In words, we say that $N1$ and $N2$ are equivalent if the same voltages on the two sets of terminals produce the same currents on the terminals, or if the same currents on the two sets of terminals produce the same voltages. Stated differently, Equation (3.16) says that $N1$ and $N2$ are equivalent if the mathematical equations relating the external voltages and currents in one are the same as the mathematical equations relating the external voltages and currents in the other.

The utility of equivalent circuits is in the simplification of circuits. If in Figure 3.11, we are interested in only the voltages and currents in M and if it is possible to find a circuit N' equivalent to N but simpler than

N in its structure and in the number of elements, then the analysis of the entire circuit will be simplified. The best-known simple example of equivalent circuit is the one in which two resistors connected in series are replaced by one resistor. We shall see more interesting examples presently.

Still other uses of equivalent circuits lie in the modeling of physical elements or circuits. Two models may be topologically different in their internal structure, but the external characterizations of the two may be the same. We regard the two as being equivalent insofar as the voltage-current relations are concerned.

Now an important question is: Suppose in $C1$ of Figure 3.11, $N1$ is replaced by its equivalent $N2$. Is it true that the internal voltages and currents of M remain unchanged? That they do is shown in the following theorem.

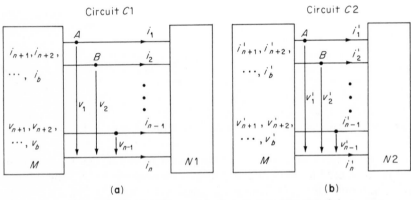

(a)   (b)

Figure 3.12   Circuits used in the proof of Theorem 3.1.

Theorem 3.1   Let $N1$ and $N2$ be equivalent circuits, each being connected to the identical circuit M as in Figures 3.12(a) and (b). Let $\{v\} = \{v_1, v_2,\ldots, v_b\}$ and $\{i\} = \{i_1, i_2,\ldots, i_b\}$ be the voltages and currents, respectively, in M

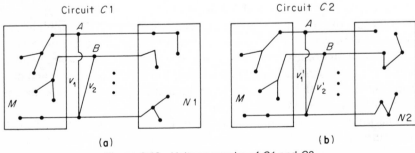

(a)   (b)

Figure 3.13   Voltage graphs of $C1$ and $C2$.

of $C1$ of Figure 3.12(a). Let $\{v'\} = \{v_1', v_2',..., v_b'\}$ and $\{i\} = \{i_1', i_2',..., i_b'\}$ be the corresponding quantities in M of $C2$ of Figure 3.12. Then

$$\{v\} = \{v'\}$$
$$\{i\} = \{i'\}$$

Proof Consider the voltage graph of $C1$. Select a tree which includes $v_1, v_2,...,$ v_{n-1} as tree branches. The tree is somewhat like that shown in Figure 3.13(a). Note that there cannot be a path from A to B that consists of only tree branches in $N1$, for otherwise we would have a loop of tree branches. The same is true for the other pairs of nodes on the terminals. Now write the set of KVL equations for the fundamental loops in M. The equations will include all the voltages v_1, $v_2,..., v_{n-1}$, but will not include any of the voltages in $N1$. Now construct the voltage graph of $C2$. Select a tree which includes $v_1', v_2',..., v_{n-1}'$ as tree branches and which has the same tree branches in M as those in M of $C1$. The tree is somewhat like that shown in Figure 3.13(b). The tree is the same as that of $C1$ except for the part in $N2$. Write the set of KVL equations for the fundamental loops of M. The equations will include all the voltages but not those in $N2$. In fact, the set of KVL equations is the same as the set written for M of $C1$.

Now consider the currents. With reference to Figure 3.12, write the KCL equations for the nodes and elements of M of $C1$. Do the same for M of $C2$. The two sets of equations are identical. Moreover, none of the current variables in $N1$ or $N2$ appear in either set of equations.

Next, consider the equations defining the elements of M. The equations relate the internal voltages $v_{n+1}, v_{n+2},..., v_b$ to the internal currents $i_{n+1}, i_{n+2},..., i_b$. The *same* equations relate the voltages $v_{n+1}', v_{n+2}',..., v_b'$ to the currents $i_{n+1}', i_{n+2}',..., i_b'$ in M of $C2$. By hypothesis, $N1$ and $N2$ are equivalent. The equations relating $v_1, v_2,..., v_n$ to $i_1, i_2,..., i_n$ are the same as those relating $v_1', v_2',..., v_n'$ to $i_1, i_2,..., i_n$. We now have two sets of variables $\{\{v\}, \{i\}\}$ and $\{\{v'\}, \{i'\}\}$. Both satisfy the *same* KVL equations, the *same* KCL equations, and the *same* equations relating the voltages and currents on the elements, including those relating the voltages and currents at the external terminals. The two sets of variables must be the same and we have

$$\{v\} = \{v'\} \quad \text{and} \quad \{i\} = \{i'\}$$

which is what is required to demonstrate. ∎

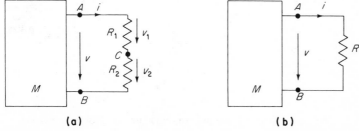

Figure 3.14 **(a)** Two resistors in series; **(b)** the equivalent resistance is the sum of the two resistances.

The basis of this theorem is again KVL, KCL, and the definitions of the elements. The theorem is valid for circuits containing any types of elements, linear or nonlinear, time invariant or time varying.

3.4 Series and Parallel Combinations of Resistors

Consider a circuit containing two resistors connected in series as shown in Figure 3.14(a). From KVL and the definition of a resistor, we have

$$v = v_1 + v_2 = R_1 i + R_2 i = (R_1 + R_2) i \tag{3.17}$$

In Figure 3.14(b), we have

$$v = Ri \tag{3.18}$$

so that the same mathematical relation holds in both circuits at the terminal AB if

$$R = R_1 + R_2 \tag{3.19}$$

As far as the voltage and current at AB are concerned, the circuit consisting of R_1 and R_2 in series is equivalent to one resistor whose value is $R_1 + R_2$. Moreover, by Theorem 3.1, if in Figure 3.14(a), the series combination is replaced by R, the voltages and currents in M remain unchanged. It is important to note that in the equivalent circuit [Figure 3.14(b)], node C has been eliminated.

The idea can be extended to n resistors connected in series. The total equivalent resistance is the sum of the n resistances, that is,

$$R_{\text{total}} = R_1 + R_2 + \cdots + R_n \tag{3.20}$$

Now consider two resistors connected in parallel as shown in Figure 3.15(a). Writing the usual equations, we have

$$i = i_1 + i_2 = (g_1 + g_2) v \tag{3.21}$$

In Figure 3.15(b), we have

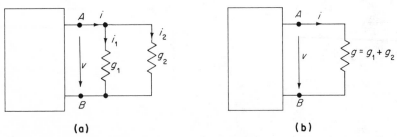

(a)   (b)

Figure 3.15 **(a)** Two resistors in parallel; **(b)** the equivalent conductance is the sum of the two conductances.

$$i = gv \qquad (3.22)$$

so that the two circuits are equivalent if

$$g = g_1 + g_2 \qquad (3.23)$$

or

$$R = \frac{1}{g} = \frac{R_1 R_2}{R_1 + R_2} \qquad (3.24)$$

The equivalent resistance is the product of the two resistances divided by the sum. Note that the equivalent circuit has one branch fewer than the original circuit.

The idea can be extended to n resistors connected in parallel. The total equivalent *conductance* is the sum of the conductances of the n resistors, that is,

$$g_{\text{total}} = g_1 + g_2 + \cdots + g_n$$

or

$$\frac{1}{R_{\text{total}}} = \frac{1}{R_1} + \frac{1}{R_2} + \cdots + \frac{1}{R_n} \qquad (3.25)$$

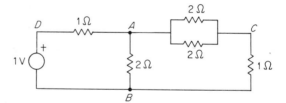

Figure 3.16 A resistor circuit. [Example 3.4]

Example 3.4 Find the voltages in the circuit of Figure 3.16.

Solution The circuit can be reduced into one which has only one loop by the following steps: (1) Combine the two 2-Ω resistors in parallel into one 1-Ω resistor [Figure 3.17(a)]. (2) Combine the two 1-Ω resistors in series into one 2-Ω resistor [Figure 3.17(b)]. (3) Combine the two 2-Ω resistors in parallel into one 1-Ω resistor [Figure 3.17(c)]. Note that in the last circuit, node C has disappeared. Applying the voltage divider formula to the last circuit, we have

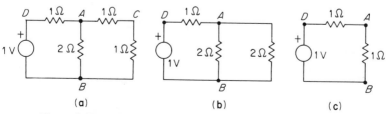

(a)   (b)   (c)

Figure 3.17 Reduction of Figure 3.16 into a one-loop circuit.

$$v_{AB} = \tfrac{1}{2} \text{ V}$$
$$v_{DA} = \tfrac{1}{2} \text{ V}$$

Now that we know v_{AB}, we get, with reference to Figure 3.17(a), the other voltages:

$$v_{CB} = (\tfrac{1}{2})(\tfrac{1}{2}) = \tfrac{1}{4} \text{ V}$$
$$v_{AC} = (\tfrac{1}{2})(\tfrac{1}{2}) = \tfrac{1}{4} \text{ V}$$

To check our results, we have in the original circuit

$$-1 + v_{DA} + v_{AC} + v_{CB} = -1 + \tfrac{1}{2} + \tfrac{1}{4} + \tfrac{1}{4} = 0$$

Example 3.5   Find the voltages and the power delivered to the circuit by the two sources in Figure 3.18.

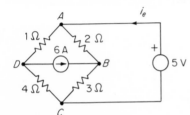

Figure 3.18 A simple nonseries-parallel circuit. [Example 3.5]

Solution   We must find i_e and v_{BD}. Applying the principle of superposition, we break up the problem into two. Figure 3.19 is one in which the excitation is the

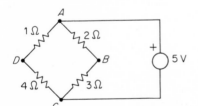

Figure 3.19 Application of superposition to Example 3.5.

voltage source alone. Using the voltage divider formula, we have

$$v_{AB} = (\tfrac{2}{5})5 = 2 \text{ V} \qquad v_{AD} = (\tfrac{1}{5})5 = 1 \text{ V}$$
$$v_{BC} = (\tfrac{3}{5})5 = 3 \text{ V} \qquad v_{DC} = (\tfrac{4}{5})5 = 4 \text{ V}$$

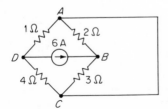

Figure 3.20 Application of superposition to Example 3.5.

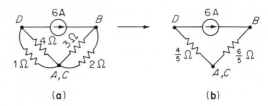

Figure 3.21 Simplification of Figure 2.20.

Figure 3.20 shows the other problem in which the excitation is the current source alone. The circuit can be redrawn as shown in Figure 3.21(a). The latter is equivalent to that shown in Figure 3.21(b). By inspection, we have

$$v_{AB} = -\left(\frac{6}{5}\right)6 = -\frac{36}{5} \text{ V} \qquad v_{AD} = \left(\frac{4}{5}\right)6 = \frac{24}{5} \text{ V}$$

$$v_{BC} = -v_{AB} = \frac{36}{5} \text{ V} \qquad v_{DC} = -v_{AD} = -\frac{24}{5} \text{ V}$$

The voltages in the original circuit with both sources present are

$$v_{AB} = 2 + \left(-\frac{36}{5}\right) = -\frac{26}{5} \text{ V} \qquad v_{AD} = 1 + \frac{24}{5} = \frac{29}{5} \text{ V}$$

$$v_{BC} = 3 + \frac{36}{5} = \frac{51}{5} \text{ V} \qquad v_{DC} = 4 - \frac{24}{5} = -\frac{4}{5} \text{ V}$$

From these voltages, we find

$$v_{BD} = v_{BC} - v_{DC} = 11 \text{ V}$$

The power delivered by the current source is $(11)(6) = 66$ W. Also,

$$i_{AD} = \frac{v_{AD}}{1} = \frac{29}{5} \text{ A}$$

$$i_{AB} = \frac{v_{AB}}{2} = -\frac{13}{5} \text{ A}$$

$$i_e = i_{AD} + i_{AB} = \frac{16}{5} \text{ A}$$

The power delivered by the voltage source is $(16/5)(5) = 16$ W. ∎

Note that in the preceding two examples, the solution was obtained more or less by inspection. The method of trees and chords was not used.

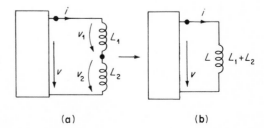

Figure 3.22 Two inductors in series.

3.5   Inductors and Capacitors in Series and Parallel

Consider two inductors in series as shown in Figure 3.22(a). Writing the usual equations, we have

$$v = v_1 + v_2 = L_1 \frac{di}{dt} + L_2 \frac{di}{dt} = (L_1 + L_2) \frac{di}{dt}$$

In Figure 3.22(b), we have

$$v = L \frac{di}{dt}$$

The two circuits will be equivalent if

$$L = L_1 + L_2 \tag{3.26}$$

In general, the total equivalent inductance of n inductors in series is the sum of the inductances of the n inductors.

Now consider two inductors connected in parallel as shown in Figure 3.23(a). Writing the usual equations, we have

$$i = i_1 + i_2 = \left(\frac{1}{L_1} + \frac{1}{L_2} \right) \int_0^t v(u) \ du + i(0)$$

In Figure 3.23(b), we have

$$i = \frac{1}{L} \int_0^t v(u) \ du + i(0)$$

The two circuits will be equivalent if

$$\frac{1}{L} = \frac{1}{L_1} + \frac{1}{L_2} \tag{3.27}$$

In general, the total equivalent inductance L_p of n inductors connected in parallel obeys the following relation:

$$\frac{1}{L_p} = \frac{1}{L_1} + \frac{1}{L_2} + \cdots + \frac{1}{L_n} \tag{3.28}$$

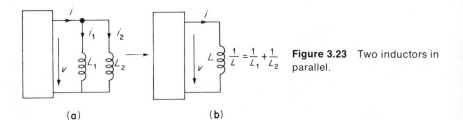

Figure 3.23 Two inductors in parallel.

(a)   (b)

Figure 3.24 Two capacitors in parallel.

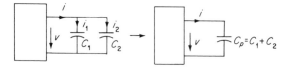

In an entirely analogous manner, we derive the equivalent capacitance of n capacitors in series and in parallel. With reference to Figure 3.24, we have, for the case of parallel connection,

$$C_p = C_1 + C_2 + \cdots + C_n \qquad (3.29)$$

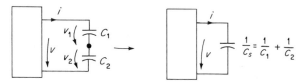

Figure 3.25 Two capacitors in series.

For the case of n capacitors in series, the equivalent capacitance C_s is given by, with reference to Figure 3.25,

$$\frac{1}{C_s} = \frac{1}{C_1} + \frac{1}{C_2} + \cdots + \frac{1}{C_n} \qquad (3.30)$$

We should note the dualism of these formulas. The dual quantities are resistance and conductance, inductance and capacitance, series connection and parallel connection.

3.6 Star-Mesh Transformation*

We noted in the last section that the combination of two resistors connected to a common node can be replaced by a single resistor. The common node is eliminated. The replacement does not, in any way, affect the voltages and the currents in the rest of the circuit. But the new circuit has one fewer node than the original. Thus the circuit has been simplified.

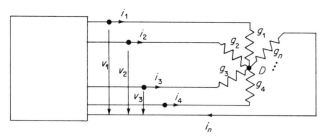

Figure 3.26 A star connection of resistors.

The natural question to ask is whether or not this simplification by the elimination of one node can be extended to the case of n resistors connected to a common node. The answer is yes.

Consider the circuit of Figure 3.26 in which we show n resistors connected to a common node D. The voltages and currents are defined as shown. Such a combination is known as a *star* connection. The circuit can be described in terms of the voltage-current relations given by the following equations:

$$i_1 = G_{11} v_1 + G_{12} v_2 + \cdots + G_{1(n-1)} v_{n-1}$$
$$i_2 = G_{21} v_1 + G_{22} v_2 + \cdots + G_{2(n-1)} v_{n-1}$$
$$i_{n-1} = G_{(n-1)1} v_1 + G_{(n-1)2} v_2 + \cdots + G_{(n-1)(n-1)} v_{n-1} \qquad (3.31)$$

Equation (3.31) is really a statement of the principle of superposition. If we regard the voltages as the excitations and the currents as the responses, the equations simply say that the responses are linear combinations of the excitations. The coefficients G_{kj} have dimension of conductance. To relate the coefficients G_{kj} to the conductances of the resistors, we use the principle of superposition. For example, suppose $v_2 = v_3 = \cdots = v_{n-1} = 0$. Then we have

$$i_k = G_{k1} v_1 \qquad k = 1, 2,\ldots, n - 1 \qquad (3.32)$$

or

$$G_{k1} = \frac{i_k}{v_1} \qquad \begin{array}{l} k = 1, 2,\ldots, n - 1 \text{ with} \\ v_2 = v_3 = \cdots = v_{n-1} = 0 \end{array} \qquad (3.33)$$

In general, we have

$$G_{kj} = \frac{i_k}{v_j}\bigg]_{v_r=0 \text{ (all } r \neq j)} \qquad (3.34)$$

Applying Equation (3.33) to the star of Figure 3.26, we have the circuit of Figure 3.27(a), which can be redrawn as that of Figure 3.27(b). By inspection of Figure 3.27(b), we obtain

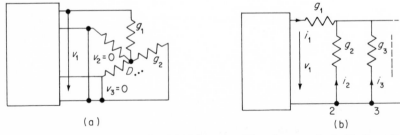

(a)   (b)

Figure 3.27 Calculation of G_{k1}. **(a)** All but one of the wires are shorted to the reference node; **(b)** simplification of the circuit of (a).

$$i_1 = \cfrac{v_1}{\cfrac{1}{g_1} + \cfrac{1}{g_2 + g_3 + \cdots + g_n}}$$

$$= \frac{g_1\,g_2 + g_1\,g_3 + \cdots + g_1\,g_n}{g_1 + g_2 + \cdots + g_n}\, v_1 \qquad (3.35)$$

so that

$$G_{11} = \frac{g_1\,g_2 + g_1\,g_3 + \cdots + g_1\,g_n}{g_1 + g_2 + \cdots + g_n} \qquad (3.36)$$

Solving for i_2, by means of the current divider formula, we obtain

$$i_2 = \frac{-g_2}{g_2 + g_3 + \cdots + g_n}\, i_1$$

$$= \frac{-g_1\,g_2}{g_1 + g_2 + \cdots + g_n}\, v_1 = G_{21} v_1 \qquad (3.37)$$

Repeating the same argument for all j, k, we get

$$G_{kj} = \frac{-g_k\,g_j}{g_1 + g_2 + \cdots + g_n} \qquad k \neq j \qquad (3.38)$$

$$G_{kk} = \frac{g_1\,g_k + g_2\,g_k + \cdots + g_{(k-1)k}\,g_k + g_{(k+1)k}\,g_k + \cdots + g_n\,g_k}{g_1 + g_2 + \cdots + g_n} \qquad (3.39)$$

Note that by Equation (3.38), $G_{kj} = G_{jk}$.

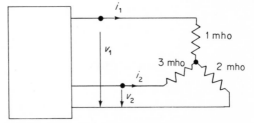

Figure 3.28 A simple example of a star of resistors.

As an example, Figure 3.28 shows a star of three resistors which, by using the formulas derived above, has the following characterization:

$$i_1 = \tfrac{5}{6}\, v_1 - \tfrac{3}{6}\, v_2 \qquad (3.40)$$
$$i_2 = -\tfrac{3}{6}\, v_1 + \tfrac{9}{6}\, v_2 \qquad (3.41)$$

Now consider the combination of resistors shown in Figure 3.29. There is a resistor connecting every pair of nodes so that the total number of resistors is $n(n-1)/2$. (Why?) Defining the voltages and currents as shown, we can write, by superposition again, the following equations:

$$i_1 = G'_{11}\, v_1 + G'_{12}\, v_2 + \cdots + G'_{1(n-1)}\, v_{n-1}$$
$$i_2 = G'_{21}\, v_1 + G'_{22}\, v_2 + \cdots + G'_{2(n-1)}\, v_{n-1}$$
$$\vdots$$
$$i_{n-1} = G'_{(n-1)1}\, v_1 + G'_{(n-1)2}\, v_2 + \cdots + G'_{(n-1)(n-1)}\, v_{n-1} \qquad (3.42)$$

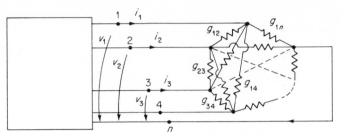

Figure 3.29   A mesh connection of resistors.

The conductances G'_{kj} can be calculated as before, that is,

$$G'_{kj} = \frac{i_k}{v_j}\Bigg]_{v_r=0\ (r\neq j)} \tag{3.43}$$

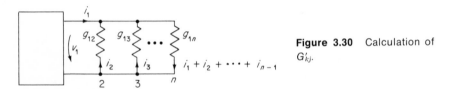

Figure 3.30   Calculation of G_{kj}.

Applying Equation (3.43) to the case of $j = 1$, we have, with reference to Figure 3.30,

$$i_1 = (g_{12} + g_{13} + \cdots + g_{1n})v_1 \tag{3.44}$$
$$i_k = -g_{1k}\,v_1 \qquad k \neq 1 \tag{3.45}$$

After obvious generalization, it follows that

$$G'_{kk} = g_{1k} + g_{2k} + \cdots + g_{(k-1)k} + g_{(k+1)k} + \cdots + g_{nk} \tag{3.46}$$
$$G'_{kj} = -g_{kj} \qquad\qquad j \neq k \tag{3.47}$$

The two sets of current-voltage relations (3.31) and (3.42) will be the same, and the two circuits will be equivalent if, and only if, $G_{kk} = G'_{kk}$ and $G_{kj} = G'_{kj}$, or

$$g_{kj} = \frac{g_k\,g_j}{g_1 + g_2 + \cdots + g_n} \tag{3.48}$$

As an example, the circuit of Figure 3.28 is equivalent to that of Figure 3.31. Applying Equations (3.46) and (3.47), we have

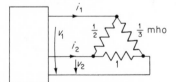

Figure 3.31   The mesh equivalent of the star of Figure 3.26.

$$G'_{11} = g_{12} + g_{13} = \tfrac{5}{6}$$
$$G'_{12} = -g_{12} = -\tfrac{1}{2}$$
$$G'_{22} = g_{21} + g_{23} = \tfrac{3}{2}$$

The equations characterizing the circuit are

$$i_1 = \tfrac{5}{6} v_1 - \tfrac{1}{2} v_2$$
$$i_2 = -\tfrac{1}{2} v_1 + \tfrac{3}{2} v_2$$

which are the same as those of the circuit of Figure 3.28.

It appears that the star-mesh transformation yields a reduction of one node at the price of increasing the number of resistors from n to $n(n-1)/2$. There does not seem to be any simplification. However, the newly created elements can sometimes be combined with elements in the rest of the circuit. By repeated application, we can reduce a circuit into one that has only one loop and two nodes.

Example 3.6 Find the voltages between pairs of nodes in the circuit of Figure 3.32(a).

Solution With reference to Figure 3.32(a), the star of four resistors at node D can be replaced by its equivalent as in Figure 3.32(b). The resistors in parallel can be combined in Figure 3.32(b), and we obtain the circuit of Figure 3.32(c).

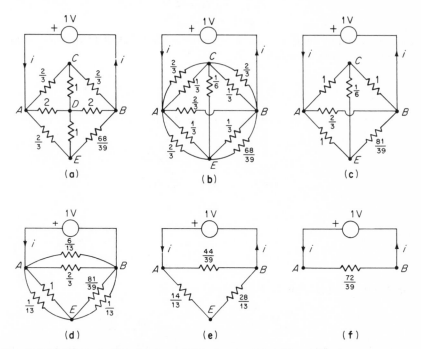

Figure 3.32 Repeated application of the star-mesh transformation to analyze a resistor circuit. (Conductances in mhos) [Example 3.6]

Node C of Figure 3.32(c) can be eliminated by replacing the star of three resistors at node C. The result is shown in Figure 3.32(d). Combining resistors in parallel again, we get Figure 3.32(e). Now node E can be eliminated by replacing the star of two resistors at node E. The result is the one-loop circuit of Figure 3.32(f).

To find the voltages and the currents in the original circuit, we go backwards. From Figure 3.32(f), we calculate the current to be

$$i = \frac{72}{39} \text{ A}$$

Restore the star of two resistors as in Figure 3.32(e). Calculate (update) the currents in the nonstar branches. We get, in this case,

$$i_{AB} = \frac{44}{39} \text{ A}$$

Calculate the currents in the branches of the star:

$$i_{AE} = i - i_{AB} = \frac{28}{39} \text{ A}$$
$$i_{EB} = i - i_{AB} = \frac{28}{39} \text{ A}$$

Calculate the voltages on the branches of the star:

$$v_{AE} = r_{AE} \, i_{AE} = \tfrac{2}{3} \text{ V}$$
$$v_{EB} = r_{EB} \, i_{EB} = \tfrac{1}{3} \text{ V}$$

Restore the star of three resistors as in Figure 3.32(c). Calculate (update) the currents in the nonstar branches:

$$i_{AB} = \frac{2}{3} v_{AB} = \frac{2}{3} \text{ A} \qquad i_{AE} = 1 \; v_{AE} = \frac{2}{3} \text{ A} \qquad i_{EB} = \frac{81}{39} v_{EB} = \frac{27}{39} \text{ A}$$

Calculate the currents in the branches of the star by KCL:

$$i_{AC} = i - i_{AB} - i_{AE} = \frac{20}{39} \text{ A}$$

$$i_{EC} = i_{AE} - i_{EB} = -\frac{1}{39} \text{ A}$$

$$i_{CB} = i - i_{EB} - i_{AB} = \frac{19}{39} \text{ A}$$

Check:

$$i_{AC} + i_{EC} - i_{CB} = 0$$

Calculate the voltages on the branches of the star:

$$v_{AC} = \frac{20}{39} \text{ V} \qquad v_{EC} = -\frac{6}{39} \text{ V} \qquad v_{CB} = \frac{19}{39} \text{ V}$$

Restore the star of four resistors as in Figure 3.32(a). Update the currents in the nonstar branches. Calculate the currents in the star branches by KCL and then the voltages on the branches. We get, finally,

$$v_{AD} = \frac{186}{351} \text{ V} \qquad\qquad v_{AC} = \frac{20}{39} \text{ V}$$

$$v_{CD} = \frac{2}{117} \text{ V} \qquad\qquad v_{EC} = -\frac{6}{39} \text{ V}$$

$$v_{ED} = -\frac{16}{117} \text{ V} \qquad\qquad v_{CB} = \frac{19}{39} \text{ V}$$

$$v_{BD} = -\frac{55}{117} \text{ V}$$

As a check, we note that

$$v_{AD} - v_{BD} - v_{EB} - v_{AE} = 186/351 + 55/117 - 1/3 - 2/3 = 0. \qquad \blacksquare$$

It is seen from this example that by repeated applications of the star-mesh transformation, we can obtain the solution of a circuit. Note the iterative nature of the algorithm that was used to find the voltages and currents. In fact, computer programs which are based on this technique have been written. The algorithm is shown by the flow chart of Figure 3.33. We first eliminate the nodes of stars one by one until we have a circuit of one loop and two nodes, the latter being those of the source. To recover the voltages and currents in the original circuit, we first calculate the current in the last element. We next restore the last star, update the resistances in the branches, calculate the currents in the nonstar branches, calculate the currents in the branches of the star by means of KCL, and finally calculate the voltages on the branches of the star. The process is repeated until all stars have been restored.

Star-mesh transformations can also be applied to capacitors or inductors connected at a common node. The formulas derived for the case of resistors are directly applicable if the conductances are replaced by capacitances in the case of capacitor stars, and by the reciprocals of inductances in the case of inductor stars.

3.7 Thévenin Equivalent Circuit

In the analysis of circuits, we are frequently interested in only the voltage and current at one or two points in the circuit. While the analysis techniques of using trees and chords or using the star-mesh transformation will always give us the solution of the circuit, these techniques also give us *all* the voltages and *all* the currents of the circuit. In order to find the voltage or current at one point, we may have to calculate all the voltages and currents even though we do not wish to know all of them. It would be desirable, therefore, to devise some way to bypass the intermediate steps and calculate only those voltages and currents of interest. The Thévenin and Norton equivalent circuits do precisely that. Another important significance of Thévenin and Norton equivalent circuits will be noted in Section 3.9.

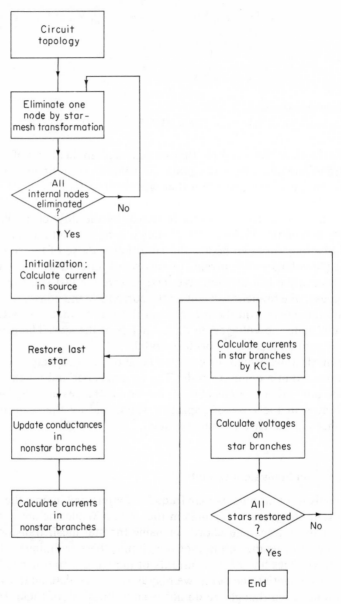

Figure 3.33 Flow chart of algorithm for the solution of circuits by the star-mesh transformation.

Consider a circuit decomposed into two parts M and N, which are connected at *two* nodes A and B as shown in Figure 3.34. Let v and i be the voltage and current at the terminals A and B. M and N may contain voltage and current sources. Suppose N is linear. We wish to find an equivalent circuit for N.

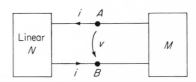

Figure 3.34 Network N is to be replaced by its equivalent circuit.

With reference to Figure 3.34, we may regard the voltage v as a response. The excitation comes from two sources, one being the external current i and the other being the collection of independent voltage and current sources in N. By the principle of superposition, v is the sum of two terms, one being due to i alone (with all independent sources in N reduced to zero) and the other due to the internal sources in N alone (with i set to zero). In equation form, we have

$$v = v_1 + v_2 \qquad (3.49)$$

where

$v_1 =$ voltage across AB under the condition that i is set to zero

$v_2 = v$ with all internal independent sources reduced to zero and the circuit N is excited by the current i alone

Equation (3.49) is the voltage-current relation describing the circuit N at the terminals AB. Schematically, the voltages v_1 and v_2 are defined as shown in Figures 3.35(a) and (b). The internal independent sources of N are exposed and are represented collectively by $e(t)$ and $j(t)$. N_0 represents the circuit N with all internal independent sources reduced to zero.

Now we assert that the circuit N is equivalent to the combination shown in Figure 3.36, where N_0 is as defined before and $e_{oc} = v_1$. To see this, we need only show that the circuit of Figure 3.36 has the same

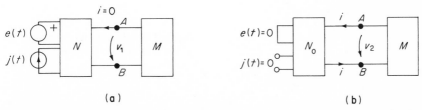

(a)   (b)

Figure 3.35 Definition of v_1 and v_2 in the derivation of the Thévenin equivalent circuit.

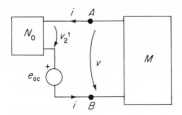

Figure 3.36   Thévenin equivalent circuit of N of Figure 3.34.

voltage-current relation at the terminal AB as N of Figure 3.34. In Figure 3.36, we have

$$v = v_2' + e_{oc} \qquad (3.50)$$

By hypothesis, $e_{oc} = v_1$. Now

$$
\begin{aligned}
v_2' &= v \qquad \text{(with } e_{oc} \text{ set to zero)} \\
&= v \qquad \text{(with internal independent sources reduced to zero)} \\
&= v_2 \qquad\qquad\qquad\qquad\qquad\qquad\qquad\qquad (3.51)
\end{aligned}
$$

Therefore, we have

$$v = v_1 + v_2$$

with is the voltage-current relation of N.

The e_{oc}–N_0 combination is known as the Thévenin equivalent circuit of N. The voltage source e_{oc} is called the *open-circuit* voltage at AB, to signify that its value is computed under the condition that the terminals A and B are opened $(i = 0)$.

The utility of Thévenin equivalent circuit is in the simplification of that part of a circuit of no interest in the analysis. The simplification frequently leads to a new circuit which can be solved by inspection.

Example 3.7   Find the voltage v_{AB} in the circuit of Figure 3.37.

Solution   Replace the circuit to the left of AB by its Thévenin equivalent circuit. We first calculate e_{oc}, which is the voltage across AB under the condition that $i = 0$. Figure 3.38(a) shows the circuit by which e_{oc} is calculated. By inspection, we have

$$e_{oc} = \tfrac{2}{3} \text{ V}$$

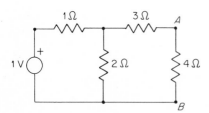

Figure 3.37   Solution of a circuit by its Thévenin equivalent.   [Example 3.7]

Next, we have N_0, which is the circuit to the left of AB under the condition that all internal independent sources are reduced to zero. Figure 3.38(b) shows the

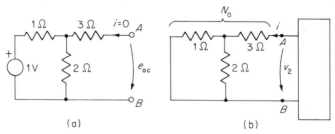

(a)   (b)

Figure 3.38 Calculation of e_{oc} and v_2 in the Thévenin equivalent circuit.

circuit N_0. Now note that N_0 can be reduced to a single resistor insofar as the voltage and current at AB are concerned. The value of the resistance is

$$R = 3 + \frac{2}{3} = \frac{11}{3}\ \Omega$$

Combining e_{oc} and N_0 in series, we get the Thévenin equivalent circuit of N as shown in Figure 3.39. By inspection, the voltage v_{AB} is given by

$$v_{AB} = \frac{4}{4 + (11/3)} \cdot \frac{2}{3} = \frac{8}{23}\ V$$

Figure 3.39 The Thévenin equivalent circuit of Figure 3.37.

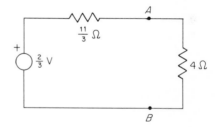

Example 3.8 Find the voltage v_{AB} in the circuit of Figure 3.40.

Figure 3.40 Solution of a circuit by use of Thévenin equivalent. [Example 3.8]

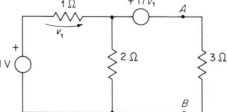

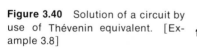

Solution   Replace the circuit to the left of AB by its Thévenin equivalent circuit. The voltage e_{oc} is found from Figure 3.41 as follows:

$$v_1 = \tfrac{1}{3} \quad \text{and} \quad v_2 = \tfrac{2}{3} \qquad \text{(by the voltage divider formula)}$$

By KVL, we have

$$e_{oc} = -17v_1 + v_2 = -5 \text{ V}$$

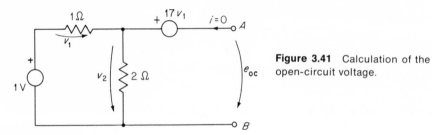

Figure 3.41 Calculation of the open-circuit voltage.

The circuit N_0 is shown in Figure 3.42, which can be simplified into a single resistor as follows. By KVL

$$v = -17v_1 - v_1 = -18v_1$$
$$v_1 = -\tfrac{2}{3} i \qquad \text{(by combining the two resistors in parallel)}$$

Therefore,

$$v = -18(-\tfrac{2}{3})i = 12i$$

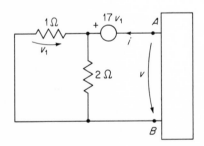

Figure 3.42 Calculation of the equivalent resistance.

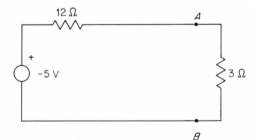

Figure 3.43 The Thévenin equivalent circuit of Figure 3.40.

The circuit N_0 is equivalent to a resistor of 12 Ω. The Thévenin equivalent circuit is shown in Figure 3.43. By inspection, $v_{AB} = (-5)(3/15) = -1$ V.

3.8 Norton Equivalent Circuit

We have yet another way of simplifying circuits. With reference to Figure 3.34, we may regard the current as the response and the voltage v and the internal sources in N as excitations. By superposition, we can write

$$i = i_1 + i_2 \qquad (3.52)$$

where

$i_1 = $ current i with $v = 0$
$i_2 = $ current i with all internal independent sources reduced to zero

Figure 3.44 shows the definition of i_1 and i_2. Now, we assert that the cir-

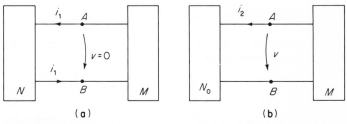

(a)   (b)

Figure 3.44 Definition of i_1 and i_2 in the derivation of the Norton equivalent circuit.

cuit shown in Figure 3.45 is equivalent to the circuit of Figure 3.34. To see this, we note that in Figure 3.45

$$i = j_{sc} + i' \qquad (3.53)$$

Now $j_{sc} = i_1$ by hypothesis, and

$$
\begin{aligned}
i' &= i && \text{(with } j_{sc} = 0) \\
&= i && \text{(with internal independent sources reduced to zero)} \\
&= i_2 && (3.54)
\end{aligned}
$$

Therefore,

$$i = i_1 + i_2$$

Figure 3.45 Norton equivalent circuit. $j_{sc} = i_1$

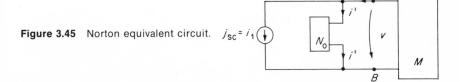

which is the current-voltage relation of Figure 3.34 at the terminals AB. The circuit to the left of AB in Figure 3.45 is known as the Norton equivalent circuit. The source j_{sc} is called the *short-circuit* current at AB to signify that its value is computed under the condition that a short circuit is placed across AB ($v = 0$).

Example 3.9 Find the voltage v_{AB} of Figure 3.37 by using the Norton equivalent circuit.

Solution The short-circuit current can be found from Figure 3.46(a). We first

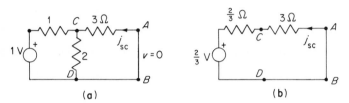

(a)   (b)

Figure 3.46   Calculation of the short-circuit current.

replace the circuit to the left of CD by its Thévenin equivalent and get the circuit of Figure 3.46(b). Now by inspection,

$$j_{sc} = -\left(\frac{2}{3}\right)\left(\frac{3}{11}\right) = -\frac{2}{11} \text{ A}$$

The circuit N_0 is the same as that of Example 3.7. (Why?) Combining the source j_{sc} in parallel with the equivalent resistor of N_0, we have the Norton equivalent circuit of N as shown in Figure 3.47. By inspection, we find the voltage v_{AB} to be $(+2/11)(44/23) = 8/23$ V as before.

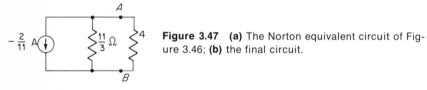

Figure 3.47   (a) The Norton equivalent circuit of Figure 3.46; (b) the final circuit.

Example 3.10 Do Example 3.8 using the Norton equivalent circuit.

Solution The short-circuit current source is found from Figure 3.48 as follows.

$$i_2 = \frac{17v_1}{2} \qquad i_1 = v_1$$

By KVL,

$$1 = v_1 + 17v_1 = 18v_1 \qquad v_1 = \frac{1}{18}$$

By KCL,

$$j_{sc} = i_2 - i_1 = \frac{17}{36} - \frac{1}{18} = \frac{5}{12} \text{ A}$$

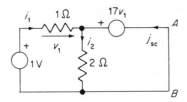

Figure 3.48 Short-circuit current of Example 3.8. [Example 3.10]

Again the circuit N_0 is the same as before. The Norton equivalent circuit is shown in Figure 3.49. By combining the two resistors into one, we get the voltage v_{AB} given by $(-5/12)(36/15) = -1$ V as before.   ∎

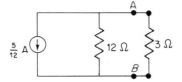

Figure 3.49 Norton equivalent of Example 3.8.

In the preceding examples, we were able to simplify each equivalent circuit into one that consists of just two elements, a source and an equivalent resistor. It is precisely this simplification that makes the Thévenin and Norton equivalent circuits useful. Can this simplification be done always? Certainly, the open-circuit voltage and the short-circuit current can always be found. Now, if N_0 contains resistors and dependent sources only, we can always reduce it into a single resistor. To see this, we note that the voltage v_2 and i in the Thévenin equivalent [Figure 3.35(b)] or the current i_2 and v in the Norton equivalent circuit [Figure 3.44(b)] must be linearly and algebraically related, since all the equations governing the voltages and currents in N_0 are linear algebraic equations. That is to say, v_2 is proportional to i, and N_0 is equivalent to a single resistor whose value is the proportionality constant. To be specific, we show in Figures 3.50(a) and (b) the Thévenin and Norton equivalent circuits, respectively, in which R_{eq} is the equivalent resistor in question. The computation of

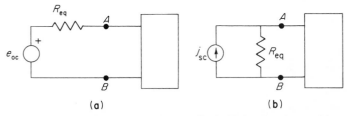

(a)   (b)

Figure 3.50 (a) Equivalent resistance R_{eq} in Thévenin circuit; (b) in Norton circuit.

R_{eq} is facilitated by the following observations. In Figure 3.50(a), suppose we were to find the Norton equivalent circuit to the left of AB. We would get, as the short-circuit current, in the direction from A to B,

$$j_{sc} = \frac{e_{oc}}{R_{eq}} \tag{3.55}$$

Similarly, if we were to find the Thévenin equivalent circuit of Figure 3.50(b), we would get, as the open-circuit voltage, in the direction from A to B,

$$e_{oc} = R_{eq} j_{sc} \tag{3.56}$$

From Equation (3.55) or (3.56), we obtain a formula for the equivalent resistance R_{eq} given by

$$R_{eq} = \frac{e_{oc}}{j_{sc}} \tag{3.57}$$

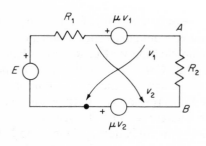

Figure 3.51 A circuit to approximate an ideal current source. [Example 3.11]

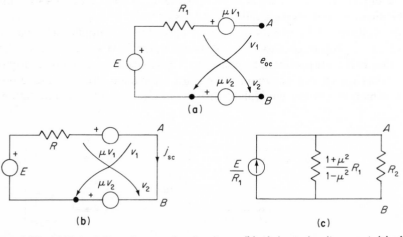

Figure 3.52 (a) Calculation of open-circuit voltage; (b) of short-circuit current; (c) of Norton equivalent circuit.

The formula is useful in situations where it is more convenient to compute the two quantities e_{oc} and j_{sc} than to compute R_{eq} directly from N_0.

Example 3.11   Find the Norton equivalent circuit to the left of AB of the circuit of Figure 3.51 in which $\mu \neq 1$.

Solution   The open-circuit voltage is found from Figure 3.52(a). By inspection, we have

$$e_{oc} = v_1 + \mu v_2$$

Also, we have

$$v_1 = -\mu v_1 + E \qquad v_1 = \frac{E}{1 + \mu}$$

$$v_2 = E + \mu v_2 \qquad v_2 = \frac{E}{1 - \mu}$$

Therefore,

$$e_{oc} = \frac{1 + \mu^2}{1 - \mu^2} E$$

The short-circuit current is found from Figure 3.52(b). Straightforward calculation shows that $v_1 = v_2 = 0$ in this case, and

$$j_{sc} = \frac{E}{R_1}$$

The equivalent resistor R_{eq} is

$$R_{eq} = \frac{e_{oc}}{j_{sc}} = \frac{1 + \mu^2}{1 - \mu^2} R_1$$

The equivalent circuit is shown in Figure 3.52(c). It will be noted that if $\mu^2 \simeq 1$, the equivalent resistance is very large, and the equivalent circuit is approximately that of an ideal current source.

Example 3.12   Figure 3.53(a) shows a simplified version of a transistor amplifier whose equivalent circuit is shown in Figure 3.53(b). Find the Thévenin equivalent circuit seen by the load resistor R_L.

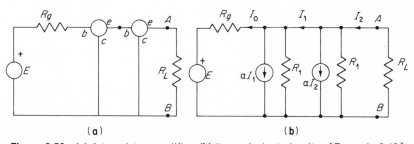

(a)   (b)

Figure 3.53   **(a)** A transistor amplifier; **(b)** its equivalent circuit.   [Example 3.12]

Solution   The circuit from which the open-circuit voltage is found is shown in Figure 3.54(a). We have

$$e_{oc} = -R_1 I_1$$

KVL: $$E - e_{oc} = -R_g I_0$$

KCL: $$I_1 = I_0 + \alpha I_1 + \frac{e_{oc}}{R_1}$$

Combining equations, we get

$$e_{oc} = \frac{E}{(R_g/R_1) + (R_g/R_1)(1 - \alpha) + 1}$$

The circuit from which the short-circuit current is found is shown in Figure 3.54(b). We have

$$j_{sc} = -I_2$$
$$I_2 - \alpha I_2 - I_1 = 0$$
$$I_1 - \alpha I_1 - I_0 = 0$$
$$E = -R_g I_0$$

Combining equations, we get

$$j_{sc} = \frac{E}{(1 - \alpha)^2 R_g}$$

The equivalent resistance is given by

$$R_{eq} = \frac{e_{oc}}{j_{sc}} = \frac{(1 - \alpha)^2 R_g}{1 + (R_g/R_1) + (R_g/R_1)(1 - \alpha)}$$

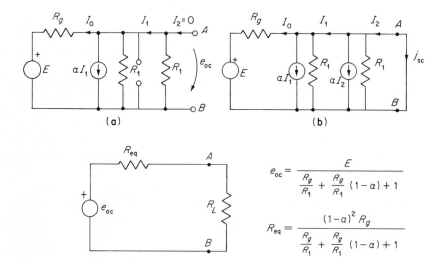

Figure 3.54   **(a)** Calculation of e_{oc}; **(b)** of j_{sc}; **(c)** of Thévenin equivalent circuit.

and the equivalent circuit is shown in Figure 3.54(c). Note that if $\alpha \simeq 1$, the equivalent resistance is very small, and the equivalent circuit seen by the load resistor R_L is approximately an ideal voltage source.

3.9 Significance of Thévenin and Norton Equivalent Circuits

The last six examples demonstrated the utility of Thévenin and Norton equivalent circuits in analysis. The most important implication of this concept of equivalent circuits, however, is that any linear two-terminal circuit, however large or small, or any physical source of electricity, can be represented, insofar as its terminal voltage and current are concerned, by a mathematical model consisting of a sourceless two-terminal circuit in series with a voltage source or in parallel with a current source. The sourceless circuit in many instances can be reduced into a single resistor. It is this concept that is the basis for regarding a battery, fuel cell, radio antenna, generator, microphone, solar cell, or thermocouple, as a nonideal voltage source (Thévenin equivalent) or a nonideal current source (Norton equivalent). The elements of the mathematical model are determined by the open-circuit voltage and short-circuit current at the terminals.

3.10 Source Conversion

A nonideal voltage source can be converted into a nonideal current source. Consider one such source shown in Figure 3.55. If we replace it

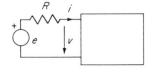

Figure 3.55 A nonideal voltage source.

by its Norton equivalent, we would get the circuit shown in Figure 3.56. The nonideal voltage source is, therefore, equivalent to a nonideal current source. Similarly, a nonideal current source can be converted into a nonideal voltage source by means of the Thévenin equivalent circuit as shown in Figure 3.57. These observations show the interchangeability of the two sources. Either a nonideal voltage source or a nonideal current source can be used to represent a source of electricity that exists in the real world. The last two examples show, however, that it is possible to construct circuits that approximate the ideal sources by suitably combining nonideal sources with *dependent sources* and resistors.

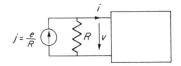

Figure 3.56 Norton equivalent of a nonideal voltage source.

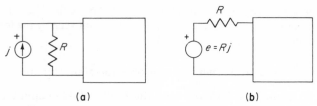

(a)   (b)

Figure 3.57 A nonideal current source **(a)** is converted into a
nonideal voltage source **(b)**.

The technique of source conversion can be used to simplify the analysis of circuits.

Example 3.13   Find v_{AB} in the circuit of Figure 3.58.

Solution   The sequence of circuits in Figure 3.58 shows the step-by-step conversion. By inspection of the last circuit, $v_{AB} = 1$ V.

3.11 Summary

This chapter deals with the analysis of simple circuits using techniques
based on the principle of superposition and the concept of equivalent circuits. The solution of the circuits is obtained by inspection, or at most
through a small number of simplification steps. One may rightfully ask
how one recognizes a "simple" circuit, and when he is confronted with
one, which of the many simplification steps he should apply first. There
is no definitive way by which simple circuits are recognized. If a circuit

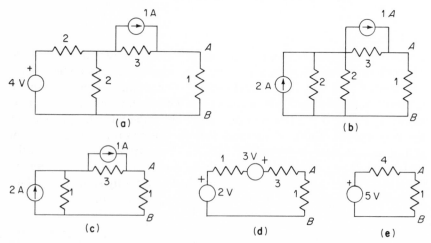

Figure 3.58 Reduction of circuit **(a)** into circuit **(e)** by repeated source conversions.
[Example 3.13]

has one or two loops, one can say it is simple. If it has more loops and has many elements in series and parallel, or has a high degree of symmetry, it can often be reduced into a simple circuit.

We conclude this chapter by an example which dramatizes the last remark.

Example 3.14   Find the resistance between A and B in the circuit of Figure 3.59(a).

Solution   It is clear if we were to write KCL, KVL, and the element equations for the circuit, there would be infinitely many of them. The solution must be found by other means. We note that since the structure is infinite in length, the resistance to the right of CD must be the same as the resistance to the right of AB. Let this resistance be R_x. We then have the equivalent circuit shown in Figure 3.59(b). R_x is simply the resistance R in series with R and R_x in parallel; or

$$R_x = R + \frac{RR_x}{R + R_x}$$

Solving for R_x, we get

$$R_x = \tfrac{1}{2}(1 + \sqrt{5})R$$

SUGGESTED EXPERIMENTS

1.   Construct a circuit with two or more sources to illustrate the homogeneity and additivity properties of the circuit. To make the experiment interesting, use sinusoidal excitations of different frequencies.
2.   Replace one or more elements of the circuit of Experiment 1 by nonlinear elements such as diodes (p–n junction diodes) and demonstrate that the principle of superposition does not hold.
3.   Using resistors and batteries, construct a circuit to illustrate Thévenin equivalent circuits. Replace a part of the circuit by its Thévenin equivalent and demonstrate that the voltages and currents in the remainder of the circuit are invariant.
4.   Experimentally verify Problems 3.23, 3.25, and 3.26.

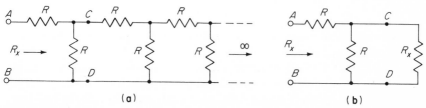

(a)   (b)

Figure 3.59   **(a)** An infinite ladder network; **(b)** its equivalent circuit for the calculation of the input resistance R_x.   [Example 3.14]

COMPUTER PROGRAMMING

Ladder Network Analysis

A class of circuits, which are amenable to analysis by inspection, is the ladder network shown in Figure 3.60. Let us calculate all the voltages and currents.

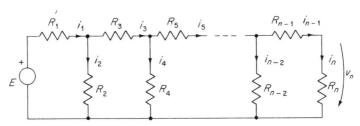

Figure 3.60   A ladder resistor network.

Assuming v_n to be 1 V for the moment, we calculate i_n from

$$i_n = \frac{v_n}{R_n}$$

By inspection, we have

$$i_{n-1} = i_n$$
$$v_{n-1} = R_{n-1}\, i_{n-1}$$
$$v_{n-2} = v_{n-1} + v_n$$
$$i_{n-2} = \frac{v_{n-2}}{R_{n-2}}$$
$$\vdots$$
$$v_3 = R_3 i_3$$
$$v_2 = v_3 + v_4$$
$$i_2 = \frac{v_2}{R_2}$$
$$v_1 = R_1 i_1$$
$$E = v_2 + v_1 = K$$

Let the numerical value of the last quantity be K. The last equation states that an excitation of K V produces a response $v_n = 1$ V. By homogeneity, an excitation of 1 V will produce a response of $v_n = 1/K$ V. v_n/E is simply the reciprocal of the last numerical quantity. Once v_n has been found, the other voltages and currents are calculated systematically in turn from the equations given above.

We note the iterative nature of the solution algorithm. The algorithm can be simply programmed. A FORTRAN version is given below.

```
C     THIS IS A PROGRAM TO COMPUTE THE CURRENTS AND VOLTAGES ON A RESISTOR
C     LADDER OF  N  SECTIONS.  INPUT DATA ARE THE BRANCH RESISTANCES R(1),
C     R(2), ...., R(N), WITH  N  EVEN.
C     THE PROGRAM ALSO COMPUTES THE SUM OF THE VOLTAGES AROUND THE PERIMETER
C     OF THE LADDER.  THE SUM, BY KVL, SHOULD BE ZERO.
      REAL I1, I2, I(100)
      DIMENSION V(100), R(100)
C     READ IN DATA
      DO 1 J=1, 100
      READ(5, 10) R(J)
   10 FORMAT(E10.3)
      IF(R(J) .EQ. .0)GO TO 5
    1 CONTINUE
    5 N = J - 1
      WRITE(6, 14)(K, R(K), K=1,N)
   14 FORMAT(1H ,'BRANCH ', I4, ' RESISTANCE', E15.3)
C     CALCULATION OF  V(N) BEGINS HERE.
      V1 = 1.
      I1 = V1/R(N)
      I2 = I1
      V2 = R(N - 1)*I2
      V1 = V2 + V1
      M = N - 1
      DO 2 K=3, M, 2
      I1 = V1/R(N + 1 - K)
      I2 = I2 + I1
      V2 = I2*R(N - K)
    2 V1 = V2 + V1
C     CURRENTS FOLLOWS.
C     WE NOW HAVE  V(N).  CALCULATION OF THE REMAINING VOLTAGES AND
      V(N) = 1./V1
      I(N) = V(N)/R(N)
      MM = N - 3
      I(N - 1) = I(N)
      DO 3 K=1, MM ,2
      V(N - K) = I(N - K)*R(N - K)
      V(N - K - 1) = V(N - K) + V(N - K + 1)
      I(N - K - 1) = V(N - K - 1)/R(N - K - 1)
    3 I(N - K - 2) = I(N - K - 1) + I(N - K)
      V(1) = I(1) *R(1)
C     CHECK KVL AROUND THE PERIMETER OF THE LADDER.
      VC = 0.
      DO 20 IC = 1, M, 2
   20 VC = VC + V(IC)
      VC = VC + V(N) - 1.
      WRITE(6, 12)
   12 FORMAT(1H , 'BRANCH NO.', 10X, ' CURRENT ', 10X, 'VOLTAGE ')
      WRITE (6, 11)(K, I(K), V(K), K=1, N)
   11 FORMAT(1H , I5, 5X, 2E20.6)
      WRITE(6, 15) VC
   15 FORMAT(1H , 'SUM OF VOLTAGES AROUND THE PERIMETER IS', E15.6)
      STOP
      END
```

```
BRANCH     1 RESISTANCE      0.100E 01
BRANCH     2 RESISTANCE      0.100E 01
BRANCH     3 RESISTANCE      0.100E 01
BRANCH     4 RESISTANCE      0.100E 01
BRANCH     5 RESISTANCE      0.100E 01
BRANCH     6 RESISTANCE      0.100E 01
BRANCH     7 RESISTANCE      0.100E 01
BRANCH     8 RESISTANCE      0.100E 01
BRANCH     9 RESISTANCE      0.100E 01
BRANCH    10 RESISTANCE      0.100E 01
BRANCH NO.          CURRENT              VOLTAGE
    1          0.617977E 00         0.617977E 00
    2          0.382022E 00         0.382022E 00
    3          0.235955E 00         0.235955E 00
    4          0.146067E 00         0.146067E 00
    5          0.898876E-01         0.898876E-01
    6          0.561798E-01         0.561798E-01
```

```
  7                0.337079E-01          0.337079E-01
  8                0.224719E-01          0.224719E-01
  9                0.112360E-01          0.112360E-01
 10                0.112360E-01          0.112360E-01
SUM OF VOLTAGES AROUND THE PERIMETER IS  -0.238419E-06
```

Input Resistance of a Ladder Network and Its Continued Fraction

The input resistance of a ladder network, namely, the resistance "seen" by the voltage source in Figure 3.60, can be expressed as a continued fraction of the branch resistances. Let this resistance be R_{in}. By inspection of Figure 3.60, we note that R_{in} is equal to R_1 in series with a circuit which consists of R_2 in parallel with a circuit which consists of R_3 in series with a circuit, and so on. In equation form, we have

$$R_{in} = R_1 + \cfrac{1}{G_2 + \cfrac{1}{R_3 + \cfrac{1}{G_4 + \cfrac{}{ \ddots + \cfrac{1}{R_{n-1} + \cfrac{1}{G_n}}}}}}$$

where $G_k = 1/R_k$. Such an expression is known as a *continued fraction*. To evaluate the fraction, we make use of the notion of "convergent." Let f be a continued fraction:

$$f = b_0 + \cfrac{a_1}{b_1 + \cfrac{a_2}{b_2 + \cfrac{a_3}{b_3 + \ddots}}}$$

We define the following fraction as the kth convergent of f:

$$f_k = b_0 + \cfrac{a_1}{b_1 + \cfrac{a_2}{b_2 + \cfrac{}{\ddots + \cfrac{a_k}{b_k}}}} = \frac{A_k}{B_k}$$

It is easy to show that A_k and B_k satisfy the recursion formula:

$$A_k = b_k A_{k-1} + a_k A_{k-2}$$
$$B_k = b_k B_{k-1} + a_k B_{k-2}$$

with $A_{-1} = 1$, $A_0 = b_0$, $B_{-1} = 0$, and $B_0 = 1$.

Continued fractions are used in numerical evaluation of functions and irrational numbers. For example, arctan x has the following expansion:

$$\arctan x = \cfrac{x}{1 + \cfrac{x^2}{3 + \cfrac{4x^2}{5 + \cfrac{9x^2}{7 + \cdots}}}}$$

The recursion formula can be easily programmed. The following is a FORTRAN program for the evaluation of a finite continued fraction or the nth convergent of an infinite continued fraction.

```
C   THIS   PROGRAM COMPUTES THE   MTH   CONVERGENT OF AN INFINITE CONTINUED
C   FRACTION IN THE FORM
C
C          F  = B(1) + A(2)/G1
C          G1 = B(2) + A(3)/G2
C          G2 = B(3) + A(4)/G3
C          ...
C   WITH A(1) = 0.
        DIMENSION A(100), B(100)
        DO 1 I=1,100
        READ(5,102)A(I), B(I)
        IF(A(I) .EQ. .0 .AND. B(I) .EQ. .0 .AND. I .GT. 1) GO TO 2
      1 CONTINUE
      2 M=I-1
    102 FORMAT(2E20.6)
        AAONE= 1.
        AATWO= B(1)
        BBONE= 0.
        BBTWO= 1.
        DO 3 J=2,M
        ATEMP= AATWO
        BTEMP= BBTWO
        AATWO = B(J)*AATWO + A(J)*AAONE
        BBTWO = B(J)*BBTWO + A(J)*BBONE
        AAONE= ATEMP
      3 BBONE= BTEMP
        F= AATWO/BBTWO
        WRITE(6,201)
    201 FORMAT(1H ,'NO.', 10X, 'A', 20X, 'B')
        WRITE(6,202)(K, A(K), B(K), K=1,M)
    202 FORMAT(1H , I3, 2E20.6/)
        WRITE(6,203)F
    203 FORMAT(1H ,'CONVERGENT = ', E15.6)
        STOP
C   EXAMPLE... COMPUTE THE  8TH CONVERGENT OF ARCTAN(1).
C   THE TRUE VALUE IS  0.7853982...
        END
```

NO.	A	B
1	0.0	0.0
2	0.100000E 01	0.100000E 01
3	0.100000E 01	0.300000E 01
4	0.400000E 01	0.500000E 01

5	0.900000E 01	0.700000E 01
6	0.160000E 02	0.900000E 01
7	0.250000E 02	0.110000E 02
8	0.360000E 02	0.130000E 02
9	0.490000E 02	0.150000E 02

CONVERGENT =   0.785397E 00

PROBLEMS

3.1 *Voltage Divider* Find the unknowns v_x and i_x in each of the following circuits.

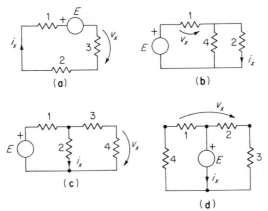

Problem 3.1   (All resistances in ohms.)

3.2 *Current Divider* Find the unknowns v_x and i_x in each of the following circuits.

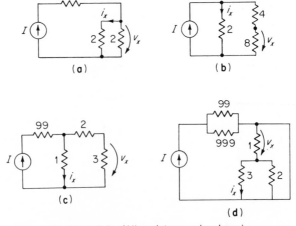

Problem 3.2   (All resistances in ohms.)

3.3 *Superposition* Express the unknowns v_x and i_x as linear combinations of the sources in each of the following circuits.

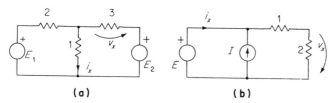

(a)   (b)

Problem 3.3 Superposition (all resistances in ohms).

3.4 *Superposition* Find v_x and i_x in each of the following circuits.

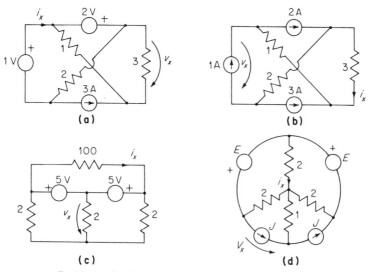

Problem 3.4 Superposition (all resistances in ohms).

3.5 *Superposition* In the circuit shown, find the value of E such that $v_1 = 5v_2$.

Problem 3.5 Superposition.

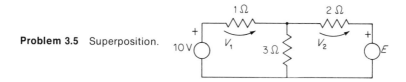

3.6 *Superposition* In the circuit of Problem 3.5, find E such that the power dissipated in the 1-Ω resistor is five times that dissipated in the 2-Ω resistor.

3.7   *Superposition* Find the current in each of the sources in the circuit shown below.

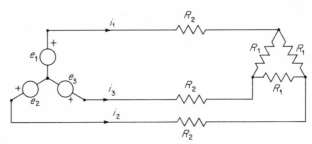

Problem 3.7 Superposition.

3.8   *Power Transmission* In the transmission of electric power from one point to another, there is always power loss in the form of heat dissipated in the resistance of the cable. To minimize this loss, the source voltage is made as high as possible and a transformer is used at the load to decrease the voltage to the required value. Part (a) of the figure shows a low-voltage transmission system. The load requires an average power of 100,000 W at 100 cos ωt V.

a.   Calculate the average power loss in the cable. Now suppose the high-voltage transmission system of Part (b) is used with the turns ratio being 100 to 1.

b.   Calculate the average power loss in the cable for the same load requirement.

c.   Which system would you use?

d.   Calculate E and E'. What can you say about them?

Note: If $v(t) = V \cos \omega t$ and $i(t) = I \cos \omega t$, then average power $= \frac{1}{2} VI$.

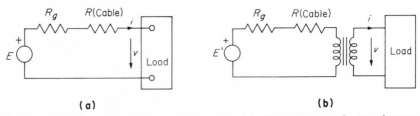

(a)   **(b)**

Problem 3.8 Power loss in transmission cable **(a)** without the use of a transformer; **(b)** with the use of a transformer.

3.9 *Thévenin Circuit* Reduce the circuit to the left of *AB* of its Thévenin equivalent and find *V/E*. Can this circuit be an amplifier?

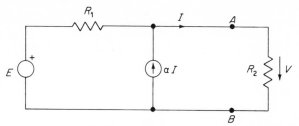

Problem 3.9 Thévenin circuit.

3.10 *Thévenin Circuit* By repeated application of Thévenin equivalent at *AB, CD,* and so on, find *V/E* in the following circuit. Also find V_{AB}, V_{CD}, and V_{EF}. Check your answers by using the ladder analysis program.

3.11 *Norton Circuit* Repeat Problem 3.10 using Norton's theorem.

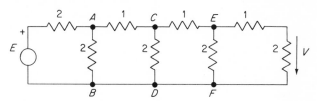

Problem 3.10 Repeated application of Thévenin equivalent circuit.

3.12 *Norton Circuit* The following circuit is a simplified transistor amplifier. Find the Norton equivalent to the left of *AB* and calculate the gain *V/E*.

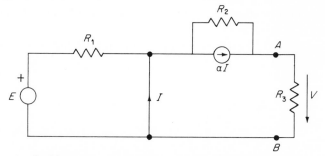

Problem 3.12 A transistor amplifier equivalent circuit.

3.13 *Fun Circuit* The following circuit is of no particular use, but it is fun. What is the value of R_X such that the resistance between AB is R_X?

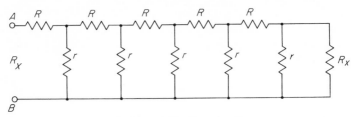

Problem 3.13 Fun circuit.

3.14 *Fun Circuit* Show that the resistance between AB is $\frac{4}{5}$ Ω.

Problem 3.14 Fun circuit (all resistor 1 Ω).

3.15 *Fun Circuit* Show that the resistance between any pair of vertices of a tetrahedron whose sides are 1-Ω resistors is $\frac{1}{2}$ Ω.

3.16 *Fun Circuit* Show that the resistance between two opposite vertices of a cube whose sides are 1-Ω resistors is $\frac{5}{6}$ Ω.

3.17 *Fun Circuit* Show that the resistance between two opposite vertices of a pentagonal dodecahedron whose 30 sides are 1-Ω resistors is $\frac{7}{6}$ Ω.

3.18 *Fun Circuit* In the following circuit, it is found that $I_1 = I_2 = \cdots = I_n$.

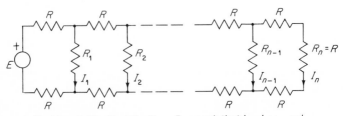

Problem 3.18 Find $R_1, R_2, ..., R_{n-1}$ such that $I_1 = I_2 = \cdots = I_n$.

a. Find $R_1, R_2,..., R_{n-1}$.

b. Let $n = 100$, find I_1. Let R be 1-Ω. Use the ladder analysis program to check your answer.

Ans.: $R_{n-k} = (k^2 + k + 1)R$; $I_1 = E/(n^2 + n + 1)R$.

3.19 *Ladder* The input resistance of a ladder network can be expressed as a continued fraction. The binomial expansion of $(1 + x)^m$ has a continued-fraction representation.

$$(1 + x)^m = 1 + \cfrac{mx}{1 + \cfrac{(1 - m)x}{2 + \cfrac{(1 + m)x}{3 + \cfrac{(2 - m)x}{2 + \cfrac{(2 + m)x}{5 + \cdot}}}}}$$

Letting $x = 1$ and $m = \frac{1}{2}$, one obtains a ladder shown below whose input resistance is $\sqrt{2}$.

Find a ladder whose input resistance is $\sqrt[3]{2}$.

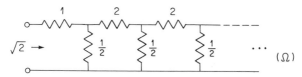

Problem 3.19 A rational approximation of an irrational number by means of continued-fraction expansion.

3.20 *Ladder* Use the continued-fraction program to check your answer in Problem 3.19. Obtain the 5th convergent.

3.21 *Ladder* Find a ladder network whose input resistance approximates the number $\pi/4$. The resistances are to be rational numbers. How many resistors must be used if the approximation is to have an error less than 0.001? Use the ladder analysis program to calculate all the voltages and currents in the circuit.

3.22 *Design* The following circuit has been proposed to simulate an ideal voltage source. Find the Thévenin circuit to the left of AB. How should you select the parameters so that the approximation would be "good"? What is the equivalent resistance seen by the source at CD? How should you select the parameters so that this equivalent resistance is high?

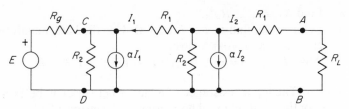

Problem 3.22 A circuit that approximates an ideal voltage source.

3.23 *Symmetry* Find the voltages V_{AD} and V_2 and currents I_1 and I_2 in the following circuit for each set of excitations: **a.** $e_1 = E_s$, $e_2 = E_s$;   **b.** $e_1 = E_s$, $e_2 = -E_s$;   **c.** $e_1 = 5$, $e_2 = 3$.

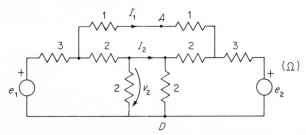

Problem 3.23 A symmetrical circuit.

3.24 *Symmetry* Repeat the last problem for the following circuit.

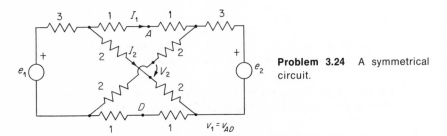

Problem 3.24 A symmetrical circuit.

3.25 *Symmetry* Show that a symmetrical network subject to symmetrical excitations can be split into two halves as shown.

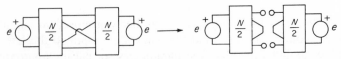

Problem 3.25 A symmetrical circuit under symmetrical excitation.

3.26 *Symmetry* Show that a symmetrical network subject to antisymmetric excitations can be split into two halves as shown.

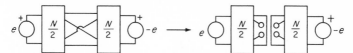

Problem 3.26 A symmetrical circuit under antisymmetrical excitation.

3.27 *Symmetry* Use the results of the last two problems to find the input resistance and the voltage V of the circuit shown.

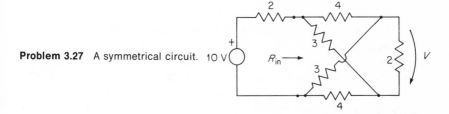

Problem 3.27 A symmetrical circuit.

4

first- and second-order circuits

In Chapter 3, we dealt mostly with simple circuits containing resistors and sources only. We now wish to examine simple circuits containing inductors and capacitors.

Specifically, we wish to consider circuits that consist of any number of resistors and sources but only one capacitor or one inductor, also circuits that consist of resistors and sources but only one capacitor and one inductor or two of the same kind. The reason for restricting our attention for the moment is that these simple circuits are used widely in practice, especially in digital computers, communication systems, and automatic control systems. More importantly once we have a good grasp of the simple cases, we shall be able to analyze the general case by extension. We shall see in Chapter 9 that the response in the general case is a linear combination of the responses of the cases considered in this chapter. At the end of the chapter, we present applications of these simple circuits. We shall see how these circuits can be used to generate waveforms for electronic display on cathode ray tubes, and to smooth a signal in a digital communication system. We shall also introduce numerical solution of differential equations for cases in which the analytic solution is difficult to obtain.

4.1 First-Order *RC* and *GL* Circuits

Consider the circuit of Figure 4.1. Choose the voltage source and the capacitor voltage as tree branches. We can write the following equations:

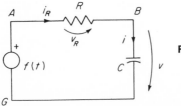

Figure 4.1 A simple *RC* circuit.

144

KVL: $\qquad v_R = f - v$

KCL: $\qquad i = i_R$

Definition of elements: $\quad i = C\dfrac{dv}{dt}$

$$v_R = Ri$$

Combining the equations, we get a first-order linear differential equation with constant coefficients:

$$\frac{dv}{dt} = -\frac{1}{RC}v + \frac{1}{RC}f(t) \tag{4.1}$$

The solution consists of two parts: a complementary solution $v_c(t)$ and a particular integral $v_p(t)$. The complementary solution satisfies the homogeneous equation

$$\frac{dv_c}{dt} = -\frac{1}{RC}v_c \tag{4.2}$$

and is given by

$$v_c(t) = Ke^{-(1/RC)t} \tag{4.3}$$

where K is an arbitrary constant. The particular integral is a solution that satisfies the nonhomogeneous equation

$$\frac{dv_p}{dt} = -\frac{1}{RC}v_p + \frac{1}{RC}f(t) \tag{4.4}$$

and is given by

$$v_p(t) = \frac{1}{RC}\int_{t_0}^{t} e^{-(1/RC)(t-x)}f(x)\,dx \tag{4.5}$$

The lower limit t_0 is the time at which we begin to describe the behavior of the circuit. The complete solution is

$$v(t) = Ke^{-(1/RC)t} + \frac{1}{RC}\int_{t_0}^{t} e^{-(1/RC)(t-x)}f(x)\,dx$$

If we begin counting time from $t = 0$, we would have

$$v(t) = Ke^{-(1/RC)t} + \frac{1}{RC}\int_{0}^{t} e^{-(1/RC)(t-x)}f(x)\,dx \tag{4.6}$$

and the constant K has the physical meaning of being the initial value of $v(t)$, since by Equation (4.6)

$$v(0) = K \tag{4.7}$$

Equation (4.6) also states that the response $v(t)$ consists of two parts, one due entirely to the initial condition on the capacitor (complementary solution) and the other to the excitation $f(t)$. We observe that the part due to the initial condition goes to zero as t approaches infinity. In other words, the response after a long time consists of essentially that part due to the excitation alone. The part due to the initial condition has by this time "died." Figure 4.2 shows the complementary solution $v_c(t)$. As a measure of the rate at which the initial value goes to zero, we define the quantity, time constant T, by the following:

$$T = RC \tag{4.8}$$

so that when $t = T$, the value of $v_c(t)$ is $v_c(0)e^{-1}$, or 36.8 percent of its initial value.

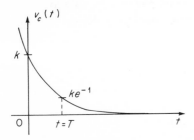

Figure 4.2 Complementary solution.

The precise nature of the response due to the excitation $f(t)$ will depend on the form of $f(t)$. It is instructive to examine the particular integral in some detail. Let

$$h(x) = \frac{1}{RC} e^{-(1/RC)x} \tag{4.9}$$

Then

$$h(t - x) = \frac{1}{RC} e^{-(1/RC)(t-x)} \tag{4.10}$$

This function is plotted in Figure 4.3 together with $f(t)$. The particular integral can now be written as

$$v_p(t) = \int_0^t h(t - x)\, f(x)\, dx \tag{4.11}$$

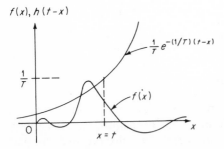

Figure 4.3 Evaluation of the particular integral.

The integral is the area under the curve defined by the product of the excitation and the function $h(t - x)$. In words, we take $h(x)$ and "fold" it over about the axis $x = 0$ to get $h(-x)$. We then shift it to the right by t to get $h(t - x)$. Multiplying $h(t - x)$ by $f(x)$ and integrating between 0 and t, we obtain the value of the integral at time t. This operation of folding, shifting, multiplying, and integrating is known as *convolution*. The integral (4.11) is known as the *convolution integral*, and we say that $f(x)$ *convolves* with the function $h(x)$ to yield $v_p(t)$. The convolution integral can also be looked upon as a statement of superposition, for we can regard $f(x)$ as composed of a series of narrow pulses of height $f(x_1)$ at $x = x_1$, and so on, as shown in Figure 4.4. The value of the integral is

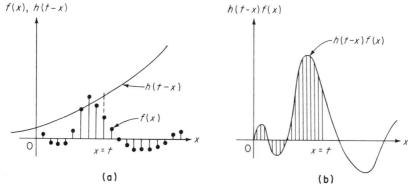

(a)   (b)

Figure 4.4   **(a)** Convolution of the characteristic function *h(t)* and the excitation *e(t)*; **(b)** the area under the curve is the value of the convolution integral at time *t*.

a weighted sum of these pulses. The weight at $x = x_1$ is $h(t - x_1)$. For this reason, the convolution integral is sometimes called the *superposition integral*.

We now observe another interesting feature. The function $h(t)$ is proportional to the complementary solution of the differential equation. It appears that at least in this case, the complementary solution determines the nature of the complete solution. Because of this, $h(t)$ is sometimes called the *characteristic function* of the circuit or the differential equation. We shall see later that in linear time-invariant circuits, this is always the case; namely, the complete solution is a linear combination of the complementary solution and the convolution of the complementary solution and the excitation.

Now consider the circuit of Figure 4.5. Writing the usual equations,

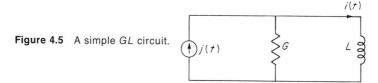

Figure 4.5   A simple *GL* circuit.

we get the following first-order linear differential equation with constant coefficients

$$\frac{di}{dt} = -\frac{1}{GL}i + \frac{1}{GL}j(t) \qquad (4.12)$$

The equation has the same form as Equation (4.4), and the complete solution is

$$i(t) = ce^{-(1/GL)t} + \frac{1}{GL}\int_0^t e^{-(1/GL)(t-x)}\,j(x)\,dx \qquad (4.13)$$

where the constant c has the physical meaning of being the initial value of the current $i(t)$ since

$$i(0) = c \qquad (4.14)$$

The time constant of the circuit in this case is

$$T = GL \qquad (4.15)$$

The dualism of this circuit and the RC circuit of Figure 4.1, and the dualism of their solutions, should be noted.

The RC and GL circuits are each governed by a first-order linear differential equation. The solution contains precisely one arbitrary constant, and the circuits are therefore called first-order circuits. They are simple circuits, but note that the RC circuit has the appearance of a Thévenin circuit and the GL circuit the appearance of a Norton equivalent circuit. As a result, we can make the following observations:

1. In a circuit that contains resistors and sources but just one capacitor, the voltage across the capacitor is given by

$$v(t) = v(0)\,e^{-(1/T)t} + \frac{1}{T}\int_0^t e^{-(1/T)(t-x)}\,e_{oc}(x)\,dx \qquad (4.16)$$

where $e_{oe}(t)$ is the open-circuit voltage of the Thévenin equivalent circuit "seen" by the capacitor, and the time constant T is RC, where R is the equivalent resistance of the Thévenin circuit. Figure 4.6 summarizes the observation.

2. In a circuit that contains resistors and sources but just one inductor, the current in the inductor is given by

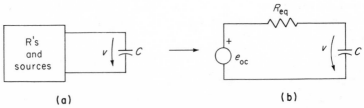

(a)   (b)

Figure 4.6 (a) A general first-order RC circuit; (b) its Thévenin equivalent.

$$i(t) = i(0) \ e^{-t/T} + \frac{1}{T} \int_0^t e^{-(1/T)(t-x)} j_{\text{sc}} \ (x) \, dx \qquad (4.17)$$

where $j_{\text{sc}}(t)$ is the short-circuit current source in the Norton equivalent circuit seen by the inductor, and the time constant T is given by GL, where G is the equivalent conductance in the Norton equivalent circuit. Figure 4.7 summarizes this observation.

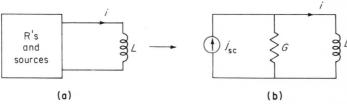

(a) **(b)**

Figure 4.7 **(a)** A general first-order *GL* circuit; **(b)** its Norton equivalent.

3. The voltage and current everywhere in a first-order circuit have the form

$$y(t) = y(0) \ e^{-1/T} + \frac{1}{T} \int_0^t e^{-(1/T)(t-x)} u(x) \, dx + z(t) \qquad (4.18)$$

where $y(t)$ is a voltage or current, and $u(t)$ and $z(t)$ are some appropriate linear combinations of the sources in the circuit.

To see this, we make reference to Figure 4.8, in which we regard the voltage $v(t)$ as an excitation to the circuit N. The response $y(t)$, by the

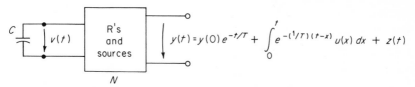

Figure 4.8 The response everywhere in *N* has the same form as *v(t)*.

principle of superposition, must be simply a linear combination of $v(t)$ and the excitations, since the equations governing the voltages and currents in N are all linear algebraic equations.

Example 4.1 Find all the voltages and currents in the circuit of Figure 4.9.

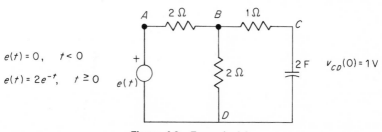

Figure 4.9 Example 4.1.

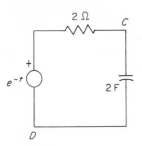

Figure 4.10   Thévenin equivalent of Example 4.1.

Solution   Replace the circuit to the left of CD by its Thévenin equivalent circuit. The resultant circuit is shown in Figure 4.10. (The details are left to the students.) The time constant of the circuit is

$$T = (2)(2) = 4 \text{ sec}$$

The voltage v_{CD} is given by

$$v_{CD}(t) = e^{-t/4} + \frac{1}{4} \int_0^t e^{-(1/4)(t-x)} e^{-x} dx = \frac{4}{3} e^{-t/4} - \frac{1}{3} e^{-t}$$

$$i_{CD}(t) = 2\frac{dv_{CD}}{dt} = -\frac{2}{3} e^{-t/4} + \frac{2}{3} e^{-t}$$

$$v_{BC}(t) = i_{CD} = \frac{2}{3} e^{-t} - \frac{2}{3} e^{-t/4}$$

$$v_{BD}(t) = v_{BC} + v_{CD} = \frac{1}{3} e^{-t} + \frac{2}{3} e^{-t/4}$$

$$i_{BD}(t) = \frac{v_{BD}}{2} = \frac{1}{6} e^{-t} + \frac{1}{3} e^{-t/4}$$

$$i_{AB}(t) = i_{BD} + i_{CD} = \frac{5}{6} e^{-t} - \frac{1}{3} e^{-t/4}$$

$$v_{AB}(t) = 2i_{AB} = \frac{5}{3} e^{-t} - \frac{2}{3} e^{-t/4}$$

$$v_{AD}(t) = v_{AB} + v_{BD} = 2e^{-t}$$

We note that the responses are all linear combinations of the voltage across the capacitor and the excitation. Observation 3 is thus verified in this example.

Example 4.2   Find the voltage $v(t)$ in the circuit of Figure 4.11(a).

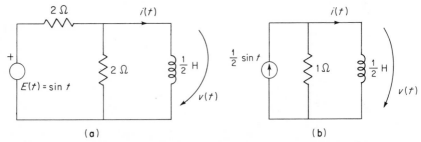

(a)   (b)

Figure 4.11   (a) A first-order inductor circuit; **(b)** its Norton equivalent. The response becomes sinusoidal for large t.   [Example 4.2]

Solution The Norton equivalent circuit is shown in Figure 4.11(b). From Equation (4.13), the current $i(t)$ is given by

$$i(t) = i(0) \ e^{-2t} + \int_0^t e^{-2(t-x)} \sin x \ dx$$

$$= i(0) \ e^{-2t} + \frac{1}{5} \ e^{-2t} + \frac{2}{5} \sin t - \frac{1}{5} \cos t$$

$$= i(0) \ e^{-2t} + \frac{1}{5} \ e^{-2t} + \frac{1}{\sqrt{5}} \sin (t + \theta)$$

with

$$\theta = -\tan^{-1} \frac{1}{2}$$

The voltage $v(t)$ is found from

$$v(t) = L \frac{di}{dt} = -i(0) \ e^{-2t} - \frac{1}{5} \ e^{-2t} + \frac{1}{2\sqrt{5}} \cos (t + \theta)$$

Note that as $t \to \infty$, the responses $i(t)$ and $v(t)$ become purely sinusoidal, as dictated by the excitation. Values of $i(t)$ and $v(t)$ are computed from the following FORTRAN program, which also calls for a computer plotting subroutine EEPLOT to plot $i(t)$ and $v(t)$ on a computer printout. The subroutine EEPLOT is described at the end of this chapter. In the following program, $i(0) = 0$ is assumed.

```
C       THIS IS EXAMPLE 4.2.
C       THE PROGRAM PLOTS THE CURRENT AND VOLTAGE VERSUS TIME.
        DIMENSION TIME(200), VOL(200), CUR(200)
        INTEGER TITLE(20)
C       DEFINE CURRENT C(T) AND VOLTAGE V(T)
        C(T) = 0.3*EXP(-2.*T) + 0.4*SIN(T) - 0.2*COS(T)
        V(T) = -0.3*EXP(-2.*T) + 0.2*COS(T) + 0.1*SIN(T)
C       READ IN TITLE CARD.
        READ(5,100)(TITLE(I),I=1,20)
    100 FORMAT(20A4)
        WRITE(6,99)
     99 FORMAT(' TIME    ',8X,'CURRENT', 8X, 'VOLTAGE')
        T=-0.1
        DO 1 I=1,200
        T=T+0.1
        TIME(I)=T
        CUR(I)=C(T)
      1 VOL(I)=V(T)
        WRITE(6,101)(TIME(I),CUR(I),VOL(I),I=1,200)
    101 FORMAT(F8.2, 2E15.6)
        CALL EEPLOT(TIME,VOL,20.,0.,,25,-.25,TITLE,200)
C       CHANGE TITLE OF PLOT.
        READ(5,100)(TITLE(I),I=1,20)
        CALL EEPLOT(TIME,CUR,20.,0.,,5,-.5,TITLE,200)
        STOP
        END
```

[Computer printout for this program follows on the next two pages.]

Computer printout for Example 4.2

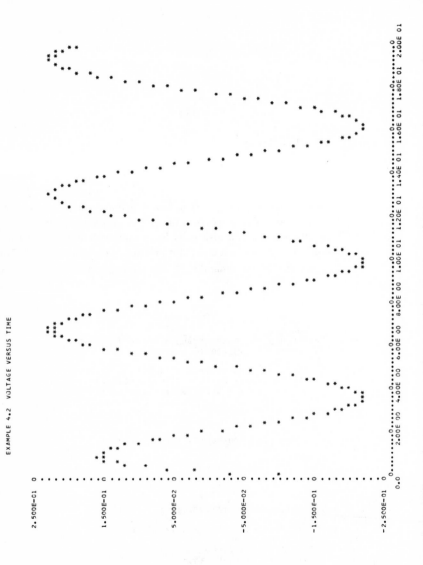

EXAMPLE 4.2   VOLTAGE VERSUS TIME

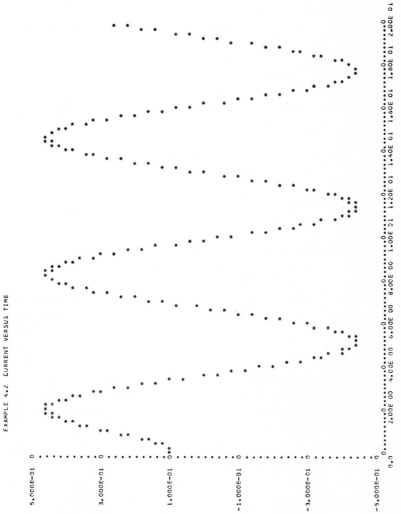

EXAMPLE 4.2 CURRENT VERSUS TIME

Computer Printout for Example 4.2 Concluded

4.2 First-Order Circuits Under Constant Excitation

In case the excitations in a first-order circuit are all constant (independent of time), the response can be found by inspection. Consider the

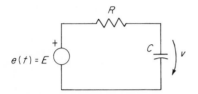

Figure 4.12 *RC* circuit with constant excitation.

circuit of Figure 4.12 where $e(t) = E$ (constant). From Equation (4.6), we have

$$v(t) = v(0) \ e^{-t/T} + \frac{1}{T} \int_0^t e^{(-1/T)(t-x)} \ E \ dx$$
$$= E + [v(0) - E] \ e^{-t/T} \tag{4.19}$$

We note that the final value of $v(t)$ is

$$v(\infty) = E \tag{4.20}$$

Equation (4.19) can, therefore, be written as

$$v(t) = v(\infty) + [v(0) - v(\infty)] \ e^{-t/T} \tag{4.21}$$

By observation 3 of the last section, we conclude that in a *first-order circuit (RC or GL) under constant excitation, the response everywhere is given by*

$$y(t) = y(\infty) + [y(0) - y(\infty)] \ e^{-t/T} \tag{4.22}$$

It is seen that the response is completely determined by three constants: the initial value $y(0)$, the final value $y(\infty)$, and the time constant T.

To facilitate the calculation of $y(0)$ and $y(\infty)$, we recall from the definition of a capacitor that

$$v(t) = \frac{1}{C} \int_0^t i(u) \, du + v(0) = v_1(t) + v(0) \tag{4.23}$$

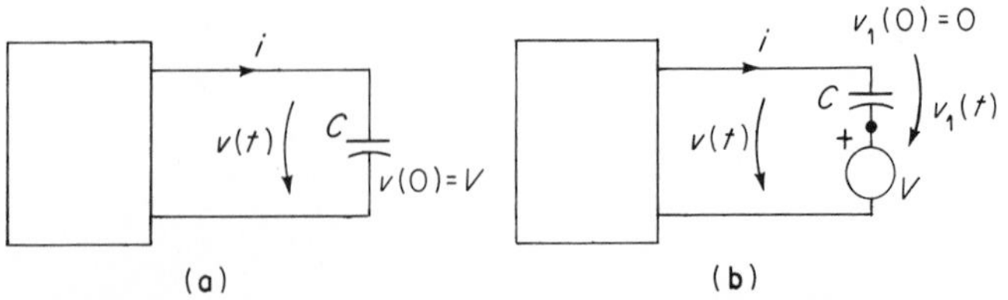

Figure 4.13 A capacitor with initial voltage V **(a)** is equivalent to a capacitor with zero initial voltage in series with a constant voltage source of value V **(b)**.

With reference to Figure 4.13(a), we may regard the charged capacitor as composed of an uncharged capacitor in series with a voltage source whose value is the initial value of the voltage across the capacitor as shown in Figure 4.13(b). At the instant $t = 0$, $v_1(0) = 0$ and the charged capacitor can be replaced by a constant voltage source of value $v(0)$.

In a dual manner, an inductor whose initial current is $i(0)$ can be replaced by an inductor with zero initial current in parallel with a current source whose value is the initial value of the current, that is, $i(0)$. This is shown in Figure 4.14. At the instant $t = 0$, an inductor can be replaced by a current source whose value is $i(0)$.

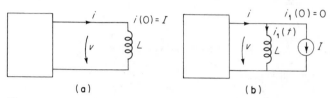

(a)   (b)

Figure 4.14 An inductor with initial current I **(a)** is equivalent to an inductor with zero initial current in parallel with a constant current source of value I **(b)**.

Next, by observations 1 and 2 of the last section and noting that the excitations are constant, we have

$$\lim_{t\to\infty} y(t) = \lim_{t\to\infty} \frac{K}{T} \int_0^t e^{-(1/T)(t-x)}\, dx = K$$

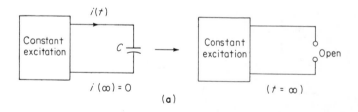

(a)

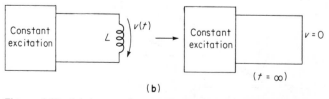

(b)

Figure 4.15 **(a)** A capacitor can be replaced by an open circuit at $t = \infty$, if the circuit is under constant excitation. **(b)** An inductor can be replaced by a short circuit at $t = \infty$, if the circuit is under a constant excitation.

where K is a constant. As a result, the final value of the voltage across a capacitor and the final value of the current in an inductor approach a constant as a limit. The rate of change of these quantities, namely, the current through the capacitor or the voltage across the inductor, is zero at $t = \infty$. Therefore, at $t = \infty$, a capacitor can be replaced by an open circuit, and an inductor can be replaced by a short circuit. This is shown in Figure 4.15.

Example 4.3   Find $v_{AB}(t)$ and $i(t)$ for $t \geq 0$ in the circuit shown in Figure 4.16(a). The initial voltage on the capacitor is -3 V.

Solution   At $t = 0$, the circuit takes the form shown in Figure 4.16(b). Using superposition, we find

$$v_{AB}(0) = 1 \text{ V} \quad \text{and} \quad i(0) = 4 \text{ A}$$

At $t = \infty$, the circuit takes the form shown in Figure 4.16(c). By inspection, we have

$$v_{AB}(\infty) = 3 \text{ V} \quad \text{and} \quad i(\infty) = 0 \text{ A}$$

The time constant is the product of the capacitance and the resistance seen by

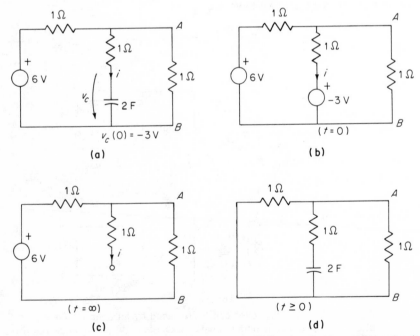

Figure 4.16   (a) The circuit; **(b)** the circuit at $t = 0$; **(c)** the circuit at $t = \infty$; **(d)** the circuit for the calculation of the time constant.   [Example 4.3]

the capacitor. The resistance can be found from the circuit of Figure 4.16(d). The time constant is

$$T = (\tfrac{3}{2})(2) = 3 \text{ sec}$$

The solutions are, by Equation (4.22),

$$v_{AB}(t) = 3 - 2e^{-t/3}$$
$$i(t) = 4e^{-t/3}$$

These are plotted in Figure 4.17.

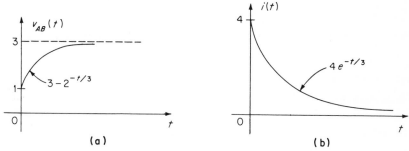

Figure 4.17   Sketch of $v_{AB}(t)$ and $i(t)$ of Example 4.3.

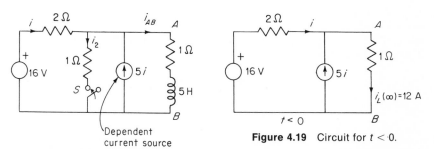

Figure 4.18   Example 4.4.

Figure 4.19   Circuit for $t < 0$.

Example 4.4   In the circuit shown in Figure 4.18, the switch S was open for $t < 0$. At $t = 0$, S closes. Find the voltage $v_{AB}(t)$ for $t \geq 0$.

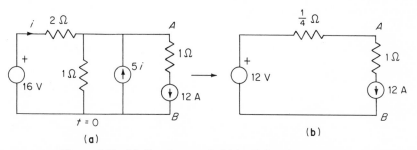

Figure 4.20   (a) Circuit at $t = 0$; (b) its Thévenin equivalent.

Solution   We must first find the initial value of the current in the inductor. Since the current in an inductor does not change instantaneously, the initial value is the final value of the circuit with the switch open as shown in Figure 4.19. By inspection, we find

$$i_L(\infty) = 12 \text{ A}$$

Now at $t = 0$, we have the circuit shown in Figure 4.20(a). Replacing the circuit to the left of AB by its Thévenin equivalent circuit, we have the circuit shown in Figure 4.20(b), from which we find

$$v_{AB}(0) = 9 \text{ V}$$

At $t = \infty$, the circuit becomes that shown in Figure 4.21(a), from which we find

$$v_{AB}(\infty) = \frac{48}{5} \text{ V}$$

The time constant can be found from the circuit shown in Figure 4.21(b), and it is given by

$$T = \left(\tfrac{4}{5}\right)(5) = 4 \text{ sec}$$

The voltage $v_{AB}(t)$ is found to be

$$v_{AB}(t) = \frac{48}{5} + \left(9 - \frac{48}{5}\right)e^{-t/4}$$

We may wish to find the other voltages and currents. For example, from Figures 4.20(b) and Figure 4.21(b), we have

$$i_{AB}(0) = 12 \qquad i_{AB}(\infty) = \frac{48}{5}$$

and

$$i_{AB}(t) = \frac{48}{5} + \left(12 - \frac{48}{5}\right)e^{-t/4}$$

From Figure 4.18, we find

$$6i(t) = i_2(t) + i_{AB} = v_{AB}(t) + i_{AB}(t)$$

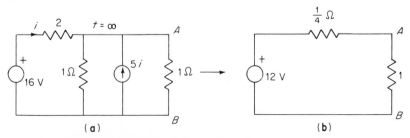

Figure 4.21   **(a)** Circuit at $t = \infty$; **(b)** its Thévenin equivalent.

so that

$$i(t) = \frac{16}{5} + \left(\frac{7}{2} - \frac{16}{5}\right)e^{-t/4}$$

To check our answers, we write KVL around the outer loop of Figure 4.18 and get

$$16 - 2i - v_{AB} = 16 - \frac{32}{5} - \left(7 - \frac{32}{5}\right)e^{-t/4} - \frac{48}{5} - \left(9 - \frac{48}{5}\right)e^{-t/4} = 0$$

4.3 Second-Order Circuits

We shall now consider circuits containing two energy-storing elements. We shall see once again that the solution consists of a complementary solution and the convolution of the complementary solution and the excitation.

Consider the circuit shown in Figure 4.22. It has one capacitor and one inductor. The remainder of the circuit consists of resistors and sources. Treating the capacitor current and the inductor voltage as responses and the capacitor voltage and the inductor current as excitations, by superposition we can write the following linear algebraic equations:

$$i_c = a'_{11} v_c + a'_{12} i_L + y'_1(t)$$
$$v_L = a'_{21} v_c + a'_{22} i_L + y'_2(t) \qquad (4.24)$$

where $y'_1(t)$ and $y'_2(t)$ are responses due to the internal sources alone. Using the definitions of the capacitor and inductor, we obtain

$$C\frac{dv_c}{dt} = a'_{11} v_c + a'_{12} i_L + y'_1(t)$$
$$L\frac{di_L}{dt} = a'_{21} v_c + a'_{22} i_L + y'_2(t) \qquad (4.25)$$

which can be written as a system of two simultaneous linear first-order differential equations with constant coefficients given below:

$$\frac{dv_c}{dt} = a_{11} v_c + a_{12} i_L + y_1(t)$$
$$\frac{di_L}{dt} = a_{21} v_c + a_{22} i_L + y_2(t) \qquad (4.26)$$

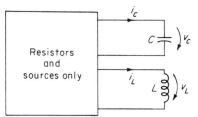

Figure 4.22 A second-order circuit.

The values of the coefficients will depend on the circuit contained in the box. We now wish to discuss the solution to this system of equations.

As in the case of first-order equations, the solution consists of a complementary solution and a particular integral. Let

$$v_c(t) = v_{co}(t) + v_{cp}(t)$$
$$i_L(t) = i_{Lo}(t) + i_{Lp}(t) \tag{4.27}$$

where v_{co} and i_{Lo} are the complementary solutions, and v_{cp} and i_{Lp} the particular integrals. To find v_{co} and i_{Lo}, we let

$$v_{co}(t) = Q_1 e^{\lambda t}$$
$$i_{Lo}(t) = Q_2 e^{\lambda t} \tag{4.28}$$

where Q_1, Q_2, and λ are constants. $v_{co}(t)$ and $i_{Lo}(t)$ satisfy the homogeneous equations. Substituting (4.28) into the differential equations with the excitations absent, we have

$$\begin{bmatrix} a_{11} - \lambda & a_{12} \\ a_{12} & a_{22} - \lambda \end{bmatrix} \begin{bmatrix} Q_1 \\ Q_2 \end{bmatrix} = \begin{bmatrix} 0 \\ 0 \end{bmatrix} \tag{4.29}$$

which is a system of linear algebraic homogeneous equations. The system has nontrivial solution if, and only if, the determinant of the matrix is zero, that is,

$$\begin{vmatrix} a_{11} - \lambda & a_{12} \\ a_{21} & a_{22} - \lambda \end{vmatrix} = 0 \tag{4.30}$$

Expanding the determinant, we get a quadratic equation in λ:

$$(a_{11} - \lambda)(a_{22} - \lambda) - a_{12}a_{21} = 0 \tag{4.31}$$

Suppose the roots are λ_1 and λ_2. There are three cases to consider:

1. λ_1 and λ_2 are real and distinct.
2. λ_1 equals λ_2.
3. λ_1 and λ_2 are complex.

The significance of the two values of λ is that the circuit, or equivalently the system of differential equations, has nontrivial complementary solution for two, and only two, values of λ. These values of λ are called the *eigenvalues* of the system (4.29) and are intrinsic to the circuit. Equation (4.31), from which λ_1 and λ_2 are determined, is called the *characteristic equation* of the circuit or the system of differential equations.

Consider the cases in which $\lambda_1 \neq \lambda_2$. Substituting λ_1 into (4.29), we get

$$\begin{bmatrix} a_{11} - \lambda_1 & a_{12} \\ a_{21} & a_{22} - \lambda_1 \end{bmatrix} \begin{bmatrix} Q_1 \\ Q_2 \end{bmatrix} = \begin{bmatrix} 0 \\ 0 \end{bmatrix}$$

Since the determinant is zero, the equations are linearly dependent. Solving for Q_2 in terms of Q_1, we have

$$Q_2 = \frac{-a_{11} + \lambda_1}{a_{12}} Q_1 \tag{4.32}$$

The complementary solution corresponding to this eigenvalue is

$$v_{co}(t) = c_1 e^{\lambda_1 t}$$
$$i_{Lo}(t) = c_1 \frac{-a_{11} + \lambda_1}{a_{12}} e^{\lambda_1 t} \tag{4.33}$$

where c_1 is an arbitrary constant. Substituting λ_2 into (4.29), we get a second complementary solution given by

$$v_{co}(t) = c_2 e^{\lambda_2 t}$$
$$i_{Lo}(t) = c_2 \frac{-a_{11} + \lambda_2}{a_{12}} e^{\lambda_2 t} \tag{4.34}$$

where c_2 is an arbitrary constant. The sum of the two complementary solutions is the complementary solution of the circuit, that is,

$$\begin{bmatrix} v_{co}(t) \\ i_{Lo}(t) \end{bmatrix} = c_1 \begin{bmatrix} 1 \\ \dfrac{-a_{11} + \lambda_1}{a_{12}} \end{bmatrix} e^{\lambda_1 t} + c_2 \begin{bmatrix} 1 \\ \dfrac{-a_{11} + \lambda_2}{a_{12}} \end{bmatrix} e^{\lambda_2 t} \tag{4.35}$$

The vectors $(1, (-a_{11} + \lambda_1)/a_{12})$ and $(1, (-a_{11} + \lambda_2)/a_{12})$ are the eigenvectors of the system (4.29). For convenience, denote the complementary solutions by

$$\begin{bmatrix} v_{co} \\ i_{Lo} \end{bmatrix} = c_1 \begin{bmatrix} x_{11} \\ x_{21} \end{bmatrix} + c_2 \begin{bmatrix} x_{12} \\ x_{22} \end{bmatrix} \tag{4.36}$$

where $x_{11} = e^{\lambda_1 t}$, $x_{21} = (-a_{11} + \lambda_1) e^{\lambda_1 t}/a_{12}$, and so on. To find the particular integrals, we use the method of variation of parameters. Let

$$\begin{bmatrix} v_{cp} \\ i_{Lp} \end{bmatrix} = u_1(t) \begin{bmatrix} x_{11} \\ x_{21} \end{bmatrix} + u_2(t) \begin{bmatrix} x_{12} \\ x_{22} \end{bmatrix} \tag{4.37}$$

The particular integrals satisfy the nonhomogeneous differential equations (4.26). Substituting (4.37) into (4.26) and simplifying by using (4.29), we get the following system of algebraic equations in the unknowns $u_1'(t)$ and $u_2'(t)$, where primes denote differentiation with respect to t.

$$\begin{bmatrix} x_{11} & x_{12} \\ x_{21} & x_{22} \end{bmatrix} \begin{bmatrix} u_1' \\ u_2' \end{bmatrix} = \begin{bmatrix} y_1 \\ y_2 \end{bmatrix} \tag{4.38}$$

Solving for $u_1'(t)$ and $u_2'(t)$ and integrating, we get $u_1(t)$ and $u_2(t)$. The complete solution of the system (4.26) is, therefore,

$$\begin{bmatrix} v_C(t) \\ i_L(t) \end{bmatrix} = c_1 \begin{bmatrix} x_{11} \\ x_{21} \end{bmatrix} + c_2 \begin{bmatrix} x_{12} \\ x_{22} \end{bmatrix} + u_1(t) \begin{bmatrix} x_{11} \\ x_{21} \end{bmatrix} + u_2(t) \begin{bmatrix} x_{12} \\ x_{22} \end{bmatrix} \tag{4.39}$$

The solution contains two arbitrary constants, which must be determined from the initial values of the capacitor voltage and the inductor current. We shall consider two numerical examples to illustrate Cases 1 and 3.

Example 4.5   Find the complete solution of the capacitor voltage and inductor current in the circuit shown in Figure 4.23.

Solution   Choose the tree and chords as shown in Figure 4.23. Writing the usual equations, we have

$$v_1 = e(t) - v_2 - v_4$$
$$v_3 = v_4$$
$$i_4 = i_1 - i_3$$
$$i_2 = i_1$$
$$v_2 = \frac{3}{2} i_2$$
$$i_3 = 4v_3$$
$$i_4 = \frac{dv_4}{dt}$$
$$v_1 = \frac{di_1}{dt}$$

Combining the equations by eliminating all variables but the capacitor voltage v_4 and inductor current i_1,

$$\frac{dv_4}{dt} = -4v_4 + i_1$$
$$\frac{di_1}{dt} = -v_4 - \frac{3}{2} i_1 + e(t) \tag{4.40}$$

Alternately, the differential equations can also be obtained as follows. Regard the capacitor voltage and inductor current as excitations and the capacitor current and inductor voltage as responses. Replace the capacitor by a voltage source of value v_4 and the inductor by a current source of value i_1 as shown in Figure 4.23(d). By superposition the capacitor current i_4 and inductor voltage v_1 are given by

$$i_4 = -4v_4 + i_1$$
$$v_1 = -v_4 - \frac{3}{2} i_1 + e(t)$$

Replacing i_4 by $(1)(dv_4/dt)$ and v_1 by $(1)(di_1/dt)$ and simplifying we obtain Equation (4.40).

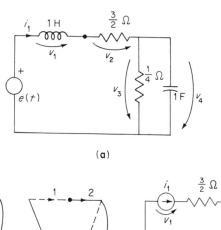

(a)

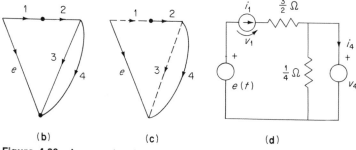

(b)   (c)   (d)

Figure 4.23 A second-order circuit with real and distinct eigenvalues.
[Example 4.5]

The characteristic equation is found from

$$\begin{vmatrix} -4 - \lambda & 1 \\ -1 & -3/2 - \lambda \end{vmatrix} = 0$$

and is given by

$$(\lambda + 2)\ (\lambda + \tfrac{7}{2}) = 0$$

The eigenvalues are real and distinct and are given by

$$\lambda_1 = -2, \qquad \lambda_2 = -\tfrac{7}{2}$$

Let the first complementary solution be

$$v'_{4o} = Q'_1\ e^{-2t}$$
$$i'_{1o} = Q'_2\ e^{-2t}$$

Substituting the above into the homogeneous differential equation, we get the following homogeneous algebraic equations:

$$-2Q'_1 + Q'_2 = 0$$
$$-Q'_1 + \tfrac{1}{2}\ Q'_2 = 0 \qquad\qquad (4.41)$$

Solving for Q'_2 in terms of Q'_1, we have

$$Q'_2 = 2Q'_1$$

Since Q_1' is arbitrary, we may set it to 1, and Q_2' is then 2. The first complementary solution is, therefore,

$$v_{4o}' = e^{-2t}$$
$$i_{1o}' = 2e^{-2t}$$

Let the second complementary solution be

$$v_{4o}'' = Q_1'' \, e^{-(7/2)t}$$
$$i_{1o}'' = Q_2'' \, e^{-(7/2)t}$$

Substitution of the above into the homogeneous differential equation yields

$$-\tfrac{1}{2} Q_1'' + Q_2'' = 0$$
$$-Q_1'' + 2Q_2'' = 0$$

Solving for Q_2'' in terms of Q_1'', we get

$$Q_2'' = \tfrac{1}{2} Q_1''$$

Setting Q_1'' to 2, we have the second complementary solution given by

$$v_{4o}'' = 2e^{-(7/2)t}$$
$$i_{1o}'' = e^{-(7/2)t}$$

The complementary solution is a linear combination of the two given by

$$v_{4o} = c_1 e^{-2t} + 2c_2 e^{-(7/2)t}$$
$$i_{1o} = 2c_1 e^{-2t} + c_2 e^{-(7/2)t} \tag{4.42}$$

where c_1 and c_2 are arbitrary constants. To find the particular integral, we set

$$v_{4p} = u_1(t) e^{-2t} + 2u_2(t) e^{-(7/2)t}$$
$$i_{1p} = 2u_1(t) e^{-2t} + u_2(t) e^{-(7/2)t} \tag{4.43}$$

Substituting the above into the nonhomogeneous differential equations, we get, after simplification by noting that the complementary solution satisfies the homogeneous equations,

$$e^{-2t} u_1'(t) + 2e^{-(7/2)t} u_2'(t) = 0$$
$$2e^{-2t} u_1'(t) + e^{-(7/2)t} u_2'(t) = e(t)$$

Solving for u_1' and u_2', we get

$$u_1'(t) = \tfrac{2}{3} e^{2t} e(t)$$
$$u_2'(t) = -\tfrac{1}{3} e^{(7/2)t} e(t)$$

Integrating, we get

$$u_1(t) = \frac{2}{3} \int_0^t e^{2x} e(x) \, dx$$

$$u_2(t) = -\frac{1}{3} \int_0^t e^{(7/2)x} e(x) \, dx$$

The particular solution is, therefore, given by

$$v_{4p}(t) = \frac{2}{3} \int_0^t e^{-2(t-x)} e(x) \, dx - \frac{2}{3} \int_0^t e^{-(7/2)(t-x)} e(x) \, dx$$

$$v_{ip}(t) = \frac{4}{3} \int_0^t e^{-2(t-x)} e(x) \, dx - \frac{1}{3} \int_0^t e^{-(7/2)(t-x)} e(x) \, dx \qquad (4.44)$$

The complete solution is

$$v_4(t) = c_1 e^{-2t} + c_2 \, 2e^{-(7/2)} + \frac{2}{3} \int_0^t e^{-2(t-x)} e(x) \, dx - \frac{2}{3} \int_0^t e^{-(7/2)(t-x)} e(x) \, dx$$

$$i_1(t) = c_1 2e^{-2t} + c_2 \, e^{-(7/2)t} + \frac{4}{3} \int_0^t e^{-2(t-x)} e(x) \, dx - \frac{1}{3} \int_0^t e^{-(7/2)(t-x)} e(x) \, dx \qquad (4.45)$$

It is seen once again that the complete solution is a linear combination of the complementary solution and the convolution of the complementary solution and the excitation. Note that the complementary solution consists of two decaying exponentials.

Example 4.6 Find $v_4(t)$ and $i_1(t)$ of the circuit of Figure 4.24.

Solution Making the obvious replacement of parameters of Example 4.5, we get the following differential equations:

$$\frac{dv_4}{dt} = -2v_4 + i_1$$

$$\frac{di_1}{dt} = -v_4 - 2i_1 + e(t) \qquad (4.46)$$

The characteristic equation is

$$\lambda^2 + 4\lambda + 5 = 0$$

whose roots are

$$\lambda_1 = -2 + j$$
$$\lambda_2 = -2 - j$$

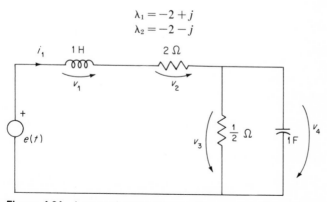

Figure 4.24 A second-order circuit with complex eigenvalues. [Example 4.6]

where $j = \sqrt{-1}$. Following the same procedure as in Example 4.5, we find the complementary solution to be

$$v_{4o}(t) = c_1 e^{(-2+j)t} + c_2 e^{(-2-j)t}$$
$$i_{1o}(t) = c_1 j e^{(-2+j)t} - c_2 j e^{(-2-j)t} \tag{4.47}$$

where c_1 and c_2 are arbitrary constants. The above can be put into a form which contains only real constants. For example, using Euler's formula for the exponential e^{jt}, we obtain

$$v_{4o}(t) = (c_1 + c_2)e^{-2t}\cos t + j(c_1 - c_2)e^{-2t}\sin t$$

Since c_1 and c_2 are arbitrary, we let

$$c_1 = \tfrac{1}{2}(A - jB) \qquad c_2 = \tfrac{1}{2}(A + jB)$$

with A and B real. We then have

$$v_{4o}(t) = Ae^{-2t}\cos t + Be^{-2t}\sin t \tag{4.48}$$

In a similar fashion, we get

$$i_{1o}(t) = Be^{-2t}\cos t - Ae^{-2t}\sin t \tag{4.49}$$

To find the particular solution, we proceed as before. Let

$$v_{4p}(t) = u_1(t)e^{(-2+j)t} + u_2(t)e^{(-2-j)t}$$
$$i_{1p}(t) = u_1(t)je^{(-2+j)t} - u_2(t)je^{(-2-j)t} \tag{4.50}$$

and $u_1(t)$ and $u_2(t)$ are found to be

$$u_1(t) = -\frac{1}{2}j\int_0^t e^{(2-j)x}e(x)\,dx$$
$$u_2(t) = \frac{1}{2}j\int_0^t e^{(2+j)x}e(x)\,dx \tag{4.51}$$

After some simplifications, the particular solution is found to be

$$v_{4p}(t) = \int_0^t e^{-2(t-x)}\sin(t-x)e(x)\,dx$$
$$i_{1p}(t) = \int_0^t e^{-2(t-x)}\cos(t-x)e(x)\,dx \tag{4.52}$$

The complete solution is given by

$$v_4(t) = Ae^{-2t}\cos t + Be^{-2t}\sin t + \int_0^t e^{-2(t-x)}\sin(t-x)e(x)\,dx$$
$$i_1(t) = Be^{-2t}\cos t - Ae^{-2t}\sin t + \int_0^t e^{-2(t-x)}\cos(t-x)e(x)\,dx \tag{4.53}$$

Once again, the complete solution is a linear combination of the complementary solution and the convolution of the complementary solution and the excitation. The complementary solution consists of two "damped" sinusoidal waves. ∎

Now we must return to Case 2 in which the two eigenvalues are equal. It is best to do a numerical example.

Example 4.7 Find $v_4(t)$ and $i_1(t)$ of the circuit shown in Figure 4.25.

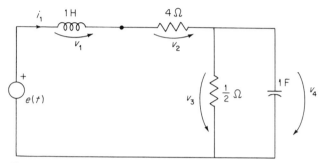

Figure 4.25 A second-order circuit with equal eigenvalues. [Example 4.7]

Solution Making the obvious replacement of element values in Example 4.5, we get the following differential equations:

$$\frac{dv_4}{dt} = -2v_4 + i_1$$

$$\frac{di_1}{dt} = -v_4 - 4i_1 + e(t) \tag{4.54}$$

The characteristic equation is

$$\lambda^2 + 6\lambda + 9 = 0$$

or

$$(\lambda + 3)^2 = 0$$

The complementary solution appears to take the form

$$v'_{4o}(t) = Q_1 e^{-3t}$$
$$i'_{1o}(t) = Q_2 e^{-3t} \tag{4.55}$$

Substituting into the homogeneous equations, we get

$$Q_1 + Q_2 = 0$$
$$-Q_1 - Q_2 = 0$$

so that by setting $Q_1 = 1$, the complementary solution appears to be

$$v'_{4o}(t) = c_1 e^{-3t}$$
$$i'_{1o}(t) = -c_1 e^{-3t} \tag{4.56}$$

Since the second eigenvalue is the same as the first, it appears that we would not be able to produce another complementary solution. In fact, however, the second complementary solution has the following form:

$$v''_{4o} = A_1 e^{-3t} + A_2 t e^{-3t}$$
$$i''_{1o} = B_1 e^{-3t} + B_2 t e^{-3t} \tag{4.57}$$

where A_1, A_2, B_1, and B_2 are constants to be determined. A_2 and B_2 may not both be zero. (Why?) Substituting (4.57) into the homogeneous differential equations, we get

$$-3A_1e^{-3t} + A_2e^{-3t} - 3A_2te^{-3t} = -2A_1e^{-3t} - 2A_2te^{-3t} + B_1e^{-3t} + B_2te^{-3t}$$
$$-3B_1e^{-3t} + B_2e^{-3t} - 3B_2te^{-3t} = -A_1e^{-3t} - A_2te^{-3t} - 4B_1e^{-3t} - 4B_2te^{-3t}$$

These equations hold for all t. Equating coefficients of like terms, we get

$$\begin{aligned}
-A_1 + A_2 - B_1 &= 0 \\
-A_2 - B_2 &= 0 \\
A_1 + B_1 + B_2 &= 0 \\
A_2 + B_2 &= 0
\end{aligned} \tag{4.58}$$

One nontrivial solution is

$$A_1 = 1, \quad B_1 = 0, \quad A_2 = 1, \quad B_2 = -1$$

The second complementary solution is then

$$\begin{aligned}
v''_{4o}(t) &= e^{-3t} + te^{-3t} \\
i''_{1o}(t) &= -te^{-3t}
\end{aligned} \tag{4.59}$$

The complementary solution is a linear combination of the first and the second complementary solutions. It is given by

$$\begin{aligned}
v_{4o} &= c_1e^{-3t} + c_2(1 + t)e^{-3t} \\
i_{1o} &= c_1(-1)e^{-3t} + c_2(-1)te^{-3t}
\end{aligned} \tag{4.60}$$

The particular integral is found by the method of variation of parameters again. Let

$$\begin{aligned}
v_{4p} &= u_1e^{-3t} + u_2(1 + t)e^{-3t} \\
i_{1p} &= u_1(-1)e^{-3t} + u_2(-1)te^{-3t}
\end{aligned} \tag{4.61}$$

Substituting the above into the nonhomogeneous differential equations, we obtain a system of two linear algebraic equations in the unknowns u'_1 and u'_2 as before. Integrating and simplifying, we get

$$\begin{aligned}
u_1(t) &= -\int_0^t (1 + x)e^{3x} e(x) \, dx \\
u_2(t) &= \int_0^t e^{3x} e(x) \, dx
\end{aligned} \tag{4.62}$$

and

$$\begin{aligned}
v_{4p}(t) &= \int_0^t (t - x)e^{-3(t-x)} e(x) \, dx \\
i_{1p}(t) &= \int_0^t e^{-3(t-x)} e(x) \, dx - \int_0^t (t - x)e^{-3(t-x)} e(x) \, dx
\end{aligned} \tag{4.63}$$

The complete solution is

$$\begin{aligned}
v_4(t) &= c_1e^{-3t} + c_2(1 + t)e^{-3t} + \int_0^t (t - x)e^{-3(t-x)} e(x) \, dx \\
i_1(t) &= -c_1e^{-3t} - c_2te^{-3t} + \int_0^t e^{-3(t-x)} e(x) \, dx - \int_0^t (t - x)e^{-3(t-x)} e(x) \, dx
\end{aligned} \tag{4.64}$$

Once again, the complete solution is a linear combination of the complementary solution and the convolution of the complementary solution with the excitation. ∎

In the preceding examples, the variables with which we describe the circuit are the inductor current and the capacitor voltage. It is the rate of change of these variables that characterizes the dynamic behavior of the circuit. Given their initial values, we can predict their values at the next instant by means of the differential equations that are obtained by combining the KVL, KCL, and the terminal equations of the circuit. Once the values of these variables are found, the voltages and currents everywhere in the circuit are determined. To see this, we may make reference to Figure 4.22. Regarding the voltage v_c and the current i_L as excitations and the voltages and currents inside the box as responses, we see that the voltages and currents are linear combinations of the capacitor voltage and the inductor current. It appears, therefore, that the capacitor voltage and the inductor current, together with the differential equations, completely describe the "state of being" of the circuit. For this reason,

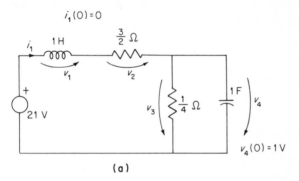

(a)

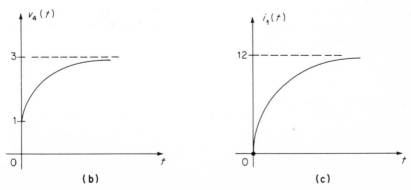

(b)　　　　　　　　　　　(c)

Figure 4.26 Response of a second-order circuit under a constant excitation. Circuit is overdamped. [Example 4.8]

these two variables are called the "state" variables of the circuit. The differential equations are called the "state" equations. If we wish to be more formal about the definitions, we say that the *state variables* of a circuit are those variables, minimum in number, whose rate of change completely characterizes the dynamic behavior of the circuit. The minimum number is called the *order* of the circuit, The differential equations governing the state variables are the *state equations,* and the "state" of the circuit at time t is the set of values of the state variables at t.

4.4 Second-Order Circuits Under Constant Excitation

It might be instructive to consider the special case in which the excitations are all constant, independent of time.

Example 4.8 If in the circuit of Example 4.5, $e(t) = 21$ V, $v_4(0) = 1$ V, and $i_1(0) = 0$ A, find all the voltages and currents. The circuit is reproduced in Figure 4.26(a).

Solution Using Equations (4.45), we first determine the constants c_1 and c_2 by means of the initial conditions.

$$v_4(0) = 1 = c_1 + 2c_2$$
$$i_1(0) = 0 = 2c_1 + c_2$$

from which we get

$$c_1 = -\tfrac{1}{3}, \; c_2 = \tfrac{2}{3}$$

Substituting into (4.45), we obtain

$$v_4(t) = 3 - \frac{22}{3} e^{-2t} + \frac{16}{3} e^{-(7/2)t}$$
$$i_1(t) = 12 - \frac{44}{3} e^{-2t} + \frac{8}{3} e^{-(7/2)t}$$

These are plotted as shown in Figures 4.26(b) and (c). To check our answers, we note that at $t = \infty$, the circuit becomes that shown in Figure 4.27. By inspection, $v_4(\infty) = 3$ V and $i_1(\infty) = 12$ A as they should be.

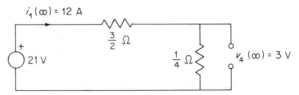

Figure 4.27 Circuit of Example 4.8 at $t = \infty$.

The other voltages and currents are

$$i_4(t) = \frac{dv_4}{dt} = \frac{44}{3} e^{-2t} - \frac{56}{3} e^{-(7/2)t}$$

$$i_3(t) = 4v_4 = 12 - \frac{88}{3} e^{-2t} + \frac{64}{3} e^{-(7/2)t}$$

$$i_1(t) = i_3 + i_4 = 12 - \frac{44}{3} e^{-2t} + \frac{8}{3} e^{-(7/2)t} \quad \text{(Check)}$$

$$v_2(t) = \frac{3}{2} i_1 = 18 - 22e^{-2t} + 4e^{-(7/2)t}$$

$$v_1(t) = \frac{di_1}{dt} = \frac{88}{3} e^{-2t} - \frac{28}{3} e^{-(7/2)t}$$

$$v_1 + v_2 + v_4 = 21 \text{ V} \quad \text{(Check)}$$

Example 4.9 Suppose in the circuit of Example 4.6, $e(t) = 25$ V for $t < 0$. At $t = 0$, $e(t)$ is suddenly reduced to zero. Find $v_4(t)$ and $i_1(t)$ for $t \geq 0$.

Solution Since the capacitor voltage and the inductor current do not change at $t = 0$, their initial values are the "final" values of the circuit with $e(t) = 25$ V. They are found from the circuit of Figure 4.28 and are given by

$$v_4(0) = 5 \text{ V} \qquad i_1(0) = 10 \text{ A}$$

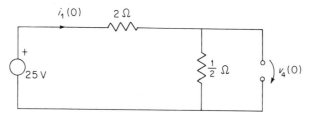

Figure 4.28 Circuit at $t = 0$. (Actual circuit is shown in Figure 4.24.)

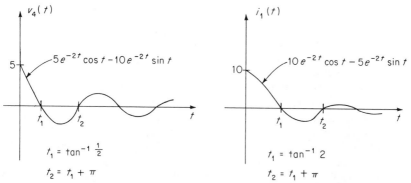

Figure 4.29 Sketch of $v_4(t)$ and $i_1(t)$ of circuit of Example 4.9 (see Figure 4.24). Circuit is underdamped.

Using these values, we have, from (4.53),

$$A = 5 \qquad B = 10$$

and the solution is

$$v_4(t) = 5e^{-2t} \cos t - 10e^{-2t} \sin t$$
$$i_1(t) = 10e^{-2t} \cos t - 5e^{-2t} \sin t$$

These are sketched in Figure 4.29.

Example 4.10   Obtain the solution of Example 4.8 for $t \geqq 0$, assuming that the initial conditions are all zero. The only contribution comes from the particular integrals. From Equations (4.64), we get, after obvious substitutions and simplifications,

$$v_4(t) = 1 - (3t + 1)e^{-3t}$$
$$i_1(t) = 2 + (3t - 2)e^{-3t}$$

These are sketched in Figure 4.30. ■

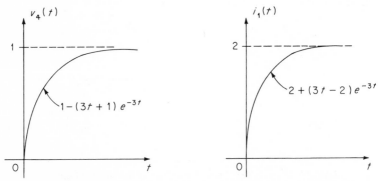

Figure 4.30   Sketch of $v_4(t)$ and $i_1(t)$ of a critically damped circuit (see Figure 4.25)   [Example 4.10]

It is time to make some observations.

1. If the eigenvalues λ_1 and λ_2 are real and distinct, the complementary solution (the response due to the initial conditions alone) is the sum of two decaying exponentials. We speak of this case as the "overdamped" case. If the eigenvalues are complex, the complementary solution is the sum of two damped sinusoidal functions. We speak of this case as the "underdamped" case. The threshold value of the eigenvalue that distinguishes the two cases is that for which $\lambda_1 = \lambda_2$, and we speak of this case as the "critically damped" case.

2. The voltages and currents everywhere in a second-order circuit are of the form given by Equations (4.45).

3. Second-order circuits are not limited to circuits containing one inductor and one capacitor only. For example, the circuits of Figure 4.31 are both second-order circuits provided that v_1 and v_2 in (a), or i_1 and in i_2 in (b), are not

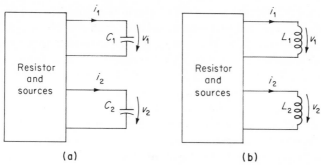

Figure 4.31 **(a)** A second-order circuit with two capacitors; **(b)** a second-order circuit with two inductors. It is assumed that no linear relation exists between the two capacitor voltages or between the two inductor currents.

linearly related. To see that the circuits are second order, we note, for example in Figure 4.31(a), that the following differential equations can be written:

$$C_1 \frac{dv_1}{dt} = i_1 = a_{11}v_1 + a_{12}v_2 + y_1(t)$$

$$C_2 \frac{dv_2}{dt} = i_2 = a_{21}v_1 + a_{22}v_2 + y_2(t)$$

where the a's are constants, and y_1 and y_2 are the responses due to internal excitations. The equations are a consequence of the principle of superposition again. See Problem 4.19.

4. The complementary solution in all of the previous examples consists of decaying exponentials of the form e^{-at} with the real part of $a = \text{Re}(a) \geq 0$. This property insures that the natural response (the response in the absence of excitations) will never grow without limit. Is this an intrinsic property of the circuits? Is it true for all circuits? The answer is yes. Things would get out of hand if the contrary were true. One can actually prove, using advanced mathematical techniques which would require a good bit of digression, that the complementary response of a circuit containing linear, time-invariant elements with all element values greater than zero is of the form e^{-at}, with $\text{Re}(a) \geq 0$. The equality holds for one, and only one, case, which is that in which resistors are completely absent. (See Problem 4.17.)

4.5 Summary

First-order circuits are those that can be reduced to one which is shown in Figure 4.1 or Figure 4.5. The response of these circuits consists of a complementary solution and the convolution of the complementary solution with the excitation. In all cases, the complementary solution is of the form $\exp(-at)$, with $a > 0$.

Second-order circuits are those that can be reduced to one containing two energy-storing elements. More precisely each of such circuits has two state variables governed by a system of two first-order differential equations. The solution consists of a complementary solution and the convolution of the complementary solution with the excitation. The nature of the complementary solution depends on the eigenvalues of the system of differential equations. Real eigenvalues lead to decaying exponential responses, and complex eigenvalues lead to damped sinusoidal responses.

We conclude this chapter by presenting a number of examples to illustrate how these circuits can be used in practice.

Example 4.11 (Sawtooth Wave Generator) In many electronic graphic display systems, a voltage that varies linearly with time is often desired. The voltage is used to deflect the electron beam of a cathode ray tube so that an image of a straight line whose length is proportional to time is produced. The voltage generally takes the form shown in Figure 4.32. A simple circuit which produces

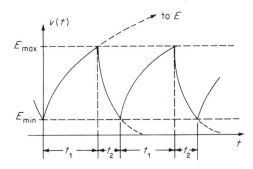

Figure 4.32 A sawtooth waveform. [Example 4.11]

such a waveform is shown in Figure 4.33, where S is a switch which opens for t_1 Sec and closes for t_2 Sec periodically. In practice, the switch is realized by a transistor or vacuum tube excited by a square wave.

Suppose E, C, R_1, and R_2 are given; find E_{max} and E_{min}.

Solution When switch S is open, the circuit becomes that shown in Figure 4.34(a). The initial capacitor voltage is E_{min}. The final value is E if the switch remains open. The capacitor voltage is given by

$$v(t) = E + (E_{min} - E)e^{-t/T_1} \qquad (4.65)$$

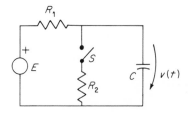

Figure 4.33 A circuit to generate a sawtooth waveform. Switch S opens and closes periodically. [Example 4.11]

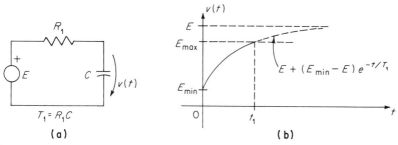

Figure 4.34 (a) The circuit of Example 4.11 when switch S is opened; (b) $v(t)$ reaches E_{max} at $t = t_1$.

where $T_1 = R_1C$. From Figure 4.34(b), it is clear that to obtain a good approximation to a straight line between E_{min} and E_{max}, we must have $E \gg E_{max}$.

When S is closed, the circuit becomes that of Figure 4.35(a). The initial voltage is E_{max} if we start counting time at t_1. The final value is $ER_2/(R_1 + R_2)$ if S is allowed to remain closed. The capacitor voltage is given by

$$v(t) = xE + (E_{max} - xE)e^{-t/T_2} \tag{4.66}$$

where $x = R_2/(R_1 + R_2)$ and $T_2 = R_1CR_2/(R_1 + R_2)$. Notice that $T_2 < T_1$. (Why?)

From Figures 4.34 and 4.35 and by Equations (4.65) and (4.66), we have

$$v(t_1) = E_{max} = E + (E_{min} - E)e^{-t_1/T_1}$$
$$v(t_2) = E_{min} = xE + (E_{max} - xE)e^{-t_2/T_2}$$

from which we obtain

$$E_{max} = \frac{E(1 - e^{-t_1/T_1}) + xE(1 - e^{-t_2/T_2})e^{-t_1/T_1}}{1 - \exp(-t_1/T_1 - t_2/T_2)} \tag{4.67}$$

$$E_{min} = \frac{xE(1 - e^{-t_2/T_2}) + E(1 - e^{-t_1/T_1})e^{-t_2/T_2}}{1 - \exp(-t_1/T_1 - t_2/T_2)} \tag{4.68}$$

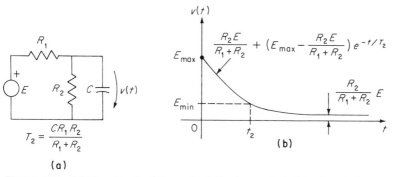

Figure 4.35 (a) The circuit of Example 4.11 when switch S is closed; (b) $v(t)$ reaches E_{min} at $t = t_2$.

In a practical design problem, E_{max}, E_{min}, t_1, and t_2 are specified, and the designer must select the element values E, R_1, R_2, and C such that the voltage will meet the specifications. Examination of the expression for E_{max} and E_{min} shows that the element values cannot be explicitly given and must be determined numerically. See Problem 4.24.

Example 4.12 (Uncompensated Attenuator) An ideal attenuator is a circuit whose output voltage is a true fractional replica of its input voltage. Figure 4.36(a) shows a simple voltage divider which is, in fact, an ideal attenuator. In practice, the output voltage is applied to a measuring device, such as a voltmeter or an oscilloscope, whose input circuit (a transistor amplifier) invariably has some "stray" capacitance connected across its terminals. The capacitance is represented by a capacitor as shown in Figure 4.36(b). The presence of this capacitor

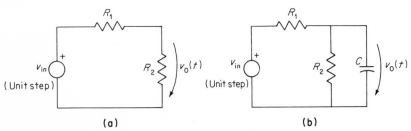

Figure 4.36 (a) A simple attenuator; **(b)** a realistic attenuator.   [Example 4.12]

causes the output voltage not to be a replica of the input, and what is observed or measured is no longer a true fraction of the input. Calculate the output voltage when the input $v_{in}(t)$ is a unit step.

Solution   For $t \leq 0$, we have $v_{in} = 0$ so that $v_0(0) = 0$. The final value is

$$v_0(\infty) = \frac{R_2}{R_1 + R_2} \, v_{in}$$

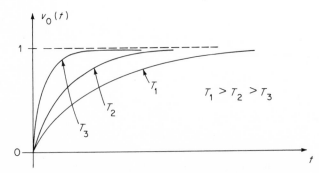

Figure 4.37   The step response of a realistic attenuator.

The time constant is

$$T = \frac{CR_1R_2}{R_1 + R_2}$$

The output voltage is

$$v_0(t) = \frac{R_2}{R_1 + R_2}\,(1 - e^{-t/T}) \qquad t \geq 0$$

Figure 4.37 shows the sketches of $v_0(t)$ for different values of T. It is seen that the smaller the time constant, the faster the output approaches its final value. ■

Is it possible to alter the circuit such that the output is a true fractional replica of the input voltage? That this is possible is demonstrated in the next example.

Example 4.13 *(Compensated Attenuator)* Figure 4.38(a) shows the circuit of a compensated attenuator. For the moment, let the input voltage be specified as follows:

$$v_{in}(t) = E(1 - e^{-kt}) \qquad t \geq 0$$
$$= 0, \qquad\qquad t < 0$$

The input becomes a step function in the limit as k approaches infinity. Let us find the output voltage $v_2(t)$.

Solution Choose a tree to include the voltage source and C_2 as shown in Figure 4.38(b). We have

$$\text{(KVL):} \qquad v_1 = v_{in} - v_2$$

$$\text{(KCL):} \qquad i_2 = i_1 + i_5 - i_4$$

$$C_2\frac{dv_2}{dt} = C_1\frac{dv_1}{dt} + \frac{v_1}{R_1} - \frac{v_2}{R_2}$$

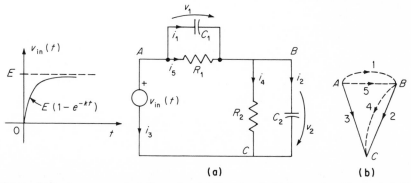

(a)   **(b)**

Figure 4.38   A compensated attenuator. **(a)** The circuit; **(b)** its graph.   [Example 4.13]

Making substitutions, we get

$$\frac{dv_2}{dt} = \frac{-1}{RC} v_2 + \frac{E}{R_1C} + E\left(\frac{kC_1}{C} - \frac{1}{R_1C}\right)e^{-kt}$$

where $C = C_1 + C_2$ and $R = R_1R_2/(R_1 + R_2)$. The initial condition $v_2(0)$ is found as follows. Note that we have a loop of capacitors and a voltage source. By KVL, at $t = 0^+$, as at all t,

$$v_{in}(0^+) = v_1(0^+) + v_2(0^+)$$

If $v_1(0^-)$ and $v_2(0^-)$ are the capacitor voltages at $t = 0^-$, then to satisfy KVL at $t = 0^+$, the voltages must change instantaneously. This may take place only if charge is transferred from one capacitor to the other instantaneously, so that the capacitor currents are unbounded. Since charge is conserved, that lost by one capacitor is gained by the other, namely,

$$C_1[v_1(0^+) - v_1(0^-)] = C_2[v_2(0^+) - v_2(0^-)]$$

from which, together with KVL, we obtain

$$v_2(0^+) = \frac{C_1}{C_1 + C_2} v_{in}(0^+) - \frac{C_1}{C_1 + C_2} v_1(0^-) + \frac{C_2}{C_1 + C_2} v_2(0^-)$$

Since $v_{in}(0^+) = 0$ and $v_1(0^-) = v_2(0^-) = 0$ in this case, we have $v_2(0^+) = 0$ V. The solution of the differential equation on $v_2(t)$ is then

$$v_2(t) = Ae^{-t/RC} + \frac{R}{R_1}E + Be^{-kt}$$

where

$$A = -\frac{R}{R_1}E - \frac{E}{C}\frac{kC_1 - 1/R_1}{1/RC - k}$$

$$B = \frac{E}{C}\frac{kC_1 - 1/R_1}{1/RC - k}$$

As $k \to \infty$, the input approaches a step and the constants

$$A \to -\frac{R}{R_1}E + \frac{C_1}{C}E = -\frac{R_2E}{R_1 + R_2} + \frac{C_1E}{C_1 + C_2}$$

$$B \to -\frac{C_1E}{C} = -\frac{C_1E}{C_1 + C_2}$$

and the output voltage becomes in the limit

$$v_2(t) = \frac{R_2E}{R_1 + R_2} - \left(\frac{R_2E}{R_1 + R_2} - \frac{EC_1}{C_1 + C_2}\right)e^{-t/RC} \qquad (4.69)$$

Figure 4.39 shows the sketches of $v_2(t)$ corresponding to the following three cases:

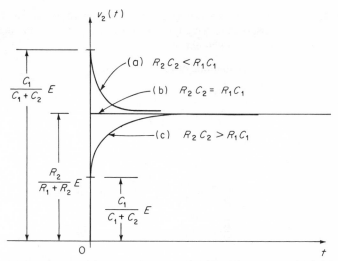

Figure 4.39 The step response of a compensated attenuator. **(a)** Overcompensated; **(b)** perfect; **(c)** undercompensated.

(a) $\dfrac{R_2}{R_1 + R_2} < \dfrac{C_1}{C_1 + C_2}$ or $R_2 C_2 < R_1 C_1$

(b) $\dfrac{R_2}{R_1 + R_2} = \dfrac{C_1}{C_1 + C_2}$ or $R_2 C_2 = R_1 C_1$

(c) $\dfrac{R_2}{R_1 + R_2} > \dfrac{C_1}{C_1 + C_2}$ or $R_2 C_2 > R_1 C_1$ (4.70)

It is seen that if we choose the element values such that $C_1 R_1 = C_2 R_2$, the circuit is a perfect attenuator for an input voltage which is a step. Is it a perfect attenuator for other types of input? The answer is yes. Using the KVL, KCL, and the terminal equations for the circuit, we obtain a differential equation in v_2 for any excitation $e(t)$. It is given below:

$$(C_1 + C_2)\,\frac{dv_2}{dt} + \frac{R_1 + R_2}{R_1 R_2}\, v_2 = C_1 \frac{de}{dt} + \frac{1}{R_1}\, e(t)$$

The element values satisfy the condition of Case (b). The equation can, therefore, be rewritten as

$$\frac{R_1 + R_2}{R_2}\left(C_1 \frac{dv_2}{dt} + \frac{v_2}{R_1}\right) = C_1 \frac{de}{dt} + \frac{e}{R_1}$$

from which it is seen that v_2 and $e(t)$ must be linearly related. In fact,

$$v_2(t) = \frac{R_2}{R_1 + R_2}\, e(t) \qquad \text{for all } t \qquad (4.71)$$

That is, the output voltage of a compensated attenuator is a true fractional replica of the input voltage, which may be any time function.

Example 4.14 (Design) A step voltage is applied to the circuit of Figure 4.40. The voltage $v(t)$ is observed to have the waveform shown. Find the value of C such that the overshoot is 10 percent of E.

Solution The differential equations of the circuit are

$$\frac{dv}{dt} = 0v + \frac{1}{C}i$$

$$\frac{di}{dt} = -\frac{1}{L}v - \frac{R}{L}i + \frac{E}{L}$$

The eigenvalues are

$$\lambda_1 = -\frac{R}{2L} + j\sqrt{\frac{1}{LC} - \left(\frac{R}{2L}\right)^2} = -\alpha + j\beta$$

$$\lambda_2 = -\frac{R}{2L} - j\sqrt{\frac{1}{LC} - \left(\frac{R}{2L}\right)^2} = -\alpha - j\beta$$

They must be complex since the circuit is underdamped. Noting that $v(0) = 0$ and $i(0) = 0$, we find $v(t)$ to be

$$v(t) = E\left[1 - \frac{1}{\sqrt{1 - \frac{CR^2}{4L}}} e^{-\alpha t} \sin(\beta t + \theta)\right] \tag{4.72}$$

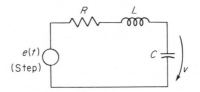

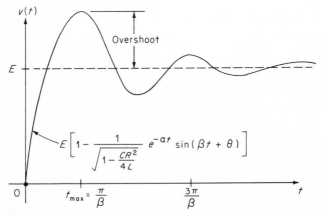

Figure 4.40 A simple second-order circuit whose step response has an overshoot. [Example 4.14]

where $\theta = \arctan{(\beta/\alpha)}$.

The maxima occur at $t_{max} = \pi/\beta,\ 3\pi/\beta,\ 5\pi/\beta,\ldots$, and they are given by

$$v(t_{max}) = E(1 + e^{-\alpha t\,max})$$

with $t_{max} = (2n + 1)\pi/\beta,\ n = 0,\ 1,\ 2,\ldots$. It is seen that the peaks of $v(t)$ follow an exponential decay.

The first maximum occurs at $t = \pi/\beta$. We must now find a value of C such that $v(\pi/\beta) = 1.1E$. Using Equation (4.72), we have

$$v\left(\frac{\pi}{\beta}\right) = 1.1E = E(1 + e^{-\alpha\pi/\beta})$$

expressing α and β in terms of the element values and solving for C, we obtain an explicit formula for C:

$$C = \frac{4L}{R^2}\ \frac{1}{1 + \left(\dfrac{\ln 10}{\pi}\right)^{-2}}$$

This example shows that, in contrast with Example 4.11, it is sometimes possible to express the element values explicitly in terms of the design parameters.

Example 4.15 (*RC Smoothing Circuit*) In a digital communication system, the signal is first sampled at equally spaced intervals, say at $T,\ 2T,\ 3T,\ldots$. Each sample is then converted into a binary number represented as a series of pulses in the time interval T, as shown in Figure 4.41. The pulses are transmitted over a coaxial cable or some other medium. The received signal is applied to a digital-to-analog converter whose output at each time interval is a constant voltage of

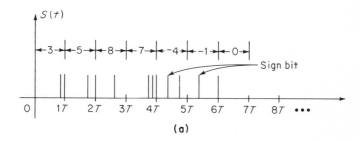

(a)

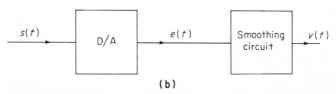

(b)

Figure 4.41 **(a)** A series of pulses which represents the sequence of binary numbers 3, 5, 8, 7, −4, −1. **(b)** The series of pulses is applied to a digital-to-analog (D/A) converter, the output of which is applied to a smoothing circuit. [Example 4.15]

amplitude proportional to the value of the binary number in that interval. The result is that the output is a "staircase" function somewhat as shown in Figure 4.42. To recover the analog signal, the output must be smoothed by applying it to a "smoothing" circuit. The simplest such circuit is an *RC* circuit shown in Figure 4.42. Compute the voltage $v(t)$.

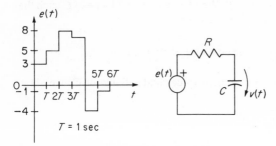

Figure 4.42 The output of a D/A converter is a staircase function. It is applied to a simple *RC* smoothing circuit to "round out" the corners. [Example 4.15]

Solution Numerical solution is clearly called for. The differential equation is

$$\frac{dv}{dt} = -\frac{1}{RC}(v - e) = f(v, t)$$

where $e(t)$ is as specified in Figure 4.42. It can be solved numerically by using the FORTRAN program RNGKT1, which is based on the Runge-Kutta algorithm described at the end of the chapter. In the computer solution, the time constant *RC* was assumed to be 3 sec, and an integration step of 0.1 sec was used. The program computes $v(t)$ for $0 \leq t \leq 10$ sec and graphs the function $v(t)$ by means of the subroutine EEPLOT, which is also described at the end of this chapter.

Figure 4.43 shows a plot of $v(t)$ based on the computer output. It is seen that

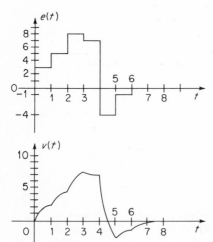

Figure 4.43 Computer solution of a simple *RC* smoothing circuit. **(a)** Input voltage; **(b)** output voltage.

the sharp corners of the staircase function are now "rounded," and the output voltage resembles the original input signal, except for a slight shift to the right (delay).

Example 4.16 *(RLC Smoothing Circuit)* Suppose that the circuit of the last example is replaced by the one shown in Figure 4.44. Compute the output voltage $v(t)$.

Figure 4.44 A *RLC* smoothing circuit. [Example 4.16]

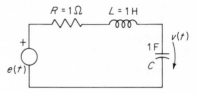

Solution The differential equations of the circuit are obtained by noting that the inductor voltage is, by **KVL**, $-(i \times 1) - v + e(t)$, and the capacitor current, by **KCL**, is i. Hence, we have

$$\frac{di}{dt} = -i - v + e(t) = f_1(i, v, t)$$

$$\frac{dv}{dt} = i = f_2(i, v, t)$$

The system of equations can be solved numerically by using the FORTRAN program RNGKT2, given at the end of the chapter. The program is also based on the fourth-order Runge-Kutta algorithm. In the computation, a step size of 0.1 sec was used. The computer output is plotted in Figure 4.45. It is seen that the second-order smoothing circuit gives much better results than the first-order circuit of the last example.

Figure 4.45 Computer solution of a *RLC* smoothing circuit. **(a)** Input voltage; **(b)** output voltage.

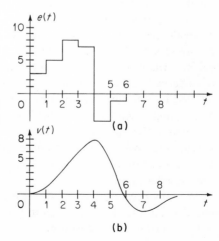

SUGGESTED EXPERIMENTS

1. Construct the circuit of Problem 4.8 and observe all waveforms. Vary the frequency of the square wave and observe the output. Explain what you see.
2. Apply a square wave to the circuit of Problem 4.3 and observe all waveforms. Decrease the time constant of the circuit by reducing the capacitance and observe all waveforms. Explain what you see.
3. Verify your design of Problems 4.9 and 4.10.
4. Apply a square wave to the circuit of Problem 4.18 and check the calculated results against the observed results.
5. Think of a way to verify the results of Problem 4.23 without actually using a frog.
6. Verify Example 4.13. Let the excitation be a square wave. Vary the frequency and observe the output. Repeat the experiment with a triangular wave. Explain the results.
7. Superimpose a series of narrow pulses on a low-frequency sine wave. Apply the composite wave to an RC smoothing circuit. Observe and explain the output.

COMPUTER PROGRAMMING

Numerical Integration of Differential Equations

In Sections 4.1 and 4.3, we gave the explicit solution of the first- and second-order circuits. The solution in both instances consists of contributions from the initial values (complementary solution) and from the convolution of the excitation and the complementary solution. If the excitation is expressible in terms of elementary functions, the convolution can be evaluated simply. In practice, however, the excitation cannot always be so specified. For example, the excitation representing a speech, a seismic wave, or an electrocardiogram can only be specified as a series of observed values taken over a period of time. When such excitation is applied to a circuit, the solution is best obtained by numerical means on a computer. (See Examples 4.15 and 4.16.)

Even in the case of an elementary excitation, there are advantages in solving the problem on a computer when we wish to study the effect on the response of a circuit as some or all of its parameters (element values) are varied.

We now present two numerical methods for solving a first-order differential equation. The methods are the Euler integration and the Runge-Kutta fourth-order algorithm.

Euler's Method

Consider a differential equation given by

$$\frac{dx}{dt} = f(x, t)$$

with an initial condition

$$x(t_0) = x_0$$

Starting with the initial value $x(t_0)$, we compute the values $x(t_1), x(t_2),\ldots$ step by step until we reach the final value $x(t_n)$. Replacing the derivative by first difference, we get

$$\frac{x(t_{k+1}) - x(t_k)}{t_{k+1} - t_k} \doteq f(x_k, t_k)$$

or

$$x(t_{k+1}) \doteq x(t_k) + f(x_k, t_k)(t_{k+1} - t_k)$$

For convenience, the increment in t is usually taken to be the same for all k, though it need not be so. Denote the increment by h and let $t_k = t_0 + kh$ and $x(t_k) = x_k$; then we have

$$x_{k+1} = x_k + f(x_k, t_k)h \tag{4.73}$$

The increment h is commonly known as the step size. Equation (4.73) is Euler's integration formula. Clearly, the solution obtained in this way is approximate. To see how large the error is at each step, we note that the Taylor series expansion of x at t_{k+1} is

$$x(t_{k+1}) = x_{k+1} = x(t_k) + hx'(t_k) + \frac{h^2}{2}x''(t_k) + \cdots$$

Euler's formula is obtained by truncating the series after the first two terms. The truncated error is seen to be of order h^2. A detailed discussion of errors will be given later.

For some applications, the Euler method is adequate if the step size is sufficiently small. If more accuracy is required, the Runge-Kutta algorithm may be used.

Fourth-Order Runge-Kutta Algorithm

There are a number of Runge-Kutta algorithms. We shall present the most popular one – known as the fourth-order algorithm. Its truncated error is of order h^5. The derivation is found in standard texts on numerical analysis and will not be given here.

Starting with the initial value x_0, we generate the subsequent values by the following procedure.

Let

$$A_1 = hf(x_k, t_k)$$

$$A_2 = hf\left(x_k + \frac{A_1}{2}, t_k + \frac{h}{2}\right)$$

$$A_3 = hf\left(x_k + \frac{A_2}{2}, t_k + \frac{h}{2}\right)$$

$$A_4 = hf(x_k + A_3, t_k + h)$$

Then

$$x_{k+1} = x_k + \frac{A_1 + 2A_2 + 2A_3 + A_4}{6}$$

The truncated error can be shown to be of order h^5. The choice of the step size is critical. If it is too large, the error will lead to meaningless results. If it is too small, the computation time will be excessive. In practice, one can monitor the incremental change in x from x_k to x_{k+1}. If the change is too large, the step size is halved. If the change is less than some small number, the step size is doubled. Another way of checking the results is to do the program twice, once with a step size h and once with $h/2$. If the results agree in the two cases, we are confident that they are correct.

The Euler and Runge-Kutta methods can be easily extended to integration of a system of n first-order differential equations. The generalized case which incorporates the feature of variable step size is given in many texts on numerical analysis. For our purpose, we list at the end the FORTRAN programs of the Runge-Kutta algorithm with fixed step size for a first-order differential equation (RNGKT1) and for a system of two first-order differential equations (RNGKT2).

Error Analysis

At each step of the Euler or Runge-Kutta algorithm, a truncated error is produced in predicting x_{k+1} from x_k. Now x_k itself is in error, and as a result so are $f(x_k, t_k)$, $f(x_k + A_1, t_k + h/2)$, and so on. As we generate the sequence $x_1, x_2, \ldots$, the truncated errors will accumulate. In addition, the truncated errors of all previous steps are carried forward in the calculation of $f(x_k, t_k)$, and so on. Let us make an analysis of the errors and see how the step size affects them. We shall do the case of Euler's method only. The analysis can be extended to the Runge-Kutta method in a straightforward way though the algebra tends to be unduly tedious.

Let $\bar{x}_k$ be the true value of $x(t_k)$. The true value of $x(t_{k+1})$ is

$$\bar{x}_{k+1} = \bar{x}_k + hf(\bar{x}_k, t_k) + \epsilon_{k,t}$$

where $\epsilon_{k,t}$ is the truncated error at step k. Now suppose the total error of x_k is ϵ_k. Then we have

$$\bar{x}_k = x_k + \epsilon_k$$

and

$$\bar{x}_{k+1} = x_k + \epsilon_k + hf(x_k + \epsilon_k, t_k) + \epsilon_{k,t} \qquad (4.74)$$

Expanding the third term in Taylor series about the point x_k, we have

$$f(x_k + \epsilon_k, t_k) = f(x_k, t_k) + \epsilon_k f_x(x_k, t_k) + O(\epsilon_k^2)$$

where $f_x(x_k, t_k)$ is the derivative with respect to x evaluated at x_k and t_k. Substituting the expansion into Equation (4.74) and neglecting higher order terms of ϵ_k we get

$$\begin{aligned} \bar{x}_{k+1} &= x_k + hf(x_k, t_k) + \epsilon_k[1 + hf_x(x_k, t_k)] + \epsilon_{k,t} \\ &= x_{k+1} + \epsilon_{k+1} \end{aligned}$$

where ϵ_{k+1} is the total error of x_{k+1} and is given by

$$\epsilon_{k+1} = \epsilon_k[1 + hf_x(x_k, t_k)] + \epsilon_{k,t}$$

Thus the total error at k "propagates" forward to $k+1$ by a propagation factor" $[1 + hf_x(x_k, t_k)]$. Proceeding for another step, we find

$$\begin{aligned} \epsilon_{k+2} = \epsilon_k[1 + hf_x(x_k, t_k)][1 + hf_x(x_{k+1}, t_{k+1})] \\ + \text{(other terms which depend on the truncated errors)} \end{aligned}$$

The propagation factor is now the product of $[1 + hf_x(x_k, t_k)]$ and $[1 + hf_x(x_{k+1}, t_{k+1})]$. In general, the total error at step k shows up in the total error in step $k + n$ as

$$\epsilon_k[1 + hf_x(x_k, t_k)][1 + hf_x(x_{k+1}, t_{k+1})] \ldots [1 + hf_x(x_{k+n}, t_{k+n})]$$

If all the derivatives of $f(x,t)$ with respect to x are positive in the various intervals, then the total error is "amplified." Unless the step size h is small, the results will not be of any value. Even if all the derivatives are negative, the step size must be chosen to be small to make the factor hf_x less than unity. Otherwise, amplification of the error will result.

As an example, consider the differential equation

$$\frac{dx}{dt} = 100x$$

The propagation factor is the same for all x. After n steps, it is

$$(1 + 100h)^n$$

On the other hand, if we have

$$\frac{dx}{dt} = -100x$$

the propagation factor after n steps is

$$(1 - 100h)^n$$

In either case, the step size must be much less than $1/100$. As a last remark, note that the eigenvalue of the last equation is 100 (time constant $=$ $1/100$). The solution is e^{-100t}, which decreases to zero very rapidly as t increases. This explains why the step size must be much less than $1/100$.

The preceding analysis can be extended to the general case of a system of first-order differential equations. We shall not do so here except to note that from the last example, it is clear that the choice of the step size is governed by the largest eigenvalue (smallest time constant) of the system.

Program RNGKT1

```
C   SEE EXAMPLE 4.15.
C   THIS IS A PROGRAM TO INTEGRATE A FIRST ORDER DIFFERENTIAL EQUATION
C   BY THE FOURTH ORDER RUNGE-KUTTA ALGORITHM.  THE EQUATION IS IN THE
C   FORM
C
C           X' = F(X,T)
C
C   THE PROGRAM GENERATES THE ARRAYS  X(I),  T(I),  I=1,2,...,N+1,
C   STARTING WITH THE INITIAL CONDITION X(1) AT T(1).  THE STEP SIZE
C   IS  H  AND IS UNCHANGED THROUGHOUT THE CALCULATION.  N IS THE
C   NUMBER OF STEPS AND IS LIMITED TO  999.
C   THE FUNCTION  F(X,T) IS DEFINED IN AN ARITHMETIC FUNCTION STATEMENT
C   OR IN A FUNCTION SUBPROGRAM.
C   THE PARAMETERS, X(1), T(1), H AND  N  ARE ASSIGNED VALUES BY MEANS
C   OF A DATA STATEMENT.
C   IN STATEMENT  200  REPLACE  'PROBLEM IDENTIFICATION'  BY YOUR OWN
C   PROBLEM NAME.
C
C
        DIMENSION X(1000), T(1000), ITITLE(20)
C   DEFINE  F(X,T)  IN THE FOLLOWING STATEMENT IF IT IS EXPRESSIBLE
C   AS AN ARITHMETIC STATEMENT.  IF NOT, USE FUNCTION SUBPROGRAM.
C   FOR EXAMPLE  4.15, THIS FUNCTION IS DEFINED IN A FORTRAN FUNCTION
C   SUBPROGRAM AS GIVEN AT THE END OF THE PROGRAM.
C   ENTER INITIAL CONDITION, STEP SIZE AND NUMBER OF INTEGRATION STEPS
C   IN THE FOLLOWING DATA STATEMENT.
C   FOR EXAMPLE  4.15, THE STATEMENT READS AS FOLLOWS.
        DATA X(1), T(1), H, N/0., 0., 0.1, 100/
C   READ IN TITLE OF COMPUTER PLOT.  SUBROUTINE EEPLOT WILL BE USED.
        READ(5,99)(ITITLE(II), II=1,20)
     99 FORMAT(20A4)
        H2 = H/2.
        M = N + 1
        DO 1000 I=1, N
        A1 = H*F(X(I), T(I))
        A2 = H*F(X(I) + A1/2. , T(I) + H2)
        A3 = H*F(X(I) + A2/2. , T(I) + H2)
        A4 = H*F(X(I) + A3,  T(I) + H)
        X(I + 1) = X(I) + (A1 + A2*2. + A3*2. + A4)/6.
   1000 T(I + 1) = T(I) + H
        WRITE(6, 200)
    200 FORMAT(1H , 'PROBLEM IDENTIFICATION'/5X, 'X', 10X, 'T')
        WRITE(6,300) (X(I), T(I), I=1, M)
    300 FORMAT(1H , 2E15.6)
C   CALL SUBROUTINE 'RANGE' TO FIND MAXIMUM AND MINIMUM VALUE OF X.
        CALL RANGE(X, YMAX, YMIN, N)
```

```
C   CALL EEPLOT TO GRAPH OUTPUT ON COMPUTER PRINTOUT.
        CALL EEPLOT(T, X, 10., 0., YMAX, YMIN, ITITLE, N)
        STOP
        END
C   THE FUNCTION F(X,T) IS DEFINED BY THE FOLLOWING FUNCTION SUBPROGRAM.
        FUNCTION F(V,T)
        IF(T .GE. 0. .AND. T .LT. 1.)GO TO 1
        IF(T .GE. 1. .AND. T .LT. 2.)GO TO 2
        IF(T .GE. 2. .AND. T .LT. 3.)GO TO 3
        IF(T .GE. 3. .AND. T .LT. 4.)GO TO 4
        IF(T .GE. 4. .AND. T .LT. 5.)GO TO 5
        IF(T .GE. 5. .AND. T .LT. 6.)GO TO 6
        E=0.
        GO TO 100
      1 E= 3.
        GO TO 100
      2 E=5.
        GO TO 100
      3 E= 8.
        GO TO 100
      4 E= 7.
        GO TO 100
      5 E= -4.
        GO TO 100
      6 E= -1.
    100 F= 3.*(-V + E)
        RETURN
        END
```

[See page 192 for the computer printout for this program.]

Program RNGKT2

```
C   SEE EXAMPLE 4.16.
C   THIS IS A PROGRAM TO INTEGRATE A SYSTEM OF TWO FIRST ORDER
C   DIFFERENTIAL EQUATIONS IN THE FORM
C
C               D (X1)/DT  =   F1(X1, X2, T)
C               D (X2)/DT  =   F2(X1, X2, T)
C
C   THE PROGRAM GENERATES THE ARRAYS  X1(I), X2(I), T(I), I=1, 2, ...,
C   N+1, STARTING WITH THE INITIAL CONDITIONS X1(1), X2(1), AT  T(1).
C   THE STEP SIZE IS  H  AND IS UNCHANGED THROUGHOUT THE CALCULATION.
C   N  IS THE NUMBER OF STEPS AND IS LIMITED TO  999.
C   THE FUNCTIONS  F1(X1, X2, T)  AND  F2(X1, X2, T)  ARE EACH DEFINED
C   IN AN ARITHMETIC FUNCTION STATEMENT, OR IN A FUNCTION SUBPROGRAM.
C   THE PARAMETERS,  X1(1), X2(1), T(1), H AND  N, ARE ASSIGNED VALUES BY
C   MEANS OF A DATA STATEMENT.
C   IN STATEMENT  200, REPLACE 'PROBLEM IDENTIFICATION' BY YOUR OWN
C   PROBLEM NAME.
C
C
        DIMENSION X1(1000), X2(1000), T(1000), ITITLE(20)
C   DEFINE THE FUNCTIONS  F1(X1, X2, T)  AND F2(X1, X2, T)  BELOW IF THEY
C   ARE EXPRESSIBLE AS ARITHMETIC FUNCTION STATEMENTS.  IF NOT, USE
C   FUNCTION SUBPROGRAMS.
C   FOR EXAMPLE 4.16,  F1(X1, X2, T)  IS DEFINED IN A FUNCTION SUBPROGRAM
C   NAMED F1(X,Y,T) AS GIVEN AT THE END OF THE MAIN PROGRAM.  F2(X1,X2,T)
C   IS DEFINED IN AN ARITHMETIC FUNCTION STATEMENT AS GIVEN BELOW.
C
        F2(X, Y, T) = X
C
C   ENTER INITIAL CONDITIONS, STEP SIZE AND NUMBER OF INTEGRATION
C   STEPS BELOW.
        DATA X1(1), X2(1), T(1), H, N/0., 0., 0., 0.1, 100/
C   READ IN TITLE OF COMPUTER PLOT  X1 VERSUS TIME.
        READ(5, 99)(ITITLE(II), II=1, 20)
```

```
   99 FORMAT(20A4)
      H2 = H/2.
      M = N + 1
      DO 100   I = 1, N
      A1 = H*F1(X1(I), X2(I), T(I))
      B1 = H*F2(X1(I), X2(I), T(I))
      A2 = H*F1(X1(I) + A1/2., X2(I) + B1/2., T(I) + H2)
      B2 = H*F2(X1(I) + A1/2., X2(I) + B1/2., T(I) + H2)
      A3 = H*F1(X1(I) + A2/2., X2(I) + B2/2., T(I) + H2)
      B3 = H*F2(X1(I) + A2/2., X2(I) + B2/2., T(I) + H2)
      A4 = H*F1(X1(I) + A3, X2(I) + B3, T(I) + H)
      B4 = H*F2(X1(I) + A3, X2(I) + B3, T(I) + H)
      X1(I + 1) = X1(I) + (A1 + 2.*A2 + 2.*A3 + A4)/6.
      X2(I + 1) = X2(I) + (B1 + 2.*B2 + 2.*B3 + B4)/6.
  100 T(I + 1) = T(I) + H
      WRITE(6,200)
  200 FORMAT(1H ,'PROBLEM IDENTIFICATION'/5X,'X1', 10X, 'X2', 10X, 'T')
      WRITE(6,300)(X1(I), X2(I), T(I), I=1, M)
  300 FORMAT(1H , 3E15.6)
C  CALL SUBROUTINE 'RANGE' TO FIND MAXIMUM AND MINIMUM VALUE OF X1.
      CALL RANGE(X1, YMAX, YMIN, N)
C  CALL SUBROUTINE 'EEPLOT' TO GRAPH  X1 VERSUS T.
      CALL EEPLOT(T, X1, 10., 0., YMAX, YMIN, ITITLE, N)
C  READ IN TITLE OF COMPUTER PLOT  X2  VERSUS TIME.
      READ(5,99)(ITITLE(II), II=1, 20)
C  CALL SUBROUTINE 'RANGE' TO FIND MAXIMUM AND MINIMUM VALUE OF X2.
      CALL RANGE(X2, YMAX, YMIN, N)
C  CALL SUBROUTINE 'EEPLOT' TO GRAPH  X2  VERSUS T.
      CALL EEPLOT(T, X2, 10., 0., YMAX, YMIN, ITITLE, N)
      STOP
      END
C  THE FOLLOWING FUNCTION SUBPROGRAM DEFINES THE FUNCTION  F1(X1,X2,T).
      FUNCTION F1(X, Y, T)
      F1 = -X -Y +E(T)
      RETURN
      END
C  THE FUNCTION E(T) IN THE ABOVE IS DEFINED BELOW.
      FUNCTION E(T)
      IF(T .GE. 0. .AND. T .LT. 1.)GO TO 1
      IF(T .GE. 1. .AND. T .LT. 2.)GO TO 2
      IF(T .GE. 2. .AND. T .LT. 3.)GO TO 3
      IF(T .GE. 3. .AND. T .LT. 4.)GO TO 4
      IF(T .GE. 4. .AND. T .LT. 5.)GO TO 5
      IF(T .GE. 5. .AND. T .LT. 6.)GO TO 6
      E=0.
      RETURN
    1 E=3.
      RETURN
    2 E=5.
      RETURN
    3 E=8.
      RETURN
    4 E=7.
      RETURN
    5 E=-4.
      RETURN
    6 E =-1.
      RETURN
      END
```

[See pages 193–194 for the computer printout for this program.]

Subroutine EEPLOT

```
      SUBROUTINE EEPLOT(X,Y,XMAX,XMIN,YMAX,YMIN,TITLE,NUM)
C     THIS SUBROUTINE PRODUCES A GRAPH OF Y VS X, WITH X BEING
C     THE HORIZONTAL AXIS.  IN THE LIST OF ARGUMENTS, X AND Y
C     ARE ARRAYS OF DIMENSION NUM.  XMAX AND YMAX ARE THE MAXIMUM VALUES
C     OF X AND Y, RESPECTIVELY.  XMIN AND YMIN ARE THE MINIMUM VALUES
```

```
C       RESPECTIVELY.  TITLE IS AN ARRAY OF DIMENSION 20 IN 20A4
C       FORMAT AND CONTAINS THE TITLE OF  THE PLOT.  IT MUST BE
C       SUPPLIED BY THE USER IN THE CALLING PROGRAM.
        DIMENSION GRAPH(52,101),YSCALE(6),XSCALE(11),X(NUM),Y(NUM)
        INTEGER TITLE(20)
        DATA DOT/1H./,ZERO/1H0/,BLANK/1H /,STAR/1H*/
        WRITE(6,100)TITLE
  100 FORMAT(1H1,20X,20A4,////)
        XSCALE(1) = XMIN
        XX=(XMAX - XMIN)/10.
        IF(XX-1.E-20)1,1,2
    1 WRITE(6,3)
    3 FORMAT('THE RANGE OF   X  IS TOO SMALL')
        GO TO 107
    2 DO 10 I=2,11
   10 XSCALE(I)=XSCALE(I-1) + XX
        YSCALE(1)=YMIN
        YY=(YMAX - YMIN)/5.
        IF(YY-1.E-20)6,6,7
    6 WRITE(6,11)YMAX
   11 FORMAT('   Y IS VIRTUALLY A CONSTANT', E15.6)
        GO TO 107
    7 DO 30 I=2,6
   30 YSCALE(I)=YSCALE(I-1)+YY
        DO 70 I=1,51
        DO 70 J=1,101
   70 GRAPH(I,J)=BLANK
        DO 50 L=1,NUM
        I=51-10.*(Y(L)-YMIN)/YY +0.2
        J=10.*(X(L)-XMIN)/XX+1.2
   50 GRAPH(I,J)=STAR
        DO 90 I=1,5
        BEGIN=ZERO
        WRITE(6,105)YSCALE(7-I),BEGIN,(GRAPH((I*10-9),J),J=1,101)
  105 FORMAT(1H  7X,1PE10.3,2X,102A1)
        BEGIN=DOT
        DO 90 K=1,9
   90 WRITE(6,104)BEGIN,(GRAPH((I*10-9+K),J),J=1,101)
  104 FORMAT(1H ,19X,102A1)
        BEGIN=ZERO
        WRITE(6,105)YSCALE(1),BEGIN,(GRAPH(51,J),J=1,101)
        DO 40 J=1,10
        M=J*10
        GRAPH(52,M)=ZERO
        DO 40 I=1,9
        K=(J-1)*10+I
   40 GRAPH(52,K)=DOT
        WRITE(6,108)BEGIN,(GRAPH(52,J),J=1,100)
  108 FORMAT(1H ,20X,101A1)
        WRITE(6,106)(XSCALE(I),I=1,11)
  106 FORMAT(1H ,16X,11(1PE9.2,1X))
  107 RETURN
        END
```

Subroutine RANGE

```
        SUBROUTINE RANGE(Y, YMAX, YMIN, N)
        DIMENSION Y(N)
        YMAX = Y(1)
        YMIN = Y(1)
        DO 2 I = 1, N
        YMAX = AMAX1(YMAX, Y(I))
    2 YMIN = AMIN1(YMIN, Y(I))
        RETURN
        END
```

Computer Printout for Example 4.15 Program

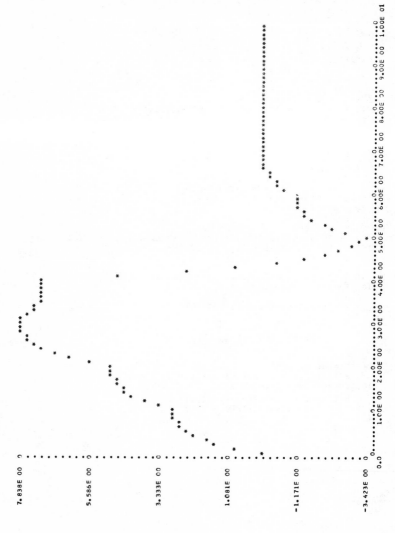

EXAMPLE 4.15.   CAPACITOR VOLTAGE VERSUS TIME

Computer Printout for Example 4.16 Program

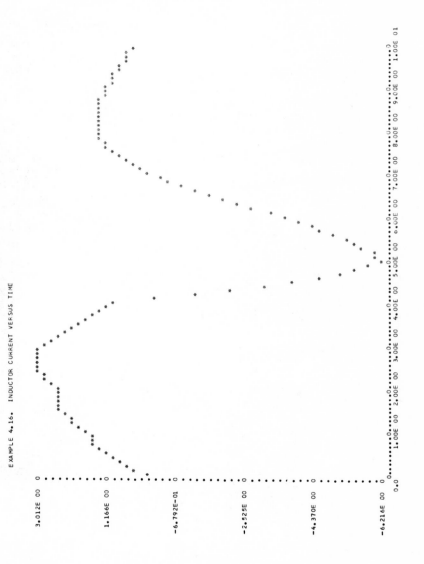

EXAMPLE 4.16. INDUCTOR CURRENT VERSUS TIME

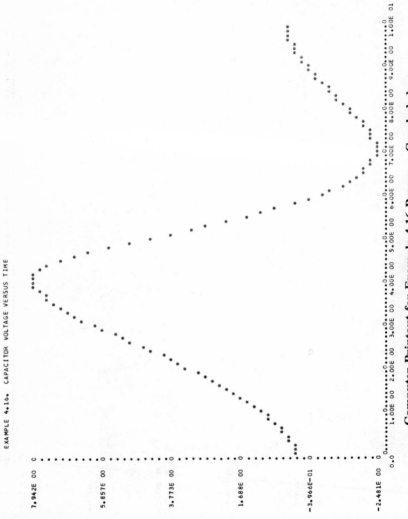

Computer Printout for Example 4.16 Program Concluded

PROBLEMS

4.1 *Exercise* Find $v_x(t)$ and $i_x(t)$ in each of the following circuits. Assume $e(t) = 0$ and $j(t) = 0$ for $t < 0$ and $e(t) = E$ and $j(t) = J$ for $t \geq 0$. Sketch $v_x(t)$ and $i_x(t)$.

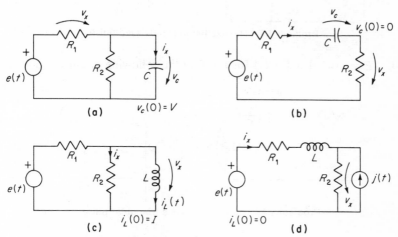

Problem 4.1 Exercises on first-order circuits.

4.2 *Exercise* In each of the following circuits, switch S is closed for $t < 0$. At $t = 0$, the switch S is suddenly opened. Find and sketch $v_x(t)$ and $i_x(t)$ in each.

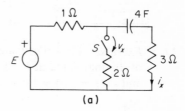

Problem 4.2 Exercises on first-order circuits.

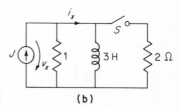

4.3   *Exercise*   The voltage $e(t)$ is a rectangular pulse as shown. Find $v_x(t)$ for $t \geq 0$.

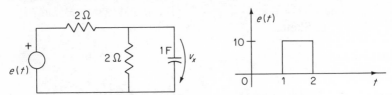

Problem 4.3   The excitation is a delayed pulse.

4.4   *Exercise*   Find $v_x(t)$ for $t \geq 0$ in the following circuit.

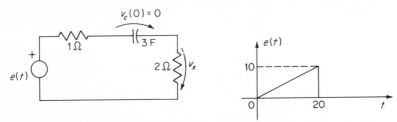

Problem 4.4   The excitation is a triangular pulse.

4.5   *Exercise*   Find the value of L such that at $t = 2$ sec $i_x(t)$ reaches 110 percent of its final value. What is this value? $e(t) = 0$ for $t < 0$ and $e(t) = E$ for $t \geq 0$.

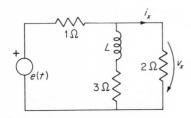

Problem 4.5   A simple design problem.

4.6   *Exercise*   Find the value of C such that $v_x(t)$ reaches 90 percent of its final value at $t = 1$ sec. Sketch $v_x(t)$. $e(t) = 10$ V for $t < 0$, and $e(t) = -10$ V for $t \geq 0$.

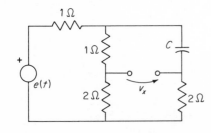

Problem 4.6   A simple design problem.

4.7 *Exercise* Find the time t at which the voltage $v_x(t)$ is zero. $e(t) =$ 100 V for $t < 0$, and $e(t) = 20$ V for $t \geq 0$. Sketch $v_x(t)$.

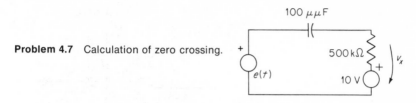

Problem 4.7 Calculation of zero crossing.

4.8 *Exercise* A square wave is applied to the circuit shown. Find and sketch $v_x(t)$. What happens as the inductance is increased? Explain.

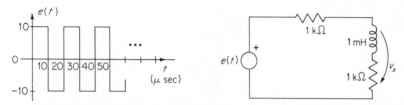

Problem 4.8 Effect of a change in time constant on the response.

4.9 *Design* Find a simple circuit which will produce an output voltage as shown when it is excited by a square-wave input.

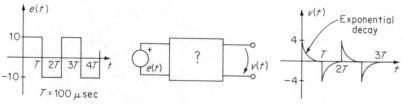

Problem 4.9 Design of a peaking circuit.

4.10 *Design* Find a simple circuit which will produce an output voltage as shown when it is excited by a series of pulses.

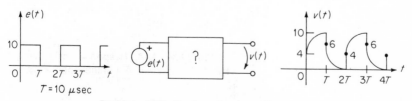

Problem 4.10 Design of a smoothing circuit.

4.11 *Design* Find the element values of the following circuit such that the voltage $v(t)$ will be as shown.

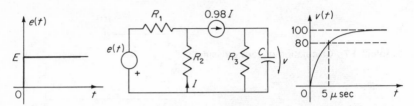

Problem 4.11 Design to meet rise time requirement.

4.12 *Design* Find the element values of the following circuit such that the current $i(t)$ will be as shown.

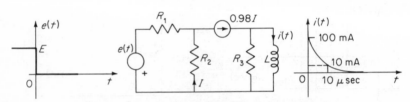

Problem 4.12 Design to meet decay requirement.

4.13 *Design* How should you select the element values in the following circuit so that the output voltage will reach 90 percent of its final value at $t = 100$ nsec.

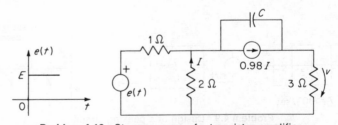

Problem 4.13 Step response of a transistor amplifier.

4.14 *Exercise* The initial voltage on the capacitor of the following circuit is 5 V, and the initial current in the inductor is zero. Find $v(t)$ for $t \geq 0$.

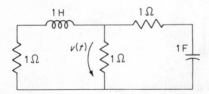

Problem 4.14 Response in a second-order circuit due to initial conditions.

4.15   The following circuit was in steady state for $t < 0$. At $t = 0$, the
switch S closes. Find and sketch the response $v(t)$. Use the computer program RNGKT2 to compute $v_c(t)$ and $i_L(t)$ for $0 \leq t \leq 10$
at 0.1 step. Compare the computer output with your results. Use
the subroutine EEPLOT to plot v_c vs. t and i_L vs. t.

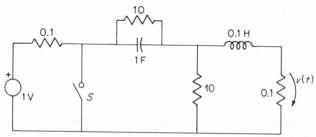

Problem 4.15   Computer solution of a second-order circuit.

4.16   In the circuit shown, $e(t) = 0$ for $t < 0$. For $t \geq 0$, $e(t) = 21 \sin t$.
Find the voltages and currents in the circuit. What becomes of the
voltages and currents at $t \to \infty$? Use the computer program
RNGKT2 to obtain v_x and the subroutine EEPLOT to plot v_x vs. t.
All initial conditions are zero.

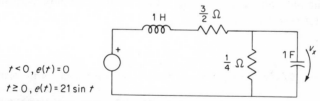

Problem 4.16   Response of a second-order circuit to a suddenly applied sine wave.

4.17   In the circuit shown, $e(t) = -1$ V for $t < 0$. At $t = 0$, it is suddenly
increased to 1 V and remains at this value for all $t > 0$. Find the
voltages and currents in the circuit. Sketch the waveforms.

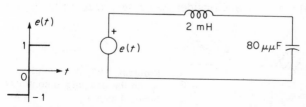

Problem 4.17   Step response of a lossless circuit.

4.18 Find $v_0(t)$ for $t \geq 0$ for three different values of the resistance R_2. $e(t) = 10$ V for $t < 0$. For $t \geq 0$, $e(t) = -10$ V; $R_1 = 470$ Ω, $L = 175$ mH; $R_3 = 680$ Ω, $C = 0.01$ μF; $R_2 = 1000, 1800, 100,000$ Ω. Use the computer program RNGKT2 to find v_0 for the three cases. Plot the results on one graph, using the subroutine EEPLOT.

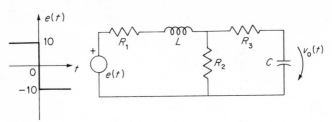

Problem 4.18 Computer solution of a circuit for three cases of damping.

4.19 Using RNGKT2 and EEPLOT, plot $v_1(t)$, $v_2(t)$, and v_2 vs. v_1 over an interval of 10 sec. Take $H = 0.05$. Use the following initial conditions:

a. $v_1(0) = 1$, $v_2(0) = -2$
b. $v_1(0) = 1$, $v_2(0) = 1$
c. $v_1(0) = 2$, $v_2(0) = -1$.

Calculate and sketch these solutions and compare with the computer results.

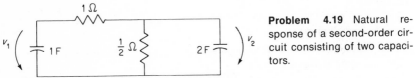

Problem 4.19 Natural response of a second-order circuit consisting of two capacitors.

4.20 In the circuit shown, $L = 0.1$ H, $C = 0.02$ F, $R_1 = 1.1$ Ω, $R_2 = 50$ Ω, $e(t) = 0$, $t < 0$, and $e(t) = 1$, $t \geq 0$. Using RNGKT2 and EEPLOT plot $v(t)$ over an interval of 1 sec. Take $H = 0.002$ and assume $v(0) = i(0) = 0$. Consider two cases: **a.** $\mu = 10$;   **b.** $\mu = 12$.

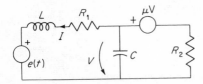

Problem 4.20 A second-order circuit in which the damping is controlled by the dependent source.

4.21 Consider circuit (a) shown below. Let steady state be established for $t < 0$. At $t = 0$, switch S opens. What is the voltage across the inductor? In order to prevent this voltage from becoming too large, a capacitor is connected across the resistance-inductor combination as shown in circuit (b). Suppose steady state has been established in (b) for $t < 0$ and at $t = 0$, S opens. What is the minimum value of C such that the inductor voltage will never exceed 10 V?

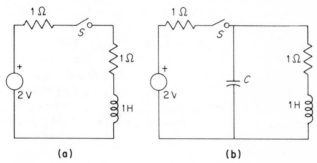

(a)   (b)

Problem 4.21 Addition of a capacitor will dampen the overshoot.

4.22 In the following circuit, S is initially open. At $t = 0$, S is suddenly closed. The current i must not exceed its final value at any time. What is the largest value of R such that i has this property? Find the time at which i reaches 90 percent of its final value.

Problem 4.22 Control of overshoot by adjusting damping in the circuit.

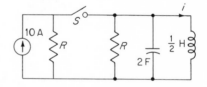

4.23 A biology student used the following circuit to study "frog kick." She observed that when switch S was suddenly closed, the frog kicked a little. When S was suddenly opened, the frog kicked 100

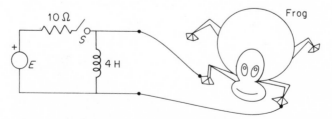

Problem 4.23 A circuit to study frog kicks.

times more violently. Explain. Assume the kick (a measurable quantity) is proportional to the voltage across the frog. What is the resistance (!?) of the frog? (A frog, for her purposes, is an element representable by a resistor.) In order to lessen the kick when the switch is suddenly opened, a friend (EE) of hers suggested that she put a capacitor across the inductor. Why? With the capacitor in the circuit, let K_0 be the amplitude of the kick when S is suddenly closed. What must be the value of the capacitance in order that the kick be less than $5K_0$ at all times when S is suddenly opened?

4.24 *Design* With reference to Example 4.11, find a set of element values of E, R_1, R_2, and C such that the sawtooth wave has $E_{max} = 2$ V and $E_{min} = 1$ V with $t_1 = 100$ μsec and $t_2 = 20$ μsec. Note: There are obviously many solutions. To guide your design, select T_1 and T_2 such that $T_1 \gg t_1$ and $T_1 \gg T_2$. A good design would require that T_1 is five to ten times t_1 and T_1 is five to ten times T_2.

5

simple nonlinear circuits

The most important property of a linear circuit is that in it the principle of superposition holds. If the excitation is multiplied by a factor, the response is multiplied by the same factor. If the excitation consists of a collection of sources, the response is a sum of the responses, each resulting from the application of one source at a time. In a nonlinear circuit, as we observed in Chapter 2, these homogeneity and additivity characteristics are no longer valid. Analysis techniques that are based on the principle of superposition cannot be used in a nonlinear circuit. Such interesting and useful ideas as Thévenin and Norton equivalent circuits cannot be applied to a nonlinear circuit as a whole, but only to that part which is linear. The differential equation that describes the dynamic behavior of a nonlinear circuit is nonlinear, and the solution cannot be expressed as the convolution of the excitation with the complementary solution. On the other hand, KCL and KVL still hold; so does Tellegen's theorem.

The formulation of circuit equations will be based, as always, on the use of trees and chords as given in Chapter 2. In contrast with the solution in a linear circuit, which can be expressed in a closed form, the solution in a nonlinear circuit, in general, must be found by numerical or graphical techniques.

In an attempt to develop a general analysis scheme which is valid for all nonlinear circuits, we find that we are dealing with too large a class of circuits. Stated differently, we find that the mathematical techniques that can be applied to all nonlinear circuits are very few in number. What we must do is to treat special cases, hoping that each case encompasses a sufficiently large class of circuits. This leads to the need for classification of nonlinear circuits, which turns out to be a very difficult task and is clearly beyond the scope of this book.

In this text, we shall be quite modest in our undertaking. We shall present an introduction to nonlinear circuit analysis and shall examine a few interesting examples of circuits that are used in practice. The principle of analysis will be emphasized. Graphical solution techniques will be given first. Computer solutions will be illustrated from time to time throughout the chapter.

5.1 Simple Nonlinear Resistor Circuits

We begin by studying circuits that consist of linear resistors, sources, and one or two nonlinear resistors. Consider the simple circuit of Figure 5.1.

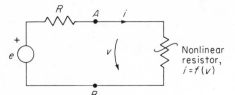

Figure 5.1 A simple nonlinear resistor circuit.

The nonlinear resistor is assumed to be a voltage-controlled element. The current in the element is related to the voltage across it by

$$i = f(v) \tag{5.1}$$

where $f(v)$ is some nonlinear function of v. The function is sometimes given analytically; more often, graphically. The graphical display of the function is known as the i-v characteristic of the element. It may take the form shown in Figure 5.2(a). Shown next to it is the i-v characteristic of a linear resistor, which is a straight line whose slope is the conductance of the resistor. We may sometimes find it convenient to speak of the conductance of a nonlinear resistor at a particular value of the voltage as the slope of the i-v curve at that point.

The solution of the circuit of Figure 5.1 is that set of branch voltages and currents that satisfy the KCL, KVL, and terminal equations. Writing these equations for the circuit, we have

$$e = iR + v \tag{5.2a}$$
$$i = f(v) \tag{5.2b}$$

These two equations completely describe the circuit. The two unknowns are i and v. Since i is expressible explicitly in terms of v, we eliminate the current variable i and get

$$e = Rf(v) + v \tag{5.3}$$

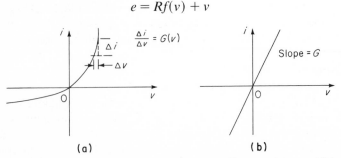

Figure 5.2 The i-v characteristic of a **(a)** nonlinear resistor and **(b)** linear resistor.

Unless the function $f(v)$ is simple, the solution of this equation must be found numerically, using an iterative scheme such as that described in Section 5.4.

If $f(v)$ is specified as a curve, we may find the solution graphically. Rewriting Equation (5.2a) as

$$i = -\frac{1}{R}v + \frac{1}{R}e \qquad (5.4)$$

we see that this is the $i\text{-}v$ characteristic of the circuit to the left of AB in Figure 5.1. We may regard the $e\text{-}R$ combination as an "element" whose $i\text{-}v$ characteristic is that given by Equation (5.4). The solution is that set of i,v which matches the current and voltage at the terminal, the "interface," or "boundary" of this element and the nonlinear resistor. Equation (5.4) describes a straight line of slope $-1/R$ with intercept at e/R as shown in Figure 5.3. The solution is the intersection of the straight line and the $i\text{-}v$ curve of the nonlinear resistor. The solution point (i_0, v_0) is commonly known as the "operating point" of the circuit. The straight line is called the "load line" as "seen" by the nonlinear resistor.

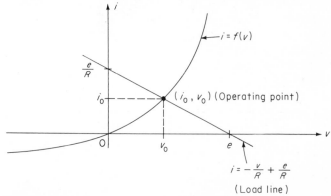

Figure 5.3 The solution, or operating, point is the intersection of the load line and the $i\text{-}v$ characteristic of the nonlinear resistor in the circuit of Figure 5.1.

Example 5.1 Find the output voltage V of the circuit of Figure 5.4. The excitation is a triangular wave, and the nonlinear resistor is a junction diode whose $i\text{-}v$ characteristic is as shown.

Solution The load line seen by the diode is

$$i = -0.1v + 0.1e$$

For each given value of e, we have a straight line with an intercept along the v-axis at e. As $e(t)$ varies periodically, the load line moves from that determined by $e = -2$ to that by $e = 2$. A family (continuum) of load lines is thus generated. The operating point moves along the curve $f(v)$ and can be read off at any instant

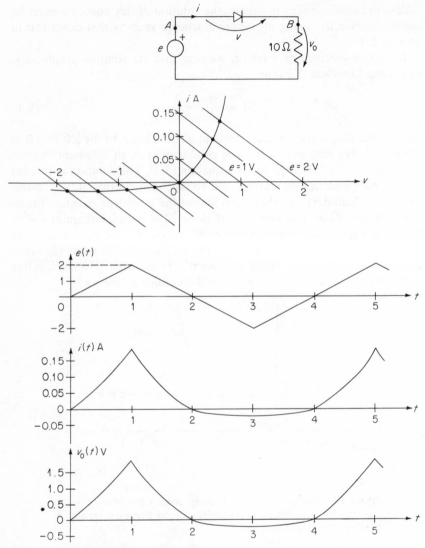

Figure 5.4 Response of a nonlinear circuit to a triangular wave. [Example 5.1]

of time. The current $i(t)$ is obtained in this way. The voltage v is simply $10i(t)$, as shown in Figure 5.4(e). Note that the circuit chops off the negative half of the input voltage.

5.2 Small Signal Analysis

Suppose in Figure 5.1, the excitation consists of a small time-varying component $e_s(t)$ superimposed on a large constant voltage E, namely,

$$e(t) = E + e_s(t) \tag{5.5}$$

with $e_s(t) \ll E$ for all t. Let the operating point in the absence of $e_s(t)$ be (V_0, I_0). As $e(t)$ takes a small excursion about E, the operating point takes a small excursion along the curve $f(v)$ and generates small deviations about (V_0, I_0) in $v(t)$ and $i(t)$. Thus we can write

$$v(t) = V_0 + v_s(t)$$
$$i(t) = I_0 + i_s(t) \tag{5.6}$$

in which $v_s(t)$ and $i_s(t)$ are small quantities for all t. Expanding the function $f(v)$ in Taylor series about the operating point, we get

$$i(t) = f(v) = f(V_0) + f'(V_0)(v - V_0) + \tfrac{1}{2} f''(V_0)(v - V_0)^2 + \cdots$$
$$i(t) = I_0 + f'(V_0)\, v_s(t) + \tfrac{1}{2} f''(V_0)\, v_s^2(t) + \cdots \tag{5.7}$$

Neglecting terms of higher-order derivatives than the first, we obtain

$$i(t) = I_0 + g(V_0)v_s(t) \tag{5.8}$$

or

$$i_s(t) = g(V_0)v_s(t) \tag{5.9}$$

It is seen that the small component of the current is linearly related to the small component of the voltage. Thus as far as the small time-varying components of the voltages and currents in the circuit are concerned, the circuit may be regarded as a linear circuit. The nonlinear resistor may be replaced by a linear one whose conductance is $g(V_0)$, which is the slope of the curve $f(v)$ evaluated at the operating point. The operating point is determined by the constant component of the voltage and current in the nonlinear element.

Focusing our attention on the small component while ignoring the large constant component of the voltages and currents in a nonlinear circuit is what is done in "small signal" analysis. Each nonlinear resistor is replaced by a linear one, with conductance equal to the slope of the i-v characteristic at the operating point. The circuit is now linear, and we can proceed with linear analysis.

In this chapter, we shall be dealing with "large signal" analysis. The voltages and currents in a circuit take on large excursions so that the elements must be treated as nonlinear.

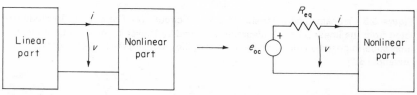

Figure 5.5 In a nonlinear circuit, the linear part may sometimes be replaced by its Thévenin equivalent, and the analysis of the circuit is simplified thereby.

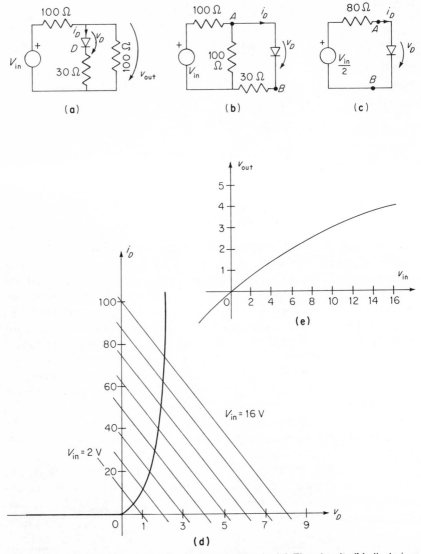

Figure 5.6 Transfer characteristic of a diode circuit. **(a)** The circuit; **(b)** diode isolated from the linear part; **(c)** linear part replaced by Thévenin equivalent; **(d)** the load lines superimposed on the i-v characteristic of the diode; **(e)** the transfer characteristic. [Example 5.2]

5.3 Decomposition of a Circuit into Linear and Nonlinear Parts

With reference to Figure 5.1, the *e-R* combination, which is the linear part of the circuit, has the appearance of a Thévenin equivalent circuit. If in a nonlinear circuit the nonlinear element can be isolated from the linear part as shown in Figure 5.5, the linear part can be replaced by its Thévenin equivalent circuit. The analysis is thus simplified.

Example 5.2   Obtain a plot of the output voltage versus the input voltage of the circuit of Figure 5.6. The plot is known as the "transfer" characteristic of the circuit.

Solution   The diode can be isolated, and the remainder can be redrawn as shown in Figure 5.6(b). Replacing the linear part by its Thévenin equivalent as in Figure 5.6(c), we find the load line to be

$$i = \frac{-v}{80} + \frac{V_{in}}{160}$$

A family of load lines is drawn in Figure 5.6(d). Consider V_{in} positive first. Reading off the operating point, we obtain the diode voltage and current as given in the following table. Noting that $V_{out} = V_D + I_D(30)$, we obtain the output voltage as given in the last column of the table. The transfer characteristic is shown in Figure 5.6(e).

V_{in}	V_D	I_D(mA)	$30I_D$(A)	V_{out}(V)
0	0	0	0	0
2	0.6	5	0.150	0.750
4	1.0	12.5	0.375	1.375
6	1.3	20	0.600	1.900
8	1.7	30	0.900	2.600
10	1.8	40	1.200	3.000
12	1.9	50	1.500	3.400
14	1.95	63	1.890	3.840
16	2.00	75	2.250	4.250

For negative values of V_{in}, the diode current is zero. The output voltage V_{out} is simply $0.5V_{in}$.

5.4 Combination of *i-v* Characteristics

By connecting nonlinear resistors in various combinations, we obtain nonlinear elements with interesting *i-v* characteristics.

1. *Series connection*   Consider the series connection of two diodes whose *i-v* characteristics are shown in Figure 5.7. The equations describing the voltage and current at the terminals of the combination are

$$v = v_1 + v_2 \tag{5.10}$$
$$i = i_1 = i_2 \tag{5.11}$$

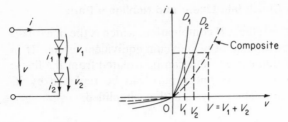

Figure 5.7 The *i-v* characteristic of two diodes in series.

The composite *i-v* characteristic is obtained by adding the two curves along the abscissa for a given value of *i*. Note that the slope of the composite curve is everywhere smaller than either of the two *i-v* curves.

2. *Parallel connection* In a parallel combination, shown in Figure 5.8, the voltages across the diodes are the same. The composite *i-v* characteristic is ob-

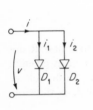
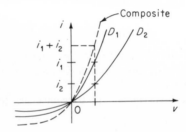

Figure 5.8 The *i-v* characteristic of two diodes in parallel.

tained by adding the two curves along the ordinate for each given value of *v*. Note that the slope of the composite curve is everywhere greater than the slope of either of the two *i-v* curves.

3. *Voltage bias* Suppose we connect a voltage source in series with a diode as shown in Figure 5.9. The equations describing the circuit are

$$v = v_D + E \qquad (5.12)$$
$$i = f(v - E) \qquad (5.13)$$

The composite *i-v* characteristic is the same as before, except that it is shifted to the right by an amount equal to *E*.

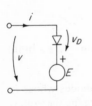
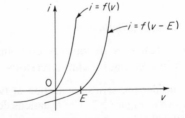

Figure 5.9 The *i-v* characteristic of a voltage-biased diode.

4. *Current bias* Suppose we connect a current source in parallel with a diode as shown in Figure 5.10. The current at the terminal is the sum of the diode

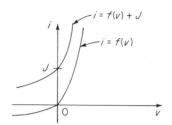

Figure 5.10 The *i-v* characteristic of a current-biased diode.

current and the source. The effect is to shift the *i-v* characteristic upward by an amount equal to J.

5. *Inversion* By inverting the direction of a diode, we obtain an *i-v* characteristic which is its reflection about the origin, as shown in Figure 5.11.

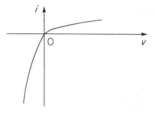

Figure 5.11 The *i-v* characteristic of an inverted diode.

We shall now present a number of examples to illustrate how diodes, linear resistors, and sources may be combined to yield interesting *i-v* and transfer characteristics.

Example 5.3 (*Limiter*) Consider the circuit of Figure 5.12 in which the diodes are assumed to have the highly unrealistic *i-v* characteristic shown. As such, the diodes may be thought of as being voltage-controlled switches. If the voltage across such a diode is positive, it may be regarded as a short circuit. If the voltage is negative, it is an open circuit.

Now the combination of D_1 and E_1 has an *i-v* characteristic shown in Figure 5.12(c). That of D_2 and E_2 is shown in Figure 5.12(d). The parallel combination has an *i-v* characteristic which is obtained by adding the two along an ordinate. The composite *i-v* is shown in Figure 5.12(e). The load line is drawn for three cases of V_{in}. For $V_{in} > E_1$ the operating point is at A and the current is finite, but the voltage $V_{out} = E_1$. For $-E_2 < V_{in} < E_1$, the operating point is at B, and the current is zero so that $V_{out} = V_{in}$. For $V_{in} < -E_2$, the operating point is at C, and the current is finite and negative, but V_{out} is held at $-E_2$. The transfer characteristic is, therefore, as shown in Figure 5.12(f).

The circuit is known as a limiter, and it is used in a frequency-modulation (FM) receiver to remove unwanted amplitude modulation. In an FM system, it is the rate of zero crossing of the carrier that conveys the information of the signal. If the amplitude of the carrier is subject to amplitude modulation, the output after detection will have components corresponding to the amplitude-modulation signal as well as components from the frequency-modulation signal. The result is that the information is distorted. To remove the amplitude modulation, the carrier is applied to a limiter before detection. The output of the limiter now has constant

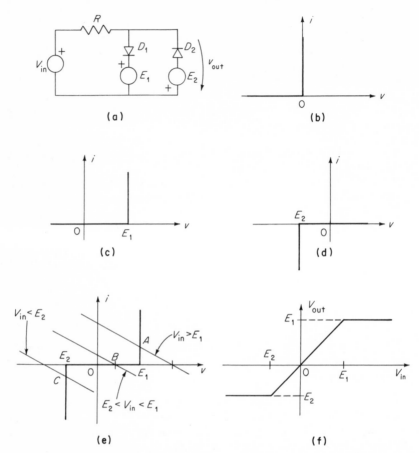

Figure 5.12 Limiter. **(a)** The circuit; **(b)** the *i-v* characteristic of an ideal diode; **(c)**, **(d)** composite characteristics of biased diodes; **(e)** composite characteristic of the two biased diodes; **(f)** the transfer characteristic.   [Example 5.3]

amplitude, and at the output of the detector, only the signal resulting from frequency modulation appears. Figure 5.13(a) summarizes what has been said.

Another application of the limiter is the conversion of a sine wave into a trapezoidal wave as depicted in Figure 5.13(b).

Example 5.4   (Approximation of a Convex Curve) In the simulation of a physical system on an analog computer, we sometimes have to approximate an arbitrary single-valued function. Using diodes and sources, we can construct an element whose *i-v* characteristic is an arbitrary convex function.

Consider the circuit of Figure 5.14(a). Let the diode characteristic be approximated by two straight line segments as shown in Figure 5.14(b). The D_1-E_1 combination has an *i-v* characteristic as shown in Figure 5.14(c), in which g_1 is the

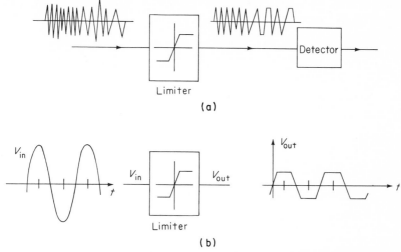

Figure 5.13 (a) Limiter used in an FM receiver; (b) as a clipper.

slope of the straight line for $v > E_1$. Similarly the *i-v* characteristic of the D_2-E_2 combination is shown in Figure 5.14(d). The parallel combination of the two is shown in Figure 5.14(e). It is seen that the composite *i-v* characteristic is a convex curve.

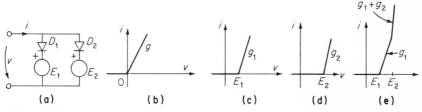

Figure 5.14 Approximation of a convex curve by a diode network. [Example 5.4]

By inversion and voltage biasing, the convex curve having the general form shown in Figure 5.15(a) can be generated by the circuit of Figure 5.15(b). The voltages E_1, E_2,... are known as break-point voltages, and the slopes g_1, g_2,... as the equivalent conductances in the various ranges of voltages.

Example 5.5 (Harmonic Generator) A nonlinear resistor can be used to generate a harmonic of a sine wave. In fact, any nonlinear element has this property. (See Problem 5.10.)

Suppose we impress a voltage $v(t)$ given by

$$v(t) = \cos \omega t \qquad (5.14)$$

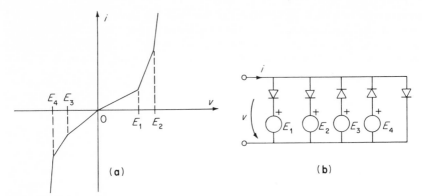

Figure 5.15 Piece-wise linear approximation of a curve by a diode network. [Example 5.4]

across a nonlinear resistor whose i-v characteristic $f(v)$ has a power series representation given by

$$i = f(v) = a_0 + a_1 v + a_2 v^2 + a_3 v^3 + \cdots \tag{5.15}$$

Then we have

$$i = a_0 + a_1 \cos t + a_2 \cos^2 t + a_3 \cos^3 t + \cdots \tag{5.16}$$

Using the identities

$$\cos^2 x = \tfrac{1}{2}(1 + \cos 2x)$$
$$\cos^3 x = \tfrac{1}{4}(3 \cos x + \cos 3x)$$
$$\cos^4 x = \tfrac{1}{8}(3 + 4 \cos 2x + \cos 4x) \tag{5.17}$$
$$\vdots$$

we find

$$i(t) = \left(a_0 + \frac{a_2}{2} + \frac{3a_4}{8} + \cdots \right)$$

$$+ \left(a_1 + \frac{3a_3}{4} + \cdots \right) \cos \omega t$$

$$+ \left(\frac{a_2}{2} + \frac{a_4}{2} + \cdots \right) \cos 2wt$$

$$+ \left(\frac{a_3}{4} + \frac{5a_5}{16} + \cdots \right) \cos 3\omega t$$

$$+ \left(\frac{a_4}{8} + \frac{3a_6}{16} + \cdots \right) \cos 4\omega t + \cdots \tag{5.18}$$

It is seen that the current consists of a sum of cosine waves of frequencies that are integral multiples of the frequency of the input voltage. The high-frequency components are called the harmonics of the input. By choosing the proper non-linearity, the coefficients $a_1, a_2, \ldots$ can be adjusted so that certain harmonics are favored over others. In engineering terminology, the harmonic generator is sometimes known as a "frequency multiplier."

Example 5.6 (Diode Logic Gates) We noted earlier that a diode may be regarded as a voltage-controlled switch. We shall now show how diodes may be connected together to realize circuits that perform logic functions.

Consider the circuit of Figure 5.16. Assume that the diodes have an *i-v* characteristic as shown in Figure 5.14(b). The *i-v* characteristics of the biased diodes are shown in Figure 5.16(b) in which the load line of the *E-R* combination is also shown. Now if $e_A < E$, then the output voltage $V_{out} < E$, regardless of whether e_B is greater or less than E. By symmetry, if $e_B < E$, then $V_{out} < E$ regardless of what e_A may be. On the other hand, if $e_A > E$ and $e_B > E$, then the diode currents are both zero and $V_{out} = E$. We summarize the observations in the following table.

e_A	e_B	V_0
0	0	0
1	0	1
0	1	1
1	1	1

In the table, "1" signifies that the voltage is less than E, and "0" signifies that the voltage is greater than or equal to E. In words, the table states that the output voltage is "1" if input A is "1," if input B is "1," or if both are "1"; the output is "0" if both are "0." This statement is precisely the description of the logical "OR" function.

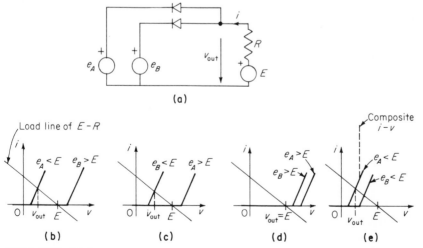

Figure 5.16 Diode OR gate: The output voltage is less than E if A, or B, or both is less than E. [Example 5.6]

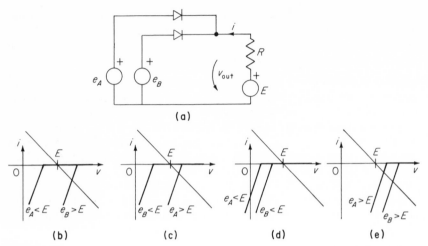

Figure 5.17   Diode AND gate: The output is equal to E if and only if, both A and B are less than, or equal to, E.   [Example 5.6]

By reversing the diodes of Figure 5.16(a), we obtain a logic circuit which performs the "AND" function. Consider Figure 5.17(a). The four possible combinations of inputs are depicted in Figures 5.17(b) to (e). The following table summarizes the observation, in which "1" denotes that the voltage in question is equal to or less than E, and "0" denotes that it is greater than E.

e_A	e_B	V_{out}
0	0	0
0	1	0
1	0	0
1	1	1

In words, the table states that the output is "1" if, and only if, both inputs are "1." Otherwise the output is "0." This is a statement of the logical "AND" function.

5.5   Newton-Raphson Algorithm

In a simple nonlinear circuit, the solution is obtained as the intersection of the $i\text{-}v$ characteristic of the nonlinear element and the load line. In all of the examples that we have presented, the intersection is found graphically. For computational purposes, graphical solution is not satisfactory, and the solution must be found numerically. There are many schemes to find the intersection of the two curves. We shall present one which is known as the Newton-Raphson algorithm.

With reference to Figure 5.3, instead of finding the intersection of the load line and the $i\text{-}v$ characteristic, we transform the problem to one of finding the root of the following equation, which is obtained by eliminating the variable i in Equation (5.2):

$$Rf(v) + v - e = 0 \qquad (5.19)$$

Given a function $g(x)$, we can find the root of the equation

$$g(x) = 0 \qquad (5.20)$$

iteratively, provided that the derivative of $g(x)$ is known or easily found.

Let $\bar{x}_0$ be the root of the equation. Starting with an initial guess of the root x_0, we obtain the next value, which is a better approximation of $\bar{x}_0$, from

$$x_1 = x_0 - \frac{g(x_0)}{g'(x_0)} \qquad (5.21)$$

Continuing, the $(n + 1)$th approximation is obtained from the nth one from

$$x_{n+1} = x_n - \frac{g(x_n)}{g'(x_n)} \qquad (5.22)$$

We terminate the iteration when the difference between the last two approximations is insignificant.

The algorithm has a very simple geometric interpretation. Writing Equation (5.22) as

$$g(x_n) = g'(x_n)(x_n - x_{n+1}) \qquad (5.23)$$

we see that this is the equation of a straight line with a slope $g'(x_n)$ and intercept at x_{n+1}. At each iteration, we pass a straight line of slope $g'(x_n)$

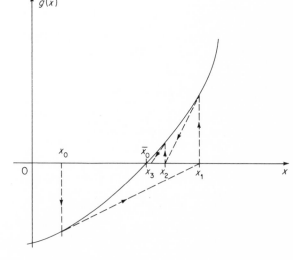

Figure 5.18 Geometric interpretation of Newton-Raphson algorithm.

through the point $(x_n, g(x_n))$. The intercept on the x-axis is the next approximation. See Figure 5.18.

The choice of the starting point is critical and depends on the characteristic of the function $g(x)$ over the interval in which the root lies. If $g(x)$ has the form shown in Figure 5.19 and the starting point is chosen

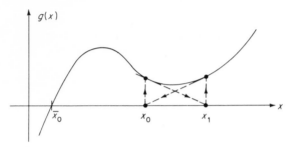

Figure 5.19 An improper choice of the starting value in the Newton-Raphson algorithm may lead itself into a trap, and the root is never reached.

as shown, the successive values of the approximation will oscillate between x_0 and x_1, and the process will never converge.

The rate of convergence of the Newton-Raphson algorithm is quadratic in the sense that if the error at the nth step is e, the error at the next step is e^2. This is shown at the end of the chapter.

Example 5.7 Suppose in Figure 5.1, the nonlinear resistor is a junction diode whose *i-v* characteristic is given by

$$i = 20 \times 10^{-6} \ (e^{38.6v} - 1)$$

Find the operating point of the circuit. Let $e = 1V$ and $R = 10^4 \Omega$.

Solution The equation whose root is the solution of the circuit is

$$g(v) = v + 0.2 \ (e^{38.6v} - 1) - 1$$

The solution is obtained from the following FORTRAN program. The operating point is found to be $v = 0.04542$ V and $i = 95.47 \times 10^{-6}$ A.

```
C THIS PROGRAM FINDS THE OPERATING POINT OF THE CIRCUIT
C      OF EXAMPLE 5.7, USING THE NEWTCN-RAPHSON ALGORITHM.
C      DEFINE THE FUNCTICN AND ITS DERIVATIVE
       G(V) = V + 0.2 * (EXP(38.6*V) - 1.) - 1.
       DG(V) = 1. + 7.72*EXP(38.6*V)
       C(V) = 20.E-06 * (EXP(38.6*V) -1)
       N=0
C      SET INITIAL VALUE OF  V  TO 0.05
       V=0.01
       CUR=C(V)
       WRITE(6,96)
   96  FORMAT(' ITERATION', 8X,'VCLTAGE',  8X, 'CURRENT')
   99  WRITE(6,98)N,V,CUR
       VNEXT=V - G(V)/DG(V)
       IF(ABS(VNEXT - V) .LE. 1.E-05)GO TO 100
       V=VNEXT
```

```
        N=N+1
        IF(N .GT. 100)GO TO 200
        CUR=C(V)
        GO TO 99
  98 FORMAT(I6,8X,2E15.6)
 200 WRITE(6,95)
  95 FORMAT('ROOT IS NOT FOUND AFTER 100 ITERATIONS.')
 100 STOP
        END
```

ITERATION	VOLTAGE	CURRENT
0	0.100000E-01	0.942169E-05
1	0.824932E-01	0.462978E-03
2	0.626870E-01	0.204853E-03
3	0.500298E-01	0.117948E-03
4	0.457990E-01	0.971635E-04
5	0.454218E-01	0.954702E-04

5.6 General Nonlinear Resistor Circuit

The underlining principle of analysis of simple nonlinear resistor circuits is the use of a load line and the combination of i-v characteristics. For computational purposes and for cases in which there is a large number of nonlinear resistors connected in an arbitrary manner, it is best to go back to first principles and formulate the circuit equations using trees and chords.

Let us do an example and see what difficulty we would encounter in an attempt to solve the general problem. Consider the circuit of Figure 5.20(a). Choose a tree as shown in Figure 5.20(b). The KVL and KCL equations are

$$v_4 = v_3 + v_2 - v_1$$
$$v_5 = v_1 - v_2$$
$$v_6 = v_3 + v_2 \tag{5.24}$$

$$i_1 = i_4 - i_5$$
$$i_2 = i_5 - i_4 - i_6$$
$$i_3 = -i_4 - i_6 \tag{5.25}$$

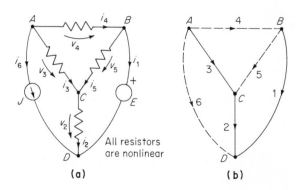

Figure 5.20 Analysis of a general nonlinear resistor network.

(a)   (b)

The terminal equations for the sources are

$$v_1 = E$$
$$i_6 = J \tag{5.26}$$

But what terminal equations should we use for the nonlinear resistors? Suppose all of them are voltage controlled. It is then possible to write

$$i_k = f_k(v_k) \qquad k = 2, 3, 4, 5 \tag{5.27}$$

Substituting the above into the two KCL equations for the resistor tree branches, we get

$$f_2(v_2) = f_5(v_5) - f_4(v_4) - J$$
$$f_3(v_3) = -f_4(v_4) - J \tag{5.28}$$

Using KVL, we have

$$f_2(v_2) = f_5(E - v_2) - f_4(v_3 + v_2 - E) - J$$
$$f_3(v_3) = -f_4(v_3 + v_2 - E) - J \tag{5.29}$$

Thus we obtain a set of simultaneous nonlinear equations in the two unknowns v_2 and v_3, which are the tree voltages. Once the tree voltages are found, the chord voltages are determined. Once all the branch voltages are found, the branch currents are obtained from the i-v characteristics, that is, Equation (5.27).

The two simultaneous equations can be put into the form:

$$g_1(v_2, v_3, E, J) = 0$$
$$g_2(v_2, v_3, E, J) = 0 \tag{5.30}$$

The solution must be found numerically.

Suppose now the nonlinear resistors are current controlled. It is then possible to write

$$v_k = h_k(i_k) \qquad k = 2, 3, 4, 5 \tag{5.31}$$

This time we substitute the above into the two KVL equations for the resistor chords. We then get

$$h_4(i_4) = h_3(i_3) + h_2(i_2) - E$$
$$h_5(i_5) = E - h_2(i_2) \tag{5.32}$$

Expressing the tree currents in terms of the chord currents, we get

$$h_4(i_4) = h_3(-i_4 - J) + h_2(i_5 - i_4 - J) - E$$
$$h_5(i_5) = E - h_2(i_5 - i_4 - J) \tag{5.33}$$

Again, we obtain a set of simultaneous nonlinear equations. The unknowns are the chord currents. Once we have the chord currents, the tree currents are determined, and from Equation (5.31) the branch voltages are then determined.

Now suppose we are not so fortunate as to be able to write either Equation (5.27) or Equation (5.31). For example, suppose we have

$$i_2 = f_2(v_2)$$
$$i_3 = f_3(v_3)$$
$$v_4 = h_4(i_4)$$
$$v_5 = h_5(i_5) \qquad\qquad (5.34)$$

If we use the first method of analysis, we would have to invert the functions $h_4(i_4)$ and $h_5(i_5)$ to get i_4 and i_5 expressed as functions of v_4 and v_5. If the chord voltages are used, then f_2 and f_3 would have to be inverted. Unfortunately the inverse of a nonlinear function does not always exist. The best that we can do is to substitute Equations (5.34) into the KVL and KCL equations and obtain a set of nonlinear simultaneous equations in the unknowns, which are the tree voltages and chord currents. In our case, we would get four equations in the form

$$g_k(v_2, v_3, i_4, i_5) = 0 \qquad k = 1, 2, 3, 4 \qquad (5.35)$$

In summary, we now can state that (1) if all the nonlinear resistors are voltage controlled, we retain the tree currents as the unknowns; (2) if all the nonlinear resistors are current controlled, we retain the chord currents as the unknowns; (3) if the resistors have mixed characterization, we retain the tree voltages and chord currents as the unknowns. In any case, we must solve a set of nonlinear simultaneous equations.

5.7 First-Order Nonlinear Circuits

A first-order nonlinear circuit is one in which there are any number of resistors but only one energy storage element, all of which may be nonlinear. It is not within the scope of this book to treat the most general case. We shall do a few examples.

We recall from Chapter 4 that the response of a linear first-order circuit consists of exponentially decaying functions. The rate of decay is determined by the time constant of the circuit, which is the product of the capacitance and the equivalent resistance seen by the capacitor, or the product of the inductance and the conductance of the resistance seen by the inductor in the circuit. In the nonlinear case, the element values are functions of voltage and current. We may intuitively think of the circuit as one in which the response is still exponentially decaying, but now the time constant changes as the voltages and currents in the circuit change.

We have already noted that the differential equation describing the dynamics of a nonlinear circuit is now nonlinear, and we can no longer find the solution by convolving the complementary solution with the excitation. In general, numerical integration of the differential equation is the only way of solving the circuit. The subject of numerical integration of general nonlinear differential equations is discussed in advanced texts on numerical analysis.

We shall confine ourselves to circuits in which the differential equation can be put in the form

$$\frac{dx}{dt} = f(x, t) \tag{5.36}$$

to which the Runge-Kutta algorithm, or any other numerical integration scheme, may be applied.

Consider the circuit of Figure 5.21(a). The capacitor is connected to a circuit containing nonlinear resistors and sources. In principle, we can

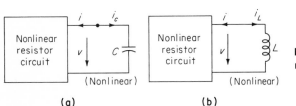

Figure 5.21 First-order nonlinear circuits.

(a)   (b)

always find an i-v characterization of the circuit at the terminals. Suppose the terminal characterization is one that expresses the current i in terms of the voltage v as

$$i = g(v, t) \tag{5.37}$$

The nonlinear capacitor, as noted in Chapter 2, is usually characterized by the charge-voltage relation:

$$q = Q(v) \tag{5.38}$$

namely, the charge q is expressible as a function of the voltage v across the capacitor. Now the current in the capacitor, i_c, is

$$i_c = \frac{dq}{dt} = \frac{dq}{dv}\frac{dv}{dt} = Q'(v)\frac{dv}{dt} \tag{5.39}$$

where $Q'(v)$ is the derivative of $Q(v)$ with respect to v. Noting $i_c = -i$, we obtain the differential equation describing the circuit at the interface

$$\frac{dv}{dt} = -\frac{1}{Q'(v)} g(v, t) \tag{5.40}$$

which is in the form of Equation (5.36).

If the capacitor is linear, then $q = Cv$ and the differential equation becomes

$$\frac{dv}{dt} = -\frac{1}{C} g(v, t) \tag{5.41}$$

Now suppose the terminal characterization of the nonlinear resistor circuit is current controlled, namely,

$$v = h(i, t) \tag{5.42}$$

In order to use the same formulation as before, we must have the inverse function of $h(i, t)$. That is, we must express the current i in terms of the voltage v before we can proceed. A formulation in terms of i is possible, as in Equation (5.54). In the most general case of nonlinearity circuits, we may not be able to obtain a differential equation in the form of Equation (5.36), and the solution becomes very difficult even with numerical techniques. To be specific, let

$$v = 1 + i + i^5 \tag{5.43}$$

and the voltage-charge relation be

$$v = q + q^2 \tag{5.44}$$

The best that we can do is to write

$$q + q^2 = 1 + \frac{dq}{dt} + \left(\frac{dq}{dt}\right)^5 \tag{5.45}$$

The equation is still a first-order differential equation, but it cannot be put in the form of Equation (5.36).

Next, suppose the capacitor is replaced by an inductor as shown in Figure 5.21(b). Let the inductor be characterized by a flux-current relation:

$$\phi = F(i_L) \tag{5.46}$$

where ϕ is the flux. What terminal characterization should we use for the nonlinear resistor circuit? Matching the voltage v with the inductor voltage at the interface, we have

$$v_L = \frac{d\phi}{dt} = \frac{d\phi}{di_L} \frac{di_L}{dt} = v \tag{5.47}$$

or

$$\frac{di_L}{dt} = \frac{1}{F'(i_L)} v \tag{5.48}$$

where $F'(i_L)$ is the derivative of $F(i_L)$ with respect to i_L. To obtain a differential equation in the form of Equation (5.36), we must have a terminal characterization for the nonlinear resistor circuit that expresses the voltage in terms of the current at the terminal; that is,

$$v = h(i, t) \tag{5.49}$$

Noting that $i = -i_L$, we find the differential equation to be

$$\frac{di_L}{dt} = \frac{1}{F'(i_L)} h(-i_L, t) \tag{5.50}$$

which is in the form of Equation (5.36).

Example 5.8   Suppose in Figure 5.21(a) the capacitor is linear with capacitance equal to 1 F. Assume the nonlinear resistor circuit is characterized by $i = v^3$. Let us compute the solution of the circuit at the interface.

The differential equation is

$$\frac{dv}{dt} = -v^3 \tag{5.51}$$

Let $v(0) = 1$. The exact solution is

$$v(t) = \frac{1}{\sqrt{2t + 1}} \tag{5.52}$$

The computer solution, obtained from the Runge-Kutta scheme with a step size of 0.05, is given below together with the exact values of $v(t)$.

t	v(computed)	v(exact)
0.	1.	1.
0.25	0.816497	0.816496
0.50	0.707107	0.707107
0.75	0.632456	0.632455
1.00	0.577350	0.577350
2.00	0.447213	0.447214
3.00	0.377964	0.377964
4.00	0.333333	0.333333
5.00	0.301510	0.301511

It is remarkable how close the computer solution is to the exact solution.

Example 5.9   Suppose in the circuit of Figure 5.21(b), the inductor is characterized by the following function:

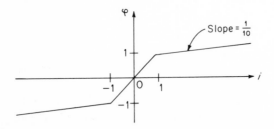

Figure 5.22   The flux-current characteristic of a nonlinear inductor. [Example 5.9]

$$\phi = i \qquad\qquad -1 \le i \le 1$$
$$= 0.1(i-1) + 1 \qquad i > 1$$
$$= 0.1(i+1) - 1 \qquad i < 1$$

The function is shown in Figure 5.22. In practice, the saturable reactor has a $\phi\text{-}i$ characteristic as shown. Let the circuit to which the inductor is connected be characterized by

$$v = 2 \sin t + i$$

The differential equation becomes

$$\frac{di_L}{dt} = \frac{1}{F'(i_L)} (2 \sin t - i_L)$$

with

$$F'(i_L) = 1 \qquad |i_L| \le 1$$
$$= 0.1 \qquad |i_L| > 1$$

Assume $i_L(0) = 0$. Let us compute the solution.

Unlike the last example, the exact solution is very difficult to find. Using the Runge-Kutta algorithm, we obtain the computer solution as plotted in Figure 5.23. A step size of 0.05 was used. Note that if the excitation is so small that $|i_L| \le 1$ for all t, then the circuit is linear. This example shows that whether or not a circuit is linear or nonlinear depends on the amplitude of the excitation. Another point to note is that the excitation is sinusoidal, but the response is not, even in the steady state.

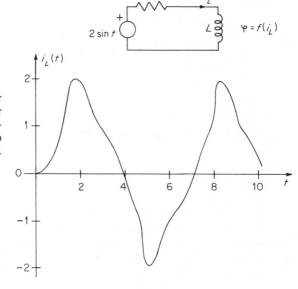

Figure 5.23 Computer solution of a nonlinear inductor circuit. Runge-Kutta was used with step size equal to 0.05 sec. [Example 5.9]

5.8 Summary

The analysis of nonlinear circuits is very different from that of linear circuits. The formulation of equations describing a circuit begins with KCL and KVL using trees and chords. However, the form and complexity of the final set of equations depend a great deal on how the nonlinear elements are characterized. It matters very much whether we can express the current in terms of the voltage on the elements or vice versa. This is particularly true in circuits containing capacitors and inductors, as we saw in the simple cases of first-order circuits.

For simple nonlinear resistor circuits, the technique of load lines and combination of the i-v characteristics are useful. For numerical computation, the Newton-Raphson scheme can be used to find the operating point. For simple first- or higher-order circuits that can be characterized by a differential equation of the form $\dot{x} = f(x, t)$, the solution can be obtained by using numerical integration schemes such as the Runge-Kutta algorithm.

SUGGESTED EXPERIMENTS

1. Measure the i-v characteristics of the following combinations: a diode in series with a voltage source; two diodes in series; two diodes in parallel in the same orientation; two diodes in parallel in the opposite orientation.
2. Construct the limiter circuit of Example 5.3. Apply a sine wave to the circuit and observe the output. Vary the amplitude of the sine wave and explain what you see.
3. Using diodes and batteries, construct a circuit whose i-v characteristic approximates the quadratic: $i = Kv^2$ ($v \geq 0$).
4. Connect a diode in series with a resistor-voltage source combination. Excite the circuit with a sine wave. Measure the harmonics generated by the circuit as the resistance varies from a low value to a high value. Explain your observations.
5. Construct a diode OR and AND gate. Verify the logic functions. Generalize the circuits of Example 5.6 to the case with N inputs. What would be the limitation on N? Why?
6. Obtain a saturable reactor (a coil wound on soft iron) and construct the circuit of Figure 5.21(b). Excite the circuit with a sine wave. Observe the output as the amplitude is increased from a small value to a large value. Explain what you see.
7. Obtain a tunnel diode and measure its i-v characteristics. Note that for certain values of voltage, the conductance is negative. Connect the diode to a resistor-voltage source combination. Vary the slope of the load line and observe the current in the circuit. Explain what you see.

COMPUTER PROGRAMMING

Rate of Convergence of the Newton-Raphson Iteration

We remarked in Section 5.4 that the Newton-Raphson iteration converges quadratically toward the root of the equation. The error at the nth step is proportional to the square of the error at the previous step. We can see this as follows.

Let the root of the equation

$$g(x) = 0$$

be $\bar{x}$. The error at the nth step is

$$e_n = \bar{x} - x_n$$

and the error at the $(n + 1)$th step is

$$\begin{aligned} e_{n+1} &= \bar{x} - x_{n+1} \\ &= \bar{x} - x_n + \frac{g(x_n)}{g'(x_n)} \\ &= e_n + \frac{g(x_n)}{g'(x_n)} \end{aligned} \tag{5.53}$$

Expanding $g(x)$ about the root, we have

$$g(x) = g(\bar{x}) + g'(\bar{x})(x - \bar{x}) + \frac{g''(\bar{x})(x - \bar{x})^2}{2} + \cdots$$

Noting that $g(\bar{x}) = 0$ and evaluating the expression at x_n, we get

$$g(x_n) = -e_n\, g'(\bar{x}) + \frac{e_n^2}{2} g''(\bar{x}) + \cdots$$

Dividing the above by $g'(\bar{x})$ and substituting into Equation (5.53), we find

$$e_{n+1} = \frac{e_n^2}{2} \frac{g''(\bar{x})}{g'(\bar{x})}$$

which is what was required to be demonstrated.

Numerical Inversion of Functions

We have mentioned on many occasions that in the solution of a nonlinear circuit, we sometimes have the need to invert a function. The problem is that given $y = f(x)$, we wish to find x as a function of y. Unless $f(x)$ is simple, analytical inversion is not always possible. However, we note that since the solution of a nonlinear circuit must almost always be found by numerical means, we should invert the function numerically. This can be done by using the Newton-Raphson algorithm, among others.

To fix ideas, suppose in the circuit of Figure 5.21(a), the nonlinear resistor circuit is current controlled:

$$v = h(i)$$

and the capacitor is voltage controlled:

$$q = Q(v)$$

The capacitor current i_c is

$$i_c = \frac{dq}{dt} = Q'(v)\frac{dv}{dt} = Q'(v)h'(i)\frac{di}{dt}$$

Noting that $i = -i_c$, we have a differential equation in the form of Equation (5.36):

$$\frac{di}{dt} = \frac{-i}{Q'(h(i))\ h'(i)} \qquad (5.54)$$

Suppose the initial voltage on the capacitor is V_0. The initial current must be found from the equation

$$V_0 = h(i)$$

and we have to invert the function $h(i)$. Now we note that the equation can be put in the form

$$g(i) = h(i) - V_0 = 0$$

The root of the equation is the initial value of the current, and the Newton-Raphson algorithm may now be applied to find it. See Problem 5.16 for a numerical example.

PROBLEMS

(Unless otherwise specified, the diode characteristic is assumed to be as shown in Figure 5.12(b).)

5.1 Sketch the i-v characteristic of each of the following circuits.

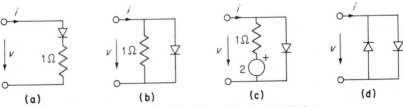

Problem 5.1 Combination of i-v characteristics.

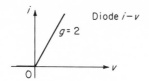

Problem 5.1   (continued).

5.2 Sketch the transfer characteristic of each of the following circuits, the transfer characteristic being v_0/v_{in}.

Problem 5.2 Transfer characteristics.

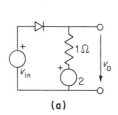

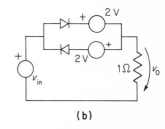

(a)   (b)

5.3 If in the circuit of Problem 5.2(b) the input voltage is $v_{in}(t) = A \sin 5t$, sketch the output voltage for $A = 1$ v. Repeat for $A = 3$ V.

5.4 Obtain the transfer characteristic of the following circuit. Let the input voltage be a triangular wave. Sketch the output voltage.

Problem 5.4 Transfer characteristic.

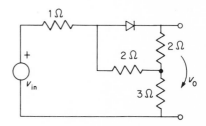

5.5 Obtain the i-v characteristic of the following circuit.

Problem 5.5 i-v characteristic.

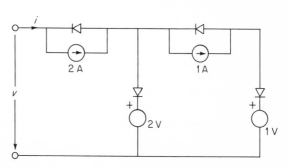

5.6   In the figure shown, the nonlinear element is the circuit of Problem 5.5. Obtain the transfer characteristic of the circuit.

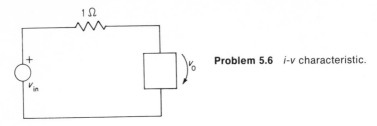

Problem 5.6   *i-v* characteristic.

5.7   The following is a logic circuit. What does it do?

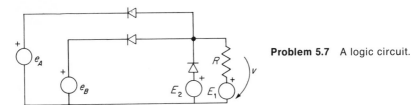

Problem 5.7   A logic circuit.

5.8   The following is a "diode matrix" encoder. As shown, the circuit encodes the presence of e_A into the binary number (1   0   1) and that of e_B the number (1   1   0). Explain the operation of the circuit.

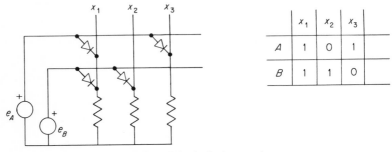

	x_1	x_2	x_3
A	1	0	1
B	1	1	0

Problem 5.8   A diode encoder.

5.9   Construct a diode matrix to encode each of the 10 decimal digits into its binary representation.

5.10   A nonlinear capacitor may be used as a harmonic generator. Let a nonlinear capacitor be characterized by a charge-voltage $(q\text{-}v)$ relation: $q = v + v^3$. Let $v(t) = \sin t$. Find the harmonics of the current in the capacitor. Choose a $q\text{-}v$ relation such that only the third harmonic appears.

5.11 Write a computer program to find all the voltages and currents in the circuit shown. Verify Tellegen's theorem.

Problem 5.11 A simple nonlinear resistor circuit.

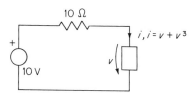

5.12 Write a computer program to compute the output voltage as a function of time in the following circuit.

Problem 5.12 A nonlinear resistor circuit under sine wave excitation.

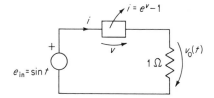

5.13 Suppose in the circuit of Figure 5.21(a), the nonlinear resistor circuit is characterized by $i = \exp(v) - 2$, and the capacitor is linear and of 1 F. Compute the terminal voltage and current for $0 \leq t \leq 5$. Assume the initial capacitor voltage to be 1 V.

5.14 Suppose in the circuit of Figure 5.21(a), the nonlinear resistor circuit is characterized by $v = (i + 1)^3$, and the capacitance is 1 F. Compute the terminal voltage and current for $0 \leq t \leq 5$ with the initial condition that $v(0) = 8$ V.

5.15 Suppose in the circuit of Figure 5.21(a), the capacitor is characterized by $v = q + q^3$, and the nonlinear resistor circuit by $v = 1 + i + i^5$. Obtain a differential equation for the current i. Indicate how you would find the numerical solution of the equation. Note that you would have to invert a function four times at every integration step if the Runge-Kutta algorithm is used.

5.16 In the circuit of Figure 5.21(a), suppose the nonlinear resistor is characterized by $v = 1 + i + i^3$, and the capacitance is 1 F. Compute the terminal voltage and current for $0 \leq t \leq 5$ with $v(0) = -2$ V.

6

formulation of circuit equations

In Chapter 2, we developed a systematic procedure of writing a set of equations that describes the behavior of a circuit completely. The set of equations consists of the KVL equations written for the fundamental loops, the KCL or current equations written for the fundamental cut sets, and the voltage-current relations of the elements. The equations were combined more or less in an *ad hoc* fashion at that time. While this *ad hoc* combination is satisfactory for simple circuits when the number of elements is small, it is not satisfactory for large circuits which we may encounter in the real world, for example, the power distribution network of an electric utility company. We now wish to inquire if it is possible to formulate circuit equations systematically from the topology of the circuit without going through the intermediate steps of combining the KVL, KCL, and terminal equations each time. In other words, is it possible to deduce formulas, which, when information about the circuit is put into them, give immediately the solution of the circuit. These formulas exist. We shall derive them in this chapter. They are, in fact, the basis on which computer programs for circuit analysis are written.

On the other hand, these formulas are unwieldy to use for hand computation. If one wishes to solve problems by "pencil and paper," one formulates the circuit equations by making physical interpretation of the various terms in the formulas. These equations can be obtained by inspection and are known as *loop equations, mesh equations,* and *node equations.*

In a circuit of b branches and n nodes, there are of course b branch voltage variables and b branch current variables. The equations that describe the circuit are the b terminal equations, the $b - n + 1$ independent voltage equations, and the $n - 1$ current equations. In all we have $b + (b - n + 1) + (n - 1) = 2b$ equations, which are what is required for the solution of the circuit.

As we shall see presently, the three sets of equations combine to yield a set of simultaneous equations in a set of variables that are either the chord currents or the tree voltages. The composite set describes the circuit completely, but the number of variables is reduced to $b - n + 1$ if chord currents are used, and to $n - 1$ if tree voltages are used.

For formulation by inspection, two sets of new variables are introduced. One is $b - n + 1$ in number, and the other $n - 1$. Both are artificial variables in that they are not associated with the voltages or currents of any set of branches. However, mathematically, equations written in terms of these describe the circuit completely and efficiently.

6.1 General Linear Resistor Circuits—Chord and Tree Analysis

We begin by considering the case of circuits containing linear resistors and independent sources only. Later, we shall also consider circuits that include inductors, capacitors, and dependent sources.

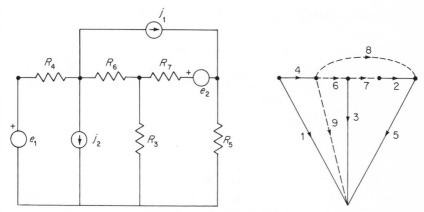

Figure 6.1 A resistor circuit.

Figure 6.1 shows a simple circuit consisting of two voltage sources and two current sources. We recall from Chapter 1 that the tree voltages determine all the voltages in a circuit. We, therefore, select a tree for the circuit shown to include all voltage sources as tree branches. We recall also that the chord currents determine all the currents in a circuit. We, therefore, select the remaining tree branches so that the current sources are chords. The selected tree is shown in the figure. We now number the branches so that the voltage sources are labelled first and the current sources are labelled last. The tree branches of the resistors follow the voltage sources, and the resistor chords follow the resistor tree branches. This numbering scheme is for our convenience, and it will simplify the expression of formulas that we derive later.

First we express the chord voltages in terms of the tree voltages by means of the KVL equations for the fundamental loops. In our case, we have

$$v_6 = v_1 - v_3 - v_4$$
$$v_7 = v_3 - v_2 - v_5$$
$$v_8 = v_1 - v_4 - v_5$$
$$v_9 = v_1 - v_4 \tag{6.1}$$

Expressing the tree currents in terms of the chord currents by means of KCL for the fundamental cut sets, we have

$$i_1 = -i_6 - i_8 - i_9$$
$$i_2 = i_7$$
$$i_3 = i_6 - i_7$$
$$i_4 = i_6 + i_8 + i_9$$
$$i_5 = i_7 + i_8 \tag{6.2}$$

Imposing the definitions of the elements, we have for sources

$$v_1 = e_1$$
$$v_2 = e_2$$
$$i_8 = j_1$$
$$i_9 = j_2 \tag{6.3}$$

and the resistors

$$v_3 = R_3 i_3$$
$$v_4 = R_4 i_4$$
$$v_5 = R_5 i_5$$
$$v_6 = R_6 i_6$$
$$v_7 = R_7 i_7 \tag{6.4}$$

Now focus our attention on the KVL equations for the resistor chords, namely, the first two equations of (6.1). Express all the variables in these equations in terms of the *resistor chord currents* i_6 and i_7 and in terms of the sources. We get

$$R_6 i_6 = e_1 - R_3(i_6 - i_7) - R_4(i_6 + j_1 + j_2)$$
$$R_7 i_7 = R_3(i_6 - i_7) - e_2 - R_5(i_7 + j_1)$$

or

$$(R_3 + R_4 + R_6)\, i_6 - R_3 i_7 = e_1 - R_4(j_1 + j_2)$$
$$-R_3 i_6 + (R_7 + R_3 + R_5)\, i_7 = -e_2 - R_5 j_1 \tag{6.5}$$

which are two linear, algebraic, simultaneous equations in the unknowns i_6 and i_7. Once i_6 and i_7 are found, the resistor tree currents are obtained from (6.2). The resistor tree voltages and resistor chord voltages are found from (6.4). The voltages across the current sources are found from (6.1).

Example 6.1   Suppose in Figure 6.1, $R_3 = 2\ \Omega$, $R_4 = 1\ \Omega$, $R_5 = 3\ \Omega$, $R_6 = 2\ \Omega$, and $R_7 = 3\ \Omega$. Also, $e_1 = 1$ V, $e_2 = -1$ V, $j_1 = 1$ A, and $j_2 = 1$ A. Substituting these values into (6.5), we get

$$
\begin{aligned}
5i_6 - 2i_7 &= -1 \\
-2i_6 + 8i_7 &= -2
\end{aligned}
\tag{6.6}
$$

Solving for the unknowns, we have

$$
i_6 = -\tfrac{1}{3}\ \text{A}, \qquad i_7 = -\tfrac{1}{3}\ \text{A}
\tag{6.7}
$$

From (6.2), the other currents are

$$
\begin{aligned}
i_1 &= \tfrac{1}{3} - 1 - 1 = -\tfrac{5}{3}\ \text{A} \\
i_2 &= -\tfrac{1}{3}\ \text{A} \\
i_3 &= -\tfrac{1}{3} + \tfrac{1}{3} = 0 \\
i_4 &= -\tfrac{1}{3} + 1 + 1 = \tfrac{5}{3}\ \text{A} \\
i_5 &= -\tfrac{1}{3} + 1 = \tfrac{2}{3}\ \text{A}
\end{aligned}
\tag{6.8}
$$

As a check, we have

$$
i_1 + i_9 + i_3 + i_5 = -\tfrac{5}{3} + 1 + 0 + \tfrac{2}{3} = 0
$$

From (6.4), the resistor voltages are

$$
\begin{aligned}
v_3 &= 2i_3 = 0 \\
v_4 &= 1i_4 = \tfrac{5}{3}\ \text{V} \\
v_5 &= 3i_5 = 2\ \text{V} \\
v_6 &= 2i_6 = -\tfrac{2}{3}\ \text{V} \\
v_7 &= 3i_7 = -1\ \text{V}
\end{aligned}
\tag{6.9}
$$

From (6.1), the voltages across the current sources are

$$
\begin{aligned}
v_8 &= 1 - \tfrac{5}{3} - 2 = -\tfrac{8}{3}\ \text{V} \\
v_9 &= 1 - \tfrac{5}{3} = -\tfrac{2}{3}\ \text{V}
\end{aligned}
\tag{6.10}
$$

As a check, we have

$$
v_8 + v_5 - v_9 = -\tfrac{8}{3} + 2 + \tfrac{2}{3} = 0 \qquad\blacksquare
$$

We now make a few observations.

CHORD ANALYSIS

1. The KVL, KCL, and the terminal equations are combined by retaining the KVL equations for the resistor chords and by expressing the variables in terms of the resistor chord currents and the sources. For this reason, the scheme is called the *chord analysis*. The number of unknowns is the number of resistor chords, which is the total number of chords less the number of current sources.

2. Once the resistor chord currents are known, all the voltages and currents in the circuit are determined. Moreover, since we have used the maximum number of independent voltage and current equations, the voltages and cur-

rents thus determined satisfy all the KVL and KCL equations and must constitute the solution of the circuit.

3. It would be desirable to deduce a way of getting (6.5) by inspection. This is possible in certain cases, and we shall present the method in Section 6.4.

TREE ANALYSIS

1. In invoking the definitions of the elements, we made the decision to express the resistor voltages in terms of the currents as in (6.4) and to retain the resistor chord currents as the only variables in the final set of equations. We could also have expressed the resistor currents in terms of the voltages and to retain another set of variables as unknowns. These variables turn out to be the *tree resistor voltages*. Let us pursue this speculation further.

Suppose, instead of (6.4), we write

$$i_3 = G_3 v_3$$
$$i_4 = G_4 v_4$$
$$i_5 = G_5 v_5$$
$$i_6 = G_6 v_6$$
$$i_7 = G_7 v_7 \tag{6.11}$$

Now focus our attention on the KCL equations for the *resistor tree currents*, that is, the last three equations of (6.2). Express all the variables in these equations in terms of the *resistor tree voltages*, v_3, v_4, and v_5, and in terms of the sources. We get

$$G_3 v_3 = G_6(e_1 - v_3 - v_4) - G_7(v_3 - e_2 - v_5)$$
$$G_4 v_4 = G_6(e_1 - v_3 - v_4) + j_1 + j_2$$
$$G_5 v_5 = G_7(v_3 - e_2 - v_5) + j_1$$

or

$$(G_3 + G_6 + G_7)v_3 + G_6 v_4 - G_7 v_5 = G_6 e_1 + G_7 e_2$$
$$G_6 v_3 + (G_4 + G_6)v_4 + 0v_5 = G_6 e_1 + j_1 + j_2$$
$$-G_7 v_3 + 0v_4 + (G_5 + G_7)v_5 = -G_7 e_2 + j_1 \tag{6.12}$$

which are three linear, algebraic, simultaneous equations in the unknowns v_3, v_4, and v_5, namely, the resistor tree voltages. Once these voltages are known, we find the chord voltages from KVL (6.1). The currents are then found from (6.11) and (6.2).

Example 6.2 Suppose the element values of the circuit of Figure 6.1 are the same as those of Example 6.1. Making substitutions into (6.12), we have

$$\tfrac{4}{3}v_3 + \tfrac{1}{2}v_4 - \tfrac{1}{3}v_5 = \tfrac{1}{6}$$
$$\tfrac{1}{2}v_3 + \tfrac{3}{2}v_4 + 0v_5 = \tfrac{5}{2}$$
$$-\tfrac{1}{3}v_4 + 0v_4 + \tfrac{2}{3}v_5 = \tfrac{4}{3}$$

Solving for the unknowns, we get

$$v_3 = 0$$
$$v_4 = \tfrac{5}{3} \text{ V}$$
$$v_5 = 2 \text{ V}$$

as before. The other voltages are found from (6.1):

$$v_6 = 1 - 0 - \tfrac{5}{3} = -\tfrac{2}{3} \text{ V}$$
$$v_7 = 0 + 1 - 2 = -1 \text{ V}$$

The currents are found from (6.11):

$$i_3 = 0$$
$$i_4 = \tfrac{5}{3} \text{ A}$$
$$i_5 = \tfrac{1}{3}(2) = \tfrac{2}{3} \text{ A}$$
$$i_6 = \tfrac{1}{2}(-\tfrac{2}{3}) = -\tfrac{1}{3} \text{ A}$$
$$i_7 = \tfrac{1}{3}(-1) = -\tfrac{1}{3} \text{ A}$$

Finally from (6.2) the currents in the voltages sources are:

$$i_1 = \tfrac{1}{3} - 1 - 1 = -\tfrac{5}{3} \text{ A}$$
$$i_2 = -\tfrac{1}{3} \text{ A}$$

as before. We continue our observations. ∎

2. The decision to keep the resistor tree voltages as variables and eliminate the other voltages and currents leads to a set of simultaneous equations in the unknowns which are the resistor tree voltages. The number of equations is the number of tree branches less the number of voltage sources. We call this scheme the *tree analysis*.

3. Chord analysis and tree analysis both lead to the same solution. As to which of the two we should use in a given circuit, it will depend on the number of variables required by each method. This is not a minor consideration, since the number of additions and multiplications grows as the third power of the number of equations in the solution of simultaneous equations.

4. In either method, voltage and current sources are present. There is no need to convert the sources into one or the other form.

6.2 Matrix Formulation—Chord Analysis*

The method of chord analysis can be formulated using matrix notations. Matrix formulation leads to formulas and computer algorithms for the computation of the resistor chord currents, and by means of KVL, KCL, and terminals equations, the remaining currents and voltages are found.

The circuit of Figure 6.1 will be used again. We select the tree as before and number the branches in the same manner. To express the chord voltages in terms of the tree voltages, we write, with reference to (6.1),

$$\begin{bmatrix} v_6 \\ v_7 \\ v_8 \\ v_9 \end{bmatrix} = \begin{bmatrix} 1 & 0 & -1 & -1 & 0 \\ 0 & -1 & 1 & 0 & -1 \\ 1 & 0 & 0 & -1 & -1 \\ 1 & 0 & 0 & -1 & 0 \end{bmatrix} \begin{bmatrix} v_1 \\ v_2 \\ v_3 \\ v_4 \\ v_5 \end{bmatrix} \tag{6.13}$$

In an abbreviated form, we have

$$\mathbf{v}_c = \mathbf{B}\mathbf{v}_t \tag{6.14}$$

where $\mathbf{v}_c$ is a column matrix of chord voltages, $\mathbf{v}_t$ a column matrix of tree voltages, and $\mathbf{B}$ a matrix of coefficients in the set of equations expressing the chord voltages in terms of the tree voltages. The rows of $\mathbf{B}$ correspond to the chords, and the columns of $\mathbf{B}$ to the tree branches.

Next, with reference to (6.2), the tree currents are related to the chord currents by

$$
\begin{bmatrix} i_1 \\ i_2 \\ i_3 \\ i_4 \\ i_5 \end{bmatrix} =
\begin{bmatrix} -1 & 0 & -1 & -1 \\ 0 & 1 & 0 & 0 \\ 1 & -1 & 0 & 0 \\ 1 & 0 & 1 & 1 \\ 0 & 1 & 1 & 0 \end{bmatrix}
\begin{bmatrix} i_6 \\ i_7 \\ i_8 \\ i_9 \end{bmatrix} \tag{6.15}
$$

or in abbreviated form

$$\mathbf{i}_t = \mathbf{C}\mathbf{i}_c \tag{6.16}$$

where $\mathbf{i}_t$ is a column matrix of tree currents, $\mathbf{i}_c$ a column of chord currents, and $\mathbf{C}$ a matrix of coefficients in the set of equations relating the tree currents to the chord currents. The rows of $\mathbf{C}$ correspond to the tree branches, and the columns to the chords.

We now make a remarkable observation, namely,

$$\mathbf{C} = -\mathbf{B}^t \tag{6.17}$$

In words, the matrix relating the tree currents to the chord currents is the negative transpose of the matrix relating the chord voltages to the tree voltages, when the columns and rows in both matrices are arranged in the same order. Note the dualism once more between voltages and currents and between trees and chords. Equation (6.17) is always true. In order not to distract ourselves from the main purpose at hand, we shall give a proof of it in the Appendix on pages 299–301.

The choice of a tree as outlined in Section 6.1 divides the set of branches of the circuit into four disjoint subsets: tree branches that are voltage sources, tree branches that are resistors, chords that are resistors, and chords that are current sources. This classification amounts to partitioning the matrices as follows:

$$
\mathbf{v}_t = \begin{bmatrix} \mathbf{v}_{tE} \\ \mathbf{v}_{tR} \end{bmatrix}
\qquad
\mathbf{v}_c = \begin{bmatrix} \mathbf{v}_{cR} \\ \mathbf{v}_{cJ} \end{bmatrix} \tag{6.18}
$$

$$
\mathbf{i}_t = \begin{bmatrix} \mathbf{i}_{tE} \\ \mathbf{i}_{tR} \end{bmatrix}
\qquad
\mathbf{i}_c = \begin{bmatrix} \mathbf{v}_{cR} \\ \mathbf{i}_{cJ} \end{bmatrix} \tag{6.19}
$$

where

$\mathbf{v}_{tE}$ is the set of tree voltages that are the voltage sources
$\mathbf{v}_{tR}$ is the set of tree voltages that are the resistor trees
$\mathbf{v}_{cR}$ is the set of chord voltages of the resistor chords
$\mathbf{v}_{cJ}$ is the set of chord voltages of the current sources
$\mathbf{i}_{tE}$ is the set of tree currents of the voltage sources
$\mathbf{i}_{tR}$ is the set of tree currents of the tree resistors
$\mathbf{i}_{cR}$ is the set of chord currents of the chord resistors, and
$\mathbf{i}_{cJ}$ is the set of chord currents of the current sources

In our case, we have

$$\mathbf{v}_{tE} = \begin{bmatrix} v_1 \\ v_2 \end{bmatrix} \qquad \mathbf{v}_{tR} = \begin{bmatrix} v_3 \\ v_4 \\ v_5 \end{bmatrix} \qquad \mathbf{v}_{cR} = \begin{bmatrix} v_6 \\ v_7 \end{bmatrix} \qquad \mathbf{v}_{cJ} = \begin{bmatrix} v_8 \\ v_9 \end{bmatrix}$$

$$\mathbf{i}_{tE} = \begin{bmatrix} i_1 \\ i_2 \end{bmatrix} \qquad \mathbf{i}_{tR} = \begin{bmatrix} i_3 \\ i_4 \\ i_5 \end{bmatrix} \qquad \mathbf{i}_{cR} = \begin{bmatrix} i_6 \\ i_7 \end{bmatrix} \qquad \mathbf{i}_{cJ} = \begin{bmatrix} i_8 \\ i_9 \end{bmatrix}$$

Correspondingly, the matrix $\mathbf{B}$ is partitioned into four submatrices:

$$\mathbf{v}_c = \begin{bmatrix} \mathbf{v}_{cR} \\ \mathbf{v}_{cJ} \end{bmatrix} = \begin{bmatrix} \mathbf{B}_{RE} & \mathbf{B}_{RR} \\ \mathbf{B}_{JE} & \mathbf{B}_{JR} \end{bmatrix} \begin{bmatrix} \mathbf{v}_{tE} \\ \mathbf{v}_{tR} \end{bmatrix} \tag{6.20}$$

where the subscripts have obvious meanings. In our case, we have, from (6.13),

$$\mathbf{B}_{RE} = \begin{bmatrix} 1 & 0 \\ 0 & -1 \end{bmatrix} \qquad \mathbf{B}_{RR} = \begin{bmatrix} -1 & -1 & 0 \\ 1 & 0 & -1 \end{bmatrix}$$

$$\mathbf{B}_{JE} = \begin{bmatrix} 1 & 0 \\ 1 & 0 \end{bmatrix} \qquad \mathbf{B}_{JR} = \begin{bmatrix} 0 & -1 & -1 \\ 0 & -1 & 0 \end{bmatrix}$$

Equation (6.15) can now be written as

$$\mathbf{i}_t = \begin{bmatrix} \mathbf{i}_{tE} \\ \mathbf{i}_{tR} \end{bmatrix} = \begin{bmatrix} -\mathbf{B}_{RE}^t & -\mathbf{B}_{JE}^t \\ -\mathbf{B}_{RR}^t & -\mathbf{B}_{JR}^t \end{bmatrix} \begin{bmatrix} \mathbf{i}_{cR} \\ \mathbf{i}_{cJ} \end{bmatrix} \tag{6.21}$$

In equation forms, (6.20) and (6.21) can be written as

$$\mathbf{v}_{cR} = \mathbf{B}_{RE}\,\mathbf{v}_{tE} + \mathbf{B}_{RR}\,\mathbf{v}_{tR} \tag{6.22}$$
$$\mathbf{v}_{cJ} = \mathbf{B}_{JE}\,\mathbf{v}_{tE} + \mathbf{B}_{JR}\,\mathbf{v}_{tR} \tag{6.23}$$
$$\mathbf{i}_{tE} = -\mathbf{B}_{RE}^t\,\mathbf{i}_{cR} - \mathbf{B}_{JE}^t\,\mathbf{i}_{cJ} \tag{6.24}$$
$$\mathbf{i}_{tR} = -\mathbf{B}_{RR}^t\,\mathbf{i}_{cR} - \mathbf{B}_{JR}^t\,\mathbf{i}_{cJ} \tag{6.25}$$

Next, the terminal equations for the elements:

$$\mathbf{v}_{tE} = \begin{bmatrix} v_1 \\ v_2 \end{bmatrix} = \begin{bmatrix} e_1 \\ e_2 \end{bmatrix} \triangleq \mathbf{E} \tag{6.26}$$

$$\mathbf{i}_{cJ} = \begin{bmatrix} i_8 \\ i_9 \end{bmatrix} = \begin{bmatrix} j_1 \\ j_2 \end{bmatrix} \triangleq \mathbf{J} \tag{6.27}$$

where $\mathbf{E}$ is the set of voltage sources, and $\mathbf{J}$ the current sources. Next, we have

$$\begin{bmatrix} v_3 \\ v_4 \\ v_5 \end{bmatrix} = \begin{bmatrix} R_3 & 0 & 0 \\ 0 & R_4 & 0 \\ 0 & 0 & R_5 \end{bmatrix} \begin{bmatrix} i_3 \\ i_4 \\ i_5 \end{bmatrix} \tag{6.28a}$$

$$\begin{bmatrix} v_6 \\ v_7 \end{bmatrix} = \begin{bmatrix} R_6 & 0 \\ 0 & R_7 \end{bmatrix} \begin{bmatrix} i_6 \\ i_7 \end{bmatrix} \tag{6.28b}$$

or in general

$$\mathbf{v}_{tR} = \mathbf{R}_t \mathbf{i}_{tR} \tag{6.29}$$
$$\mathbf{v}_{cR} = \mathbf{R}_c \mathbf{i}_{cR} \tag{6.30}$$

where $\mathbf{R}_t$ and $\mathbf{R}_c$ are diagonal matrices whose nonzero elements are the resistances of the corresponding branches in the circuit. Now take (6.22) and express all the variables in terms of $\mathbf{i}_{cR}$, $\mathbf{E}$, and $\mathbf{J}$. We get

$$\mathbf{R}_c \mathbf{i}_{cR} = \mathbf{B}_{RE} \mathbf{E} - \mathbf{B}_{RR} \mathbf{R}_t \mathbf{B}_{RR}^t \, \mathbf{i}_{cR} - \mathbf{B}_{RR} \mathbf{R}_t \mathbf{B}_{JR}^t \, \mathbf{J}$$

or

$$(\mathbf{R}_c + \mathbf{B}_{RR} \mathbf{R}_t \mathbf{B}_{RR}^t) \mathbf{i}_{cR} = \mathbf{B}_{RE} \mathbf{E} - \mathbf{B}_{RR} \mathbf{R}_t \mathbf{B}_{JR}^t \, \mathbf{J} \tag{6.31}$$

which is a system of linear algebraic equations. Treating the resistor chord currents as responses, we note the system is an expression of the principle of superposition. The formal solution of the system is

$$\mathbf{i}_{cR} = \mathbf{M}_{\bar{R}}^{1} \, (\mathbf{B}_{RE} \mathbf{E} - \mathbf{B}_{RR} \mathbf{R}_t \mathbf{B}_{JR}^t \, \mathbf{J}) \tag{6.32}$$

where

$$\mathbf{M}_R = \mathbf{R}_c + \mathbf{B}_{RR} \mathbf{R}_t \mathbf{B}_{RR}^t \tag{6.33}$$

Equation (6.32) is the formula that we seek, and it specifies the algorithm on which the computer program for analyzing resistor circuits is based. The other unknowns of the circuit are found by (6.24), (6.25), (6.29), (6.30), and (6.23), in turn,

$$\mathbf{i}_{tE} = -\mathbf{B}_{RE}^t \mathbf{i}_{cR} - \mathbf{B}_{JE}^t \, \mathbf{J} \tag{6.34}$$
$$\mathbf{i}_{tR} = -\mathbf{B}_{RR}^t \mathbf{i}_{cR} - \mathbf{B}_{JR}^t \, \mathbf{J} \tag{6.35}$$

$$\mathbf{v}_{cR} = \mathbf{R}_c \mathbf{i}_{cR} \qquad (6.36)$$
$$\mathbf{v}_{tR} = \mathbf{R}_t \mathbf{i}_{tR} \qquad (6.37)$$
$$\mathbf{v}_{cJ} = \mathbf{B}_{JE}\mathbf{E} + \mathbf{B}_{JR}\mathbf{v}_{tR} \qquad (6.38)$$

The computer algorithm can be described as follows.

1. Read in the topology of the circuit. The input data consists of a list. Each entry consists of two node names to which a branch is connected, the element type, and the value of the branch.
2. Find a tree such that all the voltage sources are tree branches and all the current sources are chords. The computer program given in Chapter 1 can be modified to find such a tree, commonly known as the "proper" tree.
3. Identify the chords and construct the matrices $\mathbf{R}_c$, $\mathbf{R}_t$, $\mathbf{E}$, and $\mathbf{J}$.
4. Construct the matrix $\mathbf{B}$. [See Equation (6.14).] The rows of $\mathbf{B}$ correspond to the chords, and the columns of $\mathbf{B}$ correspond to tree branches. The element of $\mathbf{B}$, $\mathbf{B}_{ij}$, is nonzero if, and only if, tree branch j is contained in the loop associated with chord i; otherwise it is zero.
5. Partition $\mathbf{B}$ into four submatrices: $\mathbf{B}_{RE}$, $\mathbf{B}_{RR}$, $\mathbf{B}_{JE}$, and $\mathbf{B}_{JR}$.
6. Obtain the transpose of the four submatrices of $\mathbf{B}$.
7. Calculate $\mathbf{M}_{\bar{R}}^{-1}$ [Equation (6.33)].
8. The resistor chord currents are computed from Equation (6.32).
9. The other voltages and currents from Equations (6.34) to (6.38).

A computer program based on the tree-chord analysis has been written. The program, called RNET, computes all the branch voltages and currents of a linear resistor network excited by independent voltage and current sources. A reference node is chosen by the program, and as a check, the sum of the currents entering this node is computed and printed. A complete listing of RNET and the instructions for its use are given at the end of the chapter.

6.3 Matrix Formulation—Tree Analysis*

The method of tree analysis will now be formalized. We express the resistor currents in terms of the resistor voltages as follows:

$$\mathbf{i}_{tR} = \mathbf{G}_t \mathbf{v}_{tR} \qquad (6.39)$$
$$\mathbf{i}_{cR} = \mathbf{G}_c \mathbf{v}_{cR} \qquad (6.40)$$

where $\mathbf{G}_t$ and $\mathbf{G}_c$ are diagonal matrices whose nonzero elements are the conductances of the tree and chord resistor branches, respectively. In the example of Figure 6.1, we have

$$\begin{bmatrix} i_3 \\ i_4 \\ i_5 \end{bmatrix} = \begin{bmatrix} G_3 & 0 & 0 \\ 0 & G_4 & 0 \\ 0 & 0 & G_5 \end{bmatrix} \begin{bmatrix} v_3 \\ v_4 \\ v_5 \end{bmatrix}$$

$$\begin{bmatrix} i_6 \\ i_7 \end{bmatrix} = \begin{bmatrix} G_6 & 0 \\ 0 & G_7 \end{bmatrix} \begin{bmatrix} v_6 \\ v_7 \end{bmatrix}$$

Now take the KCL equation (6.25) and express all the variables in terms of the resistor tree voltages and the sources. We get

$$\mathbf{G}_t \mathbf{v}_{tR} = -\mathbf{B}_{RR}^t \mathbf{G}_c (\mathbf{B}_{RE} \mathbf{v}_{tE} + \mathbf{B}_{RR} \mathbf{v}_{tR}) - \mathbf{B}_{JR}^t \mathbf{i}_{cJ}$$

or

$$(\mathbf{G}_t + \mathbf{B}_{RR}^t \mathbf{G}_c \mathbf{B}_{RR}) \mathbf{v}_{tR} = -\mathbf{B}_{RR}^t \mathbf{G}_c \mathbf{B}_{RE} \mathbf{E} - \mathbf{B}_{JR}^t \mathbf{J} \qquad (6.41)$$

which is a system of linear algebraic equations in the unknowns $\mathbf{v}_{tR}$, the resistor tree voltages. The formal solution is

$$\mathbf{v}_{tR} = \mathbf{M}_G^{-1} (- \mathbf{B}_{RR}^t \mathbf{G}_c \mathbf{B}_{RE} \mathbf{E} - \mathbf{B}_{JR}^t \mathbf{J}) \qquad (6.42)$$

where

$$\mathbf{M}_G = \mathbf{G}_t + \mathbf{B}_{RR}^t \mathbf{G}_c \mathbf{B}_{RR} \qquad (6.43)$$

Equation (6.42) is the formula on which the computer program of the method of tree analysis is based. The other variables in the circuit can also be explicitly given in terms of the sources, but from the computational point of view, they should be calculated from the following in turn.

$$\mathbf{v}_{cR} = \mathbf{B}_{RE} \mathbf{E} + \mathbf{B}_{RR} \mathbf{v}_{tR} \qquad (6.44)$$
$$\mathbf{i}_{tR} = \mathbf{G}_t \mathbf{v}_{tR} \qquad (6.45)$$
$$\mathbf{i}_{cR} = \mathbf{G}_c \mathbf{v}_{cR} \qquad (6.46)$$
$$\mathbf{i}_{tE} = -\mathbf{B}_{RE}^t \mathbf{i}_{cR} - \mathbf{B}_{JE}^t \mathbf{i}_{cJ} \qquad (6.47)$$
$$\mathbf{v}_{cJ} = \mathbf{B}_{JE} \mathbf{E} + \mathbf{B}_{JR} \mathbf{v}_{tR} \qquad (6.48)$$

6.4 Loop Analysis

The method of chord analysis is applicable to any circuit containing resistors and both voltage and current sources. If a circuit contains only voltage sources, the method can be simplified, and the equations in the unknowns of the resistor chord currents can be obtained by inspection. We must keep in mind that writing equations by inspection is for the convenience of analysis by "pencil and paper."

With reference to Figure 6.1, suppose the current sources are absent. The circuit becomes that shown in Figure 6.2(a). Choose the same tree as before and number the branches in the same way, as given in Figure 6.2(b). The two chords define two fundamental loops to which we may assign directions to coincide with the directions of the chords. With reference to Figure 6.2(b), the fundamental loops are:

First loop: 3,1,4,6   clockwise
Second loop: 2,5,3,7   clockwise

Now following the same derivation as in Section 6.1 but with the current sources missing ($j_1 = j_2 = 0$), we obtain a system of linear equations in the unknowns i_6 and i_7, the chord currents. [See Equations (6.5).]

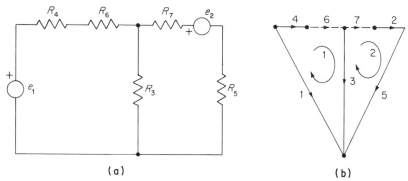

(a)   (b)

Figure 6.2 A resistor circuit containing only voltage sources.

$$(R_3 + R_4 + R_6)i_6 - R_3 i_7 = e_1$$
$$-R_3 i_6 + (R_5 + R_3 + R_7)i_7 = -e_2 \tag{6.49}$$

We now interpret the terms in the equations. Each term on the left-hand side of the first equation represents a voltage – the voltage across a resistor. For example, $R_3 i_6$ is the voltage across R_3 due to a current whose value is i_6. We can make similar interpretations for the other terms and for the second equation. We may, therefore, imagine (yes, imagine) as if there were a circulating current of magnitude i_6 flowing around the fundamental loop defined by the chord R_6 in a clockwise direction. Similarly, we may imagine there were a circulating current of magnitude i_7 circulating around the fundamental loop defined by the chord R_7. Note that R_3 is shared by both of the fundamental loops, and the current in it is the algebraic sum of the two circulating currents, namely, $i_6 - i_7$, in our case. We recall that each equation in (6.49) is a KVL equation. The first equation states that the sum of the voltages across the resistor in the fundamental loop equals the sum of the voltage sources which cause the circulating current to flow. The voltages are expressed in terms of the chord currents. Now note that the coefficient on i_6 in the first equation is the sum of the resistances in the fundamental loop. The coefficient on i_7 is the negative of the sum of the resistances shared by the two loops. The right-hand side of the equation consists of the sum of the voltage sources in the loop, each with a sign which is determined as follows:

Figure 6.3 Sign convention for a voltage source in loop analysis.

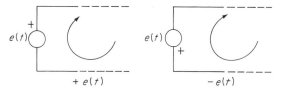

The sign is "+" if the circulating current is in the opposite direction of the source. Otherwise it is "−." See Figure 6.3.

Having made these observations, we can now deduce the following rules of writing the equations, by inspection, in the unknowns which are the loop currents.

1. Choose a tree. Place all voltage sources in a tree.
2. Identify the directions of the "loop" currents, which are the circulating currents in the fundamental loops.
3. The equations in the loop currents are

$$R_{11}i_{c1} + R_{12}i_{c2} + \cdots + R_{1m}i_{cm} = E_1$$
$$R_{21}i_{c1} + R_{22}i_{c2} + \cdots + R_{2m}i_{cm} = E_2$$
$$\vdots$$
$$R_{m1}i_{c1} + R_{m2}i_{c2} + \cdots + R_{mm}i_{cm} = E_m \qquad (6.50)$$

where $i_{c1}, i_{c2},\ldots, i_{cm}$ are the loop currents

$R_{ii} =$ the sum of the resistances in loop i

$R_{ij} = \pm$ (the sum of the resistances shared by loop i and loop j). The sign is "+" if i_{ci} and i_{cj} go through the common resistors in the same direction; otherwise it is "−"

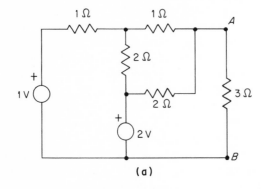

(a)

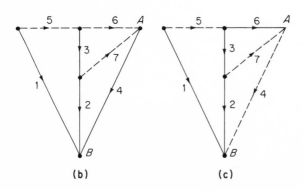

(b)   (c)

Figure 6.4 Loop analysis for two different trees. [Example 6.3]

$$R_{ij} = R_{ji}$$

E_r = the sum of the voltage sources in loop r

m = the number of chords in the circuit = the number of equations

Equations (6.50) are known as the *loop equations* of the circuit for the choice of the tree. If a different tree is chosen, the equations will of course be different, but the solution of the circuit is, and must be, the same.

Example 6.3   Find the voltage v_{AB} in the circuit of Figure 6.4(a).

Solution   The choice of tree is shown in Figure 6.4(b). There are three fundamental loops. Assign the loop currents as follows:

$$i_{c1} : \text{loop } 5,3,2,1 \quad \text{clockwise}$$
$$i_{c2} : \text{loop } 6,4,2,3 \quad \text{clockwise}$$
$$i_{c3} : \text{loop } 7,4,2 \quad \text{clockwise}$$

Following the rules stated above, we have

$$3i_{c1} - 2i_{c2} - 0i_{c3} = -1$$
$$-2i_{c1} + 6i_{c2} + 3i_{c3} = 2$$
$$0i_{c1} + 3i_{c2} + 5i_{c3} = 2$$

Since we are interested only in v_{AB}, which is $3i_4 = 3(i_6 + i_7) = 3(i_{c2} + i_{c3})$, we need to solve for i_{c2} and i_{c3} only. Doing just that, we have

$$i_{c2} = \frac{2}{43} \text{ A} \qquad i_{c3} = \frac{16}{43} \text{ A}$$

and

$$v_{AB} = 3\left(\frac{2}{43} + \frac{16}{43}\right) = \frac{54}{43} \text{ V}$$

Alternate solution   Suppose we choose a different tree as shown in Figure 6.4(c). Define the loop currents as follows:

$$i_{c1} : \text{loop } 5,3,2,1 \quad \text{clockwise}$$
$$i_{c2} : \text{loop } 7,6,3 \quad \text{counterclockwise}$$
$$i_{c3} : \text{loop } 4,2,3,6 \quad \text{clockwise}$$

The loop equations are

$$3i_{c1} + 2i_{c2} - 2i_{c3} = -1$$
$$2i_{c1} + 5i_{c2} - 3i_{c3} = 0$$
$$-2i_{c1} - 3i_{c2} + 6i_{c3} = 2$$

In this case, $v_{AB} = 3i_{c3}$, and we need to solve for i_{c3} only, which is found to be

$$i_{c3} = \frac{18}{43} \text{ A}$$

and

$$v_{AB} = 3i_{c3} = \frac{54}{43} \text{ V}$$

as before. It should be obvious that if we are interested only in the voltage or current at a particular element, we should force the element to be a chord in the choice of the tree.

6.5 Planar and Nonplanar Graphs

A short digression is now taken in order to introduce a very interesting and useful concept in graph theory. We define a *planar graph* as a graph that can be mapped on a plane such that the branches do not cross one another (more formally, such that the branches meet only at points that are the nodes of the graph). If the graph cannot be so mapped, it is called a nonplanar graph. Figure 6.5 shows a number of planar and nonplanar graphs.

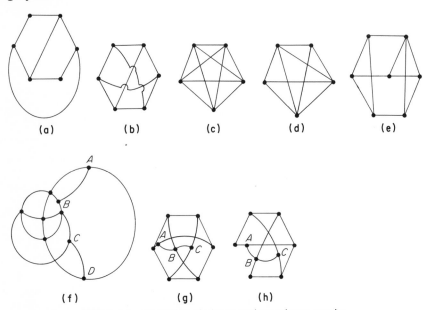

Figure 6.5 Examples of planar and nonplanar graphs.

An interesting problem in graph theory, which has been solved in various degrees of sophistication, is to determine, in a systematic way, whether or not a graph is planar. The problem has obvious applications in the wiring of electronic circuits and the layout of pipeline networks. One solution, known as the Kuratowski condition,[1] is the following:

> *Kuratowski's Theorem* A necessary and sufficient condition that a graph be planar is that it does not contain a subgraph that can be reduced to either the graph of Figure 6.5(b) or the graph of Figure 6.5(c) by replacing, in the subgraph, each series connection of branches by a single branch.

For example, in Figure 6.5, graphs (f) and (g) are nonplanar. In graph (f), the removal of branches AB and CD leaves a subgraph that can be re-

[1] Kuratowski, C., "Sur le Problème des Courbes Gauches en Topologie." *Fund. Math.,* vol. 15, pp. 271–283, 1930.

duced to graph (c). In graph (g), the removal of branches AB and BC leaves a subgraph that can be reduced to graph (b).

Planar graphs have many interesting properties. We define a *mesh* in a planar graph as a loop whose interior is empty. If we map a planar graph on a plane such that the branches do not cross one another, the "windows," or the "regions," are the meshes. In Figure 6.6, the meshes are $ABGFA$, $BCDHGB$, $GHEFG$, and $HDEH$. We define the *boundary* of a planar graph as the set of branches that form the perimeter of the planar map. In Figure 6.6, $ABCDEFA$ is the boundary. The set of meshes associated with a planar graph is a function of the mapping of the graph.

Figure 6.6 A planar graph. ABCDEFA is the boundary mesh.

If a graph is mapped differently, we would get a different set of meshes. For example, the graphs of Figures 6.6 and 6.7 are the same, but in Figure 6.7, the meshes are $ABGFA$, $ABCDEFA$, $GFEHG$, and $HDEH$. The boundary is $GBCDHG$. The point to note here is that *any* mesh can be made into the boundary of some map, and vice versa. A better way to see this is to map the planar graph on the surface of a sphere. Then every region can be considered as a mesh or as a boundary, depending on how the map is viewed. See Figure 6.8.

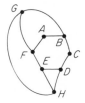

Figure 6.7 The graph of Figure 6.6 mapped differently. GBCDHG is the boundary mesh.

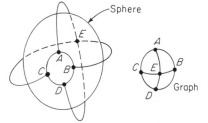

Figure 6.8 A graph mapped on a sphere as viewed from the "back."

An important observation about meshes is in the following.

Remark 6.1 The number of meshes is $b - n + 1$ in a connected graph of n nodes and b branches.

Proof We shall prove the remark by induction on b. The assertion is certainly true for a graph of two branches $(b = 2)$. Assume the assertion is true for all connected graphs with k or less branches and n nodes. Consider a planar graph of $b = k + 1$ and focus our attention on a branch x which belongs to the boundary as shown in Figure 6.9. Now, suppose we remove the branch x. By hypothesis, the remaining planar graph G has $k - n + 1$ meshes. The addition of x to G as a boundary branch increases the number of meshes exactly by one, while preserving the planarity of the overall graph. The number of meshes is therefore $k - n + 1 + 1$, which is $(k + 1) - n + 1$. The assertion is, therefore, true for a graph of $k + 1$ branches and n nodes. Since any branch in a planar graph can be made into a boundary branch, we conclude that the assertion is true for any number of branches.

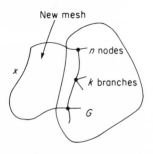

Figure 6.9 *G* is a planar graph. The addition of a boundary branch increases the number of meshes by one.

Remark 6.2 Suppose we write the KVL equations for the $b - n + 1$ meshes of a planar circuit and suppose we append to the set of equations another KVL equation that is written for the boundary of the graph. Then the voltage variable of every branch appears exactly twice in the appended set of equations, once with a coefficient of $+1$ and once -1, provided that we assign to all the meshes the same orientation and to the boundary the opposite orientation.

The remark is obvious if we consider a typical planar graph shown in Figure 6.10. The meshes are assigned a clockwise orientation, and the boundary counterclockwise. The conclusion of the remark follows if we note that every branch is shared by exactly two meshes, or by a mesh and a boundary mesh, for example,

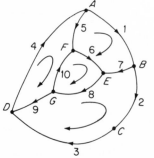

Figure 6.10 Every branch in a planar graph is shared either by exactly two meshes or by a mesh and the boundary.

branch 6 by $ABEFA$ and $FEGF$, branch 1 by $ABEFA$ and $ADCBA$. We also note that each voltage variable bears a coefficient $+1$ in the KVL equation for one mesh, and -1 in the KVL equation of the other mesh. To be specific, we write the KVL equations for the meshes and the boundary of Figure 6.10 as follows:

$$
\begin{array}{lll}
\text{Mesh} & ABEFA & v_1 + v_7 - v_6 - v_5 = 0 \\
& BCDGEB & v_2 + v_3 - v_9 - v_8 - v_7 = 0 \\
& DAFGD & v_4 + v_5 - v_{10} + v_9 = 0 \\
& FEGF & v_6 + v_8 + v_{10} = 0 \\
\text{Boundary} & ADCBA & -v_4 - v_3 - v_2 - v_1 = 0
\end{array} \tag{6.51}
$$

∎

An immediate consequence of Remark 6.2 is that the set of appended KVL equations is linearly dependent. What is the maximum number of independent equations and which of the KVL equations in the appended set are independent? The answers are given in the following theorem.

Theorem 6.1 The maximum number of independent KVL equations written for the $b - n + 1$ meshes and the boundary of a connected planar graph is $b - n + 1$. Moreover, the KVL equations for the $b - n + 1$ meshes constitute a maximal set of independent equations.

Proof Let $m = b - n + 2$. Let the appended set of KVL equations be $\phi_1, \phi_2,\ldots, \phi_m$, corresponding to the meshes $M_1, M_2,\ldots, M_{m-1}$ and the boundary, which will be denoted as M_m. By Remark 6.2, we have

$$
\phi_1 + \phi_2 + \cdots + \phi_m = 0 \tag{6.52}
$$

We wish to show that this is the only linear relation among the m equations. Assume the contrary. Let there be another linear relation

$$
c_1\phi_1 + c_2\phi_2 + \cdots + c_m\phi_m = 0 \tag{6.53}
$$

in which not all the c's are zero. For convenience, let $c_{k+1} = c_{k+2} = \cdots = c_m = 0$ and the other c's are not zero. We would have

$$
c_1\phi_1 + c_2\phi_2 + \cdots + c_k\phi_k = 0 \tag{6.54}
$$

Consider the subgraph of the graph consisting of the meshes $M_1, M_2,\ldots, M_k$ as shown in Figure 6.11. Since the graph is connected, there is at least one mesh, say M_{k+1}, not in the set $M_1,\ldots, M_k$, which is adjacent to at least one mesh, say M_2, in the set $M_1,\ldots, M_k$. M_{k+1} and M_2 share at least one voltage variable, say v_x. In the set of KVL equations $\phi_1, \phi_2,\ldots, \phi_k$, v_x appears exactly once, so that the relation (6.54) is not possible, thus contradicting the assumption. We con-

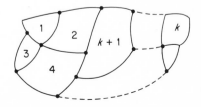

Figure 6.11 Graph used in the proof of Theorem 6.1.

clude, therefore, that the relation (6.52) is the only linear relation. It follows that the maximum number of independent KVL equations among the set $\phi_1, \phi_2, \ldots, \phi_m$ is $m - 1 = b - n + 1$. Since any mesh can be made into a boundary and vice versa, we conclude that the KVL equations of the $b - n + 1$ meshes are independent and constitute a maximal set of independent equations. ■

We should note the similarity, in fact dualism, of this theorem and the theorem on the maximal number of independent KCL equations.

As a final remark about meshes, we note that meshes are not in general fundamental loops. In other words, it is not always possible to find a tree such that the fundamental loops and the meshes coincide. An example of such a graph is shown in Figure 6.12.

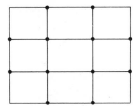

Figure 6.12 A planar graph whose meshes cannot be made to coincide with any set of fundamental loops.

6.6 Mesh Equations

Having established that the KVL equations written for the meshes constitute a maximal set of independent voltage equations, we now introduce another scheme of formulating equations for *planar* circuits. The scheme is very much like that of the loop equations except that it is easier to apply by inspection. However, the scheme for loop equations is ap-

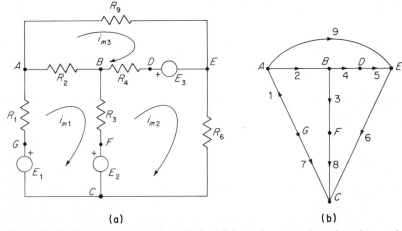

(a)   (b)

Figure 6.13 Illustration of mesh analysis. (a) A mesh current is assigned to each mesh; (b) the graph of the circuit.

plicable to planar and nonplanar graphs, whereas the scheme for the meshes is applicable to planar graphs only.

Consider Figure 6.13. Let the meshes be numbered and oriented as follows:

Mesh 1: *ABFCGA*   clockwise
Mesh 2: *BDECFB*   clockwise
Mesh 3: *AEDBA*   clockwise

Now we introduce a set of fictitious current variables known as the *mesh currents*. We imagine (yes, imagine) that in each of the meshes there is a circulating current flowing in the elements that constitute the mesh. The mesh currents are denoted as i_{m1}, i_{m2}, and i_{m3} in this example and are assigned the same orientations as those of the meshes. We must emphasize that the mesh currents are not physical currents and, in general, cannot be measured directly. Because of the last remark of the last section, mesh currents, in general, are not loop currents.

Suppose we write the KVL equations for the meshes. For example, in mesh 1, we have

$$v_1 + v_2 + v_3 + v_8 - v_7 = 0 \tag{6.55}$$

Now, express all the resistor voltages in terms of the mesh currents. For example,

$$v_1 = R_1 i_{m1}$$
$$v_2 = R_2(i_{m1} - i_{m3})$$

Note that R_2 is shared by both mesh 1 and mesh 3. The current in R_2 is the algebraic sum of the mesh currents i_{m1} and i_{m3}. Following this procedure, we obtain, for the three meshes, the following three KVL equations:

$$R_1 i_{m1} + R_2(i_{m1} - i_{m3}) + R_3(i_{m1} - i_{m2}) - E_1 + E_2 = 0$$
$$R_4(i_{m2} - i_{m3}) + R_6 i_{m2} - R_3(i_{m1} - i_{m2}) + E_3 - E_2 = 0$$
$$R_9 i_{m3} - R_4(i_{m2} - i_{m3}) - R_2(i_{m1} - i_{m3}) - E_3 = 0 \tag{6.56}$$

Rearranging the terms, we have

$$(R_1 + R_2 + R_3)i_{m1} - R_3 i_{m2} - R_2 i_{m3} = E_1 - E_2$$
$$-R_3 i_{m1} + (R_3 + R_4 + R_6)i_{m2} - R_4 i_{m3} = E_2 - E_3$$
$$-R_2 i_{m1} - R_4 i_{m2} + (R_2 + R_4 + R_9)i_{m3} = E_3 \tag{6.57}$$

which is a set of simultaneous algebraic equations in the unknowns i_{m1}, i_{m2}, and i_{m3}. The set of equations is of the form

$$R_{11} i_{m1} + R_{12} i_{m2} + \cdots + R_{1m} i_{mm} = e_1$$
$$R_{21} i_{m1} + R_{22} i_{m2} + \cdots + R_{2m} i_{mm} = e_2$$
$$\vdots$$
$$R_{m1} i_{m1} + R_{m2} i_{m2} + \cdots + R_{mm} i_{mm} = e_m \tag{6.58}$$

The coefficients can be readily interpreted as follows:

R_{rr} = sum of the resistances in mesh r

R_{rs} = −(resistance shared by meshes r and s) = R_{sr}

e_r = sum of voltage sources in mesh r, with
     the algebraic signs to be determined as
     in Figure 6.3

m = $b - n + 1$

Equations (6.58) are known as the *mesh equations*. They are merely the KVL equations expressed in terms of the fictitious variables, the mesh currents. Now an interesting question is: Does the set of mesh equations describe the circuit completely? We have used a maximal set of KVL equations and the definitions of the elements. What happens to KCL?

Consider a typical node in a planar graph as shown in Figure 6.14.

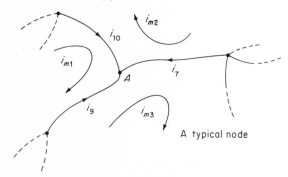

Figure 6.14 In mesh analysis, KCL is automatically satisfied at every node.

A typical node

By our convention, we have

$$i_{10} = i_{m1} - i_{m2}$$
$$i_9 = i_{m3} - i_{m1}$$
$$i_7 = i_{m2} - i_{m3} \qquad (6.59)$$

The sum of the branch currents at the node is

$$i_{10} + i_9 + i_7 = (i_{m1} - i_{m2}) + (i_{m3} - i_{m1}) + (i_{m2} - i_{m3}) = 0 \quad (6.60)$$

so that KCL is satisfied at the node. We conclude that the convention of expressing the branch currents in terms of the mesh currents guarantees that the KCL equations at all nodes are satisfied. Therefore, the description of the circuit by the mesh equations is complete.

Example 6.4 Solve the problem of Example 6.3 by mesh equations.

Solution The circuit is reproduced in Figure 6.15. Assigning the mesh currents as shown, we have

$$\begin{bmatrix} 3 & -2 & 0 \\ -2 & 5 & -2 \\ 0 & -2 & 5 \end{bmatrix} \begin{bmatrix} i_{m1} \\ i_{m2} \\ i_{m3} \end{bmatrix} = \begin{bmatrix} -1 \\ 0 \\ 2 \end{bmatrix}$$

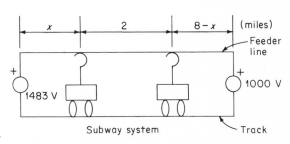

Figure 6.15 Mesh analysis. [Example 6.4]

Solving for i_{m3}, we get

$$i_{m3} = \frac{18}{43}$$

and

$$v_{AB} = 3i_{m3} = \frac{54}{43} \text{ V}$$

as before.

Example 6.5 A single-track subway line is powered by two generators located separately at the two ends of the line. The line has two trains, which are always kept 2 mi apart. The length of the line is 10 mi. The track has a resistance of 0.4 Ω/mi, and the feeder line 0.1 Ω/mi. Assume each train is equivalent to a load of 1 Ω. At what point along the line will the two generators supply the same power to the system? At this point, (1) what is the power? (2) What is the power taken by the two trains? (3) What is the power lost in transmission?

Solution The circuit of the line is shown in Figure 6.16. Assign mesh currents as shown. The mesh equations are

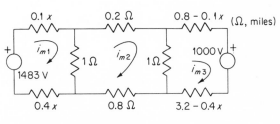

Figure 6.16 A two-car subway system. [Example 6.5]

Equivalent circuit

$$\begin{bmatrix} 1+0.5x & -1 & 0 \\ -1 & 3 & -1 \\ 0 & -1 & 5-0.5x \end{bmatrix} \begin{bmatrix} i_{m1} \\ i_{m2} \\ i_{m3} \end{bmatrix} = \begin{bmatrix} 1483 \\ 0 \\ -1000 \end{bmatrix}$$

Solving for the unknowns, we get

$$i_{m1} = \frac{(14-1.5x)(1483) - 1000}{9+6x-0.75x^2}$$

$$i_{m2} = \frac{1483(5-0.5x) - 1000(1+0.5x)}{9+6x-0.75x^2}$$

$$i_{m3} = \frac{1483 - 1000(2+1.5x)}{9+6x-0.75x^2}$$

The power supplied by the 1483-V generator is

$$P_1 = 1483 i_{m1}$$

The power supplied by the 1000-V generator is

$$P_2 = -1000 i_{m3}$$

Setting $P_1 = P_2$ and solving for x, we get

$$x = 6 \text{ mi}$$

At this value of x, we find

$$P_1 = P_2 = 528,000 \text{ W}$$

The powers taken by the trains P_{t1} and P_{t2} are, respectively,

$$P_{t1} = (1)(i_{m1} - i_{m2})^2 = 171,000 \text{ W}$$
$$P_{t2} = (1)(i_{m2} - i_{m3})^2 = 222,000 \text{ W}$$

The power lost in transmission is

$$P_{\text{lost}} = (3)(i_{m1})^2 + (1)(i_{m2})^2 + (1)(i_{m3})^2$$
$$= 663,000 \text{ W}$$

As a check, we note that

$$P_1 + P_2 - P_{t1} - P_{t2} - P_{\text{lost}} = (528 + 528 - 171 - 222 - 663)10^3 = 0$$

as expected.   ∎

As a final remark on mesh equations, we note from the two examples that they are much easier to use than the loop equations. There is no need to find a tree and to identify the fundamental loops. For "pencil and paper" analysis, mesh equations are clearly superior. However, one must remember that meshes are defined for planar graphs only, and that solution by pencil and paper is only practical for simple circuits containing less than five meshes. For computer solution, the chord or tree analysis is still better even for planar graphs. It is very difficult to identify the meshes in a graph by a computer.

6.7 Node Equations

Another scheme of writing circuit equations by inspection that is applicable to both planar and nonplanar graphs is that of *node analysis,* provided that the circuit contains only current sources as the excitations. In the case of mesh equations, we write KVL equations for the meshes in terms of a set of new current variables. In the case of node analysis, we write KCL equations at each and every node except one in terms of a set of new *voltage* variables.

Consider the circuit of Figure 6.17. Number the nodes as shown. Define four voltage variables, one from each of the four nodes 1, 2, 3, and 4 to node 5, which shall be called the reference node. The voltage graph is a tree with node 5 as the "root" as shown in Figure 6.17. Denote the four voltages as v_{1r}, v_{2r}, v_{3r}, and v_{4r}. We now note that all the branch voltages can be expressed in terms of the four voltages as follows:

$$v_1 = v_{1r}$$
$$v_2 = v_{1r} - v_{4r}$$
$$v_3 = v_{2r} - v_{4r}$$
$$v_4 = v_{2r} - v_{3r}$$
$$v_5 = v_{3r} - v_{4r}$$
$$v_6 = v_{3r}$$
$$v_7 = v_{1r} - v_{2r}$$
$$v_8 = v_{1r} \qquad\qquad (6.61)$$

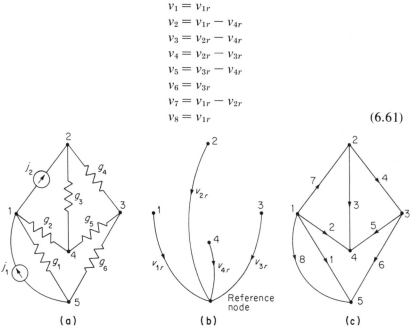

Figure 6.17 Node analysis. (**a**) The circuit; (**b**) its voltage graph with node 5 chosen as the reference node; (**c**) the graph of the circuit.

The branch currents can also be expressed in terms of the node-to-reference voltages. For example,

$$i_1 = g_1 v_1 = g_1 v_{1r}$$
$$i_2 = g_2 v_2 = g_2(v_{1r} - v_{4r}) \qquad\qquad (6.62)$$
$$\vdots$$

Next we write the KCL equations for each of the four nodes 1, 2, 3, and 4. For example, at node 1, we have

$$i_1 + i_2 + i_7 + i_8 = 0 \tag{6.63}$$

Expressing the currents in terms of the node-to-reference voltages and the sources, we have

$$g_1 v_{1r} + g_2(v_{1r} - v_{4r}) + j_2 - j_1 = 0$$

or

$$(g_1 + g_2)v_{1r} + 0v_{2r} + 0v_{3r} - g_2 v_{4r} = j_1 - j_2 \tag{6.64}$$

Repeating the procedure for all nodes, we obtain the following system of equations in the unknowns which are the node-to-reference voltages.

$$
\begin{aligned}
(g_1 + g_2)v_{1r} + 0v_{2r} + 0v_{3r} - g_2 v_{4r} &= j_1 - j_2 \\
0v_{1r} + (g_3 + g_4)v_{2r} - g_4 v_{3r} - g_3 v_{4r} &= j_2 \\
0v_{1r} - g_4 v_{2r} + (g_4 + g_5 + g_6)v_{3r} - g_5 v_{4r} &= 0 \\
-g_2 v_{1r} - g_3 v_{2r} - g_5 v_{3r} + (g_2 + g_3 + g_5)v_{4r} &= 0
\end{aligned} \tag{6.65}
$$

The equations are of the form

$$
\begin{bmatrix}
G_{11} & G_{12} & \cdots & G_{1q} \\
G_{21} & G_{22} & \cdots & G_{2q} \\
\vdots & \vdots & \ddots & \vdots \\
G_{q1} & & \cdots & G_{qq}
\end{bmatrix}
\begin{bmatrix}
v_{1r} \\
v_{2r} \\
\vdots \\
v_{qr}
\end{bmatrix}
=
\begin{bmatrix}
J_1 \\
J_2 \\
\vdots \\
J_q
\end{bmatrix} \tag{6.66}
$$

where $q = n - 1 =$ number of nodes less one. The coefficients can be interpreted physically and found by inspection as follows:

$G_{ii} =$ sum of conductances at node i
$G_{ij} = -$(conductance connected directly between node i
      and node j) $= G_{ji}$
$J_i\quad =$ sum of current source entering into node i

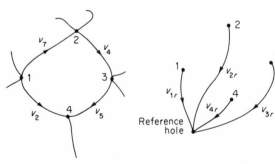

A typical loop   Voltage graph

Figure 6.18 In node analysis, KVL is automatically satisfied at every loop.

These equations are known as the node equations. They are KCL equations in terms of the node-to-reference voltages. Moreover, they give a complete description of the circuit. To see this, we note that we have used $n - 1$ KCL equations and the definitions of the elements. As to the KVL equations, they are satisfied at every loop automatically. Consider Figure 6.18, which shows a typical loop. The sum of the voltages around the loop is

$$v_7 + v_4 + v_5 - v_2 = (v_{1r} - v_{2r}) + (v_{2r} - v_{3r}) + (v_{3r} - v_{4r}) - (v_{1r} - v_{4r}) = 0$$

This is true for all loops, and we conclude that the definition of the node-to-reference voltages guarantees that the KVL equations are satisfied everywhere.

Example 6.6   Find the current I in the circuit of Figure 6.19.

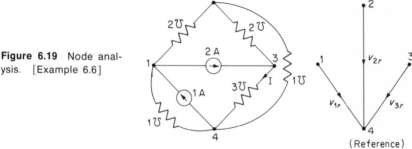

Figure 6.19 Node analysis. [Example 6.6]

Solution   Choose the reference node and number the other nodes as shown. Define the node-to-reference voltages as indicated. The node equations are

$$\begin{bmatrix} 3 & -2 & 0 \\ -2 & 5 & -2 \\ 0 & -2 & 5 \end{bmatrix} \begin{bmatrix} v_{1r} \\ v_{2r} \\ v_{3r} \end{bmatrix} = \begin{bmatrix} -1 \\ 0 \\ 2 \end{bmatrix}$$

Solving for v_{3r}, we get

$$v_{3r} = \frac{18}{43} \text{ V}$$

The current I is given by

$$I = 3v_{3r} = \frac{54}{43} \text{ A} \qquad \blacksquare$$

We now make an interesting observation. The node equations of this example are mathematically the same as the mesh equations of Example 6.4. The variables in this example are the node voltages, and the variables in Example 6.4 are the mesh currents. The two sets of variables satisfy the same equations and hence have the same numerical values. We further note that the branch *voltages* in the case of *node* equations and the branch *currents* in the case of the *mesh* equations are expressed in the same ways in terms of the variables of the equations. It follows that the branch voltages in the node analysis are numerically the same as the branch currents in the mesh analysis. Dualism is present again. But an interesting question is: How are the two circuits topologically related in order that the mesh equations of one are the node equations of the other and in order that the branch currents of one have the same numerical values as the branch voltages of the other? The two circuits are known as *duals* of each other. We shall answer these questions in Section 6.9.

Example 6.7   A generator in the form of an imperfect current source supplies power to three electroplating tanks. The first tank requires a current of 1 A, the second 2 A, and the third 3 A. The circuit of the system is shown in Figure 6.20 in which the resistors represent the resistances of the buses that connect the tanks to the generator. The tanks are represented by current sources of the appropriate values. What is the minimum value of the generator current in order that none of the voltages across the tanks be less than 100 V?

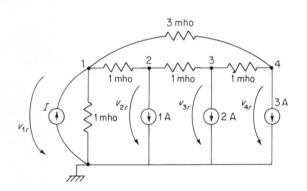

Figure 6.20 A system of electroplating tanks represented by current sources. [Example 6.7]

Solution   Number the nodes and define the node voltages as shown. With reference to Figure 6.20, the node equations are

$$
\begin{bmatrix}
5 & -1 & 0 & -3 \\
-1 & 2 & -1 & 0 \\
0 & -1 & 2 & -1 \\
-3 & 0 & -1 & 4
\end{bmatrix}
\begin{bmatrix}
v_{1r} \\ v_{2r} \\ v_{3r} \\ v_{4r}
\end{bmatrix}
=
\begin{bmatrix}
I \\ -1 \\ -2 \\ -3
\end{bmatrix}
$$

Solving for the unknowns, we get

$$
\begin{aligned}
v_{1r} &= I - 6.0 \\
v_{2r} &= I - 7.8 \\
v_{3r} &= I - 8.6 \\
v_{4r} &= I - 7.4
\end{aligned}
$$

The minimum voltage is v_{3r}. Setting it to 100 V yields $I = 108.6$ A, and we find the other voltages to be

$$v_{1r} = 108.6 - 6.0 = 102.6 \text{ V}$$
$$v_{2r} = 108.6 - 7.8 = 100.8 \text{ V}$$
$$v_{4r} = 108.6 - 7.4 = 101.2 \text{ V}$$

As a check, let us sum the voltages around the loop of resistors. We have

$$v_{1r} - v_{2r} = 1.8 \text{ V}$$
$$v_{2r} - v_{3r} = 0.8 \text{ V}$$
$$v_{3r} - v_{4r} = -1.2 \text{ V}$$
$$v_{4r} - v_{1r} = -1.4 \text{ V}$$

The sum of these quantities is zero. ∎

6.8 Circuits Containing Resistors, Inductors, and Capacitors

The loop, mesh, and node equations given in the last several sections were written for circuits containing resistors and sources only. The extension to circuits containing the three types of elements — resistor, inductor, and capacitor — is straightforward.

The KVL and KCL equations are independent of the type of elements. In the more general case, the terminal equations would include those for the inductors and capacitors. Specifically, in addition to the terminal equations for the resistors of the form

$$v = Ri \quad \text{or} \quad i = Gv$$

we have, for the inductors, relations of the form

$$v = \frac{Ldi}{dt} = LDi \quad \text{or} \quad i = \frac{1}{LD}v \qquad (6.67)$$

and for the capacitors, relations of the form

$$i = \frac{Cdv}{dt} = CDv \quad \text{or} \quad v = \frac{1}{C}\left(\frac{1}{D}\right)i \qquad (6.68)$$

The symbol D is the derivative operator defined by

$$D(x) = \frac{dx}{dt}$$

and the inverse operator $1/D$ is

$$\frac{1}{D}(x) = \int_t x(y)\, dy$$

It is seen that in terms of the operators, the current-voltage relations for the inductors and capacitors are in the same form as the current-voltage relations for the resistors. In place of R we have LD and $1/CD$, and in place of g we have $1/LD$ and CD. As a result, we can deduce immediately the form of the loop and mesh equations for the general case as follows.

Loop equations now take the form:

$$Z_{11}(D)i_{c1} + Z_{12}(D)i_{c2} + \cdots Z_{1m}(D)i_{cm} = E_1$$
$$Z_{21}(D)i_{c1} + Z_{22}(D)i_{c2} + \cdots Z_{2m}(D)i_{cm} = E_2$$
$$\vdots$$
$$Z_{m1}(D)i_{c1} + Z_{m2}(D)i_{c2} + \cdots Z_{mm}(D)i_{cm} = E_m \qquad (6.69)$$

where $Z_{ij}(D) = R_{ij} + L_{ij}D + \dfrac{1}{C_{ij}D}$, with

$R_{ii} = $ the sum of resistances in loop i
$L_{ii} = $ the sum of inductances in loop i
$C_{ii} = $ the equivalent series capacitance in loop i
$R_{ij} = \pm($the sum of resistances shared by loops i and $j)$
$L_{ij} = \pm($the sum of inductances shared by loops i and $j)$
$C_{ij} = \pm($the equivalent series capacitance shared by
            loops i and $j)$
$E_i = $ the sum of the voltages sources in loop i

The algebraic sign of an off-diagonal term is "+" if both loop currents go through the element in the same direction; otherwise, the sign is "−."

As an example, consider the circuit of Figure 6.21. Select the tree as shown. We have

$$\left(3 + \frac{1}{2D}\right)i_{c1} - \frac{1}{2D}i_{c2} - \left(2 + \frac{1}{2D}\right)i_{c3} = e_1 - e_2$$

$$-\frac{1}{2D}i_{c1} + \left(D + \frac{1}{D}\right)i_{c2} + \frac{1}{D}i_{c3} = 0$$

$$-\left(2 + \frac{1}{2D}\right)i_{c1} + \frac{1}{D}i_{c2} + \left(\frac{2}{D} + 2D + 3\right)i_{c3} = e_2$$

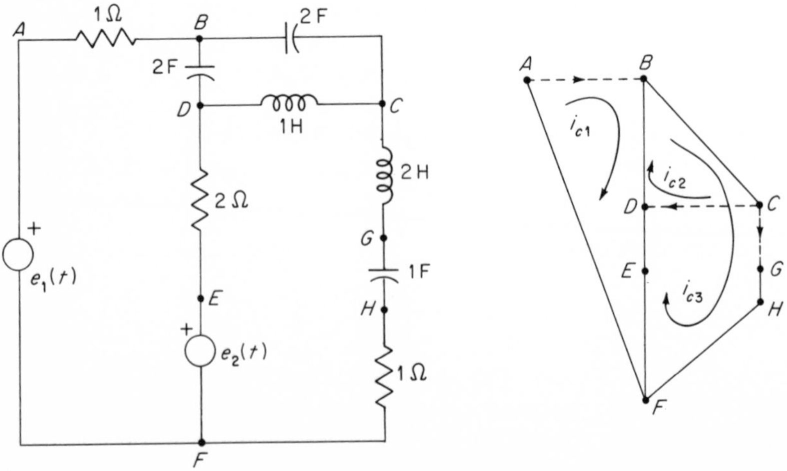

Figure 6.21 Loop equations in a circuit containing R, L, and C elements.

The set of loop equations is a set of simultaneous integral-differential equations with constant coefficients in the unknowns i_{c1}, i_{c2}, and i_{c3}, the loop currents. We shall discuss the solution of these equations later.

For mesh equations, we have, similar to the loop equations, the following:

$$Z_{11}(D)i_{m1} + Z_{12}(D)i_{m2} + \cdots + Z_{1m}(D)i_{mm} = E_1$$
$$Z_{21}(D)i_{m1} + Z_{22}(D)i_{m2} + \cdots + Z_{2m}(D)i_{mm} = E_2$$
$$\vdots$$
$$Z_{m1}(D)i_{m1} + Z_{m2}(D)i_{m2} + \cdots + Z_{mm}(D)i_{mm} = E_m \qquad (6.70)$$

where $Z_{ij}(D) = R_{ij} + L_{ij}D + \dfrac{1}{C_{ij}D}$, with

R_{ii} = the sum of the resistances in mesh i
L_{ii} = the sum of the inductances in mesh i
C_{ii} = the equivalent series capacitance in mesh i
R_{ij} = −(the resistance shared by mesh i and mesh j)
L_{ij} = −(the inductance shared by mesh i and mesh j)
C_{ij} = −(the capacitance shared by mesh i and mesh j)
E_i = the sum of the voltage sources in mesh i

As an example, when all the mesh currents are assigned the same orientation, the mesh equations of the circuit of Figure 6.21 are

$$\left(3 + \frac{1}{2D}\right)i_{m1} - \frac{1}{2D}i_{m2} - 2i_{m3} = e_1 - e_2$$
$$-\frac{1}{2D}i_{m1} + \left(D + \frac{1}{D}\right)i_{m2} - Di_{m3} = 0$$
$$-2i_{m1} - Di_{m2} + \left(3 + 3D + \frac{1}{D}\right)i_{m3} = e_2$$

Again the set of equations is a set of simultaneous integral-differential equations with constant coefficients in the unknowns which are the mesh currents.

Finally, the node equations become, in the general case, the following:

$$Y_{11}(D)v_{1r} + Y_{12}(D)v_{2r} + \cdots + Y_{1q}(D)v_{qr} = J_1$$
$$Y_{21}(D)v_{1r} + Y_{22}(D)v_{2r} + \cdots + Y_{2q}(D)v_{qr} = J_2$$
$$\vdots$$
$$Y_{q1}(D)v_{1r} + Y_{q2}(D)v_{2r} + \cdots + Y_{qq}(D)v_{qr} = J_q \qquad (6.71)$$

where $Y_{ij}(D) = G_{ij} + C_{ij}D + \dfrac{1}{L_{ij}D}$, $q = n - 1$, with

G_{ii} = the sum of conductances connected to node i
C_{ii} = the sum of capacitances connected to node i

L_{ii} = the equivalent parallel inductance of all inductances connected to node i

G_{ij} = −(the conductance connected between node i and node j)

C_{ij} = −(the capacitance connected between node i and node j)

L_{ij} = −(the equivalent parallel inductance connected between node i and node j)

J_i = the sum of the current sources entering into node i

As an example, consider the circuit of Figure 6.22. The node equations are

$$(2 + D)v_{1r} - Dv_{2r} - 0v_{3r} - v_{4r} = j_1$$

$$-Dv_{1r} + \left(D + \frac{1}{D}\right)v_{2r} - \frac{1}{2D} v_{3r} - \frac{1}{2D} v_{4r} = 0$$

$$-0v_{1r} - \frac{1}{2D} v_{2r} + \left(4D + \frac{1}{2D}\right)v_{3r} - 2Dv_{4r} = 0$$

$$-v_{1r} - \frac{1}{2D} v_{2r} - 2Dv_{3r} + \left(1 + 2D + \frac{1}{2D}\right)v_{4r} = j_2$$

We see from Equations (6.69), (6.70), and (6.71) that the loop, mesh, and node equations can all be obtained by inspection. Moreover, as noted in Sections 6.4, 6.6, and 6.7, each of the three sets independently describes the circuit completely. On the other hand, these equations are not natural in the sense that the variables are artificial and are only indirectly related to the branch voltages and branch currents. The dynamic behavior of the circuit cannot be directly inferred from these equations. To solve for the unknowns, one would eliminate all but one of the unknowns and obtain a differential equation for this unknown. In general,

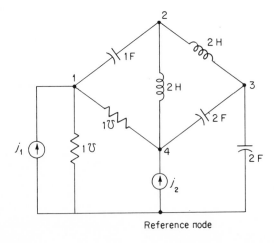

Figure 6.22 Node equations in a circuit containing R, L, C elements.

the differential equation will be of high order, say, n. The complete solution will contain n undetermined coefficients, which can be found by imposing the initial values of the zeroth, the first, and so on, up to the $(n - 1)$st derivatives of the unknown. But these initial values are, in general, not specified and must be found from the circuit in terms of the initial currents in the inductors and the initial voltages across the capacitors. It is the currents in the inductors and the voltages across the capacitors that are the natural variables of the circuit. We shall formulate equations in terms of these variables in Chapter 9 when we present formally the idea of *state variables* and *state equations*.

Nevertheless, the mesh equations, the loop equations, and the node equations do lead to a complete solution of the circuit, and for the special case in which the excitations are sinusoidal functions of time, they are attractive and convenient to use. More will be said about this case in Chapter 7. We shall now show by an example what we mean by saying that the mesh currents, for example, are not natural.

Example 6.8 Find $i_{m2}(t)$ for $t > 0$ in the circuit of Figure 6.23(a).

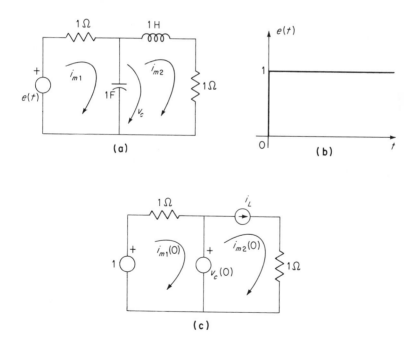

Figure 6.23 Illustration of the fact that mesh currents are not the natural variables to use in the description of the dynamics of a circuit. Their initial values must be expressed in terms of the inductor current and capacitor voltage. [Example 6.8]

Solution   The mesh equations are

$$\left(1 + \frac{1}{D}\right)i_{m1} - \frac{1}{D}i_{m2} = 1$$

$$-\frac{1}{D}i_{m1} + \left(1 + D + \frac{1}{D}\right)i_{m2} = 0$$

Solving for i_{m1} in terms of i_{m2} from the second equation, we get

$$i_{m1} = (D^2 + D + 1)i_{m2}$$

Substitute this relation into the first equation, and we get

$$(D^2 + 2D + 2)i_{m2} = 1$$

or

$$\frac{d^2 i_{m2}}{dt^2} + 2\frac{di_{m2}}{dt} + 2i_{m2} = 1$$

which is a second-order differential equation in i_{m2}. The characteristic equation is

$$\lambda^2 + 2\lambda + 2 = 0$$

and the eigenvalues are

$$\lambda_1 = -1 + j$$
$$\lambda_2 = -1 - j$$

The complementary solution is

$$i_{m2c} = c_1 e^{(-1+j)t} + c_2 e^{(-1-j)t}$$

where c_1 and c_2 are arbitrary constants. The particular integral, in this case, is simply

$$i_{m2p} = \tfrac{1}{2}$$

The complete solution is

$$i_{m2}(t) = c_1 e^{(-1+j)t} + c_2 e^{(-1-j)t} + \tfrac{1}{2}$$

The constants are determined from the initial conditions. At $t = 0$, the circuit takes the form shown in Figure 6.23(c). By inspection, we find

$$i_{m2}(0) = i_L(0)$$

The derivative of i_{m2} is related to the voltage across the inductor:

$$\frac{di_{m2}}{dt} = \frac{1}{L}v_L$$

From Figure 6.23(c), we find

$$\left.\frac{di_{m2}}{dt}\right]_{t=0} = v_c(0) - i_L(0)$$

We see that the initial values are given in terms of those of the inductor current and capacitor voltage. For the special case where $v_c(0) = 0$ and $i_L(0) = 0$, we find the constants to be

$$c_1 = -\frac{1}{2(1+j)}$$
$$c_2 = -\frac{j}{2(1+j)}$$

The solution is

$$i_{m2}(t) = \tfrac{1}{2} - \tfrac{1}{2}e^{-t}\cos t - \tfrac{1}{2}e^{-t}\sin t$$

To find $i_{m1}(t)$, we use

$$i_{m1} = (D^2 + D + 1)i_{m2}$$

and

$$i_{m1} = \tfrac{1}{2} + \tfrac{1}{2}e^{-t}\cos t - \tfrac{1}{2}e^{-t}\sin t$$

6.9 Duality

We have remarked on many occasions that dualism exists among circuit quantities. If we make a statement about a situation in a circuit, we can also make a "dual" statement about a "similar" situation in another circuit, provided that certain key words such as *voltage* and *current, loop* and *cut set* are interchanged. Thus, we write voltage equations for loops and current equations for cut sets. The voltage across an inductor is proportional to the derivative of the current, and the current in a capacitor is proportional to the derivative of the voltage, and so on. In addition, we have observed dualism among topological quantities as well. Thus, we speak of the fundamental cut sets with respect to a tree on one hand, and the fundamental loops with respect to a set of chords on the other. We write KVL equations for the meshes and KCL equations for the nodes. Let us now formalize these observations.

Let N and N' be two circuits of b branches and n nodes. Let the branch voltages and currents of N be $v_1, v_2,..., v_b$ and $i_1, i_2,..., i_b$, respectively. Let those of N' be $v_1', v_2',..., v_b'$ and $i_1', i_2',..., i_b'$, respectively. Now if it is possible to establish a one-to-one correspondence between branches of N and N' such that numerically

$$v_k = i_k'$$
$$i_k = v_k' \qquad (k = 1, 2,..., b) \tag{6.72}$$

then we say that N' is the dual of N, and by symmetry N is the dual of N'.

For example, the two circuits of Figure 6.24 are duals. The mesh equations of N are

$$\left(2 + D + \frac{1}{D}\right)i_{m1} - \left(D + \frac{1}{D}\right)i_{m2} = 5 \sin 6t$$

$$-\left(D + \frac{1}{D}\right)i_{m1} + \left(1 + 4D + \frac{1}{D}\right)i_{m2} = 0 \qquad (6.73)$$

On the other hand, the node equations of N' are

$$\left(2 + D + \frac{1}{D}\right)v_{1r} - \left(D + \frac{1}{D}\right)v_{2r} = 5 \sin 6t$$

$$-\left(D + \frac{1}{D}\right)v_{1r} + \left(1 + 4D + \frac{1}{D}\right)v_{2r} = 0 \qquad (6.74)$$

The mesh currents and the node voltages satisfy the same equations and, therefore, must have the same numerical values if the initial conditions are the same in both circuits. Moreover, the branch currents of N can be expressed as a linear combination of the mesh currents. The branch voltages in N' can be expressed as the *same* linear combination of the node voltages.

How does one construct the dual of a circuit? The scheme outlined below is what we must follow.

1. Corresponding to each mesh of N we assign a node in N'. In addition, we assign a node corresponding to the boundary mesh of N.
2. In N every element is shared by two meshes, or by a mesh and the boundary mesh. To each element shared by mesh i and mesh j in N, there corresponds an element connected between node i and node j in N'. The type and value of the element are obtained as in the following table.

N	N'
Voltage source $e(t)$	Current source $j'(t) = e(t)$
Resistor $R = r \ \Omega$	Resistor of conductance $G = r$ mhos
Inductor $L = l$ H	Capacitor $C' = l$ F
Capacitor $C = c$ F	Inductor $L' = c$ H

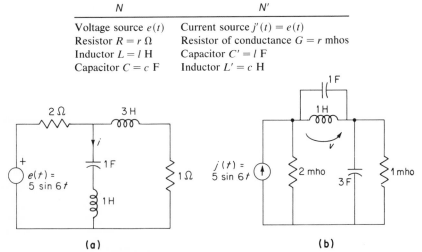

Figure 6.24   (a) A circuit; and (b) its dual.

3. The resulting circuit N' is the dual of N. Examination of Figure 6.24 shows that the dual circuit is obtained in a manner outlined above.

To show that the construction scheme indeed leads to the dual of a circuit, we must show that N' satisfies the definition of a dual. First, we note that there are as many nodes in N' as there are meshes in N. Now assign a clockwise direction to all meshes in N. Let the node corresponding to the boundary mesh be chosen as the reference node in N'.

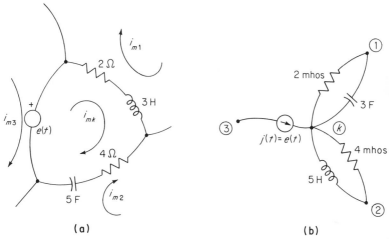

Figure 6.25 (a) A typical mesh; (b) its dual.

Consider a typical mesh in N as shown in Figure 6.25(a). By construction, the corresponding node in N' is shown in Figure 6.25(b). The mesh equation is

$$\left(6 + 3D + \frac{1}{5D}\right) i_{mk} - (2 + 3D)\, i_{m1} - \left(4 + \frac{1}{5D}\right) i_{m2} = e(t)$$

The node equation in N' is

$$\left(6 + 3D + \frac{1}{5D}\right) v'_{kr} - (2 + 3D) v'_{lr} - \left(4 + \frac{1}{5D}\right) v'_{2r} = j(t) = e(t)$$

We conclude that the mesh equations of N are the same as the node equations of N'. Moreover, the numerical values of the mesh currents and their corresponding node voltages in N' are the same. Now, let i_q be the branch current in element q which is shared by mesh p and mesh s. We have

$$i_q = i_{mp} - i_{ms}$$

The branch voltage v'_q of the corresponding element in N' is

$$v'_q = v'_{pr} - v'_{sr}$$

Since the mesh currents and the node voltages of N' have the same value, we conclude that the branch currents of N are the same as the branch voltages of N'. Next, we note that by construction, if the element q is a resistor, we have, in N,

$$v_q = R_q i_q$$

and in N',

$$i'_q = G'_q v'_q = R_q v'_q$$

If the element q is an inductor, we have, in N,

$$v_q = L_q D i_q$$

and in N',

$$i'_q = C'_q D v'_q = L_q D v'_q$$

If the element is a capacitor, we have, in N,

$$v_q = \frac{1}{C_q} \frac{1}{D} i_q$$

and in N',

$$i'_q = \frac{1}{L'_q} \frac{1}{D} v'_q = \frac{1}{C_q} \frac{1}{D} v'_q$$

Since i_q and v'_q have the same value, we see from the above that v_q and i'_q have the same value. We conclude, therefore, that the branch voltages in N and the branch currents of N' have the same value. By definition, N' is the dual of N.

In the construction of N', we speak of the meshes of N. In addition, we assume that the elements of the circuits are two-terminal elements. We conclude, therefore, that it is only meaningful to speak of the dual of a planar circuit with two-terminal elements. The dual of a nonplanar graph or the dual of a multiterminal element does not exist.

Example 6.9 Find the dual of the circuit of Figure 6.26(a). Suppose the current in the inductor is $5 \sin 6t$ for $t \geq 0$. Find the current in the corresponding branch in the dual.

Solution The dual is shown in Figure 6.26(b). The voltage across the capacitor is the same as the current in the inductor·in the circuit of Figure 6.26(a). The current in the capacitor is

$$i'_c(t) = C \frac{dv'}{dt} = (3)(5)(6) \cos 6t = 90 \cos 6t$$

Example 6.10 Repeat the problem of Example 6.9 for the circuit of Figure 6.26(c).

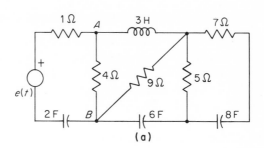

(a)

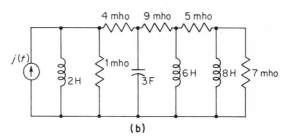

Figure 6.26 (a) A circuit;
(b) its dual; (c) another circuit whose dual is also (b).
[Example 6.9]

(b)

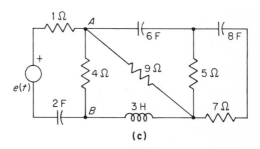

(c)

Solution By construction, the dual of (c) is the same as that of (a). The point to
note is that two seemingly different circuits have the same dual. Stated in a different way, the dual of a circuit may not be unique. If in (a), we disconnect the
circuit at *AB* and turn the part on the right 180° and reconnect the circuit, we get
the circuit (c). It should be noted that the two circuits have the same mesh equations but different node equations.   ∎

We now summarize the dual quantities in the following table.

Circuit *N*	Dual of *N*
Branch voltage	Branch current
Mesh current	Node-to-reference voltage
KCL	KVL
Cut sets	Loops
Mesh equations	Node equations
Series connection	Parallel connection
Resistor	Resistor
Capacitor	Inductor
Voltage source	Current source
Short circuit	Open circuit

6.10 Circuits Containing Dependent Sources

The methods of loop, mesh, and node analyses will now be extended to circuits containing dependent sources. We recall that a dependent source is one whose value is proportional to the voltage or current at some part of the circuit other than the terminals of the source. To write the circuit equations, we proceed as before, first treating the dependent sources as if they were independent sources, then expressing their values in terms of the variables of the equations. The variables are the loop currents, mesh currents, or the node voltages, depending on the method of analysis. These techniques are illustrated in the following three examples.

Example 6.11 The transistor is a three-terminal device which under certain conditions has an equivalent circuit shown in Figure 6.27. The letter "b" designates

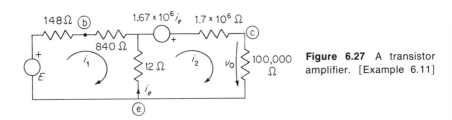

Figure 6.27 A transistor amplifier. [Example 6.11]

the "base" terminal, "c" the collector, and "e" the emitter. Typical values of the elements are as shown. Find the gain V_0/E of the circuit.

Solution Writing mesh equations for the mesh currents i_1 and i_2 as shown in Figure 6.27, we have, approximately,

$$1000i_1 - 12i_2 = E$$
$$-12i_1 + 1,800,000i_2 = 1,670,000i_e$$

The dependent voltage source is now expressed in terms of the mesh currents by noting

$$i_e = i_2 - i_1$$

Substituting this into the mesh equations and simplifying, we get, approximately,

$$1000i_1 - 12i_2 = E$$
$$1,670,000i_1 - 130,000i_2 = 0$$

Solving for i_2, we get

$$i_2 = \frac{E}{66}$$

and

$$V_0 = 100,000i_2 = 1515E \quad \text{or} \quad V_0/E = 1515$$

Example 6.12 The circuit of Figure 6.28 is a differential amplifier whose output is proportional to the difference of the two input currents. Find the proportionality constant.

Solution We first note the symmetry of the circuit. Define I_s and I_u as, respectively, the sum and difference of I_A and I_B:

$$I_s = I_A + I_B$$
$$I_u = I_A - I_B$$

or

$$I_A = \frac{1}{2}(I_s + I_u) = \frac{I_s}{2} + \frac{I_u}{2}$$
$$I_B = \frac{1}{2}(I_s - I_u) = \frac{I_s}{2} - \frac{I_u}{2}$$

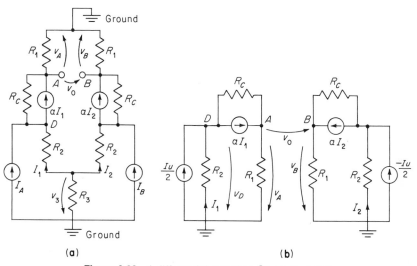

Figure 6.28 A differential amplifier. [Example 6.12]

If we apply the symmetrical excitations $I_s/2$ to the circuit, $V_A = V_B$ and therefore $V_0 = 0$. The only response results from the antisymmetric excitation $I_A = I_u/2$ and $I_B = -I_u/2$. By superposition and symmetry, $V_A = -V_B$ and $V_3 = 0$. Therefore, V_A can be calculated from the circuit of Figure 6.28(b). Writing node equations, we get

$$(G_2 + G_c)V_D - G_cV_A = I_u/2 - \alpha I_1$$
$$-G_cV_D + (G_c + G_1)V_A = \alpha I_1$$

where the G's are the conductances of the resistors. We must express I_1 in terms of V_A and V_D. By inspection of Figure 6.28(b), we have

$$I_1 = -G_2V_D$$

Substituting this into the node equations and solving for V_A, we get

$$V_A = \frac{1}{2} \frac{G_c - \alpha G_2}{G_c G_1 + G_c G_2 + (1 - \alpha) G_2 G_1} I_u$$

Noting that $I_u = I_A - I_B$ and $V_A = -V_B$ under the antisymmetric excitation, we find

$$V_0 = \frac{G_c - \alpha G_2}{G_c G_1 + G_c G_2 + (1 - \alpha) G_2 G_1} (I_A - I_B)$$

The difference amplifier is useful as a *comparator*. One of the inputs, say I_B, is a reference source. If I_A is the same as I_B, the output is zero; otherwise the output is finite.

Example 6.13 An ideal voltmeter is one that gives a visual indication of the voltage existing between two points without disturbing the circuit being measured in any way. A realistic voltmeter, however, must draw a small but finite amount of current from the circuit in order to convert the electrical quantity into an observable quantity, for example, a deflection of the needle in a meter. A circuit that "isolates" the circuit being measured from the measuring circuit is shown in Figure 6.29. Calculate the current i_x in terms of the voltage v_x; that is, find the input resistance of the circuit.

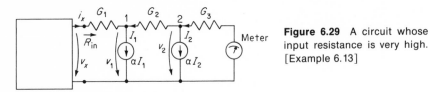

Figure 6.29 A circuit whose input resistance is very high. [Example 6.13]

Solution Number the nodes as shown. Treating i_x as a source for the moment, we have the following node equations:

$$G_2 v_1 - G_2 v_2 = i_x - \alpha I_1$$
$$-G_2 v_1 + (G_2 + G_3) v_2 = -\alpha I_2$$

Expressing the dependent currents and i_x in terms of v_x, v_1, and v_2, we get

$$i_x = (v_x - v_1) G_1$$
$$I_1 = G_2 (v_2 - v_1)$$
$$I_2 = -G_3 v_2$$

Making substitutions and solving for v_1, we find

$$v_1 = \frac{G_1 G_2 + G_1 G_3 (1 - \alpha)}{G_1 G_2 + G_1 G_3 (1 - \alpha) + G_2 G_3 (1 - \alpha)^2} v_x$$

and

$$i_x = (v_x - v_1) G_1 = \frac{(1 - \alpha)^2}{R_3 + R_2 (1 - \alpha) + R_1 (1 - \alpha)^2} v_x$$

Thus by choosing α to be close to unity, the current i_x can be made very small. The input resistance is very large, and the circuit is effectively isolated from the meter.

6.11 Circuits Containing Multiterminal Elements

Multiterminal elements are characterized by a set of equations relating the voltages and currents at the terminals. When these elements are imbedded in a circuit, the mesh, loop, or node equations of the circuit are written in the usual way except that the voltages and currents at the terminals of the multiterminal elements are first treated as sources and then replaced by the equations that define the elements. The following two examples illustrate the technique.

Example 6.14 Write a set of mesh equations for the circuit shown in Figure 6.30.

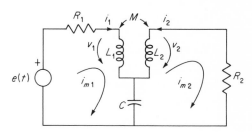

Figure 6.30 Mesh equations in a circuit containing couple coils. [Example 6.14]

Solution Assign mesh currents as shown. We first regard v_1 and v_2 as voltage sources and write the mesh equations given by

$$\left(R_1 + \frac{1}{DC}\right)i_{m1} - \frac{1}{DC}i_{m2} = e - v_1$$

$$-\frac{1}{DC}i_{m1} + \left(\frac{1}{DC} + R_2\right)i_{m2} = v_2$$

The coupled coil is characterized by

$$v_1 = DL_1 i_1 + DM i_2 = DL_1 i_{m1} - DM i_{m2}$$
$$v_2 = DM i_1 + DL_2 i_2 = DM i_{m1} - DL_2 i_{m2}$$

Substituting these into the mesh equations, we get

$$\left(R_1 + \frac{1}{DC} + DL_1\right)i_{m1} - \left(\frac{1}{DC} + DM\right)i_{m2} = e$$

$$-\left(\frac{1}{DC} + DM\right)i_{m1} + \left(\frac{1}{DC} + R_1 + DL_2\right)i_{m2} = 0$$

which is a system of integral-differential equations in the unknowns i_{m1} and i_{m2}. In Chapter 7, we shall discuss its solution for a special class of excitations.

Example 6.15   (Linear Sweep) The circuit of Figure 6.31 is used to generate a voltage which varies linearly with time when the input is a step function. The three-terminal element is characterized by the following terminal equations:

$$i_1 = g_{11}v_1 + g_{12}v_2$$
$$i_2 = g_{21}v_1 + g_{22}v_2$$

Find the output voltage $v_2(t)$ and obtain the condition on the element values such that $v_2(t)$ varies linearly with time.

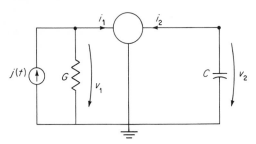

Figure 6.31 A circuit containing a three-terminal device whose parameters may be chosen so that the output voltage is proportional to time, when the input is a step. [Example 6.15]

Solution   Treating the currents i_1 and i_2 as sources for the moment, we have the following node equations:

$$Gv_1 + 0v_2 = j - i_1$$
$$0v_1 + DCv_2 = -i_2$$

Substituting the expressions for i_1 and i_2 into the above, we get

$$(G + g_{11})v_1 + g_{12}v_2 = j$$
$$g_{21}v_1 + (g_{22} + DC)v_2 = 0$$

Solving for v_2, we obtain a differential equation on v_2:

$$C(G + g_{11})\frac{dv_2}{dt} + (g_{11}g_{22} - g_{12}g_{21} + Gg_{22})v_2 = -g_{21}$$

The solution to this equation, subject to the initial condition that $v_2(0) = 0$, is

$$v_2(t) = \frac{-g_{21}}{g_{11}g_{22} - g_{12}g_{21} + Gg_{22}}\,(1 - e^{-t/T})$$

where

$$T = \frac{C(G + g_{11})}{g_{11}g_{22} - g_{12}g_{21} + Gg_{22}}$$

The voltage $v_2(t)$ will vary linearly with t if it is possible to select elements such that $T = \infty$, or

$$\frac{g_{21}g_{21}}{(G + g_{11})g_{22}} = 1$$

6.12　Summary

We have presented a number of systematic procedures for writing circuit equations. Some are applicable to all circuits, and some are suitable for hand computation. It is well to summarize our observations.

　　1. For computer solution, use the tree or chord analysis. The method is applicable to any circuit containing two-terminal elements.

　　2. For hand computation, we refer to the following table in which E stands for voltage sources and J for current sources.

Circuit	Sources		
	E only	J only	E and J
Planar	Mesh	Node	Tree or chord
Nonplanar	Loop	Node	Tree or chord

Thus, if a circuit is nonplanar and contains voltage sources only, one should use loop equations. When both voltage and current sources are present, one should use tree or chord analysis. However, in some instances, the sources are nonideal, and they can all be converted into voltage or current sources. (See Section 3.8.)

　　3. When dependent sources or multiterminal elements are present, the equations are written in the same way except that the terminal voltages are regarded as sources when we write mesh or loop equations, and the terminal currents are regarded as sources when we write node equations.

SUGGESTED EXPERIMENTS

1. Construct a resistor circuit with several voltage sources. Measure all the branch voltages and currents. Check your results by hand calculation or by using the computer program RNET.
2. Construct a planar circuit and its dual. Verify that the branch currents in one have the same value as the branch voltages in the other. Check your answers by using the computer program RNET.

COMPUTER PROGRAMMING

Computer Analysis of a General Resistor Network—RNET*

RNET[2] is a computer program for the numerical analysis of linear, resistive circuits excited by independent voltage and current sources. Written in FORTRAN IV for the IBM 360 system, this program computes, in single precision, all branch voltages and currents for up to 50

---

[2] This program was written by Andrew Gaspar in 1968, then a junior in the Department of Electrical Engineering, Columbia University.

elements. A ground or reference node is chosen if not already specified, and as a check, KCL is computed at this reference node. A complete listing of the program is given at the end of the chapter. The program contains many "comment" statements which explain the functions of the various parts of the program. Essentially it is based on the tree-chord analysis of Section 6.1.

Operation   The *input* of the program consists of the topological description of the circuit. The *output* of the program consists of the following:

1. voltage across each branch (E12.5 format)
2. current through each branch (E12.5 format)
3. computation of KCL at reference node (E20.10 format)
4. an error message diagnosing the trouble, since the program is equipped with numerous checks for errors in format or topology

Instructions for use   The input parameters are:

N – the number of nodes contained in the network
B – the number of branches contained in the network
V – the number of voltage sources contained in the network
C – the number of current sources contained in the network
R – the number of resistors contained in the network
GR – the node number of the reference. This node is used by the program to perform a KCL check. If not specified, the program will choose as a reference that node which joins the most branches.

1. *Numbering nodes*   Number all nodes in the network in random order using consecutive numbers 1 to N.
2. *Numbering branches*   In order to keep track of all of the elements, each branch (element) will be given a number 1 to B. This time the numbering is not completely arbitrary. The voltage sources must be numbered first, then the resistors, and then the current sources. This sequencing must be kept even if one of the three is not present (that is, if there were no voltage sources, then the resistors would be numbered 1 to R, and then the current sources $R + 1$ to B).
3. *Entering data*   All input data must be entered in the format indicated, and the cards must appear in proper order.

a. **FIRST CARD** (All values are integer)

Columns	1–2	5–6	9–10	13–14	17–18	21–22	23–60	61–80
Parameter name	N	B	R	V	C	GR	Blank	Name

N, B, R, V, C, and GR were previously defined. In 61–80 enter either your name or the program name.

b. SUBSEQUENT CARDS

Columns	1–2	3–4	11–20	26–35	41–50
	From node	To node	Voltage sources	Resistors	Current sources

The first two fields serve to define current directions as well as to determine where the element is connected. Integer values are used for the node numbers. The value of the voltage source, resistor source, or current source is entered in the appropriate columns. These values are expressed in real (floating point) numbers and may be entered anywhere in the field. The real number must include a decimal point (that is, 1.0, 3.14, 5.).

The elements must be entered in the same order that they were numbered. *Each and every element must be entered on a separate card.* One last instruction: The polarity of the source must be entered as below.

	From node	To node	Source value
$+$ $\bigcirc$ 6V 01 ── 02	01	02	$+6.0$ V
$+$ $\bigcirc$ 12V 07 ── 11	11	07	-12.0 V
$\bigcirc$ 2A 14 ← 02	02	14	$+2.0$ A
$\bigcirc$ 3A 19 ← 18	19	18	-3.0 A

4. *Multiple data sets*   As many problems as desired can be run at the same time. Simply punch up all the data for each problem according to the prescribed format and place them one after another in the appropriate place. The program will automatically identify and process each problem. The last card of the data set must be a blank card. Only *one* blank card must appear, and it is to be at the end of the data set.

5. *Examples*   Two examples are given at the end of the program listing. The computer outputs are also shown. Note that the sum of currents at the reference node in one case is small, but not identically zero. (Why?)

PROBLEMS

6.1   *Exercise*   Find the voltages v_1 and v_2 in the following circuit.

Problem 6.1 Mesh equations.

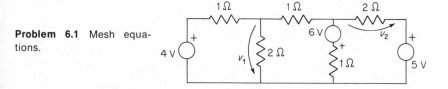

6.2   *Exercise*   The voltage v_x is required to be 5 V. Find the value of R_1 that is necessary.

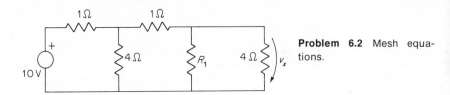

Problem 6.2   Mesh equations.

6.3   *Exercise*   Find the voltages v_{AB}, v_{BC}, and v_{CA} in the following circuit.

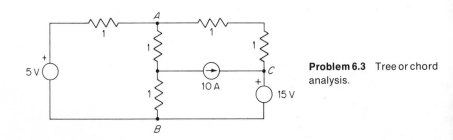

Problem 6.3   Tree or chord analysis.

6.4   *Exercise*   What must be the value of E such that $v_{AB} = 5$ V in the following circuit?

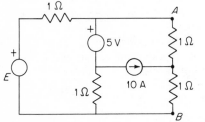

Problem 6.4   Tree or chord analysis.

6.5   *Exercise*   Select a set of values for R_1, R_2, and R_3 such that $i_1 : i_2 : i_3 = 9 : 3 : 1$.

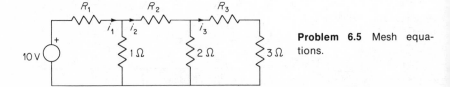

Problem 6.5   Mesh equations.

6.6   *Exercise* Find v_{AB} in the following figure. Use **RNET** to check your results.

Problem 6.6   Node equations.

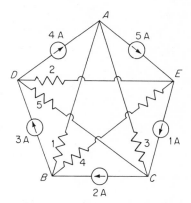

6.7   *Exercise* Find i_1 and i_2 in the following figure. Use **RNET** to check your results.

Problem 6.7   Loop equations.

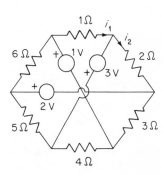

6.8   *Bridged-T* Show that the resistance of the circuit to the right of AB is R and that $V_0/E = \frac{1}{2} R/(R + R_1)$. Find the Thévenin equivalent circuit to the left of CD.

Problem 6.8   A bridged-*T* circuit.

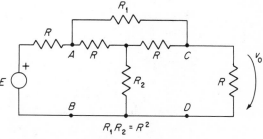

6.9 *Bridged-T* Show that V_0/E is given by $V_0/E = \frac{1}{2} \pi_i^N 1/(1 + R_{ai}/R)$.

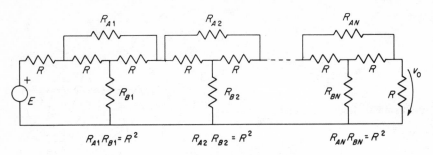

Problem 6.9 A cascade of bridged-*T* circuits.

6.10 *Design* Using the bridged-*T* structure of Problem 6.9, design a circuit with 5 taps as shown below such that $v_1 = 1$ V, $v_2 = 2$ V, $v_3 = 3$ V, $v_4 = 4$ V, and $v_5 = 5$ V.

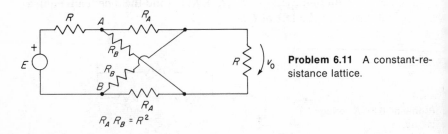

Problem 6.10 A simple design problem.

6.11 *Lattice* Shown below is a constant-resistance lattice. Show that the resistance of the circuit to the right of *AB* is *R* and that $V_0/E = \frac{1}{2}(R - R_A)/(R + R_A)$. See also Problem 6.29.

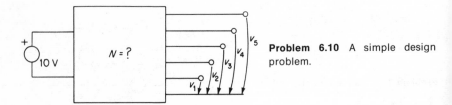

Problem 6.11 A constant-resistance lattice.

6.12 *Lattice* Show that V_0/E is given by $V_0/E = \frac{1}{2} \pi_i^N (R - R_{Ai})/(R + R_{Ai})$.

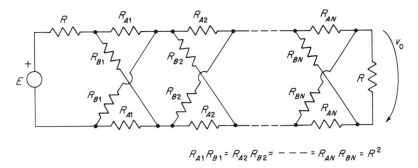

$$R_{A1}R_{B1} = R_{A2}R_{B2} = \; - \; - \; - \; = R_{AN}R_{BN} = R^2$$

Problem 6.12 A cascade connection of lattices.

6.13 *Design* Using the lattice structure of Problem 6.12, design a circuit with 10 taps as shown below such that $v_1 = 2^{-1}$, $v_2 = 2^{-2}$,..., and $v_{10} = 2^{-10}$.

Problem 6.13 A simple design problem.

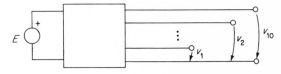

6.14 *Exercise* Calculate the total power delivered to the circuit by the current source.

Problem 6.14 Node equations and calculation of power.

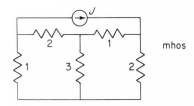

mhos

6.15 *Exercise* Find the value of the conductance such that the current i_x is 5 A. (Compare with Problem 6.2.)

Problem 6.15 Node equations and duality.

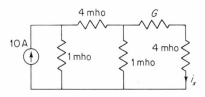

6.16 *Exercise* Find i_1, i_2, and i_3 in the following figure. Use RNET to check your answer. See also Problem 6.30.

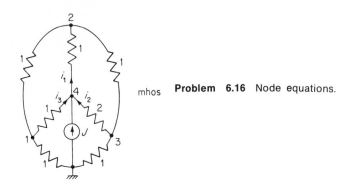

mhos   **Problem 6.16** Node equations.

6.17 *Exercise* Find $v(t)$. The input current is a step, and the initial voltage on the capacitor is zero.

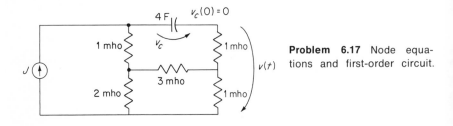

Problem 6.17 Node equations and first-order circuit.

6.18 *Exercise* Find $v(t)$. The input current is a step. Assume all initial conditions are zero.

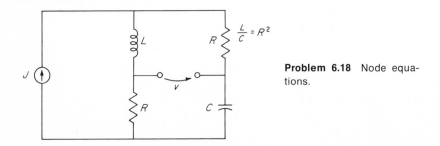

Problem 6.18 Node equations.

6.19 *Exercise* Find the value of R such that the following circuit is **a.** overdamped, **b.** underdamped, and **c.** critically damped. See also Problem 6.33.

Problem 6.19 Mesh equations in a second-order circuit.

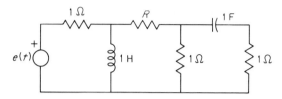

6.20 *Exercise* Find the power in R_2.

Problem 6.20 Calculation of power.

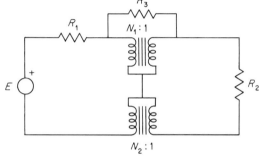

6.21 *Design* The equivalent circuit of a transistor is shown below. Find the value of R_1 and R_2 such that the gain is 100.

Problem 6.21 Design of an amplifier.

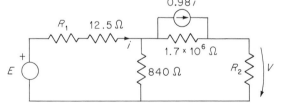

6.22 *Application* A fault occurred somewhere along a 20-mi cable, which has a resistance of 2 Ω/mi. To locate the fault, the following experiment was performed: A 90-V battery was connected at one end of the cable and an ammeter at the other. The reading of the ammeter was 2 A. The current in the battery was 4 A. Find the location and the resistance of the fault.

6.23 *Application* Show that in a circuit consisting of resistors, the input resistance between any two nodes increases if the resistance of any of the resistors is increased.[3] (*Hint:* Write the loop equations for the circuit and express the input resistance in terms of the resistance of any one resistor.)

6.24 *Application* Show that in a circuit consisting of resistors, the input resistance between any two nodes decreases if a short circuit is placed across any other pair of nodes. (See Problem 6.23.)

6.25 *Application* Show that in a circuit consisting of resistors, the input resistance between any two nodes increases if any resistor is removed from the circuit. (See Problem 6.23.)

6.26 *Application* A circuit has four nodes. Each node is connected to the other three by a resistor. The six resistors have conductances G_{12}, G_{13}, G_{14}, G_{23}, G_{24}, and G_{34}, respectively. Show that the input conductance between nodes 1 and 2 lies between the limits $G_{12} + (G_{13}^{-1} + G_{23}^{-1})^{-1} + (G_{14}^{-1} + G_{24}^{-1})^{-1}$ and $G_{12} + [(G_{13} + G_{14})^{-1} + (G_{23} + G_{24})^{-1}]^{-1}$.

6.27 *Application* The equivalent circuit of a transistor amplifier is as shown. Find the power gain $-V_0 I_2 / EI$.

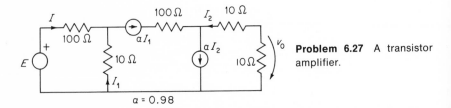

Problem 6.27 A transistor amplifier.

$a = 0.98$

6.28 *Application* A circuit that approximates an integrator is shown. Show that the output voltage is nearly proportional to the integral of the input voltage. How should you select the elements to make the circuit as good an approximation as possible?

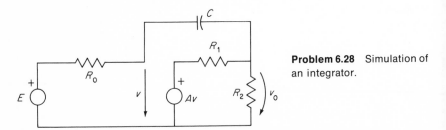

Problem 6.28 Simulation of an integrator.

6.29 *Duality* Find the dual of the lattice of Problem 6.11. What can you say about it? Can you find other circuits whose duals are topologically the same as the original circuits?

---

[3] Lord Rayleigh, "On the Theory of Resonance," *Phil. Trans.*, 1871, p. 77.

6.30 *Duality* Find the dual of the circuit of Problem 6.16. Use RNET to verify that the branch voltages and currents in the two circuits correspond as expected.

6.31 *Duality* Construct a circuit which is the dual of that of Problem 6.4. Compute the branch voltages and currents in the two circuits. What is the dual of v_{AB} of Problem 6.4?

6.32 *Duality* What can you say about Tellegen's theorem in the dual of a circuit?

6.33 *Duality* Find the dual of the circuit of Problem 6.19. What are the answers to Parts a, b, c of the problem?

6.34 *Duality* What can you say about the dual of a circuit containing dependent sources? Obtain the dual of the circuit of Figure 6.27.

Program RNET

```
C   THIS IS A PROGRAM FOR THE NUMERICAL ANALYSIS  OF LINEAR, RESISTOR NETWORKS
C   DRIVEN BY  INDEPENDENT VOLTAGE AND CURRENT SOURCES AND CONSISTING OF NO MORE
C   THAN 50 ELEMENTS.   THE PROGRAM CHOOSES A REFERENCE NODE IF NOT GIVEN ONE AND
C   AS A CHECK COMPUTES 'KCL' AT THAT NODE.   COMMENTS ARE PROVIDED THROUGHOUT
C   THE PROGRAM TO EXPLAIN THE PURPOSE OF EACH SECTION.  COMPUTATION IS IN
C   SINGLE PRECISION.
C
C   STEPS FOR INPUT SHOULD BE AS FOLLOWS,
C                  1. NUMBER ALL NODES IN RANDOM ORDER.
C                  2. NUMBER ALL ELEMENTS IN THE FOLLOWING SEQUENCE -
C                     VOLTAGE SOURCES,RESISTORS,CURRENT SOURCES.
C                  3. DEFINE CURRENT DIRECTIONS IN THE NETWORK.
C                  4. ENTER DATA AS FOLLOWS -
C
C                  FIRST CARD     (ALL VALUES ARE INTEGERS)
C
C                  COLUMN->   | 1 2| 5 6|9 10|1314|1718|21  22|23  60|61  80|
C                             | N |  B |  R | V  |  C |GROUND|BLANK | NAME |
C
C                  SUBSEQUENT CARDS (FIRST TWO FIELDS ARE INTEGER, OTHERS ARE REAL)
C
C                  COLUMN->   | 1-2| 3-4| 11-20 |  26-35  | 41-50 |
C                             |FROM| TO |VOLTAGE|RESISTORS|CURRENT|
C                             |NODE|NODE|SOURCES|         |SOURCES|
C                  5. EACH ELEMENT MUST HAVE ITS OWN CARD OR LINE.
C    N=NUMBER OF NODES
C    B=NUMBER OF BRANCHES
C    R=NUMBER OF RESISTORS
C    V=NUMBER OF VOLTAGE SOURCES
C    C=NUMBER OF CURRENT SOURCES
C    GROUND=NUMBER OF REFERENCE NODE
C
C
C        SAMPLE PROBLEM:
C
C             J6    2.0
C             JJJJ
C         +_____J<-J_____+
C         |     JJJJ      |
C         |              |
C         |     VV       |
C         |     V V    |2)
C      1)*_____V V_____*___+
C         | E1 V V     |   |
C         | 12.0 VV    |   |
C         |           |   |
C         -          / VVV
C         /         R3 - V  V E2
C         - R5      / V  V 2.0
C         / 3.0   4.0 - VVV
C         |           |  |
C         |   R4      |  |
C         |   7.0     |  |
C         *_____/|/|_____*___+
C         4)          3)
C
C  ...........................................................
C
C        INPUT:
C  4      6      3    2     1                              SAMPLE PROBLEM
C1 2            -12.0
C2 3             2.0
C2 3                          4.0
C3 4                          7.0
C4 1                          3.0
C2 1                                              2.0
C        BLANK CARD TO DENOTE END OF INPUT.
```

```
C     ...................................................
C           OUTPUT:
C     BRANCH NUMBER            VOLTAGE ACROSS ELEMENT      CURRENT THROUGH BRANCH
C     1   (VOLTAGE SOURCE)        -1.20000E 01 VOLTS          3.00000E 00 AMPS
C     2   (VOLTAGE SOURCE)         2.00000E 00 VOLTS          5.00000E-01 AMPS
C     3   (RESISTOR)               2.00000E 00 VOLTS          5.00000E-01 AMPS
C     4   (RESISTOR)               7.00000E 00 VOLTS          1.00000E 00 AMPS
C     5   (RESISTOR)               3.00000E 00 VOLTS          1.00000E 00 AMPS
C     6   (CURRENT SOURCE)         1.20000E 01 VOLTS          2.00000E 00 AMPS
C
C     REFERENCE NODE = 2
C     SUM OF CURRENTS AT REFENCE NODE =     9.5367431641E-07
C
C     ...............................
      SUBROUTINE RNET
      INTEGER N, B, R, V, C, D, T, GROUND, CR, TR              RNET  10
      REAL JJ,MA,MB,IR                                        RNET  20
      DIMENSION ALPHA(9)                                      RNET  30
      DIMENSION TBTREE(49,49)                                 RNET  40
      DIMENSION APRIME(20,50)                                 RNET  50
      DIMENSION FMT(7)                                        RNET  60
      DIMENSION TBJE(20,20), TBRCE(20,49)                     RNET  70
      DIMENSION YE(49,49), YO(49,49), L(20), M(20)            RNET  80
      DIMENSION A(20,50), ATREE(19,49), ACHORD(19,49), BTREE(49,49),   RNET  90
     1 AARRAY(2500), NODEI(50), NODEJ(50), RR(50), E(50), JJ(50),      RNET 100
     2 INDEX(50), RTREE(50), RCHORD(50), GTREE(50), GCHORD(50), VR(50),RNET 110
     3 IR(50), TEMP(50), X(19,50), XX(19,49), MA(19,49), MB(19,49),    RNET 120
     4 XT(19,49), XD(19,49), XE(19,49), XF(49,49), XG(49,49), BJE(19,49)RNET 130
     5. BRCE(19,49), BRCRT(19,49), BJRT(19,49), SUM(50), SUM1(50),     RNET 140
     6SUM2(50), SUM3(50), SUM4(50), TBRCRT(49,49), TBJRT(49,49)        RNET 150
C                                                             RNET 160
C                                                             RNET 170
100   READ(5,101)N, B, R, V, C, GROUND, N1, N2, N3, N4, N5    RNET 180
101   FORMAT(6(I2,2X),36X,5A4)
      IF(N .EQ. 0) GO TO 30000                                RNET 200
      WRITE(6,123)N1,N2,N3,N4, N5                             RNET 210
123   FORMAT(1H1,40X,      5A4)                               RNET0220
      WRITE(6,102) N, B, R, V, C, GROUND                      RNET 230
102   FORMAT(1H0,10X, 'N=',I2, 5X, 'B=',I2, 5X, 'R=',I2, 5X, 'V=',I2, RNET 240
     1 5X, 'C=',I2, 15X, 'GROUND=',I3)                        RNET 250
      IF( V .EQ. 0.0 ) GO TO 2                                RNET 260
1     IF( C .EQ. 0.0 ) GO TO 4                                RNET 270
      GO TO 6                                                 RNET 280
2     WRITE(6,3)                                              RNET 290
3     FORMAT(1H0,//15X,'NO VOLTAGE SOURCES ARE PRESENT')      RNET 300
      GO TO 1                                                 RNET 310
4     WRITE(6,5)                                              RNET 320
5     FORMAT(1H0,//15X,'NO CURRENT SOURCES ARE PRESENT')      RNET 330
6     IF(N .LT. 2) GO TO 9002                                 RNET 340
      IF(B .LT. 2) GO TO 9004                                 RNET 350
      IF(R .EQ. 0) GO TO 9006                                 RNET 360
      IF(B .LT. N)GO TO 9008                                  RNET 370
      NM = N-1                                                RNET 380
      NB = B-(N-1)                                            RNET 390
      TR = NM - V                                             RNET 400
      KV = V + 1                                              RNET 410
      CR = R - (NM-V)                                         RNET 420
      NP = N + 1                                              RNET 430
      KVR = V + R                                             RNET 440
      WRITE(6,203)                                            RNET 450
203   FORMAT(1H0,// 10X, 'NODEI',5X,'NODEJ',10X, 'VOLTAGE SOURCES',    RNET 460
     1 20X, 'RESISTORS',10X, 'CURRENT SOURCES' /)             RNET 470
      DO 200 K = 1,B                                          RNET 480
      READ(5,201) NODEI(K),NODEJ(K),E(K),RR(K),JJ(K)          RNET 490
201   FORMAT(2I2,6X, 3(F10.2, 5X))                            RNET 500
      WRITE(6,202)NODEI(K),NODEJ(K),E(K),RR(K),JJ(K)          RNET 510
202   FORMAT(1H0,10X,I3,6X,I3,15X,F12.3,15X,F12.3,10X,F12.3)  RNET 520
200   CONTINUE                                                RNET 530
C                                                             RNET 540
C     SET UP INCIDENCE MATRIX A(I,K)                          RNET 550
C                                                             RNET 560
      DO 300 K = 1,B                                          RNET 570
      INDEX(K) = K                                            RNET 580
      DO 300 I = 1,N                                          RNET 590
      IF(NODEI(K) .EQ. I) GO TO 10                            RNET 600
```

```
      IF(NODEJ(K) .EQ. I) GO TO 20                        RNET 610
      A(I,K) = 0.0                                        RNET 620
      GO TO 300                                           RNET 630
   10 A(I,K) = 1.0                                        RNET 640
      GO TO 300                                           RNET 650
   20 A(I,K) = -1.0                                       RNET 660
  300 CONTINUE                                            RNET 670
C                                                         RNET 680
C         CHECKING TOPOLOGY                               RNET 690
      DO 360 I = 1,N                                      RNET 700
      CHECK = 0.0                                         RNET 710
      DO 350 K = 1,B                                      RNET 720
      IF(A(I,K) .EQ. 0.0) GO TO 350                       RNET 730
      CHECK = CHECK + 1.0                                 RNET 740
  350 CONTINUE                                            RNET 750
      IF(CHECK .LT. 2.0) GO TO 9010                       RNET 760
  360 CONTINUE                                            RNET 770
C     CHOOSING GROUND IF ONE IS NOT GIVEN                 RNET 780
      IF(GROUND .NE.0) GO TO 30                            RNET 790
      BIG = 0.0                                           RNET 800
      DO 400 I = 1,N                                       RNET 810
      SUM(I) = 0.0                                        RNET 820
      DO 500 K = 1,B                                       RNET 830
  500 SUM(I) = ABS(A(I,K)) + SUM(I)                       RNET 840
      IF(SUM(I) .LE. BIG) GO TO 400                       RNET 850
      BIG = SUM(I)                                        RNET 860
      GROUND = I                                          RNET 870
  400 CONTINUE                                            RNET 880
C                                                         RNET 890
C     DELETING GROUND ROW FROM INCIDENCE MATRIX           RNET 900
   30 DO 600 K = 1,B                                      RNET 910
      DO 600 I = 1,NM                                      RNET 920
      IF(I .GE. GROUND) GO TO 35                          RNET 930
      A(I,K) = A(I,K)                                     RNET 940
      GO TO 600                                           RNET 950
   35 IP = I + 1                                          RNET 960
      A(I,K) = A(IP,K)                                    RNET 970
  600 CONTINUE                                            RNET 980
      DO 650 I = 1,NM                                      RNET 990
      DO 650 J = 1,B                                      RNET1000
  650 APRIME(I,J) = A(I,J)                                RNET1010
      CALL TRIANG(A,NM,B,INDEX)                           RNET1020
C         CHECKING TOPOLOGY FOR VOLTAGE LOOPS AND CURRENT CUTSETS.   RNET1030
C                                                         RNET1040
      KVR = V + R                                         RNET1050
      DO 805 K = 1,NM                                      RNET1060
      IF(INDEX(K) .GT. KVR) GO TO 9014                    RNET1070
  805 CONTINUE                                            RNET1080
      DO 810 K = N,B                                      RNET1090
      IF(INDEX(K) .LE. V) GO TO 9000                      RNET1100
  810 CONTINUE                                            RNET1110
C         CHECKING TOPOLOGY FOR DISCONNECTED LOOPS.       RNET1120
      DO 820 J = 1,NM                                      RNET1130
      IF( A(J,J) .EQ. 0.0) GO TO 9012                     RNET1140
  820 CONTINUE                                            RNET1150
C         REARRANGING THE COLUMNS OF A.                   RNET1160
      DO 800 K = 1,B                                      RNET1170
      DO 800 I = 1,NM                                      RNET1180
  800 A(I,K) = APRIME(I,INDEX(K))                         RNET1190
C                                                         RNET1200
C     DEFINING CHORD AND TREE SUBSETS                     RNET1210
      DO 69 K = 1, NM                                      RNET1220
      DO 69 I = 1, NM                                      RNET1230
      ATREE(I,K) = A(I,K)                                 RNET1240
   69 CONTINUE                                            RNET1250
      DO 1600 K=1,NB                                       RNET1260
      DO 1600 I = 1, NM                                    RNET1270
      KH=K+NM                                             RNET1280
      ACHORD(I,K) = A(I,KH)                               RNET1290
 1600 CONTINUE                                            RNET1300
C                                                         RNET1310
C     CHANGING ATREE TO ONE DIMENSIONAL ARRAY             RNET1320
      DO 88 J = 1, NM                                      RNET1330
      DO 88 I = 1, NM                                      RNET1340
      NU = I + NM * ( J-1 )                               RNET1350
```

```
        AARRAY(NU) = ATREE(I,J)                                       RNET1360
     88 CONTINUE                                                      RNET1370
        CALL MINV(AARRAY,NM,D,L,M)                                    RNET1380
        DO 25 J = 1,NM                                                RNET1390
        DO 25 I = 1,NM                                                RNET1400
        NU = I + NM * ( J-1 )                                         RNET1410
        ATREE(I,J) = AARRAY(NU)                                       RNET1420
     25 CONTINUE                                                      RNET1430
C                                                                     RNET1440
C       SOLVING FOR TREE SUBSET OF FUNDEMENTAL LOOP MATRIX (BTREE)    RNET1450
        DO 1660 L = 1,NM                                              RNET1460
        DO 1660 J = 1,NB                                              RNET1470
        BTREE(I,J) = 0.0                                              RNET1480
        DO 1665 K = 1,NM                                              RNET1490
   1665 BTREE(I,J) = -ATREE(I,K) * ACHORD(K,J) + BTREE(I,J)           RNET1500
   1660 CONTINUE                                                      RNET1510
        IF(TR .EQ. 0) GO TO 6000                                      RNET1520
        IF(CR .EQ. 0) GO TO 6600                                      RNET1530
        GO TO 1644                                                    RNET1540
C        SOME CIRCUITS HAVE NO TREE RESISTORS, OTHERS HAVE NO CHORD RESISRNET1550
C        THESE WILL BE CONSIDERED AS SPECIAL CASES AND WILL BE DEALT WITHRNET1560
C        SEPERATELY                                                   RNET1570
C                                                                     RNET1580
C        NO   TREE   RESISTORS                                        RNET1590
C                                                                     RNET1600
   6000 DO 6100 I = 1,R                                               RNET1610
        IPV = I + V                                                   RNET1620
   6100 GCHORD(I) = 1.0 / RR(INDEX(IPV))                              RNET1630
C          VOLTAGE ACROSS VOLTAGE SOURCES.                            RNET1640
        DO 6200 I = 1,NM                                              RNET1650
        VR(INDEX(I)) = E(I)                                           RNET1660
        DO 6200 J = 1,NB                                              RNET1670
   6200 TBTREE(J,I) = BTREE(I,J)                                      RNET1680
C          VOLTAGE ACROSS CURRENT SOURCES.                            RNET1690
        DO 6300 I = 1,NB                                              RNET1700
        IPV = I + V                                                   RNET1710
        VR(INDEX(LPV)) = 0.0                                          RNET1720
        DO 6225 J = 1,NM                                              RNET1730
   6225 VR(INDEX(IPV)) = TBTREE(I,J) * E(J) + VR(INDEX(IPV))          RNET1740
        IF(I .LE. R) GO TO 6250                                       RNET1750
        GO TO 6300                                                    RNET1760
   6250 IR(INDEX(IPV)) = GCHORD(I) * VR(INDEX(IPV))                   RNET1770
   6300 CONTINUE                                                      RNET1780
        IF(C .EQ. 0) GO TO 6450                                       RNET1790
        DO 6400 K = 1,C                                               RNET1800
        KVR = K + R + V                                               RNET1810
   6400 IR(INDEX(KVR)) = JJ(KVR)                                      RNET1820
C          CURRENT THROUGH VOLTAGE SOURCES                            RNET1830
   6450 DO 6500 J = 1,NM                                              RNET1840
        IR(INDEX(J)) = 0                                              RNET1850
        DO 6500 I = 1,NB                                              RNET1860
        IPV = I + V                                                   RNET1870
   6500 IR(INDEX(J)) = BTREE(J,I) * IR(INDEX(IPV)) + IR(INDEX(J))     RNET1880
        GO TO 10000                                                   RNET1890
C                                                                     RNET1900
C        NO   CHORD   RESISTORS                                       RNET1910
C                                                                     RNET1920
   6600 DO 6650 I = 1,R                                               RNET1930
        IPV = I + V                                                   RNET1940
   6650 RTREE(I) = RR(INDEX(IPV))                                     RNET1950
        DO 6700 K = 1,NB                                              RNET1960
        KVR = K + V + R                                               RNET1970
        IR(INDEX(KVR)) = JJ(KVR)                                      RNET1980
        DO 6700 I = 1,NM                                              RNET1990
   6700 TBTREE(K,I) = BTREE(I,K)                                      RNET2000
C          CURRENT THROUGH VOLTAGE SOURCES                            RNET2010
        DO 6800 I = 1,NM                                              RNET2020
        IPV = I + V                                                   RNET2030
        IR(INDEX(I)) = 0.0                                            RNET2040
        DO 6725 K = 1,C                                               RNET2050
        KVR = K + V + R                                               RNET2060
   6725 IR(INDEX(I)) = BTREE(I,K) * JJ(KVR) + IR(INDEX(I))            RNET2070
        IF(I .GT. V) GO TO 6750                                       RNET2080
        GO TO 6800                                                    RNET2090
   6750 VR(INDEX(IPV)) = RTREE(I-V) * IR(INDEX(IPV))                  RNET2100
```

```
      6800 CONTINUE                                                  RNET2110
           IF(V .EQ. 0) GO TO 6875                                   RNET2120
    C           VOLTAGE ACROSS VOLTAGE SOURCES.                      RNET2130
           DO 6850 I = 1,V                                           RNET2140
      6850 VR(INDEX(I)) = E(I)                                       RNET2150
    C           VOLTAGE ACROSS CURRENT SOURCES.                      RNET2160
      6875 DO 6900 K = 1,NB                                          RNET2170
           KVR = K + V + R                                           RNET2180
           VR(INDEX(KVR)) = 0.0                                      RNET2190
           DO 6900 J = 1,NM                                          RNET2200
      6900 VR(INDEX(KVR)) = TBTREE(K,J) * VR(INDEX(J)) + VR(INDEX(KVR)) RNET2210
           GO TO 10000                                               RNET2220
    C      TRANSPOSING BTREE                                         RNET2230
      1644 DO 1700 I = 1, NM                                         RNET2240
           DO 1700 J = 1,NB                                          RNET2250
           A(I,J) = BTREE(I,J)                                       RNET2260
      1700 CONTINUE                                                  RNET2270
           DO 1705 I = 1,NM                                          RNET2280
           DO 1705 J = 1,NB                                          RNET2290
      1705 BTREE(J,I) = A(I,J)                                       RNET2300
    C                                                                RNET2310
    C      DEFINING SUBMATRICES OF BTREE                             RNET2320
    C           NOTE THAT B-RC-E, FOR EXAMPLE, REFERS TO THE INTERSECTION OF RNET2330
    C      CHORD RESISTORS AND VOLTAGE SOURCES IN MATRIX 'B'.        RNET2340
           IF(V .EQ. 0) GO TO 1850                                   RNET2350
           DO 1800 I = 1, CR                                         RNET2360
           DO 1800 K = 1,V                                           RNET2370
           BRCE(I,K) = BTREE(I,K)                                    RNET2380
      1800 CONTINUE                                                  RNET2390
      1850 DO 1900 I = 1, CR                                         RNET2400
           DO 1900 K = 1,TR                                          RNET2410
           JKV = K + V                                               RNET2420
           BRCRT(I,K) = BTREE(I, JKV)                                RNET2430
      1900 CONTINUE                                                  RNET2440
           IF(C .EQ. 0.0) GO TO 2197                                 RNET2450
           IF(V .EQ. 0) GO TO 2001                                   RNET2460
           MCR = CR + 1                                              RNET2470
           JCR = CR + C                                              RNET2480
           DO 2000 I = 1,C                                           RNET2490
           DO 2000 K = 1,V                                           RNET2500
           KP = I + CR                                               RNET2510
           BJE(I,K) = BTREE(KP,K)                                    RNET2520
      2000 CONTINUE                                                  RNET2530
      2001 DO 2100 I = 1,C                                           RNET2540
           DO 2100 K = 1,TR                                          RNET2550
           IT = I + CR                                               RNET2560
           KV = K + V                                                RNET2570
           BJRT(I,K) = BTREE(IT,KV)                                  RNET2580
      2100 CONTINUE                                                  RNET2590
    C      DEFINING MATRICES R66, G66, R33, G33                      RNET2600
      2197 DO 2200 I = 1,TR                                          RNET2610
           IPV = I + V                                               RNET2620
           RTREE(I) = RR(INDEX(IPV))                                 RNET2630
           GTREE(I) = 1.0/RR(INDEX(IPV))                             RNET2640
      2200 CONTINUE                                                  RNET2650
           DO 2300 I = 1, CR                                         RNET2660
           INX = I + NM                                              RNET2670
           RCHORD(I) = RR(INDEX(INX))                                RNET2680
           GCHORD(I) = 1.0/RR(INDEX(INX))                            RNET2690
      2300 CONTINUE                                                  RNET2700
    C                                                                RNET2710
    C      TRANSPOSE MATRIX BJRT                                     RNET2720
           IF(C .EQ. 0.0) GO TO 2401                                 RNET2730
           DO 2400 I = 1,C                                           RNET2740
           DO 2400 K = 1,TR                                          RNET2750
      2400 TBJRT(K,I) = BJRT(I,K)                                    RNET2760
    C      TRANSPOSE MATRIX BRCRT                                    RNET2770
      2401 DO 2500 I = 1,CR                                          RNET2780
           DO 2500 K = 1,TR                                          RNET2790
      2500 TBRCRT(K,I) = BRCRT(I,K)                                  RNET2800
    C                                                                RNET2810
    C  V3=(1+BRCRT*RTREE*TBRCRT*GCHORD)-|(BRCRT*RTREE*TBJRT)JJ-(BRCE)E RNET2820
    C  V6=(1+RTREE*TBRCRT*GCHORD*BRCRT)-|(RTREE*TBJRT)JJ-(RTREE*TBRCRT*GCHORRNET2830
    C                                                                RNET2840
```

```
C       MATRIX MULTIPLICATION                                            RNET2850
C                                                                        RNET2860
C       X = RTREE * TBRCRT                                               RNET2870
        DO 2800 I = 1, TR                                                RNET2880
        DO 2800 J = 1, CR                                                RNET2890
        X(I,J) = 0                                                       RNET2900
        DO 2900 K = 1, TR                                                RNET2910
        IF(I .NE. K)GO TO 2900                                           RNET2920
        X(I,J) = RTREE(I) * TBRCRT(K,J) + X(I,J)                         RNET2930
 2900 CONTINUE                                                           RNET2940
 2800 CONTINUE                                                           RNET2950
C                                                                        RNET2960
C       XX = RTREE*TBRCRT*GCHORD                                         RNET2970
        DO 3000 I = 1, TR                                                RNET2980
        DO 3000 J = 1, CR                                                RNET2990
        XX(I,J) = 0                                                      RNET3000
        DO 3100 K = 1, CR                                                RNET3010
        IF(J .NE. K)GO TO 3100                                           RNET3020
        XX(I,J) = X(I,K) * GCHORD(K) + XX(I,J)                           RNET3030
 3100 CONTINUE                                                           RNET3040
 3000 CONTINUE                                                           RNET3050
C                                                                        RNET3060
C       DEFINING M31 = MB                                                RNET3070
        DO 3400 J = 1, CR                                                RNET3080
        DO 3400 I = 1, CR                                                RNET3090
        MB(I,J) = 0                                                      RNET3100
        DO 3500 K = 1, TR                                                RNET3110
 3500 MB(I,J) = BRCRT(I,K) * XX(K,J) + MB(I,J)                           RNET3120
 3400 CONTINUE                                                           RNET3130
        DO 3600 K = 1, CR                                                RNET3140
        DO 3600 I = 1, CR                                                RNET3150
        IF(I .EQ. K) GO TO 95                                            RNET3160
        MB(I,K) = MB(I,K)                                                RNET3170
        GO TO 3600                                                       RNET3180
   95 MB(I,K) = MB(I,K) + 1                                              RNET3190
 3600 CONTINUE                                                           RNET3200
        DO 74 I = 1, CR                                                  RNET3210
        DO 74 J = 1, CR                                                  RNET3220
        NU = I + CR * (J-1)                                              RNET3230
   74 AARRAY(NU) = MB(I,J)                                               RNET3240
        CALL MINV(AARRAY,CR,D,L,M)                                       RNET3250
        DO 76 I = 1, CR                                                  RNET3260
        DO 76 J = 1, CR                                                  RNET3270
        NU = I + CR * (J-1)                                              RNET3280
   76 MB(I,J) = AARRAY(NU)                                               RNET3290
C                                                                        RNET3300
C       XE REPRESENTS THE LOWER LEFT ELEMENT                             RNET3310
        IF(C .EQ. 0.0) GO TO 4750                                        RNET3320
        DO 4200 I = 1, TR                                                RNET3330
        DO 4200 J = 1, C                                                 RNET3340
        XE(I,J) = 0                                                      RNET3350
        DO 4300 K = 1, TR                                                RNET3360
        IF(I .NE. K) GO TO 4300                                          RNET3370
        XE(I,J) = RTREE(K) * TBJRT(K,J) + XE(I,J)                        RNET3380
 4300 CONTINUE                                                           RNET3390
 4200 CONTINUE                                                           RNET3400
C                                                                        RNET3410
C       XF = COEFFICIENT OF JJ FOR V3                                    RNET3420
        DO 4400 I = 1, CR                                                RNET3430
        DO 4400 J = 1, C                                                 RNET3440
        XF(I,J) = 0                                                      RNET3450
        DO 4500 K = 1, TR                                                RNET3460
 4500 XF(I,J) = BRCRT(I,K) * XE(K,J) + XF(I,J)                           RNET3470
 4400 CONTINUE                                                           RNET3480
C                                                                        RNET3490
C       XG = MB * XF                                                     RNET3500
C       XG REPRESENTS THE UPPER LEFT ELEMENT                             RNET3510
C       NOTE THAT THE UPPER RIGHT IS ALREADY STORED AS BRCE             RNET3520
        IF(C .EQ. 0.0) GO TO 4750                                        RNET3530
        DO 4600 I = 1, CR                                                RNET3540
        DO 4600 J = 1, C                                                 RNET3550
        XG(I,J) = 0                                                      RNET3560
        DO 4700 K = 1, CR                                                RNET3570
 4700 XG(I,J) = MB(I,K) * XF(K,J) + XG(I,J)                              RNET3580
```

```
 4600 CONTINUE                                                      RNET3590
C                                                                   RNET3600
C     CALCULATING CHORD RESISTOR VOLTAGES                           RNET3610
 4750 IF(V .EQ. 0) GO TO 4880                                       RNET3620
      DO 4637 I = 1,CR                                              RNET3630
      DO 4637 J = 1,V                                               RNET3640
      YE(I,J) = 0.0                                                 RNET3650
      DO 4638 K = 1,CR                                              RNET3660
 4638 YE(I,J) = MB(I,K) * BRCE(K,J) + YE(I,J)                       RNET3670
 4637 CONTINUE                                                      RNET3680
 4880 DO 4800 I = 1,CR                                              RNET3690
      SUM1(I) = 0                                                   RNET3700
      IF(C .EQ. 0.0) GO TO 4901                                     RNET3710
      DO 4900 K = 1, C                                              RNET3720
      KRV = K + R + V                                               RNET3730
 4900 SUM1(I) = XG(I,K) * JJ(KRV) + SUM1(I)                         RNET3740
 4901 SUM2(I) = 0.0                                                 RNET3750
      IF(V .EQ. 0) GO TO 4870                                       RNET3760
      DO 5000 K = 1, V                                              RNET3770
 5000 SUM2(I) =   YE(I,K) * E(K) + SUM2(I)                          RNET3780
 4870 INX = I + NM                                                  RNET3790
      VR(INDEX(INX)) =-SUM1(I) - SUM2(I)                            RNET3800
 5400 IR(INDEX(INX)) = GCHORD(I) * VR(INDEX(INX))                   RNET3810
 4800 CONTINUE                                                      RNET3820
C     DEFINING M68 = MA                                             RNET3830
C     MA = RTREE * TBRCRT * GCHORD * BRCRT                          RNET3840
      DO 3200 J = 1, TR                                             RNET3850
      DO 3200 I = 1, TR                                             RNET3860
      MA(I,J) = 0                                                   RNET3870
      DO 3300 K = 1, CR                                             RNET3880
 3300 MA(I,J) = XX(I,K) * BRCRT(K,J) + MA(I,J)                      RNET3890
 3200 CONTINUE                                                      RNET3900
      DO 3700 K = 1, TR                                             RNET3910
      DO 3700 I = 1, TR                                             RNET3920
      IF(I .EQ. K) GO TO 85                                         RNET3930
      MA(I,K) = MA(I,K)                                             RNET3940
      GO TO 3700                                                    RNET3950
   85 MA(I,K) = MA(I,K) + 1                                         RNET3960
 3700 CONTINUE                                                      RNET3970
      DO 49 I = 1, TR                                               RNET3980
      DO 49 J = 1, TR                                               RNET3990
      NU = I + TR * (J-1)                                           RNET4000
   49 AARRAY(NU) = MA(I,J)                                          RNET4010
      CALL MINV(AARRAY,TR,D,L,M)                                    RNET4020
      DO 65 I = 1, TR                                               RNET4030
      DO 65 J = 1, TR                                               RNET4040
      NU = I + TR * (J-1)                                           RNET4050
   65 MA(I,J) = AARRAY(NU)                                          RNET4060
C                                                                   RNET4070
C     XT =(MA)-I * RTREE * TBRCRT * GCHORD                          RNET4080
      DO 3800 I = 1, TR                                             RNET4090
      DO 3800 J = 1, CR                                             RNET4100
      XT(I,J) = 0                                                   RNET4110
      DO 3900 K = 1, TR                                             RNET4120
 3900 XT(I,J) = MA(I,K) * XX(K,J) + XT(I,J)                         RNET4130
 3800 CONTINUE                                                      RNET4140
C                                                                   RNET4150
C     XD = COEFFICIENT OF E FOR V6                                  RNET4160
      IF(V .EQ. 0) GO TO 4041                                       RNET4170
      DO 4000 I = 1, TR                                             RNET4180
      DO 4000 J = 1, V                                              RNET4190
      XD(I,J) = 0                                                   RNET4200
      DO 4100 K = 1, CR                                             RNET4210
 4100 XD(I,J) = XT(I,K) * BRCE(K,J) + XD(I,J)                       RNET4220
 4000 CONTINUE                                                      RNET4230
C     XE = COEFFICIENT OF JJ FOR V6                                 RNET4240
C     YO = MA * XE                                                  RNET4250
 4041 IF(C .EQ. 0.0) GO TO 5150                                     RNET4260
      DO 4037 I = 1,TR                                              RNET4270
      DO 4037 J = 1,C                                               RNET4280
      YO(I,J) = 0.0                                                 RNET4290
      DO 4039 K = 1,TR                                              RNET4300
 4039 YO(I,J) = MA(I,K) * XE(K,J) + YO(I,J)                         RNET4310
 4037 CONTINUE                                                      RNET4320
C                                                                   RNET4330
```

```
C       CALCULATING TREE RESISTOR VOLTAGES                          RNET4340
 5150 DO 5100 I = 1, TR                                             RNET4350
      SUM3(I) = 0.0                                                 RNET4360
      IF(C .EQ. 0.0) GO TO 5201                                     RNET4370
      DO 5200 K = 1, C                                              RNET4380
      KRV=K+R+V
 5200 SUM3(I) = YO(I,K) * JJ(KRV) + SUM3(I)                         RNET4390
 5201 SUM4(I) = 0.0                                                 RNET4400
      IF(V .EQ. 0) GO TO 5310                                       RNET4410
      DO 5300 K = 1, V                                              RNET4420
 5300 SUM4(I) = XD(I,K) * E(K) + SUM4(I)                            RNET4430
 5310 IPV = I + V                                                   RNET4440
      VR(INDEX(IPV)) =  SUM3(I) - SUM4(I)                           RNET4450
 5500 IR(INDEX(IPV)) = GTREE(I) * VR(INDEX(IPV))                    RNET4460
 5100 CONTINUE                                                      RNET4470
C                                                                   RNET4480
C          CALCULATING THE VOLTAGE DROP ACROSS CURRENT SOURCES.     RNET4490
C                                                                   RNET4500
      IF(C .EQ. 0) GO TO 5480                                       RNET4510
      DO 5430 K = 1,C                                               RNET4520
      SUM3(K) = 0.0                                                 RNET4530
      DO 5430 I = 1,TR                                              RNET4540
      IPV = I + V                                                   RNET4550
 5430 SUM3(K) = - BJRT(K,I) * VR(INDEX(IPV)) + SUM3(K)              RNET4560
      DO 5450 I = 1,C                                               RNET4570
      SUM4(I) = 0.0                                                 RNET4580
      IF (V .EQ. 0)GO TO 5450
      DO 5451 J=1,V
 5451 SUM4(I)=-BJE(I,J)*E(J) + SUM4(I)
 5450 CONTINUE
      DO 5490 K = 1,C                                               RNET4610
      KRV = K + R + V                                               RNET4620
 5490 VR(INDEX(KRV)) = SUM3(K) + SUM4(K)                            RNET4630
C                                                                   RNET4640
C       CALCULATING VOLTAGE DROP ACROSSS VOLTAGE SOURCE             RNET4650
 5480 IF(V .EQ. 0) GO TO 5665                                       RNET4660
      DO 5600 I = 1,V                                               RNET4670
 5600 VR(INDEX(L)) = E(I)                                           RNET4680
C       CALCULATING VOLTAGE SOURCE CURRENTS                         RNET4690
C       TRANSPOSING MATRICES                                        RNET4700
      IF(C .EQ. 0.0) GO TO 5640                                     RNET4710
      DO 5630 J = 1,V                                               RNET4720
      SUM1(J) = 0.0                                                 RNET4730
      DO 5630 K = 1,C                                               RNET4740
      TBJE(J,K) = BJE(K,J)                                          RNET4750
      KRV = K + R + V                                               RNET4760
 5630 SUM1(J) = TBJE(J,K) * JJ(KRV)  + SUM1(J)                      RNET4770
 5640 DO 5650 J = 1,V                                               RNET4780
      SUM2(J) = 0.0                                                 RNET4790
      DO 5650 I = 1,CR                                              RNET4800
      TBRCE(J,I) = BRCE(I,J)                                        RNET4810
      INX = I + NM                                                  RNET4820
 5650 SUM2(J) = TBRCE(J,I) * IR(INDEX( INX))  + SUM2(J)             RNET4830
      DO 5660 I = 1,V                                               RNET4840
 5660 IR(INDEX(I)) = SUM1(I) + SUM2(I)                              RNET4850
 5665 IF(C .EQ. 0.0) GO TO 10000                                    RNET4860
C       CALCULATING CURRENT SOURCE CURRENTS                         RNET4870
      DO 5670 K = 1,C                                               RNET4880
      KRV = K + R + V                                               RNET4890
 5670 IR(INDEX(KRV)) = JJ(KRV)                                      RNET4900
      GO TO 10000                                                   RNET4910
C                                                                   RNET4920
C       THIS SECTION CONTAINS ERROR ANALYSIS OF CIRCUIT CONSTRUCTION RNET4930
 9000 WRITE(6,9001)                                                 RNET4940
 9001 FORMAT(1H0,//5X,'**ERROR**  VOLTAGE SOURCES ARE IN A LOOP.')  RNET4950
      GO TO 1102                                                    RNET4960
 9002 WRITE(6,9003)                                                 RNET4970
 9003 FORMAT(1H0,//5X,'**ERROR**  THERE MUST BE AT LEAST TWO NODES TO FORNET4980
     1RM A NETWORK. CHECK INPUT, IT''S PROBABLY MISPUNCHED')        RNET4990
      GO TO 1102                                                    RNET5000
 9004 WRITE(6,9005)                                                 RNET5010
 9005 FORMAT(1H0,//5X, '**ERROR**  THERE MUST BE AT LEAST TWO BRANCHES TRNET5020
     10 FORM A NETWORK. CHECK YOUR INPUT, IT IS PROBABLY MISPUNCHED.')RNET5030
      GO TO 1102                                                    RNET5040
 9006 WRITE(6,9007)                                                 RNET5050
```

```
9007 FORMAT(1H0,//5X,'**ERROR**  YOU MUST HAVE RESISTORS TO MAKE A RESIRNET5060
    1STIVE NETWORK. SHAPE UP. ALSO CHECK YOUR INPUT,IT IS PROBABLY MISPRNET5070
    2UNCHED.')                                                      RNET5080
     GO TO 1102                                                     RNET5090
9008 WRITE(6,9009)                                                  RNET5100
9009 FORMAT(1H0,//5X,'**ERROR**   YOUR NETWORK HAS MORE NODES THAN BRANCHRNET5110
    1ES. ')                                                         RNET5120
     GO TO 1102                                                     RNET5130
9010 WRITE(6,9011)I                                                 RNET5140
9011 FORMAT(1H0, //5X,'**ERROR**   NODE',I3,1X, 'IS DANGLING FROM CIRCUIRNET5150
    1T.')                                                           RNET5160
     GO TO 1102                                                     RNET5170
9012 WRITE(6,9013)                                                  RNET5180
9013 FORMAT(1H1, '**ERROR**   PROGRAM IS UNABLE TO CHOSE A PROPER TREE.RNET5190
    1   CHECK NETWORK FOR DISCONNECTED LOOP.')                      RNET5200
     GO TO 1102                                                     RNET5210
9014 WRITE(6,9015)                                                  RNET5220
9015 FORMAT(1H0,/5X,'**ERROR** CURRENT SOURCES ARE IN A CUTSET')     RNET5230
     GO TO 1102                                                     RNET5240
C                                                                   RNET5250
C    PRINT OUT VALUES OF BRANCH CURRENTS AND VOLTAGES               RNET5260
10000 WRITE(6,10001)                                                RNET5270
10001 FORMAT(1H1,2X,'BRANCH NUMBER',20X,'VOLTAGE ACROSS ELEMENT',30X,RNET5280
    1 'CURRENT THROUGH BRANCH')                                     RNET5290
     DO 11111 K = 1,B                                               RNET5300
     IF(K .LE. V) GO TO 11000                                       RNET5310
     IF(K .LE. V+R) GO TO 11002                                     RNET5320
     IF(K .LE. B) GO TO 11004                                       RNET5330
11000 WRITE(6,11001) K, VR(K), IR(K)                                RNET5340
11001 FORMAT(1H0,5X,I3,3X,'(VOLTAGE SOURCE)',10X,1PE12.5,1X,'VOLTS', RNET5350
    135X,1PE12.5,1X,'AMPS')                                         RNET5360
     GO TO 11111                                                    RNET5370
11002 WRITE(6,11003) K, VR(K), IR(K)                                RNET5380
11003 FORMAT(1H0,5X,I3,3X,'(RESISTOR)',16X,1PE12.5,1X,'VOLTS',35X,  RNET5390
    1 1PE12.5,1X,'AMPS')                                            RNET5400
     GO TO 11111                                                    RNET5410
11004 WRITE(6,11005) K, VR(K), IR(K)                                RNET5420
11005 FORMAT(1H0,5X,I3,3X,'(CURRENT SOURCE)',10X,1PE12.5,1X,'VOLTS', RNET5430
    1 35X, 1PE12.5, 1X, 'AMPS')                                     RNET5440
11111 CONTINUE                                                      RNET5450
C    A BRIEF CHECK ON KCL AT GROUND NODE                            RNET5460
     WRITE(6,401) GROUND                                            RNET5470
 401 FORMAT(1H0,7HGROUND=, I3)                                      RNET5480
C    A CHECK OF 'KCL' AT REFERENCE NODE                             RNET5490
     SUMKCL = 0.0                                                   955 55
     DO 12000K=1,B                                                  5510
     IF(NODEI(K) .EQ. GROUND) GO TO 12005                           RNET5520
     IF(NODEJ(K) .EQ. GROUND) GO TO 12010                           RNET5530
     GO TO 12000                                                    RNET5540
12005 SUMKCL = IR(K) + SUMKCL                                       RNET5550
     GO TO 12000                                                    RNET5560
12010 SUMKCL = - IR(K) + SUMKCL                                     RNET5570
12000 CONTINUE                                                      RNET5580
     WRITE(6,5708)GROUND,SUMKCL                                     RNET5590
 5708 FORMAT(1H0, // 10X,'THE SUM OF THE CURRENTS AT REFERENCE NODE(NODERNET5600
    1 #',I2,') =',1PE25.8)                                          RNET5610
 1102 WRITE(6,30008)N1, N2, N3, N4, N5                              RNET5620
30008 FORMAT(1H0,///// 25X, 'THIS IS THE END OF PROGRAM ''', 5A4, '''') RNET5630
     GO TO 100                                                      RNET5640
30000 RETURN
     END                                                            RNET5660
     SUBROUTINE TRIANG(A,N,M,INDEX)                                 TRIA  10
     DIMENSION A(20,50), INDEX(50)                                  TRIA  20
C                                                                   TRIA  30
C        THIS SUBROUTINE REDUCES THE FIRST N ROWS AND N COLUMNS OF AN TRIA  40
C        MATRIX TO TRIANGULAR FORM.                                 TRIA  50
C                                                                   TRIA  60
C        LOOK AT THE DIAGONAL ELEMENTS                              TRIA  70
     DO 1000 LL = 1,N                                               TRIA  80
     IMP = LL + 1                                                   TRIA  90
 100 IF( A(LL,LL) .EQ. 0.0) GO TO 400                               TRIA 100
C        ELIMINATING THE NONZERO ELEMENTS IN THE COLUMN            TRIA 110
     DO 300 I = IMP,N                                               TRIA 120
     IF( A(I,LL) .EQ. 0.0) GO TO 300                                TRIA 130
     DUMMY = A(LL,LL) / A(I,LL)                                     TRIA 140
     DO 200 K = LL,M                                                TRIA 150
```

```
  200 A(I,K) = DUMMY * A(I,K) - A(LL,K)                              TRIA 160
  300 CONTINUE                                                       TRIA 170
      GO TO 1000                                                     TRIA 180
C            INTERCHANGING ROWS TO PUT NONZERO ELEMENT ON DIAGONAL.  TRIA 190
  400 IF(LL .EQ. N) GO TO 800                                        TRIA 200
      DO 700 I = IMP,N                                               TRIA 210
      IF(A(I,LL) .EQ. 0.0) GO TO 600                                 TRIA 220
      DO 500 K = LL,M                                                TRIA 230
      TEMP = A(I,K)                                                  TRIA 240
      A(I,K) = A(LL,K)                                               TRIA 250
      A(LL,K) = TEMP                                                 TRIA 260
  500 CONTINUE                                                       TRIA 270
      GO TO 100                                                      TRIA 280
  600 IF(I .EQ. N) GO TO 800                                         TRIA 290
  700 CONTINUE                                                       TRIA 300
C            INTERCHANGING COLUMNS                                   TRIA 310
  800 IF(LL .EQ. M) GO TO 1000                                       TRIA 320
      DO 900 J = IMP,M                                               TRIA 330
      IF( A(LL,J) .EQ. 0.0) GO TO 900                                TRIA 340
      DO 850 IT = 1,N                                                TRIA 350
      TEMP = A(IT,J)                                                 TRIA 360
      A(IT,J) = A(IT,LL)                                             TRIA 370
      A(IT,LL) = TEMP                                                TRIA 380
  850 CONTINUE                                                       TRIA 390
C            KEEPING TRACK OF INDEX                                  TRIA 400
      KEEP = INDEX(LL)                                               TRIA 410
      INDEX(LL) = INDEX(J)                                           TRIA 420
      INDEX(J) = KEEP                                                TRIA 430
      GO TO 100                                                      TRIA 440
  900 CONTINUE                                                       TRIA 450
 1000 CONTINUE                                                       TRIA 460
      RETURN                                                         TRIA 470
      END                                                            TRIA 480
      SUBROUTINE MINV(A,N,D,L,M)                                     MINV 033
      DIMENSION A(1),L(1),M(1)                                       MINV 034
C         SEARCH FOR LARGEST ELEMENT                                 MINV 054
C                                                                    MINV 055
      D=1.0                                                          MINV 056
      NK=-N                                                          MINV 057
      DO 80 K=1,N                                                    MINV 058
      NK=NK+N                                                        MINV 059
      L(K)=K                                                         MINV 060
      M(K)=K                                                         MINV 061
      KK=NK+K                                                        MINV 062
      BIGA=A(KK)                                                     MINV 063
      DO 20 J=K,N                                                    MINV 064
      IZ=N*(J-1)                                                     MINV 065
      DO 20 I=K,N                                                    MINV 066
      IJ=IZ+I                                                        MINV 067
   10 IF( ABS(BIGA)- ABS(A(IJ))) 15,20,20                           MINV 068
   15 BIGA=A(IJ)                                                     MINV 069
      L(K)=I                                                         MINV 070
      M(K)=J                                                         MINV 071
   20 CONTINUE                                                       MINV 072
C                                                                    MINV 073
C         INTERCHANGE ROWS                                           MINV 074
C                                                                    MINV 075
      J=L(K)                                                         MINV 076
      IF(J-K) 35,35,25                                               MINV 077
   25 KI=K-N                                                         MINV 078
      DO 30 I=1,N                                                    MINV 079
      KI=KI+N                                                        MINV 080
      HOLD=-A(KI)                                                    MINV 081
      JI=KI-K+J                                                      MINV 082
      A(KI)=A(JI)                                                    MINV 083
   30 A(JI) =HOLD                                                    MINV 084
C                                                                    MINV 085
C         INTERCHANGE COLUMNS                                        MINV 086
C                                                                    MINV 087
   35 I=M(K)                                                         MINV 088
      IF(I-K) 45,45,38                                               MINV 089
   38 JP=N*(I-1)                                                     MINV 090
      DO 40 J=1,N                                                    MINV 091
      JK=NK+J                                                        MINV 092
      JI=JP+J                                                        MINV 093
```

```
      HOLD=-A(JK)                                                  MINV 094
      A(JK)=A(JI)                                                  MINV 095
   40 A(JI) =HOLD                                                  MINV 096
C                                                                  MINV 097
C         DIVIDE COLUMN BY MINUS PIVOT (VALUE OF PIVOT ELEMENT IS  MINV 098
C         CONTAINED IN BIGA)                                       MINV 099
C                                                                  MINV 100
   45 IF(BIGA) 48,46,48                                            MINV 101
   46 D=0.0                                                        MINV 102
      RETURN                                                       MINV 103
   48 DO 55 I=1,N                                                  MINV 104
      IF(I-K) 50,55,50                                             MINV 105
   50 IK=NK+I                                                      MINV 106
      A(IK)=A(IK)/(-BIGA)                                          MINV 107
   55 CONTINUE                                                     MINV 108
C                                                                  MINV 109
C         REDUCE MATRIX                                            MINV 110
C                                                                  MINV 111
      DO 65 I=1,N                                                  MINV 112
      IK=NK+I                                                      MINV 113
      HOLD=A(IK)                                                   MINV M01
      IJ=I-N                                                       MINV 114
      DO 65 J=1,N                                                  MINV 115
      IJ=IJ+N                                                      MINV 116
      IF(I-K) 60,65,60                                             MINV 117
   60 IF(J-K) 62,65,62                                             MINV 118
   62 KJ=IJ-I+K                                                    MINV 119
      A(IJ)=HOLD*A(KJ)+A(IJ)                                       MINV M02
   65 CONTINUE                                                     MINV 121
C                                                                  MINV 122
C         DIVIDE ROW BY PIVOT                                      MINV 123
C                                                                  MINV 124
      KJ=K-N                                                       MINV 125
      DO 75 J=1,N                                                  MINV 126
      KJ=KJ+N                                                      MINV 127
      IF(J-K) 70,75,70                                             MINV 128
   70 A(KJ)=A(KJ)/BIGA                                             MINV 129
   75 CONTINUE                                                     MINV 130
C                                                                  MINV 131
C         PRODUCT OF PIVOTS                                        MINV 132
C                                                                  MINV 133
      D=D*BIGA                                                     MINV 134
C                                                                  MINV 135
C         REPLACE PIVOT BY RECIPROCAL                              MINV 136
C                                                                  MINV 137
      A(KK)=1.0/BIGA                                               MINV 138
   80 CONTINUE                                                     MINV 139
C                                                                  MINV 140
C         FINAL ROW AND COLUMN INTERCHANGE                         MINV 141
C                                                                  MINV 142
      K=N                                                          MINV 143
  100 K=(K-1)                                                      MINV 144
      IF(K) 150,150,105                                            MINV 145
  105 I=L(K)                                                       MINV 146
      IF(I-K) 120,120,108                                          MINV 147
  108 JQ=N*(K-1)                                                   MINV 148
      JR=N*(I-1)                                                   MINV 149
      DO 110 J=1,N                                                 MINV 150
      JK=JQ+J                                                      MINV 151
      HOLD=A(JK)                                                   MINV 152
      JI=JR+J                                                      MINV 153
      A(JK)=-A(JI)                                                 MINV 154
  110 A(JI) =HOLD                                                  MINV 155
  120 J=M(K)                                                       MINV 156
      IF(J-K) 100,100,125                                          MINV 157
  125 KI=K-N                                                       MINV 158
      DO 130 I=1,N                                                 MINV 159
      KI=KI+N                                                      MINV 160
      HOLD=A(KI)                                                   MINV 161
      JI=KI-K+J                                                    MINV 162
      A(KI)=-A(JI)                                                 MINV 163
  130 A(JI) =HOLD                                                  MINV 164
      GO TO 100                                                    MINV 165
  150 RETURN                                                       MINV 166
      END                                                          MINV 167
```

Sample Output of RNET—Problem 6.6

                                        PROBLEM 6.6

N= 5   B=10   R= 5   V= 0   C= 5   GROUND= 5

        NO VOLTAGE SOURCES ARE PRESENT

NODEI	NODEJ	VOLTAGE SOURCES	RESISTORS	CURRENT SOURCES
2	5	0.0	1.000	0.0
1	3	0.0	2.000	0.0
2	4	0.0	3.000	0.0
3	5	0.0	4.000	0.0
1	4	0.0	5.000	0.0
1	2	0.0	0.0	4.000
2	3	0.0	0.0	5.000
3	4	0.0	0.0	1.000
4	5	0.0	0.0	2.000
1	5	0.0	0.0	-3.000

BRANCH NUMBER		VOLTAGE ACROSS ELEMENT	CURRENT THROUGH BRANCH
1	(RESISTOR)	-1.00000E 00 VOLTS	-1.00000E 00 AMPS
2	(RESISTOR)	-4.00000E 00 VOLTS	-2.00000E 00 AMPS
3	(RESISTOR)	1.90735E-06 VOLTS	6.35783E-07 AMPS
4	(RESISTOR)	8.00000E 00 VOLTS	2.00000E 00 AMPS
5	(RESISTOR)	5.00000E 00 VOLTS	1.00000E 00 AMPS
6	(CURRENT SOURCE)	5.00000E 00 VOLTS	4.00000E 00 AMPS
7	(CURRENT SOURCE)	-9.00000E 00 VOLTS	5.00000E 00 AMPS
8	(CURRENT SOURCE)	9.00000E 00 VOLTS	1.00000E 00 AMPS
9	(CURRENT SOURCE)	-1.00000E 00 VOLTS	2.00000E 00 AMPS
10	(CURRENT SOURCE)	4.00000E 00 VOLTS	-3.00000E 00 AMPS

GROUND= 5

    THE SUM OF THE CURRENTS AT REFERENCE NODE(NODE # 5) =   2.86102295E-06

                THIS IS THE END OF PROGRAM 'PROBLEM 6.6

Sample Output of RNET—Problem 6.7

PROBLEM 6.7

N= 6   B= 9   R= 6   V= 3   C= 0   GROUND=   6

NO CURRENT SOURCES ARE PRESENT

NODE I	NODE J	VOLTAGE SOURCES	RESISTORS	CURRENT SOURCES
1	4	1.000	0.0	0.0
6	3	2.000	0.0	0.0
5	2	3.000	0.0	0.0
1	2	0.0	1.000	0.0
2	3	0.0	2.000	0.0
3	4	0.0	3.000	0.0
4	5	0.0	4.000	0.0
5	6	0.0	5.000	0.0
6	1	0.0	6.000	0.0

BRANCH NUMBER		VOLTAGE ACROSS ELEMENT	CURRENT THROUGH BRANCH
1	(VOLTAGE SOURCE)	1.00000E 00 VOLTS	-7.79280E-01 AMPS
2	(VOLTAGE SOURCE)	2.00000E 00 VOLTS	2.25230E-02 AMPS
3	(VOLTAGE SOURCE)	3.00000E 00 VOLTS	-9.59460E-01 AMPS
4	(RESISTOR)	9.54956E-01 VOLTS	9.54956E-01 AMPS
5	(RESISTOR)	-9.00918E-03 VOLTS	-4.50459E-03 AMPS
6	(RESISTOR)	5.40538E-02 VOLTS	1.80179E-02 AMPS
7	(RESISTOR)	-3.04505E 00 VOLTS	-7.61262E-01 AMPS
8	(RESISTOR)	9.90993E-01 VOLTS	1.98199E-01 AMPS
9	(RESISTOR)	1.05405E 00 VOLTS	1.75676E-01 AMPS

GROUND=   6

THE SUM OF THE CURRENTS AT REFERENCE NODE(NODE # 6) =   0.0

THIS IS THE END OF PROGRAM 'PROBLEM 6.7

Proof of Equation (6.17)

Consider a graph G. Assign orientations to its branches, loops, and cut sets. Construct a *cut set matrix* $\mathbf{K}$ whose columns correspond to the branches and whose rows to the cut sets. The elements of $\mathbf{K}$, k_{ij}, are determined as follows.

$$k_{ij} = 1 \text{ if branch } j \text{ is in cut set } i \text{ and has}$$
$$\text{the same orientation as cut set } i$$
$$= -1 \text{ if branch } j \text{ is in cut set } i \text{ and has}$$
$$\text{the opposite orientation as cut set } i$$
$$= 0 \text{ if branch } j \text{ is not in cut set } i$$

For example, the cut set matrix $\mathbf{K}$ of the graph of Figure A.1 is

$$
\mathbf{K} = \begin{array}{c} \\ \\ \end{array}
\begin{bmatrix}
1 & 0 & 0 & 0 & 1 & 1 & 0 \\
0 & 1 & 0 & 0 & -1 & 0 & -1 \\
0 & 0 & 1 & 0 & 0 & 1 & -1 \\
0 & 0 & 0 & 1 & 0 & 0 & -1 \\
1 & 1 & 0 & -1 & 0 & 1 & 0 \\
0 & 0 & 1 & -1 & 0 & 1 & 0 \\
\vdots & & & & & &
\end{bmatrix}
\begin{array}{c}
1 \\ 2 \\ 3 \\ 4 \\ 5 \\ 6 \\ \vdots
\end{array}
\tag{A.1}
$$

Branch columns: 1 2 3 4 5 6 7; rows labeled Cut set.

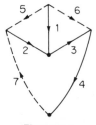

Figure A.1

Next we construct a *loop matrix* $\mathbf{L}$ whose columns correspond to the branches of G and the rows to the loops of G. The elements of $\mathbf{L}$, l_{ij}, are determined as follows.

$$l_{ij} = 1 \text{ if branch } j \text{ is in loop } i \text{ and has}$$
$$\text{the same orientation as loop } i$$
$$= -1 \text{ if branch } j \text{ is in loop } i \text{ and has}$$
$$\text{the opposite orientation as loop } i$$
$$= 0 \text{ if branch } j \text{ is not in loop } i$$

For example, the loop matrix $\mathbf{L}$ of G is

$$\begin{array}{c}
\text{Branch}
\end{array}$$

$$\mathbf{L} = \begin{array}{c}
 1 2 3 4 5 6 7 \\
\left[\begin{array}{ccccccc}
-1 & 1 & 0 & 0 & 1 & 0 & 0 \\
-1 & 0 & -1 & 0 & 0 & 1 & 0 \\
0 & 1 & 1 & 1 & 0 & 0 & 1 \\
0 & 0 & 0 & -1 & 1 & -1 & -1 \\
0 & 1 & 1 & 0 & 1 & -1 & 0 \\
& & & \vdots & & & \\
\end{array}\right]
\begin{array}{c}
1 \\ 2 \\ 3 \\ 4 \\ 5 \\ \vdots
\end{array}
\end{array} \tag{A.2}$$

Lemma $$\mathbf{K L}^t = 0 \tag{A.3}$$

In words, we say that the product of **K** and the transpose of **L**, when their columns are ordered the same, is a matrix whose elements are all zero.

Proof Each row in **K** corresponds to a cut set, and each column of $\mathbf{L}^t$ corresponds to a loop. Consider a cut set K_i shown in Figure A.2. K_i cuts the graph into two disjoint parts G_1 and G_2. Let L_j be a loop. If all the branches of L_j are in G_1 or G_2, L_j and K_i do not have any branches in common. The product of row i in **K** and column j in $\mathbf{L}^t$ is zero. If L_j has branches in G_1 and G_2, then L_j and K_i must have an even number of branches in common. Moreover, the number of common branches having the same orientation equals the number of branches having the opposite orientation. This can be seen from the sketches of Figure A.3. The product $K_i L_j^t$ is the sum of products of corresponding elements and is therefore zero.

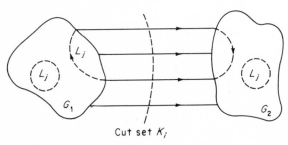

Cut set K_i

Figure A.2

Theorem Let $\mathbf{K}_f$ be the cut set matrix consisting of the fundamental cut sets with respect to a tree. Let $\mathbf{L}_f$ be the loop matrix consisting of the fundamental loops with respect to the same tree. Let $\mathbf{K}_f$ and $\mathbf{L}_f$ be written as

$$\begin{array}{cc}
\text{Trees} & \text{Chords} \\
\mathbf{K}_f = \begin{bmatrix} \mathbf{I} & \mathbf{C} \end{bmatrix} \\
\mathbf{L}_f = \begin{bmatrix} \mathbf{B} & \mathbf{I} \end{bmatrix}
\end{array}$$

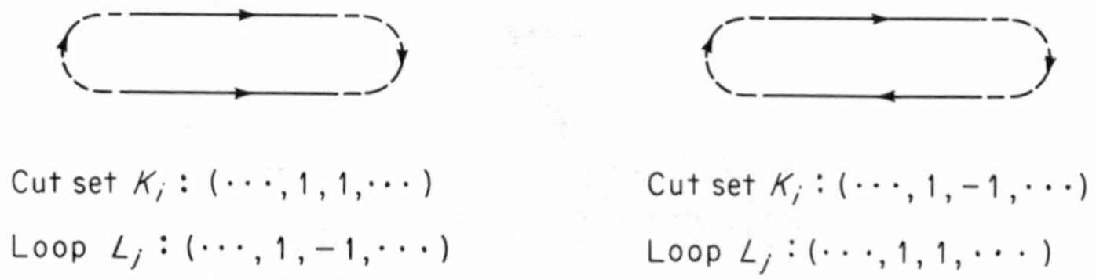

Cut set K_j : $(\cdots, 1, 1, \cdots)$ Cut set K_j : $(\cdots, 1, -1, \cdots)$

Loop L_j : $(\cdots, 1, -1, \cdots)$ Loop L_j : $(\cdots, 1, 1, \cdots)$

Figure A.3

Then

$$C = -B^t \qquad (A.4)$$

Proof From the lemma, we have

$$K_f L_f^t = \begin{bmatrix} I & C \end{bmatrix} \begin{bmatrix} B^t \\ I \end{bmatrix} = 0 \qquad (A.5)$$

Expanding the equation, we obtain immediately

$$C = -B^t$$

which is Equation (6.17), if we recall that the matrix C expresses the tree currents in terms of the chord currents, and the matrix B expresses the chord voltages in terms of the tree voltages.

7

circuits in the
sinusoidal steady state

In this chapter we consider circuits under a class of excitations which are sinusoidal functions of time and develop analysis techniques to study their behavior in the steady state, namely, at times long after the application of the excitations. The techniques of loop, mesh, and node equations of Chapter 6 will be used to describe these circuits, and the analysis is the same as for resistor circuits, provided that the concept of "impedance" is employed.

7.1 Motivation

There are several reasons for studying circuits in the sinusoidal steady state. Electric power systems in almost all parts of the world generate and transmit electricity in the form of sinusoidal voltages and currents (60 Hz in the United States and many parts of the world, and 50 Hz in some parts). Second, all physically produced excitations, speech, music, television signals, radar pulses, computer signals, seismic waves, and electrocardiographs can be resolved into a linear combination of sine and cosine functions. The mathematical procedure for doing this is known as Fourier analysis. It is a simple concept. We shall digress for a moment to present it.

Consider a periodic function $g(t)$ with period T, that is,

$$g(t) = g(t + T)$$

If $g(t)$ meets certain continuity and boundedness conditions which in practice can always be satisfied, it can be represented as a series of orthogonal functions as follows:

$$g(t) = a_0 + \sum_{n=1}^{\infty} \left(a_n \cos n \frac{2\pi}{T} t + b_n \sin n \frac{2\pi}{T} t \right) \qquad (7.1)$$

To find the coefficients a_k, we first multiply both sides by $\cos (2\pi\, kt/T)$ and integrate over a period. Because of the orthogonal property of the cosine functions, all terms of the right-hand side are zero except that which corresponds to $n = k$. Solving for a_k, we get

$$a_k = \frac{2}{T} \int_0^T g(t) \cos k \frac{2\pi}{T} t \, dt \qquad k = 1, 2,\ldots \qquad (7.2)$$

In a similar manner, we get

$$b_k = \frac{2}{T} \int_0^T g(t) \sin k \frac{2\pi}{T} t \, dt \qquad k = 1, 2,\ldots \qquad (7.3)$$

and

$$a_0 = \frac{1}{T} \int_0^T g(t) \, dt \qquad (7.4)$$

The expression on the right-hand side of Equation (7.1) is known as the Fourier series representation of $g(t)$. The constant a_0 is the average value of $g(t)$ and is called the "dc" component. Each nonconstant component of the series is a sine or cosine wave. We speak of the *frequency* of a component as the number of cycles it has in a second (hertz). Thus the dc component has a frequency of zero hertz; the first nonconstant component has a frequency of

$$f_1 = \frac{1}{T} \text{ Hz}$$

and the kth component has a frequency

$$f_k = \frac{k}{T} \text{ Hz}$$

The factor 2π is always associated with $1/T$, and we define another term ω, known as "radian frequency," as

$$\omega = 2\pi f \text{ rad/sec}$$

where f is the frequency of the given sine or cosine wave. The Fourier series now can be written as

$$g(t) = a_0 + \sum_{n=1}^{\infty} (a_n \cos \omega_n t + b_n \sin \omega_n t) \qquad (7.5)$$

with

$$\omega_k = k \frac{2\pi}{T} = k\omega_1$$

The kth component has a frequency k times that of the first. The sinusoidal waves are, therefore, *harmonically* related.

As an example, the square wave of Figure 7.1 has a Fourier series representation given by

$$g(t) = \frac{1}{2} + \frac{2}{\pi} \left[\sin \omega t + \frac{1}{3} \sin 3\omega t + \frac{1}{5} \sin 5\omega t + \cdots \right] \qquad (7.6)$$

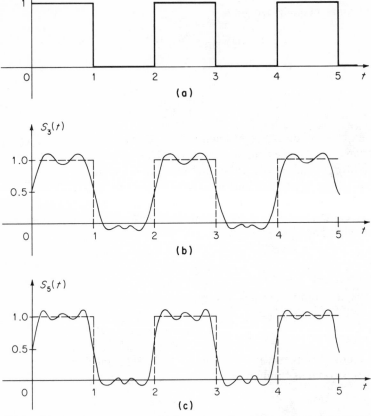

Figure 7.1 **(a)** A square wave; **(b)** three-term Fourier series approximation; **(c)** five-term approximation.

where $\omega = \pi$ rad/sec or $f = 0.5$ Hz. The wave has components at the *discrete* frequencies of 0.5, 1.5, 2.5, 3.5,... Hz. The spacing between components is 1 Hz.

Now consider the pulse train of Figure 7.2. It has a Fourier series given by

$$g(t) = 0.1 + \frac{1}{\pi}\left[\sin \omega \cos \omega t + \frac{1}{2}\sin 2\omega \cos 2\omega t + \cdots\right]$$
$$+ \frac{1}{\pi}\left[(1 - \cos \omega)\sin \omega t + \frac{1}{2}(1 - \cos 2\omega)\sin 2\omega t + \cdots\right]$$

$$(7.7)$$

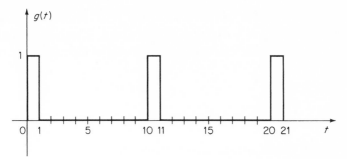

Figure 7.2 A series of pulses of height 1, width 1, and period 10.

where $\omega = \pi/5$ rad/sec or $f = 0.1$ Hz. The pulse train therefore has frequency components at 0.1, 0.2, 0.3,... Hz. The spacing between components is 0.1 Hz. It is clear that if we further increase the period, the spacing between components is further reduced. In the limit as the period goes to infinity, the spacing becomes zero and the pulse train has continuous frequency "components." At the same time, the pulse train becomes a *nonperiodic* function since it never repeats itself, and the Fourier series becomes the Fourier integral:

$$g(t) = \int_0^\infty A(\omega) \cos \omega t \, d\omega + \int_0^\infty B(\omega) \sin \omega t \, d\omega \qquad (7.8)$$

where

$$A(\omega) = 2 \int_0^\infty g(t) \cos \omega t \, dt$$

$$B(\omega) = 2 \int_0^\infty g(t) \sin \omega t \, dt \qquad (7.9)$$

We should note the similarity of these expressions and those of the Fourier series. It is beyond the scope of this book to present a detailed discussion of Fourier integral. The digression of the last several paragraphs is to convince ourselves that any "well-behaved" time functions can be regarded as a sum of sinusoidal functions. The components have frequencies which take on discrete values in the case of periodic functions and on continuous values in the case of nonperiodic functions.

There is yet another reason for studying circuits in the sinusoidal steady state. It turns out that the steady-state response under sinusoidal excitations completely determines the character of the "natural" response, that is, the response in the absence of any excitation. (See "Characteristic Function" of Chapter 4.) This point will be discussed in Chapter 8.

7.2 A Simple Example

Consider the circuit of Figure 7.3, which has a cosine excitation of frequency ω rad/sec. The differential equations of the circuit are

$$\frac{di}{dt} = -\frac{R}{L}i - \frac{1}{L}v + \frac{1}{L}E\cos\omega t$$

$$\frac{dv}{dt} = \frac{1}{C}i \qquad\qquad (7.10)$$

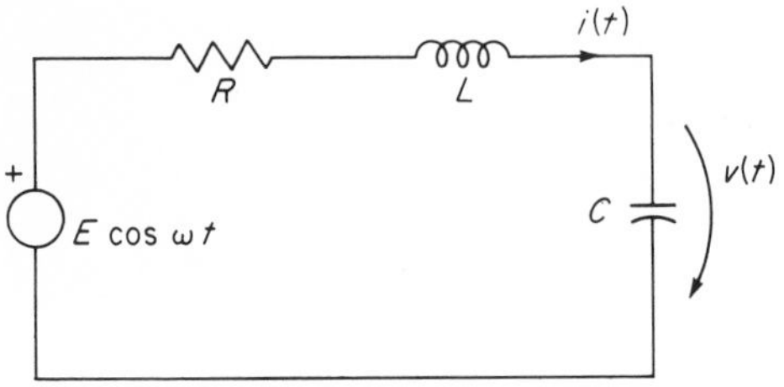

Figure 7.3 A simple circuit under sinusoidal excitation.

The solution $i(t)$ consists of the complementary part $i_c(t)$ and the particular integral $i_p(t)$:

$$i(t) = i_c(t) + i_p(t)$$

where, from Chapter 4,

$$i_c(t) = Ae^{\lambda_1 t} + Be^{\lambda_2 t}$$

with

$$\lambda_1, \lambda_2 = -\frac{R}{2L} \pm \sqrt{\left(\frac{R}{2L}\right)^2 - \frac{1}{LC}}$$

and with A, B as arbitrary constants. We note that in the steady state $(t \to \infty)$, the complementary solution becomes zero, and the only contribution to $i(t)$ comes from the particular solution. The particular solution can be found by convolution as we did in Chapter 4. However, for our present purpose, we take a different approach.

The particular integral is a solution to the inhomogeneous differential equation. We assume

$$i_p(t) = I_p \cos(\omega t + \theta)$$

Substituting it into Equation (7.10), we find

$$\left(\omega L - \frac{1}{\omega C}\right) I_p \cos(\omega t + \theta) + RI_p \sin(\omega t + \theta) = E \sin \omega t$$

Expanding the trigonometric functions, we get

$$\left[\left(\omega L - \frac{1}{\omega C}\right)\cos\theta + R\sin\theta\right] I_p \cos\omega t$$

$$+ \left[-\left(\omega L - \frac{1}{\omega C}\right)\sin\theta + R\cos\theta\right] I_p \sin\omega t = E \sin\omega t$$

Equating coefficients on $\cos \omega t$ and $\sin \omega t$ or both sides, we obtain

$$I_p = \frac{E}{\sqrt{R^2 + \left(\omega L - \dfrac{1}{\omega C}\right)^2}} \tag{7.11}$$

$$\theta = -\tan^{-1} \frac{\omega L - \dfrac{1}{\omega C}}{R} \tag{7.12}$$

and the steady-state response is

$$i_p(t) = \frac{E}{\sqrt{R^2 + \left(\omega L - \dfrac{1}{\omega C}\right)^2}} \cos(\omega t + \theta)$$

which is seen to be a sinusoidal wave of the same frequency. The quantity I_p is known as the *amplitude,* and θ as the *phase* of the response $i_p(t)$. Figure 7.4 shows a sketch of $i_p(t)$. It is seen that the phase is a measure of the relative delay of $i_p(t)$ with respect to the excitation $e(t)$.

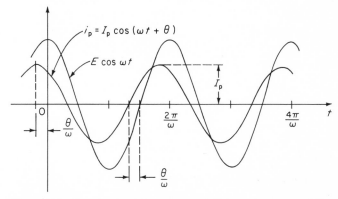

Figure 7.4 The steady-state response $i_p(t)$ is a sine wave of the same frequency as the excitation, but is out of phase with respect to the excitation.

We see in this example that the process of finding the steady-state response is rather cumbersome. We must assume a solution, substitute it into the differential equations, expand the trigonometric functions, equate the coefficients, and simplify the answer. Can we not be more sophisticated? Yes, but at the cost of having to use complex algebra.

7.3 Exponential Excitation

Suppose in the circuit of Figure 7.3, the excitation is

$$e'(t) = Ee^{j\omega t}$$

where $j = \sqrt{-1}$. What becomes of $i_p(t)$? Before we proceed to find the solution, we note that $e'(t)$ is not physical in the sense that it cannot be produced. It contains an imaginary part as seen from the expansion of the exponential:

$$e'(t) = Ee^{j\omega t} = E \cos \omega t + jE \sin \omega t$$

Nevertheless, there is nothing to prevent us from using it mathematically. We note further that the original excitation $E \cos \omega t$ is simply the real part of the exponential excitation, that is,

$$e(t) = E \cos \omega t = \text{Re}[e'(t)]$$

where $\text{Re}[x(t)]$ stands for the real part of the function $x(t)$.

Let the steady-state response of $i(t)$ be $i_p'(t)$. We assume

$$i_p'(t) = I_p' e^{j\omega t} \tag{7.13}$$

Substituting it into the differential equations, we find

$$\left(R + j\omega L + \frac{1}{j\omega C} \right) I_p' e^{j\omega t} = E e^{j\omega t}$$

or

$$I_p' = \frac{E}{R + j\omega L + \dfrac{1}{j\omega C}} \tag{7.14}$$

and

$$i_p'(t) = \frac{E}{R + j\omega L + \dfrac{1}{j\omega C}} e^{j\omega t}$$

Using the identity that for A, B real,

$$A + jB = \sqrt{A^2 + B^2} \; e^{j \tan^{-1} B/A}$$

we express $i_p'(t)$ as

$$i_p'(t) = \frac{E}{\sqrt{R^2 + \left(\omega L - \dfrac{1}{\omega C} \right)^2}} e^{j(\omega t + \theta)} \tag{7.15}$$

where

$$\theta = -\tan^{-1} \frac{\left(\omega L - \dfrac{1}{\omega C} \right)}{R}$$

Now suppose we take the real part of $i_p'(t)$. We get

$$\text{Re}[i'_p(t)] = \frac{E}{\sqrt{R^2 + \left(\omega L - \dfrac{1}{\omega C}\right)^2}} \cos(\omega t + \theta) \qquad (7.16)$$

which is precisely the steady-state response $i_p(t)$ when the excitation is $E \cos \omega t = \text{Re}[e'(t)]$. It appears that the steady-state response under a cosine excitation can be found by taking the real part of the response under an exponential excitation. This observation is summarized as follows.

Excitation	Response
$e(t) = E \cos \omega t$	$i(t) = I \cos(\omega t + \theta)$
$e'(t) = E e^{j\omega t}$	$i'(t) = I' e^{j\omega t}$
$\text{Re}[e'(t)] = e(t)$	$\text{Re}[I' e^{j\omega t}] = i(t)$

The use of the exponential excitation eliminates the algebraic manipulations of the trigonometric functions. This simplification is particularly desirable in the case of large circuits where we have a system of integral-differential equations (loop, mesh, node).

We must now show that in a general circuit under cosine excitation, the response can be found by taking the real part of the response under an exponential excitation.

We first observe that an excitation, which is a real function of time, produces a response which is real. This follows from the fact that the coefficients of the differential equations describing a circuit are real. Now, with reference to Figure 7.5, let

$$v(t) = \text{response to excitation } e^{j\omega t}$$
$$v_r(t) = \text{response to excitation } \cos \omega t$$
$$v_i(t) = \text{response to excitation } \sin \omega t$$

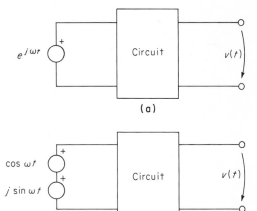

Figure 7.5 An exponential excitation is equal to the sum of the two excitations $\cos \omega t$ and $j \sin \omega t$.

By homogeneity, $jv_i(t)$ is the response to excitation $j \sin \omega t$. By additivity,

$$v_r(t) + jv_i(t) = \text{response to } (\cos \omega t + j \sin \omega t)$$
$$= \text{response to } e^{j\omega t} = v(t)$$

Since $v_r(t)$ and $v_i(t)$ are real functions of t, we have

$$v_r(t) = \text{Re}[v(t)]$$
$$v_i(t) = \text{Im}[v(t)]$$

In words, *the response to an excitation cos ωt can be found by taking the real part of the response to the complex excitation $e^{j\omega t}$. The response to an excitation sin ωt can be found by taking the imaginary part of the response to the excitation $e^{j\omega t}$.* More generally, we may state as follows: The response to $A \cos(\omega t + \theta) = \text{Re}[Ae^{j\theta} e^{j\omega t}]$ can be found by taking the real part of the response to $Ae^{j\theta} e^{j\omega t}$. The response to $A \sin(\omega t + \theta)$ can be found by taking the imaginary part of the response to $Ae^{j\theta} e^{j\omega t}$.

Example 7.1   Find the steady-state response $v_1(t)$ and $v_2(t)$ of the circuit of Figure 7.6(a).

Solution   Instead of the circuit of Figure 7.6(a), we consider the circuit of Figure 7.6(b) in which the excitation is $5e^{j2t}$. Let the mesh currents be $I_1 e^{j2t}$ and $I_2 e^{j2t}$. The mesh equations are

$$(1 + D) I_1 e^{j2t} - I_2 e^{j2t} = 5e^{j2t}$$

$$-I_1 e^{j2t} + \left(1 + \frac{4}{D}\right) I_2 e^{j2t} = 0$$

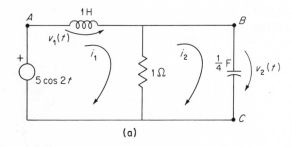

(a)

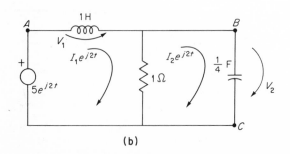

(b)

Figure 7.6 (a) A circuit under sinusoidal excitation; (b) the circuit under exponential excitation. [Example 7.1].

Carrying out the operations and dropping e^{j2t}, we get

$$(1 + j2)I_1 - I_2 = 5$$
$$-I_1 + (1 - j2)\,I_2 = 0$$

which is a system of two algebraic equations in the unknowns I_1 and I_2. Solving for them, we find

$$I_1 = \frac{5(1 - j2)}{4}$$

$$I_2 = \frac{5}{4}$$

The mesh currents are

$$I_1 e^{j2t} = \frac{5(1 - j2)}{4}\,e^{j2t}$$

$$I_2 e^{j2t} = \frac{5}{4}\,e^{j2t}$$

The voltages $v_{AB}(t)$ and $v_{BC}(t)$ are

$$v_{AB}(t) = D(I_1 e^{j2t}) = \frac{j5(1 - j2)}{2}\,e^{j2t}$$

$$v_{BC}(t) = \frac{4}{D}(I_2 e^{j2t}) = \frac{-j5}{2}\,e^{j2t}$$

Now referring to the circuit of Figure 7.6(a), we find $v_1(t)$ and $v_2(t)$ to be

$$v_1(t) = \text{Re}[v_{AB}(t)] = \text{Re}\left[\frac{j5(1 - j2)}{2}\right]e^{j2t}$$

$$= 5\cos 2t - \frac{5}{2}\sin 2t$$

$$v_2(t) = \text{Re}[v_{BC}(t)] = \text{Re}\left[-j\frac{5}{2}\,e^{j2t}\right]$$

$$= \frac{5}{2}\sin 2t$$

As a check, we note

$$v_1(t) + v_2(t) = 5\cos 2t = e(t)$$

7.4 Circuit Equations in the Sinusoidal Steady State

We saw in the last example that a complex excitation of the form $Xe^{j\omega t}$ produces a response of the form $Ye^{j\omega t}$, where X and Y are in general complex. More importantly, the set of differential equations, after the factor $e^{j\omega t}$ has been dropped, becomes a system of algebraic equations in the unknowns which are the complex amplitudes of the circuit variables. It appears that the factor $e^{j\omega t}$ may be dropped at the outset, that algebraic equations on the complex circuit variables may be written, and that the steady-state solution is obtained by taking the real part of the solution multiplied by the factor $e^{j\omega t}$. We now formalize these observations.

Let $v_1, v_2,..., v_b$ be branch voltages and $i_1, i_2,..., i_b$ be branch currents. If the excitations are of the form $Xe^{j\omega t}$, the voltages and currents take the form

$$v_k(t) = V_k e^{j\omega t}$$
$$i_k(t) = I_k e^{j\omega t} \qquad k = 1, 2,..., b$$

A typical KCL equation may be

$$i_1 + i_7 - i_8 = 0$$

or

$$I_1 e^{j\omega t} + I_7 e^{j\omega t} - I_8 e^{j\omega t} = 0$$

or

$$I_1 + I_7 - I_8 = 0 \qquad (7.17)$$

That is, KCL equations hold for the complex amplitudes of the currents. A typical KVL equation may be

$$v_2 + v_3 - v_5 = 0$$

or

$$V_2 e^{j\omega t} + V_3 e^{j\omega t} - V_5 e^{j\omega t} = 0$$

or

$$V_2 + V_3 - V_5 = 0 \qquad (7.18)$$

That is, KVL equations hold for the complex amplitudes of the voltages. Now consider the terminal equations of the elements.

Resistor:
$$v_k = R i_k$$
$$V_k e^{j\omega t} = R I_k e^{j\omega t}$$
$$\text{or} \qquad V_k = R I_k \qquad (7.19)$$

Inductor:
$$v_k = L \frac{di_k}{dt}$$
$$V_k e^{j\omega t} = j\omega L I_k e^{j\omega t}$$
$$\text{or} \qquad V_k = j\omega L I_k \qquad (7.20)$$

Capacitor:
$$i_k = C \frac{dv_k}{dt}$$
$$I_k e^{j\omega t} = j\omega C V_k e^{j\omega t}$$
$$\text{or} \qquad I_k = j\omega C V_k \qquad (7.21)$$

From Equations (7.19), (7.20), and (7.21), we note that the complex voltages and complex currents are simply proportional to each other as if the inductor were a resistor of value $j\omega L$ and the capacitor a resistor of value $1/j\omega C$. Proceeding in a similar manner, we find that for a set of mutually coupled coils,

$$V_1 = j\omega L_1 I_1 + j\omega M I_2$$
$$V_2 = j\omega M I_1 + j\omega L_2 I_2$$

where L_1 and L_2 are the self-inductances, and M is the mutual inductance. For a transformer,

$$V_1 = \frac{n_1}{n_2} V_2$$

$$I_1 = \frac{n_2}{n_1} I_1$$

For dependent sources

$$V_k = \mu V_m \qquad k \neq m$$
$$I_k = \alpha I_m$$
$$V_k = r I_m$$
$$I_k = g V_m$$

where μ, α, r, and g are constants.

In summary, KVL and KCL equations hold for the complex voltages V_1, V_2,..., V_b and currents I_1, I_2,..., I_b. The terminal equations become algebraic equations in V_k and I_k. The solution is obtained by treating the circuit as if it contained only resistors after each inductor L is replaced by a "complex resistance" of $j\omega L$, and each capacitor C by $1/j\omega C$. The factor $e^{j\omega t}$ is dropped, but is understood to be associated with every complex voltage and current in the form $V e^{j\omega t}$ and $I e^{j\omega t}$.

Example 7.2   Redo Example 7.1, using the concept of complex resistance.

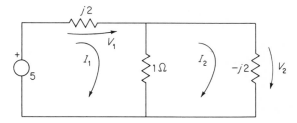

Figure 7.7 The circuit of Figure 7.6 under exponential excitation, with the factor $e^{j\omega t}$ dropped.

Solution   Dropping the factor e^{j2t} on all variables and replacing the inductor by $j2$, the capacitor by $[1/j(2)]\,(1/4) = -j2$, we obtain the circuit of Figure 7.7. Writing mesh equations, we get

$$(1 + j2)I_1 - I_2 = 5$$
$$-I_1 + (1 - j2)I_2 = 0$$

and

$$I_1 = \left(\frac{1 - j2}{4}\right) 5$$

$$I_2 = \frac{5}{4}$$

as before. To find $v_1(t)$ and $v_2(t)$, let

$$v_1(t) = Re[V_1 e^{j2t}]$$
$$v_2(t) = Re[V_2 e^{j2t}]$$

The complex voltages are found from

$$V_1 = (j2)I_1 = j\frac{5(1 - j2)}{2}$$

$$V_2 = (-j2)I_2 = -j\frac{5}{2}$$

and

$$v_1(t) = Re\left[j\frac{5(1 - j2)}{2} e^{j2t}\right]$$

$$= 5 \cos 2t - \frac{5}{2} \sin 2t$$

$$v_2(t) = Re\left[-j\frac{5}{2} e^{j2t}\right] = \frac{5}{2} \sin 2t$$

as before.

7.5 Impedance and Admittance

We saw in Example 7.2 the convenience of describing the circuit using the complex voltage and current and the notion of complex resistance. We shall further generalize this notion by defining the concept of *impedance* and *admittance*.

Consider Figure 7.8(a) in which the excitation is $Ee^{j\omega t}$. Let $Ie^{j\omega t}$ and $Ve^{j\omega t}$ be the response at the terminals shown. Dropping the factor $e^{j\omega t}$, we consider the circuit of Figure 7.8(b). We define the impedance across the terminals AB as

$$Z(j\omega) = \frac{V}{I} \tag{7.22}$$

$Z(j\omega)$ is a function of $j\omega$, and its unit is ohm. The reciprocal of $Z(j\omega)$ is *admittance*.

$$Y(j\omega) = \frac{I}{V} = \frac{1}{Z(j\omega)} \tag{7.23}$$

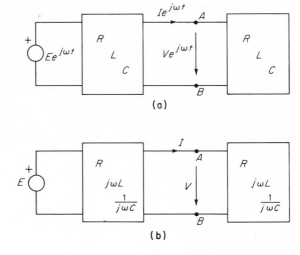

Figure 7.8 In a circuit with exponential excitation, the analysis can be simplified by dropping the factor $e^{j\omega t}$ and replacing every inductor by $j\omega L$ and every capacitor by $\frac{1}{j\omega C}$.

We note the impedance has all the properties of a resistance, and the admittance has those of a conductance. Therefore, we can state immediately that *the total impedance of two impedances connected in series is the sum of the impedances, and the total admittance of two admittances connected in parallel is the sum of the admittances.* Thus, with reference to Figure 7.9, we have

$$Z = Z_1 + Z_2$$
$$Y = Y_1 + Y_2$$

Figure 7.9 (a) Impedances in series are equivalent to one whose value is the sum of the impedances; **(b)** admittances in parallel are equivalent to one whose value is the sum of the admittances.

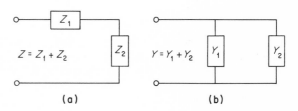

Using the concept of impedance, we note from Equations (7.19), (7.20), and (7.21) that the impedance and admittance of a resistor are, respectively

$$Z_R = R \qquad Y_R = \frac{1}{R} = G$$

of an inductor are

$$Z_L = j\omega L \qquad Y_L = \frac{1}{j\omega L}$$

and of a capacitor are

$$Z_C = \frac{1}{j\omega C} \qquad Y_C = j\omega C$$

Example 7.3   Redo Example 7.2, using the concept of impedance.

Solution   Referring to Figure 7.7, combining the two impedances in parallel into one, we get

$$Z_p = \frac{1}{1 + \dfrac{1}{-j2}} = \frac{-j2}{1 - j2} = \frac{4 - j2}{5}$$

Using the voltage divider formula, we get

$$V_1 = \frac{j2}{j2 + Z_p}\,(5) = \frac{j10}{j2 + \dfrac{4}{5} - j\dfrac{2}{5}} = j\,\frac{5(1 - j2)}{2}$$

$$V_2 = \frac{Z_p}{j2 + Z_p}\,(5) = \frac{(4 - j2)}{j2 + \dfrac{4}{5} - j\dfrac{2}{5}} = -j\,\frac{5}{2}$$

as before.

Example 7.4   Find $v(t)$ in the circuit of Figure 7.10.

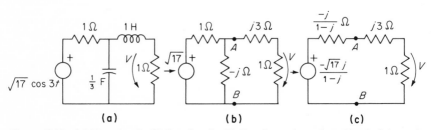

(a)   (b)   (c)

Figure 7.10   (a) The circuit; (b) the circuit with the elements replaced by their imped-ances; (c) the part to the left of AB replaced by its Thévenin equivalent. [Example 7.4]

Solution   Using complex voltages and currents and dropping the factor $e^{j\omega t}$, we have the circuit of Figure 7.10(b). Replacing the circuit to the left of AB by its Thévenin equivalent, we get the circuit of Figure 7.10(c). Using the voltage divider formula, we obtain

$$V = \frac{1}{1 + j3 + \dfrac{-j}{1 - j}} \cdot \frac{-j\sqrt{17}}{1 - j}$$

$$V = \frac{-j\sqrt{17}}{4 + j} = e^{j\theta}$$

where $\theta = -\pi/2 - \tan^{-1} 1/4 = -1.815$ radians or $-104.03°$. The voltage $v(t)$ is

$$v(t) = \mathrm{Re}[Ve^{j3t}] = \mathrm{Re}[e^{j\theta}e^{j3t}] = \cos(3t - 1.815)$$

Figure 7.11 shows the sketches of $e(t)$ and $v(t)$. It is seen that $v(t)$ lags $e(t)$ by a phase difference of $104.03°$.

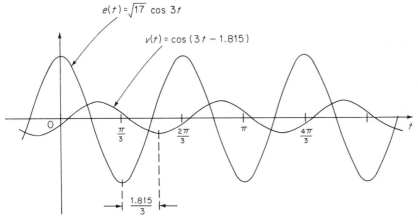

Figure 7.11 Sketch of the excitation $e(t)$ and the response $v(t)$. [Example 7.4]

7.6 Superposition in the Sinusoidal Steady State

We recall that a realistic signal is composed of infinitely many sinusoidal components, each of different frequency. The response to the signal is the sum of the responses to the components. We must note, however, that the impedances in a circuit are functions of frequency so that in calculating the responses due to the components, the correct values of the impedances must be used.

Example 7.5   (Low-Pass Filter) The circuit of Figure 7.12(a) is a simple "filter" which passes certain components of the signal while it attenuates certain others.
    Find the output voltage $v(t)$ if $e(t) = \cos t + \cos 10t + \cos 100t$.

Solution Suppose the excitation is $e(t) = \cos \omega t$. Replacing the elements by their impedances and combining them as shown in Figure 7.12(b), we get

$$V_0(j\omega) = \frac{\dfrac{1}{1+j\omega}}{1+j\omega+\dfrac{1}{1+j\omega}} = \frac{1}{2-\omega^2+j2\omega}$$

$$V_0(j\omega) = \frac{1}{\sqrt{(2-\omega^2)^2 + 4\omega^2}} e^{j\theta}$$

$$\theta = -\tan^{-1}\frac{2\omega}{2-\omega^2}$$

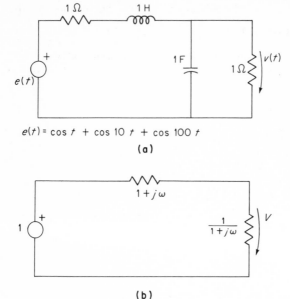

$$e(t) = \cos t + \cos 10\, t + \cos 100\, t$$

(a)

(b)

Figure 7.12 (a) A low-pass filter; (b) the circuit with the elements replaced by their impedances and the latter simplified. [Example 7.5]

Let $v(t) = v_1(t) + v_2(t) + v_3(t)$ where v_1, v_2, and v_3 are the responses to $\cos t$, $\cos 10t$, and $\cos 100t$, respectively. Then we have

$$v_1(t) = \mathrm{Re}[V_0(j)e^{jt}] = \frac{1}{\sqrt{(2-1)^2 + 4}}\cos(t + \theta_1)$$
$$v_1(t) = 0.447 \cos(t + \theta_1)$$

where

$$\theta_1 = -\tan^{-1} 2 = -63.45°$$
$$v_2(t) = \mathrm{Re}[V_0(j10)e^{j10t}] = 0.01 \cos(10t + \theta_2)$$

where

$$\theta_2 = -\tan^{-1}\frac{20}{(-98)} = -168.47°$$
$$v_3(t) = \mathrm{Re}[V_0(j100)e^{j100t}] \cong 0.0001 \cos(100t + \theta_3)$$

where

$$\theta_3 = -\tan^{-1}\frac{200}{(-9998)} = -178.85°$$

Finally,

$$v(t) = 0.447 \cos(t + \theta_1) + 0.01 \cos(10t + \theta_2) + 0.0001 \cos(100t + \theta_3)$$

We note that the amplitudes of the three components of the input signal are equal, but those of the output are drastically different. The circuit has "filtered" out the high-frequency component ($\omega = 100$) while passes the low-frequency component ($\omega = 1$) relatively unaffected. The circuit is commonly known as a simple *low-pass filter*. It is used in communication systems to separate low-frequency signals from high-frequency signals, for example, and it is used to remove the high-frequency noise superimposed on an electrocardiograph.

7.7 Loop and Mesh Equations

If we note that the impedance has all the properties of the resistance, we deduce immediately the mesh and loop equations for circuits in sinusoidal steady state. With reference to Section 6.6, we can write the loop equations in the form:

$$Z_{11}I_1 + Z_{12}I_2 + \cdots + Z_{1l}I_l = E_1$$
$$Z_{21}I_1 + Z_{22}I_2 + \cdots + Z_{2l}I_l = E_2$$
$$\vdots$$
$$Z_{l1}I_1 + Z_{l2}I_2 + \cdots + Z_{ll}I_l = E_l \qquad (7.24)$$

where

$Z_{ii} =$ sum of impedances in loop i
$Z_{ij} = \pm$(sum of impedances shared by loop i and loop j)
$E_j =$ sum of voltage sources in loop j
$I_j =$ loop current j

The algebraic sign of the off-diagonal terms is plus if loop currents i and j go through the impedance Z_{ij} in the same direction; otherwise, it is minus.

In a similar manner, we have the mesh equations of the form:

$$Z_{11}I_1 + Z_{12}I_2 + \cdots + Z_{1m}I_m = E_1$$
$$Z_{21}I_1 + Z_{22}I_2 + \cdots + Z_{2m}I_m = E_2$$
$$\vdots$$
$$Z_{m1}I_1 + Z_{m2}I_2 + \cdots + Z_{mm}I_m = E_m \qquad (7.25)$$

where

$Z_{ii} =$ sum of impedances in mesh m
$Z_{ij} = -$(sum of impedances shared by mesh i and j)
$E_j =$ sum of voltage sources in mesh j
$I_j =$ mesh current j

The algebraic sign is always minus provided that the mesh currents are all assigned the same direction.

Example 7.6 *(Balanced Bridge)* In Figure 7.13, what condition must the impedances Z_1, Z_2, Z_3, and Z_4 satisfy in order that $I = 0$?

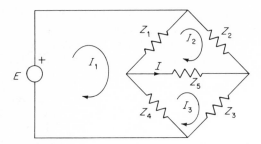

Figure 7.13 A balanced bridge, if the condition $Z_1Z_3 = Z_2Z_4$ is satisfied. [Example 7.6]

Solution   Assign mesh currents as shown. The mesh equations are

$$(Z_1 + Z_4)I_1 - Z_1I_2 - Z_4I_3 = E$$
$$-Z_1I_1 + (Z_1 + Z_2 + Z_5)I_2 - Z_5I_3 = 0$$
$$-Z_4I_1 - Z_5I_2 + (Z_3 + Z_4 + Z_5)I_3 = 0$$

Solving for I_2 and I_3, we have

$$I_2 = \frac{Z_4Z_5 + Z_1(Z_3 + Z_4 + Z_5)}{\Delta} E$$

$$I_3 = \frac{Z_1Z_5 + Z_4(Z_1 + Z_2 + Z_5)}{\Delta} E$$

where Δ is the determinant of the set of coefficients of the simultaneous equations. Now

$$I = I_3 - I_2 = \frac{Z_2Z_4 - Z_1Z_3}{\Delta} E$$

Hence $I = 0$ if, and only if,

$$Z_2Z_4 = Z_1Z_3$$

Under this condition, the bridge is said to be *balanced*. The balanced bridge can be used to measure unknown capacitance, inductance, and frequency of excitation. (See Problems 7.22–7.24.) As a simple illustration, Figure 7.14 shows a bridge to determine the unknown inductance L_x. Z_5 of Figure 7.13 is replaced by a detector (ammeter). R_1 and C are fixed. R_2 is varied until the current in the detector is zero. The unknown L_x is then found from

$$(j\omega L_x)\frac{1}{j\omega C} = R_1R_2$$

or

$$L_x = R_1R_2C$$

Note that the condition is independent of frequency.

Example 7.7   (High-Pass Filter) Figure 7.15 shows a circuit which is used in a communication system to suppress the low-frequency components of a signal.

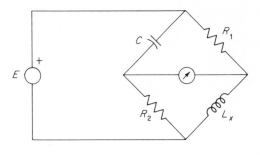

Figure 7.14 A bridge to determine an unknown inductance.

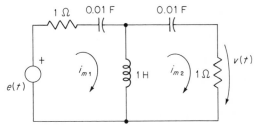

Figure 7.15 High-pass filter. [Example 7.7]

Suppose $e(t) = \cos t + \cos 10t + \cos 100t$. Find the output voltage $v(t)$.

Solution Let the excitation be $e^{j\omega t}$. Writing the mesh equations, we have

$$\left(1 + j\omega + \frac{1}{j0.01\omega}\right) I_1 - j\omega I_2 = 1$$

$$-j\omega I_1 + \left(1 + j\omega + \frac{1}{j0.01\omega}\right) I_2 = 0$$

Solving for I_2 and then V, we get

$$V(j\omega) = 1 I_2 = \frac{j\omega}{\left(201 - \dfrac{10^4}{\omega^2}\right) + j\omega\left(2 - \dfrac{200}{\omega^2}\right)}$$

Let $v(t) = v_1(t) + v_2(t) + v_3(t)$, where $v_1(t)$ is the response to $\cos t$, $v_2(t)$ to $\cos 10t$, and $v_3(t)$ to $\cos 100t$. Then

$$v_1(t) = \mathrm{Re}[V(j)e^{jt}] = \mathrm{Re}\left[\frac{je^{jt}}{-9799 - j198}\right]$$

$$v_1(t) \cong 10^{-4} \cos\left(t - \frac{\pi}{2}\right)$$

$$v_2(t) = \mathrm{Re}[V(j10)e^{j10t}] = \mathrm{Re}\left[\frac{j10e^{j10t}}{101}\right]$$

$$v_2(t) \cong 0.099 \cos\left(10t + \frac{\pi}{2}\right)$$

$$v_3(t) = \mathrm{Re}[V(j100)e^{j100t}] = \mathrm{Re}\left[\frac{j100e^{j100t}}{200 + j198}\right]$$

$$v_3(t) \cong 0.354 \cos\left(100t + \frac{\pi}{4}\right)$$

It is seen that the low-frequency components are suppressed at the output, while the high-frequency component ($\omega = 100$) is relatively unaffected. This circuit is one of the many varieties of high-pass filter.

Example 7.8   (Excitations in Different Parts of Circuit) Figure 7.16 shows a circuit in which the sources are of different frequencies and are placed in different parts of the circuit. Calculate the output voltage $v(t)$.

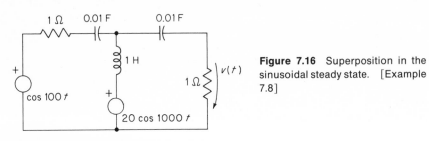

Figure 7.16   Superposition in the sinusoidal steady state.   [Example 7.8]

Solution   Recalling the process of transforming the differential equations into algebraic equations and of dropping the factor $e^{j\omega t}$, we note that the mesh (or loop) and node equations are written for excitations of one particular frequency. To use the complex notations for the voltages and currents, we must be certain that they are associated with excitations of the same frequency. If the excitations are of different frequencies, we must use superposition and solve the problem using one excitation at a time.

With reference to Figure 7.16, let $v(t) = v_1(t) + v_2(t)$, where $v_1(t)$ is the response to cos 100t, and $v_2(t)$ to 20 cos 1000t. We write the mesh equations for the circuit with the source cos 100t present only. We have

$$\left(1 + j100 + \frac{1}{j}\right)I_1 - j100I_2 = 1$$

$$-j100I_1 + \left(1 + j100 + \frac{1}{j}\right)I_2 = 0$$

The solution is the same as in Example 7.7 for $\omega = 100$, namely,

$$v_1(t) \cong 0.354 \cos\left(100t + \frac{\pi}{4}\right)$$

The mesh equations for the case in which only 20 cos 1000t is present are

$$\left(1 + j1000 + \frac{1}{j10}\right)I_1 - j1000I_2 = -20$$

$$-j1000I_1 + \left(1 + j1000 + \frac{1}{j10}\right)I_2 = 20$$

Solving for I_2 and then V, we get

$$V \cong \frac{20}{j2000} \cong 0.01e^{j\theta}, \qquad \theta = -84.3°$$

so that

$$v_2(t) \cong 0.01 \cos(1000t + \theta)$$

The solution is then

$$v(t) \cong 0.354 \cos\left(100t + \frac{\pi}{4}\right) + 0.01 \cos(1000t + \theta)$$

Note that the response to the high-frequency ($\omega = 1000$) component is much less than that to the low-frequency ($\omega = 100$) component. The circuit is a high-pass or low-pass filter depending upon where the excitation is placed and where the response is taken.

7.8 Node Equations

Recalling that an admittance has all the properties of a conductance, we can write immediately the node equations for a circuit in sinusoidal steady state. With reference to Section 6.7, we have, with conductances replaced by admittances

$$Y_{11}V_1 + Y_{12}V_2 + \cdots + Y_{1n}V_n = J_1$$
$$Y_{21}V_1 + Y_{22}V_2 + \cdots + Y_{2n}V_n = J_2$$
$$\vdots$$
$$Y_{n1}V_1 + Y_{n2}V_2 + \cdots + Y_{nn}V_n = J_n \qquad (7.26)$$

where

Y_{ii} = the sum of admittances connected to node i
Y_{ij} = $-$ (the admittance connected between node i and node j)
J_i = the sum of current sources entering into node i
V_i = node-to-reference voltage of node i
n = number of nodes less one.

If the circuit contains only resistors, inductors, and capacitors, then $Y_{ij} = Y_{ji}$. The case of circuits containing dependent sources is treated in the next section.

Example 7.9 (Band-Pass Filter) Figure 7.17 shows a circuit which is used to

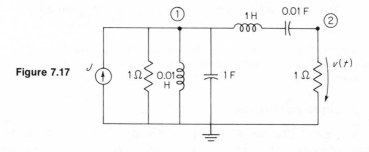

Figure 7.17

select one signal among a number of signals of different frequencies. Suppose $j(t) = \cos t + \cos 10 + \cos 100t$. Find $v(t)$.

Solution Denote the admittance in parallel with the current source by Y_1, and the admittance between nodes 1 and 2 by Y_2. Then, since Y_1 and Y_2 are functions of frequency, we have

$$Y_1(j) = 1 + j + \frac{1}{j0.01} = 1 - j99$$

$$Y_2(j) = \frac{1}{j + \dfrac{1}{j0.01}} = j\frac{1}{99} \qquad Z_2(j) = -j99$$

$$Y_1(j10) = 1 + j10 + \frac{1}{j0.1} = 1$$

$$Y_2(j10) = \frac{1}{j10 + \dfrac{1}{j10}} = j\infty \qquad Z_2(j10) = 0$$

$$Y_1(j100) = 1 + j100 + \frac{1}{j} = 1 + j99$$

$$Y_2(j100) = \frac{1}{j100 + \dfrac{1}{j}} = -j\frac{1}{99} \qquad Z_2(j100) = j99$$

The node equations for an excitation $Je^{j\omega t}$ are

$$(Y_1 + Y_2)V_1 - Y_2V_2 = J$$
$$-Y_2V_1 + (1 + Y_2)V_2 = 0$$

Solving for V_2, we get

$$V_2 = \frac{J}{1 + Y_1 + Y_1Z_2}$$

Substituting the values of Y_1 and Z_2 for each of the frequencies, we get, approximately

$$V_2(j) = 10^{-4}e^{j\pi}$$
$$V_2(j10) = 0.5$$
$$V_2(j100) = 10^{-4}e^{j\pi}$$

The steady-state response $v(t)$ is, by superposition,

$$v(t) = \text{Re}[V_2(j)e^{jt}] + \text{Re}[V_2(j10)e^{j10t}] + \text{Re}[V_2(j100)e^{j100t}]$$
$$= 10^{-4}\cos(t + \pi) + 0.5\cos(10t) + 10^{-4}\cos(100t + \pi)$$

It is seen that the circuit attenuates the low ($\omega = 1$) and high ($\omega = 100$) components and passes the medium component. The circuit is one of the many varieties of band-pass filters.

7.9 Circuits Containing Dependent Sources

The techniques that we developed in Chapter 6 for handling resistor circuits containing dependent sources apply to the case of circuits in the sinusoidal steady state. Again, we treat the dependent sources as if they were independent sources in writing the loop, mesh, and node equations. The values of the dependent sources are then expressed in terms of the variables of the equations. The equations are then simplified to obtain the solution.

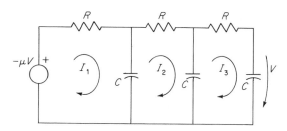

Figure 7.18 Phase-shift oscillator. [Example 7.10]

Example 7.10 (Phase-Shift Oscillator) Figure 7.18 shows a circuit known as a phase-shift oscillator which is used as a sine wave generator. The circuit has no external excitation. Finite voltages and currents exist in the circuit by virtue of the dependent source μV. Show that voltages and currents can exist only at one particular frequency and find the value of μ for which this is possible.

Solution The mesh equations are

$$\begin{bmatrix} R + \dfrac{1}{j\omega C} & -\dfrac{1}{j\omega C} & 0 \\[2mm] -\dfrac{1}{j\omega C} & R + \dfrac{2}{j\omega C} & -\dfrac{1}{j\omega C} \\[2mm] 0 & -\dfrac{1}{j\omega C} & R + \dfrac{2}{j\omega C} \end{bmatrix} \begin{bmatrix} I_1 \\[2mm] I_2 \\[2mm] I_3 \end{bmatrix} = \begin{bmatrix} -\mu V \\[2mm] 0 \\[2mm] 0 \end{bmatrix}$$

Expressing V in terms of I_3, we have

$$V = \frac{I_3}{j\omega C}$$

Substituting the above into the mesh equations and transposing terms, we get a set of *homogeneous* equations:

$$\begin{bmatrix} R + \dfrac{1}{j\omega C} & -\dfrac{1}{j\omega C} & \dfrac{\mu}{j\omega C} \\[2mm] -\dfrac{1}{j\omega C} & R + \dfrac{2}{j\omega C} & -\dfrac{1}{j\omega C} \\[2mm] 0 & -\dfrac{1}{j\omega C} & R + \dfrac{2}{j\omega C} \end{bmatrix} \begin{bmatrix} I_1 \\[2mm] I_2 \\[2mm] I_3 \end{bmatrix} = \begin{bmatrix} 0 \\[2mm] 0 \\[2mm] 0 \end{bmatrix}$$

Nontrivial solution exists if the determinant is zero. Setting it to zero and expanding it, we get

$$R\left(R^2 - \frac{6}{\omega^2 C^2}\right) + \frac{1}{j\omega C}\left(5R^2 - \frac{1+\mu}{\omega^2 C^2}\right) = 0$$

Equating the real part and imaginary part separately to zero, we get

$$R^2 = \frac{6}{\omega^2 C^2} \quad \text{or} \quad \omega = \frac{\sqrt{6}}{RC}$$

$$5R^2 = \frac{1+\mu}{\omega^2 C^2} \quad \text{or} \quad \mu = 29$$

In other words, finite voltages and currents exist if, and only if, $\mu = 29$. The voltages and currents are sinusoidal functions of frequency $\omega = \sqrt{6}/RC$ rad/sec. The circuit sustains sinusoidal voltages and currents in the *steady state* without any excitation. Under this condition, the circuit is said to be in *oscillation*. The frequency of oscillation is determined by the RC product. We recall in Chapter 4 that sustained oscillation is possible in a circuit consisting of an inductor and a capacitor. This example shows that it is possible to produce a sinusoidal oscillation in a circuit consisting of resistors and capacitors *together with a dependent source*. In practice, the dependent source is realized by an amplifier.

7.10 Circuits Containing Mutual Inductors

It is best to do an example.

Example 7.11   Figure 7.19 shows a circuit representing the input circuit of a radio receiver. The source and resistor combination represents the signal source as seen by the antenna. The output voltage $v(t)$ is applied to an amplifier which is not shown. Suppose the signal is $\sin 10^4 t + \sin 10^5 t + \sin 10^6 t$. Calculate $v(t)$.

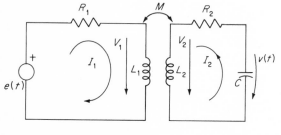

Figure 7.19 The input circuit of a radio receiver. [Example 7.11]

$$L_1 = 2 \times 10^{-3}\,\text{H} \qquad C = 1.0 \times 10^{-7}\,\text{F}$$
$$L_2 = 1 \times 10^{-3}\,\text{H} \qquad R_1 = 300\,\Omega$$
$$M = 1 \times 10^{-3}\,\text{H} \qquad R_2 = 50\,\Omega$$

Solution   Let the excitation be $Ee^{j\omega t}$, assigning mesh currents as shown. Note that they have different directions. Regarding the voltages V_1 and V_2 as sources for the moment, we find the mesh equations to be

$$R_1 I_1 = E - V_1 = E - (j\omega L_1 I_1 + j\omega M I_2)$$

$$\left(-R_2 - \frac{1}{j\omega C}\right) I_2 = V_2 = j\omega M I_1 + j\omega L_2 I_2$$

Simplifying, we get

$$(R_1 + j\omega L_1)I_1 + j\omega M I_2 = E$$

$$j\omega M I_1 + \left(j\omega L_2 + \frac{1}{j\omega C} + R_2\right)I_2 = 0$$

Solving for I_2 and then V, we get

$$V = \frac{(M/C)E}{R_1 R_2 - \omega^2 L_1 L_2\left(1 - \dfrac{1}{\omega^2 L_2 C}\right) + \omega^2 M^2 + j\omega L_1 R_2 \\ + j\omega L_2 R_1\left(1 - \dfrac{1}{\omega^2 L_2 C}\right)}$$

Substituting the values for the elements, we find

$$V(j10^4) \cong 0.03 e^{j\pi/2}$$
$$V(j10^5) \cong 0.371 e^{j\theta_2} \qquad \theta_2 = -21.8°$$
$$V(j10^6) \cong 0.0093 e^{j\theta_3} \qquad \theta_3 = -158.2°$$

and

$$v(t) = 0.03 \sin\left(10^4 t + \frac{\pi}{2}\right) + 0.371 \sin(10^5 t + \theta_2)$$
$$+ 0.0093 \sin(10^6 t + \theta_3)$$

It is seen that the circuit is a band-pass filter that suppresses the low- (10^4) and high- (10^6) frequency components while passes the medium-frequency component.

7.11 Power in the Sinusoidal Steady State

With reference to Figure 7.20, let the excitations be contained in N_1.

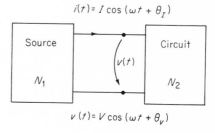

Figure 7.20 The power taken by N_2 is $v(t) \cdot i(t)$.

Let the voltage across the terminals of N_2 be $v(t) = |V| \cos(\omega t + \theta_V)$ and the current be $i(t) = |I| \cos(\omega t + \theta_I)$. The power delivered to N_2 is

$$p(t) = v(t)i(t) = |VI| \cos(\omega t + \theta_V) \cos(\omega t + \theta_I)$$
$$= \tfrac{1}{2}|VI| \cos(\theta_V - \theta_I) + \tfrac{1}{2}|VI| \cos(2\omega t + \theta_V + \theta_I) \qquad (7.27)$$

Figure 7.21 shows a sketch of $v(t)$, $i(t)$, and $p(t)$. We can make the following observations:

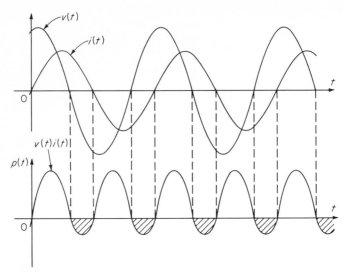

Figure 7.21 Sketch of voltage $v(t)$, current $i(t)$, and power $p(t)$. The shaded area indicates that a portion of $p(t)$ is negative and is returned to the source.

1. The power has an average component, denoted by P_{av}, given by

$$P_{av} = \tfrac{1}{2}|VI| \cos(\theta_V - \theta_I) \tag{7.28}$$

2. Superimposed on the average component is a sinusoidal component whose frequency is twice that of the voltage or current.
3. The power is negative for part of the time (shaded area), indicating that power is returned to the source during this time, and it is not used to do any useful work at the load which is represented by N_2.
4. The average power P_{av} depends on the phase difference between the voltage and current. It is maximum when $\theta_V = \theta_I$, namely, when the voltage and currents are in phase. It is zero when $\theta_V - \theta_I = \pm\pi/2$, namely, when the voltage and current are 90° out of phase. To summarize,

$$\max P_{av} = \tfrac{1}{2}|VI| \qquad \theta_V = \theta_I$$

$$\min P_{av} = 0 \qquad |\theta_V - \theta_I| = \frac{\pi}{2}$$

The two cases are illustrated in Figure 7.22. It is seen that when $\theta_V = \theta_I$, the power delivered to the circuit is positive at all times. None is returned to the source. When $|\theta_V - \theta_I| = \pi/2$, the power is positive half of the time and negative half of the time. The average power is zero.

7.12   Power in Terms of Complex Voltage and Current

It is convenient to express the average power in terms of the complex voltage and current and, in particular, of the impedance across the ter-

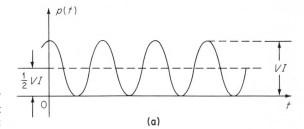

Figure 7.22 (a) Power with voltage and current in phase with each other: $p(t) \geq 0$ and average power is maximum; (b) power with voltage and current 90° out of phase with each other. Average power is zero.

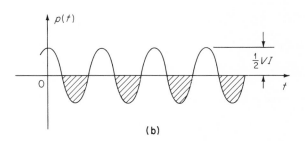

minals of N_2 in Figure 7.20. Using complex notations and noting that $\cos X$ is the real part of e^{jX}, we have, from Equation (7.28),

$$P_{av} = \tfrac{1}{2} \, \mathrm{Re}[|V|e^{j\theta_V} \, |I|e^{-j\theta_I}]$$
$$= \tfrac{1}{2} \, \mathrm{Re}[VI^*]$$

where

$$V = |V|e^{j\theta_V} \qquad I = |I|e^{j\theta_I} \qquad (7.29)$$

and I^* is the complex conjugate of I. Since $V/I = Z =$ input impedance across the terminals of N_2, we also have

$$P_{av} = \tfrac{1}{2} \, \mathrm{Re}\!\left[\frac{VV^*}{Z^*}\right] = \tfrac{1}{2}|V|^2 \, \mathrm{Re}\!\left[\frac{1}{Z^*}\right]$$
$$P_{av} = \tfrac{1}{2}|V|^2 \, \mathrm{Re}[Y] \qquad (7.30)$$

and

$$P_{av} = \tfrac{1}{2} \, \mathrm{Re}[ZII^*] = \tfrac{1}{2}|I|^2 \, \mathrm{Re}[Z] \qquad (7.31)$$

From Equations (7.30) and (7.31), we see that the average power is zero if the real part of the impedance is zero.

Let us examine three special cases. Suppose N_2 is a resistor as shown in Figure 7.23(a). We have

$$Z = R$$

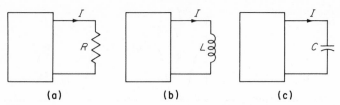

Figure 7.23 (a) Average power taken by a resistor is $\frac{1}{2}I^2R$; (b) average power taken by an inductor is 0; (c) average power taken by a capacitor is 0.

and

$$P_{av} = \tfrac{1}{2}|I|^2 \, \text{Re}[Z] = \tfrac{1}{2}|I|^2 \, R \qquad (7.32)$$

If N_2 is an inductor, we have

$$Z = j\omega L$$

$$P_{av} = \tfrac{1}{2}|I|^2 \, \text{Re}[j\omega L] = 0 \qquad (7.33)$$

If N_2 is a capacitor, we have

$$Z = \frac{1}{j\omega C}$$

$$P_{av} = \tfrac{1}{2}|I|^2 \, \text{Re} \, \frac{1}{j\omega C} = 0 \qquad (7.34)$$

Thus, the average power in an inductor and capacitor is zero. The resistor is the only element that dissipates power in the steady state.

7.13 Tellegen's Theorem

We recall from Chapter 2 Tellegen's theorem which states that the sum of the powers of the elements in a circuit is zero at all times. Let us apply the theorem to the case of sinusoidal steady state.

With reference to Figure 7.24, let the branch voltages be $v_1(t)$, $v_2(t)$,..., $v_b(t)$. Let the corresponding currents be $i_1(t)$, $i_2(t)$,..., $i_b(t)$. The theorem states that

$$\sum_{k=1}^{b} v_k(t)i_k(t) = 0$$

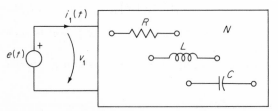

Figure 7.24 Pertaining to Tellegen's theorem.

In Figure 7.24, the excitation is $v_1(t) = e(t)$. We have

$$-e(t)i_1(t) = v_2(t)i_2(t) + \cdots + v_b(t)i_b(t)$$

Taking the average value on both sides and noting that the average of a sum of quantities is the sum of the averages of the quantities, we get

$$-P_1 = P_2 + P_3 + \cdots + P_b \qquad (7.35)$$

where P_k is the average power taken by element k. Now from the last section, we note that if the element is an inductor or capacitor, the average power is zero. Therefore,

P_1 = average power delivered by the source
  = sum of the average powers taken by
     resistors in the circuit

Example 7.12   Compute the total average power taken by the resistors and the average power delivered by the source in the circuit of Figure 7.25.

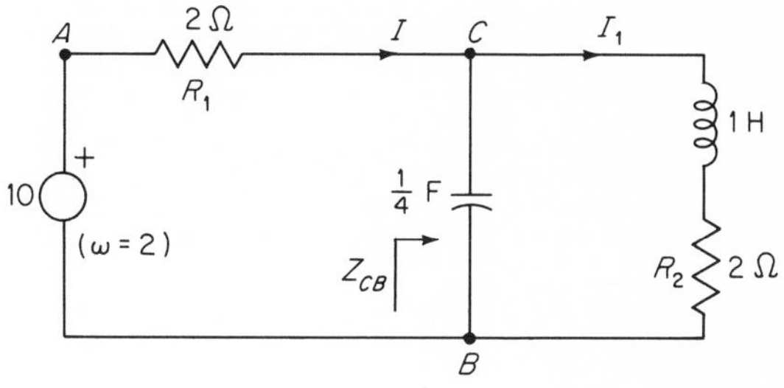

Figure 7.25 Computation of power.   [Example 7.12]

Solution   The impedance to the right of AB is

$$Z_{AB} = 2 + \cfrac{1}{j\dfrac{1}{2} + \cfrac{1}{j2 + 2}} = 4 - j2, \ \Omega$$

$$I = \frac{10}{Z} = \frac{5}{2 - j}$$

$$|I|^2 = 5$$

The average power delivered to the circuit by the source is

$$P_{\text{source}} = \tfrac{1}{2}|I|^2 \ \text{Re}[Z_{AB}] = (\tfrac{1}{2}) \ (5) \ (4) = 10 \ \text{W}$$

From the figure, we have

$$I_1 = \frac{1}{j\frac{1}{2}} \frac{I}{\dfrac{1}{j\frac{1}{2}} + j2 + 2} = -jI$$

The power taken by R_2 is

$$P_2 = \tfrac{1}{2}|I_1|^2(2) = 5 \ \text{W}$$

The power taken by R_1 is

$$P_1 = \tfrac{1}{2}|I|^2(2) = 5 \text{ W}$$

We note that $P_{\text{source}} = P_1 + P_2$.

7.14   Real Part of an Impedance

An important theorem about the real part of an impedance can be deduced from Tellegen's theorem in sinusoidal steady state.

With reference to Figure 7.24, if the circuit marked N contains only resistors, inductors, capacitors, mutual inductors, and ideal transformers, the average power P_{av} taken by N is, by Tellegen's theorem, the sum of the powers dissipated in the resistors. As long as the resistances are positive, P will be nonnegative. P_{av} is identically equal to zero if the circuit does not contain any resistance. From Equation (7.31), we have

$$P_{\text{av}} = \tfrac{1}{2}|I|^2 \,\text{Re}[Z]$$

The argument given above leads us to write

$$P_{\text{av}} \geq 0$$

or

$$\text{Re}[Z] \geq 0 \qquad\qquad (7.36)$$

The relation (7.36) is valid at *any* value of frequency. We, therefore, have

> *Theorem 7.1*   The real part of an impedance of a circuit containing resistors, inductors, capacitors, mutual inductors, and ideal transformers is nonnegative at any frequency. The real part is identically zero if the circuit contains no resistors.

The imaginary part of an impedance can be positive or negative. Let the impedance Z be written as

$$Z = \text{Re}[Z] + j\,\text{Im}[Z] = |Z|e^{j\theta_Z}$$
$$= \frac{V}{I} = \frac{|V|}{|I|}\,e^{j(\theta_V - \theta_I)}$$

with

$$\theta_Z = \tan^{-1}\frac{\text{Im}[Z]}{\text{Re}[Z]}$$

Because $\text{Re}[Z] \geq 0$, the phase angle θ_Z or the phase difference $\theta_V - \theta_I$ must lie between $-\pi/2$ and $\pi/2$. We, therefore, have an alternate form of Theorem 7.1.

Theorem 7.2 The phase angle of an impedance of a circuit described in Theorem 7.1 lies between $-\pi/2$ and $\pi/2$.

Example 7.13 Compute the real part of the impedance Z_{CB} of the circuit of Figure 7.25 as a function of the frequency of excitation ω.

Solution By inspection, we have

$$Z_{CB}(j\omega) = \cfrac{1}{j\dfrac{\omega}{4} + \cfrac{1}{2 + j\omega}}$$

$$= \cfrac{2 + j\omega}{1 - \dfrac{\omega^2}{4} + \dfrac{j\omega}{2}}$$

$$\text{Re}[Z_{CB}(j\omega)] = \cfrac{2}{\left(1 - \dfrac{\omega^2}{4}\right)^2 + \dfrac{\omega^2}{4}}$$

which is positive for all finite frequency ω.

Example 7.14 Compute the real part of the impedance Z of Figure 7.26.

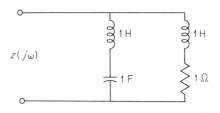

Figure 7.26 The real part of $Z(j\omega)$ is always nonnegative at all ω. [Example 7.14]

Solution By inspection, we have

$$Z(j\omega) = \cfrac{1}{\cfrac{1}{j\omega + \dfrac{1}{j\omega}} + \cfrac{1}{1 + j\omega}}$$

$$= \frac{(1 + j\omega)(1 - \omega^2)}{(1 - 2\omega^2) + j\omega}$$

$$\text{Re}[Z(j\omega)] = \frac{(1 - \omega^2)^2}{(1 - 2\omega^2)^2 + \omega^2}$$

which is positive for all frequencies except at $\omega = 1$, where it is zero. ∎

The significance of Theorem 7.1 (or 7.2) is that the impedance of a circuit belongs to a rather special class of functions of the variable ω. It is beyond the scope of this book to define this class and to present its properties. It is sufficient to say that this theorem is the starting point of the theoretical study of network analysis and synthesis.

7.15 Power Factor

We have seen that in the sinusoidal steady state, the average power is proportional to the cosine of the phase difference between the voltage and current. The average power is maximum if the voltage and current are in phase. An electric power company, which supplies electricity to its customers, would like to do so in the most efficient manner. The efficiency depends critically on this phase difference.

With reference to Figure 7.20, we have the average power given by

$$P_{\mathrm{av}} = \tfrac{1}{2}|V|\,|I|\,\cos(\theta_V - \theta_I)$$

We define the *power factor* of the circuit N_2 as

$$\text{Power factor} = \cos(\theta_V - \theta_I) \tag{7.37}$$

From Theorem 7.2 of the last section, $|\theta_V - \theta_I| \le \pi/2$. Therefore, the power factor has values between 0 and 1. In particular,

$$\text{Power factor} = 1 \quad \text{if } \theta_V = \theta_I, \qquad \text{and } P_{\mathrm{av}} = \tfrac{1}{2}|V|\,|I|$$

$$\text{Power factor} = 0 \quad \text{if } |\theta_V - \theta_I| = \frac{\pi}{2}, \quad \text{and } P_{\mathrm{av}} = 0$$

The significance of the power factor is illustrated in the following example.

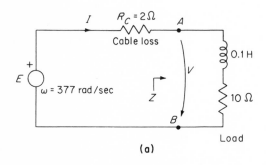

(a)

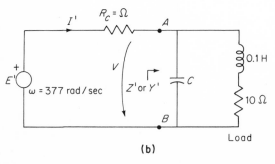

(b)

Figure 7.27 (a) The cable loss is 6.57 W; (b) the addition of a capacitor reduces the cable loss to 0.433 W. [Example 7.15]

Example 7.15 *(Power Factor Correction)*
  Case 1: Figure 7.27(a) shows a circuit that represents, in a very crude fashion, a power system. The voltage source represents the generator in a power station. R_C is the resistance of the transmission cable. The load consists of a resistor in series with an inductor. Calculate the average power taken by the load and the power dissipated in R_C. The voltage across AB is maintained at 100 V ($v(t) = 100 \cos 377t$, $f = 60$ Hz).
  Case 2: In Figure 7.27(b) the same circuit is shown with a capacitor C added across AB. Find the value of C such that the power factor of the load including C is 1. Calculate the power taken by the load and the power dissipated in R_C. The voltage V is unchanged, but E is changed to some other value E'.

Solution *(Case 1):* The impedance Z of the load is given by

$$Z = 10 + j37.7 = 39e^{j\theta}$$
$$\theta = 75.15°$$

and the power factor is $\cos \theta = 0.256$. The current is

$$I = \frac{100}{Z} = \frac{100}{(10 + j37.7)}$$
$$|I|^2 = 6.57$$

The power taken by the load is

$$P_{\text{load}} = \tfrac{1}{2}|I|^2 \, \text{Re}[Z] = \tfrac{1}{2}\,(6.57)(10) = 32.9 \text{ W}$$

The power taken by the cable resistance is

$$P_{\text{cable}} = \tfrac{1}{2}|I|^2 \, R_C = \tfrac{1}{2}\,(6.57)(2) = 6.57 \text{ W}$$

Solution *(Case 2):* The admittance is

$$Y' = j377C + \frac{1}{10 + j37.7} = j377C + \frac{10 - j37.7}{1521}$$

The current I' is

$$I' = Y'V = \left(j377C - \frac{j37.7}{1521} + \frac{10}{1521}\right)(100)$$

The voltage V and I' will be in phase (power factor = 1) if

$$377C = \frac{37.7}{1521} \quad \text{or} \quad C = 65.7 \ \mu\text{F}$$

and the current becomes

$$I' = 0.657 \text{ A}$$

The power taken by the load is

$$P_{\text{load}} = \frac{1}{2}\,|V|^2 \, \text{Re}[Y'] = \frac{1}{2}\,(100)^2 \left(\frac{10}{1521}\right) = 32.9 \text{ W}$$

as before and as expected. (Why?) The power dissipated (wasted) in the cable resistance is

$$P'_{\text{cable}} = \tfrac{1}{2}|I'|^2\, R_C = \tfrac{1}{2}(0.657)^2(2) = 0.432 \text{ W}$$

Thus the powers delivered to the load are the same in the two cases. However, the cable loss in the second case is substantially less than that in the first case. In practice, the power factor (or phase angle) of the load is constantly monitored. When it becomes too small, a capacitor or its equivalent is connected across the load automatically to "correct" the power factor, so as to make it as close to unity as possible.

7.16 Maximum Power Transfer

Consider the circuit of Figure 7.28 in which the E–Z_S combination may be considered as the Thévenin equivalent circuit of the circuit to the left of AB. Z_L represents the load impedance. Under what condition will the load receive maximum power?

The average power taken by Z_L is P_L, given by

$$P_L = \frac{1}{2}\left|\frac{E}{Z_S + Z_L}\right|^2 \text{Re}[Z_L]$$

$$= \frac{1}{2}\frac{\text{Re}[Z_L]E^2}{(\text{Re}[Z_S] + \text{Re}[Z_L])^2 + (\text{Im}[Z_S] + \text{Im}[Z_L])^2}$$

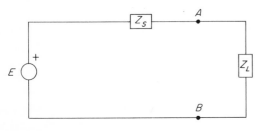

Figure 7.28 The average power taken by the load impedance Z_L is a maximum if it is related to the source impedance Z_S by $Z_L = Z_s^*$|

As the real part of Z_S and Z_L must be nonnegative, P_L is maximum if

$$\text{Re}[Z_S] = \text{Re}[Z_L]$$

and

$$\text{Im}[Z_S] = -\text{Im}[Z_L]$$

that is, if

$$Z_L = Z_S^* \tag{7.38}$$

the load impedance should be the complex conjugate of the source impedance. The maximum power is found to be

$$P_{L(\text{max})} = \frac{1}{8}\frac{E^2}{\text{Re}[Z_S]}$$

Under the same condition, the power dissipated in Z_S is

$$P_{S(max)} = \frac{(\frac{1}{2})\,E}{|Z_S + Z_L|^2}\,\mathrm{Re}[Z_S] = P_{L(max)}$$

$$= \frac{1}{8}\frac{E^2}{\mathrm{Re}[Z_S]} \tag{7.39}$$

which is the same as the power delivered to the load. The total power available from the source E is simply

$$P_{avail} = P_{L(max)} + P_{S(max)} = \frac{1}{4}\frac{E^2}{\mathrm{Re}[Z_S]} \tag{7.40}$$

and it is independent of the load impedance.

Example 7.16 In the circuit of Figure 7.29(a), the E–R_g combination represents a realistic signal source. The C–R_L combination represents a load. The capacitance C is the unavoidable stray capacitance associated with the load resistance R_L. Calculate the power taken by the load.

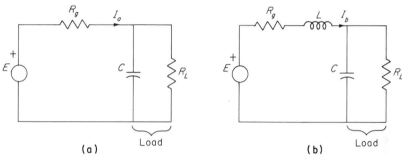

Figure 7.29 The power taken by the load in (**a**) is increased by the addition of an inductor L as in (**b**). [Example 7.16]

Now an inductor is added to the circuit as shown in Figure 7.29(b). Find the value of L and R_g such that maximum power is received in R_L.

Solution In the circuit of Figure 7.29(a), we have

$$I_a = \frac{E}{R_g + \dfrac{R_L}{1 + j\omega R_L C}} = \frac{E(1 + j\omega R_L C)}{R_g + R_L + j\omega R_L C R_g}$$

The average power taken by the load is

$$P_a = \frac{1}{2}\,|I_a|^2\,\mathrm{Re}[Z_L]$$

$$= \frac{1}{2}\frac{E^2(1 + \omega^2 R_L^2 C^2)}{(R_g + R_L)^2 + \omega^2 R_L^2 C^2 R_g^2}\,\mathrm{Re}\!\left[\frac{R_L}{1 + j\omega C R_L}\right]$$

$$= \frac{1}{2}\frac{E^2 R_L}{(R_g + R_L)^2 + \omega^2 R_L^2 C^2 R_g^2}$$

In the second case [Figure 7.29(b)], maximum power transfer takes place if

$$R_g + j\omega L = \frac{R_L}{1 - j\omega C R_L}$$

Equating the real and imaginary parts on both sides, we get

$$R_g = \frac{R_L}{1 + \omega^2 C^2 R_L^2}$$

$$L = \frac{R_L^2 C}{1 + \omega^2 C^2 R_L^2}$$

If R_g and L are chosen to satisfy these conditions at a given ω, the current I_b and the average power P_b become

$$I_b = \frac{E}{R_g + j\omega L + \dfrac{R_L}{1 + j\omega C R_L}} = \frac{E}{2R_g}$$

$$P_b = \frac{E^2}{8R_g^2} \operatorname{Re}\left[\frac{R_L}{1 + j\omega C R_L}\right] = \frac{E^2}{8R_g}$$

The load receives more power from the source in the second case than in the first case by a ratio given by

$$\frac{P_b}{P_a} = \frac{(R_g + R_L)^2 + \omega^2 R_L^2 C^2 R_g^2}{4 R_g R_L}$$

Thus, the addition of an inductor increases the power delivered to the load at a given ω. This is particularly important in the design of communication circuits where it is desired that the signal power be as large as possible in relation to the noise power generated locally at the load.

7.17 Average Power in a Periodic Signal

If the excitation of a circuit is a periodic signal, it can be resolved into a sum of sinusoidal functions of frequencies that are harmonically related. The response is a sum of sinusoids. Let us calculate the average power taken by the circuit under this condition.

Consider Figure 7.30. The voltage source is assumed to have a Fourier representation given by

$$e(t) = A_1 \cos \omega t + A_2 \cos 2\omega t + \cdots + A \cos k\omega t + \cdots$$
$$+ B_1 \sin \omega t + B_2 \sin 2\omega t + \cdots + B \sin k\omega t + \cdots$$

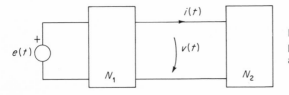

Figure 7.30 Calculation of power when the excitation is a sum of sinusoidal waves.

By superposition, the responses $v(t)$ and $i(t)$ can be written as

$$v(t) = V_1 \cos(\omega t + \theta_1) + V_2 \cos(2\omega t + \theta_2) + \cdots$$
$$+ V_k \cos(k\omega t + \theta_k) + \cdots$$
$$i(t) = I_1 \cos(\omega t + \phi_1) + I_2 \cos(2\omega t + \phi_2) + \cdots$$
$$+ I_k \cos(k\omega t + \phi_k) + \cdots$$

The power received by N_2 is

$$p(t) = v(t)i(t) = \sum_{n=1}^{\infty} \sum_{m=1}^{\infty} V_m I_n \cos(m\omega t + \theta_m) \cos(n\omega t + \phi_n)$$

The average value of $p(t)$ is the sum of the averages of the terms. A typical term is

$$p_{mn} = V_m I_n \cos(m\omega t + \theta_m) \cos(n\omega t + \phi_n)$$

It has an average value given by

$$\frac{\omega}{2\pi} \int_0^{2\pi/\omega} p_{mn}(t) \, dt = \frac{\omega}{2\pi} \int_0^{2\pi/\omega} V_m I_m \cos(m\omega t + \theta_m) \cos(n\omega t + \phi_n) \, dt$$

$$= 0 \qquad \text{if } m \neq n$$
$$= \tfrac{1}{2} V_m I_m \cos(\theta_m - \phi_m) \qquad \text{if } m = n$$

Therefore, the averages of the terms are zero except those for which $m = n$. The average of $p(t)$ is then

$$P_{\text{av}} = \frac{1}{2} \sum_{m=1}^{\infty} V_m I_m \cos(\theta_m - \phi_m) \tag{7.41}$$

But each term is recognized to be the average power taken by N_2 when the excitation is $A_m \cos m\omega t + B_m \sin m\omega t$ alone. We, therefore, have the "additivity" principle of power:

$$P_{\text{av}} = P_1 + P_2 + \cdots + P_m \cdots \tag{7.42}$$

where P_m is the average power due to the excitation $A_m \cos m\omega t + B_m \sin m\omega t$ alone. It must be noted that each term in the sum is contributed by a sinusoid of a frequency *different* from the frequencies of the rest. The principle is valid also if the circuit contains two or more excitations located at different parts of the circuit, provided they are of different frequencies. The principle is *not valid* if the excitations are of the same frequency. We shall demonstrate these observations by examples.

Example 7.17 Calculate the average power taken by the resistor in the circuit of Figure 7.31. The excitation is the square wave of Figure 7.1 which has a Fourier representation given by

$$e(t) = \frac{1}{2} + \frac{2}{\pi} \left[\sin \pi t + \frac{1}{3} \sin 3\pi t + \frac{1}{5} \sin 5\pi t + \cdots \right]$$

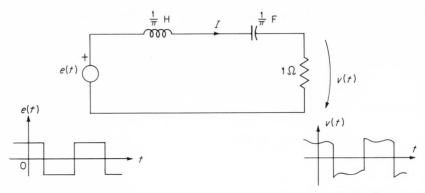

Figure 7.31 Calculation of average power in a circuit when the excitation is square wave. [Example 7.17]

Solution Let the excitation be $E(\omega)e^{j\omega t}$. The complex current is found to be

$$I(\omega) = \frac{E(\omega)}{1 + j\dfrac{\omega}{\pi} + \dfrac{\pi}{j\omega}} = \frac{E(\omega)}{\left[1 + \left(\dfrac{\omega}{\pi} - \dfrac{\pi}{\omega}\right)^2\right]^{1/2}} e^{j\theta}$$

$$\theta = -\tan^{-1}\left(\frac{\omega}{\pi} - \frac{\pi}{\omega}\right)$$

The values of $I(\omega)$ at $\omega = 0, \pi, 3\pi, 5\pi,\dots$ are

$$I(0) = 0$$
$$I(\pi) = \frac{2}{\pi} = 0.637$$
$$I(3\pi) = 0.0745e^{j\theta_3} \qquad \theta_3 = -69.45°$$
$$I(5\pi) = 0.026e^{j\theta_5} \qquad \theta_5 = -78.23°$$
$$\vdots$$

The total average power is then

$$P_{\text{total}} = \tfrac{1}{2}|I(0)|^2(1) + \tfrac{1}{2}|I(\pi)|^2(1) + \tfrac{1}{2}|I(3\pi)|^2(1) + \dots$$
$$= 0 + 0.203 + 0.00278 + 0.000338 + \cdots$$
$$= 0.206$$

It is instructive to calculate the current $i(t)$ also. From the expressions for $I(\omega)$ [$\omega = 0, \pi, 3\pi, \cdots$], we have

$$i(t) = \text{Im}[I(0)] + \text{Im}[I(\pi)e^{j\pi t}] + \text{Im}[I(3\pi)e^{j3\pi t}] + \cdots$$

$$= 0.637 \sin \pi t + 0.0745 \sin(3\pi t + \theta_3) + 0.026 \sin(5\pi t + \theta_5) + \cdots$$

The response consists of principally that due to $\sin \pi t$. The other components are much suppressed. This example shows that a circuit can be used to shape a waveform. The excitation is a square wave, but the response is not.

Example 7.18   Find the average power taken by the resistor in Figure 7.32.

Solution   Using complex notation, we have the mesh equations given by

$$(1 + j)I_1 - I_2 = 3$$
$$-I_1 + (1 - j)I_2 = -5$$
$$I_1 = -2 - 3j$$
$$I_2 = -2 - 5j$$

Figure 7.32 Additivity of power does not hold when the excitations are of the same frequency. [Example 7.18]

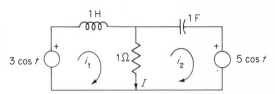

The total current in the resistor is

$$I = I_1 - I_2 = 2j$$

The average power taken by the resistor is

$$P = \tfrac{1}{2}\,|I|^2(1) = 2 \text{ W}$$

The average power delivered to the circuit by the source 3 cos t and that by 5 cos t are, respectively,

$$P_1 = \tfrac{1}{2}\,\text{Re}[(3)(I_1^*)] = -3 \text{ W}$$
$$P_2 = -\tfrac{1}{2}\,\text{Re}[(5)(I_2^*)] = 5 \text{ W}$$

We note that $P_1 + P_2 = P$ as expected.

Now suppose we apply additivity to find I. Let I' be I due to 3 cos t alone, and I'' be I due to 5 cos t alone. We find

$$I' = 3j$$
$$I'' = 5j$$

If we apply superposition to calculate the power in this case, we would get

$$P = \frac{1}{2}\,|I'|^2(1) + \frac{1}{2}\,|I''|^2(1) = \frac{9}{2} + \frac{25}{2}$$
$$= 17 \text{ W}$$

which is wrong. The point to note is that the excitations are of the same frequency and that additivity of power does not hold in this case.

7.18   Energy in the Sinusoidal Steady State

We noted earlier that the average power in an inductor or capacitor is zero in the sinusoidal steady state. The power is positive half of the time and negative the other half. The energy stored in an inductor and capacitor, whose rate of change is the power, on the other hand, has an average value which is not identically zero.

Let $i(t) = I \cos(\omega t + \theta)$ be the current on an inductor. Then from Section 2.18, we have that the energy stored in an inductor L is given by

$$
\begin{aligned}
w_L(t) &= \tfrac{1}{2} L i^2 \\
&= \tfrac{1}{2} L I^2 \cos^2(\omega t + \theta) \\
&= \tfrac{1}{4} L I^2 + \tfrac{1}{4} L I^2 \cos(2\omega t + 2\theta)
\end{aligned}
\tag{7.43}
$$

which has an average value W_L given by

$$
W_L = \tfrac{1}{4} L I^2 \tag{7.44}
$$

In a similar manner, if $v(t) = V \cos(\omega t + \theta)$ is the voltage on a capacitor C, then the energy stored in it is

$$
\begin{aligned}
w_C(t) &= \tfrac{1}{2} C v^2 \\
&= \tfrac{1}{4} C V^2 + \tfrac{1}{4} C V^2 \cos(2\omega t + 2\theta)
\end{aligned}
\tag{7.45}
$$

and the average value of the energy W_C is

$$
W_C = \tfrac{1}{4} C V^2 \tag{7.46}
$$

Note that the energy is a sinusoid, whose frequency is twice that of the voltage and current on the inductor and capacitor.

Example 7.19 Calculate the total average power and the total average energy stored in the inductors and capacitors in the circuit of Figure 7.33.

Solution The mesh equations are

$$
\left(1 + j2 + \frac{1}{j2}\right)I_1 - \frac{1}{j2}I_2 = 1
$$

$$
-\frac{1}{j2}I_1 + \left(1 + j2 + \frac{1}{j}\right)I_2 = 0
$$

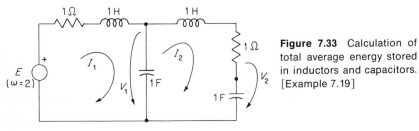

Figure 7.33 Calculation of total average energy stored in inductors and capacitors. [Example 7.19]

Solving for I_1 and I_2, we get

$$
I_1 = \frac{1 + j}{-\frac{1}{4} + j\frac{5}{2}} = 0.563 e^{j\theta_1} \qquad \theta_1 = -50.7°
$$

$$
I_2 = \frac{-j\frac{1}{2}}{-\frac{1}{4} + j\frac{5}{2}} = 0.2 e^{j\theta_2} \qquad \theta_2 = 174.3°
$$

The total average power is

$$P = \tfrac{1}{2}\,|I_1|^2(1) + \tfrac{1}{2}\,|I_2|^2(1) = 0.18 \text{ W}$$

The total average energy stored in the inductors is

$$W_L = \tfrac{1}{4}\,|I_1|^2(1) + \tfrac{1}{4}\,|I_2|^2(1) = 0.09 \text{ J}$$

The total average energy stored in the capacitors is

$$W_C = \frac{1}{4}\,C_1|V_1|^2 + \frac{1}{4}\,C_2|V_2|^2$$

$$= \frac{1}{4}\,\frac{|I_1 - I_2|^2}{\omega^2 C_1} + \frac{1}{4}\,\frac{|I_2|^2}{\omega^2 C_2}$$

$$= \frac{1}{4}\,\frac{0.519}{(4)(1)} + \frac{1}{4}\,\frac{0.04}{(4)(1)}$$

$$= 0.035 \text{ J}$$

In Section 7.20, we shall see that the average energies W_L and W_C are related to the instantaneous power $p(t)$ in a circuit.

7.19 Complex Power, Reactive Power

Let $v(t) = |V| \cos \omega t$ and $i(t) = |I| \cos(\omega t + \theta)$ be the voltage and current, respectively, that exist at the terminals of a circuit. The instantaneous power $p(t)$ can be written as

$$p(t) = v(t)i(t) = |VI| \cos \omega t \cos(\omega t + \theta)$$

$$= \frac{|VI|}{2} \cos \theta + \frac{|VI|}{2} \cos(2\omega t + \theta)$$

$$= \frac{|VI|}{2} \cos \theta + \frac{|VI|}{2} \cos \theta \cos 2\omega t$$

$$- \frac{|VI|}{2} \sin \theta \sin 2\omega t$$

It is convenient to define the term *complex power* as

$$\boldsymbol{P} = \tfrac{1}{2}\,VI^* \tag{7.47}$$

Then

$$\text{Re}[\boldsymbol{P}] = \frac{1}{2}\,\text{Re}[VI^*] = \frac{|VI|}{2} \cos \theta = \text{average power} = P$$

$$\text{Im}[\boldsymbol{P}] = \frac{1}{2}\,\text{Im}[VI^*] = \frac{|VI|}{2} \sin \theta$$

The last term is known as the *reactive* power and is given the symbol Q:

$$Q = \text{reactive power} = \frac{|VI|}{2} \sin \theta \tag{7.48}$$

The instantaneous power then becomes

$$p(t) = P + P \cos 2\omega t - Q \sin 2\omega t \qquad (7.49)$$

and it is seen that $p(t)$ consists of a constant term and two time-varying terms, one of which is 90° out of phase with respect to the other. Figure 7.34 shows a sketch of the three terms. It is noted that when $v(t)$ and $i(t)$ are in phase, P is maximum and Q is zero. When they are 90° out of place, P is zero and Q is maximum. The physical significance of Q is given in the next section.

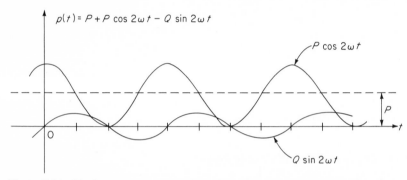

Figure 7.34 The instantaneous power $p(t)$ consists of an average value P and two sinusoidal components that are 90° out of phase with respect to each other. Q is the reactive power.

7.20 Tellegen's Theorem Again

We recall that if $v_1, v_2,..., v_b$ are branch voltages and $i_1, i_2,..., i_b$ are branch currents, then

$$v_1 i_1 + v_2 i_2 + \cdots + v_b i_b = 0$$

We recall also that the set of voltages can be *any* set that satisfies KVL, and the set of currents can be *any* set that satisfies KCL.

Now in the sinusoidal steady state with complex notations, the complex voltages and currents satisfy KVL and KCL, respectively. Let $V_1, V_2,..., V_b$ be the complex branch voltages and $I_1, I_2,..., I_b$ be the complex branch currents. Then $I_1^*, I_2^*, \cdots, I_b^*$, that is, the complex conjugates of the currents, satisfy KCL. By Tellegen's theorem, we have

$$V_1 I_1^* + V_2 I_2^* + \cdots + V_b I_b^* = 0$$

Let the voltages and currents be those of the circuit of Figure 7.35, in which the only excitation is the voltage source E. From the above, we have

$$\begin{aligned} \tfrac{1}{2} E I_1^* &= \tfrac{1}{2} V_2 I_2^* + \tfrac{1}{2} V_3 I_3^* + \cdots + \tfrac{1}{2} V_b I_b^* \\ &= P + jQ \end{aligned} \qquad (7.50)$$

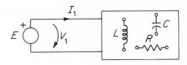

Figure 7.35 Pertaining to Tellegen's theorem.

The right-hand side is the complex power taken by the circuit. The real part is the average power P, and the imaginary part the reactive power Q. Each term on the right corresponds to the complex power taken by a particular branch. If the branch is a resistor R_k, we have

$$\tfrac{1}{2} V_k I_k^* = \tfrac{1}{2} R_k |I_k|^2 = \text{average power in } R_k$$

If the branch is an inductor L_k, we have

$$\begin{aligned}\tfrac{1}{2} V_k I_k^* &= \tfrac{1}{2} j\omega L_k |I_k|^2 = j2\omega(\tfrac{1}{4} L_k |I_k|^2)\\ &= j2\omega \text{ (average energy stored in } L_k)\end{aligned}$$

If the branch is a capacitor C_k, we have

$$\tfrac{1}{2} V_k I_k^* = -\tfrac{1}{2} j\omega C_k |V_k|^2 = -j2\omega(\tfrac{1}{4} C_k |V_k|^2)$$

If we group all the terms corresponding to the resistor branches and group those corresponding to the inductors and capacitors, we obtain

$$P + jQ = \text{(sum of average powers in resistors)} + j2\omega(W_L - W_C) \quad (7.51)$$

where

$$W_L = \text{total average energy stored in inductors}$$
$$W_C = \text{total average energy stored in capacitors}$$

The reactive power, therefore, has the physical significance that it is proportional to the difference of the total average energy stored in the inductors and the total average energy stored in the capacitors of a circuit. If the power factor is unity, the two energies are the same. (Why?)

To verify Equation (7.51), we note from Example 7.19,

$$\tfrac{1}{2} E I_1^* = \tfrac{1}{2} (1)(0.563 e^{-j\theta_1}) = 0.18 + j0.22$$
$$P = \text{Re}[\tfrac{1}{2} E I_1^*] = 0.18$$
$$Q = \text{Im}[\tfrac{1}{2} E I_1^*] = 0.22$$
$$2\omega(W_L - W_C) = (2)(2)(0.09 - 0.035) = 0.22 = Q$$

as expected.

7.21 Summary

This chapter deals with the analysis of circuits under sinusoidal excitation. We have seen that the response in the steady state is sinusoid of the same frequency, but possibly being out of phase with respect to the excitation. To facilitate the computation of the response, we introduced the idea of exponential excitation of the form $Xe^{j\omega t}$. The response is then obtained by taking the real part of the answer. More importantly, the use of exponential excitation transforms the circuit equations from differential equations to algebraic equations in the unknowns that are the complex amplitudes of the voltages and currents.

To further facilitate the computation, we introduced the concept of impedance and admittance. Using these quantities, we found the analysis of circuits in sinusoidal steady state to be similar in every aspect to the analysis of resistive circuits.

We presented a number of examples of circuits to illustrate that circuits can be constructed to separate the low-, high-, or medium-frequency components of a signal. The idea that under sinusoidal excitation there is an average value of the instantaneous power taken by a circuit was introduced. The average power depends on the relative phase between the voltage and current and is equal to the total power dissipated in the resistors of the circuit. The term power factor was defined as a measure of the efficiency of transmission of power from a generator to a load.

In sinusoidal steady state, the inductors and capacitors store magnetic and electric energy, respectively. Each has an average value. The difference between the total average energy stored in the inductors and the total average energy stored in the capacitors is related to the reactive power of the circuit. The reactive power is the imaginary part of the complex power, but it has the physical significance that it is the amplitude of the time-varying component of the instantaneous power, that is 90° out of phase with respect to the other component whose amplitude is the average power in the circuit.

SUGGESTED EXPERIMENTS

1. Construct a circuit with resistors, inductors, and capacitors. Excite the circuit with a sinusoidal signal and observe all responses. Note that all the voltages and currents are sinusoidal. Carefully sketch the voltage and current in an inductor. Note the phase difference. Repeat for a capacitor.

2. Excite the circuit of Experiment 1 with a square wave. Note that the responses are no longer square waves. Why? Explain what you see.

3. Verify Example 7.1 experimentally. Sketch the waveforms of the input and output voltages. Use a phase meter to measure the phase difference.

4. Using realistic elements, construct the low-pass circuit of Example 7.12. For example, use 10,000-Ω resistors, a 1-H inductor, and a 0.01-μF capacitor. Vary the frequency of the sinusoidal source and observe the output voltage.
5. Construct the simple band-pass filter of Figure 7.19. Experimentally verify the frequency selective property of the circuit.
6. Construct the impedance bridge of Figure 7.13. Verify the balanced condition.
7. Construct Maxwell's bridge of Problem 7.22 and verify the condition for balance.
8. Construct the Wien bridge of Problem 7.23 and use it to determine the frequency of the source.
9. Construct the bridged-T network of Problem 7.16. How would you verify the fact that the input impedance is constant, independent of frequency?
10. Construct the lattice network of Problem 7.14. Measure the output voltage. Verify the fact that its amplitude is constant, independent of frequency. Measure the phase of the output voltage with respect to the input. Explain what you see.

COMPUTER PROGRAMMING

1. Write a program to compute the Fourier series approximation of a square wave of Figure 7.1 using three, four, five, and six terms of the series. Plot the results carefully.
2. Write a program to compute the Fourier series approximation of the pulse train of Figure 7.2 using three, four, five, and six terms of the series. Plot the results carefully.
3. Write a program to calculate all the voltages and currents in the steady state of the circuit of Example 7.5. Compute the product of branch voltage and current for every branch. Verify Tellegen's theorem for every instant of time.
4. Compute the instantaneous power delivered to the load for each of the two cases of Example 7.15. Sketch the power as a function of time and note the instants at which the power is negative. Compute and plot the voltage and current at the terminals AB. Note the phase difference in the two cases.
5. Modify the program LADDER of Chapter 3 so that it can be used to compute the input impedance of a ladder network.

PROBLEMS

7.1   *Exercise*   Find the impedance and admittance of each of the following circuits ($\omega = 2$ rad/sec).

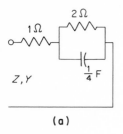

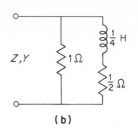

Problem 7.1 Computation of impedance and admittance.

(a)   (b)

7.2   Find the impedance and admittance of each of the following ladder circuits. (See Chapter 3.) Extend the ladder analysis technique to circuits with impedances.

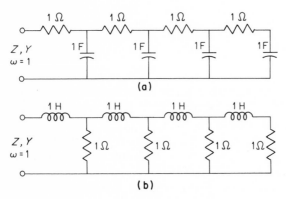

(a)

(b)

Problem 7.2 Computation of impedance and admittance of a ladder network.

7.3   *Exercise*   Find the steady-state response $v(t)$ of the following circuit. Sketch $v(t)$ carefully. What is $v(t)$ if the input voltage is $\sin(2t + \pi/4)$?

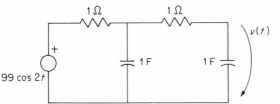

Problem 7.3 Steady-state response.

7.4   *Exercise*   Find the current $i(t)$ in the following circuit. What can you say about this circuit with respect to the circuit of Problem 7.3? What is $i(t)$ if the excitation is $\cos(2t + \pi/3)$?

Problem 7.4 Steady-state response and duality.

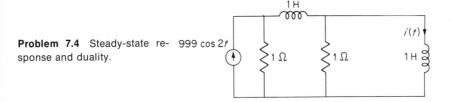

7.5   *Exercise*   Find the output voltage $v(t)$ of the following circuit. What is the impedance of the *L–C* combination? Are you surprised at the answer?

Problem 7.5 Steady-state
response.

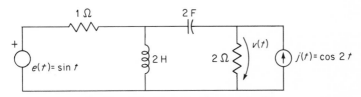

7.6   *Superposition*   Find the steady-state response of the circuit shown. Find the dual of the circuit and compute the current in the capacitor in the dual.

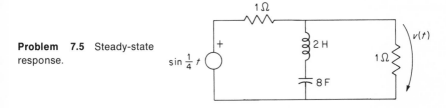

Problem 7.6   Superposition and duality.

7.7   *Superposition*   Is it possible to adjust the phase of the source $e_2(t)$ in the circuit below such that the output voltage $v(t)$ has a phase difference of $\pi/3$ with respect to the source $e_1(t)$?

Problem 7.7 Superposition.

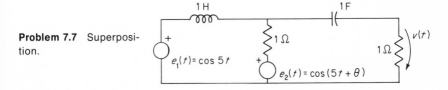

7.8   *Superposition*   Find the complex voltage V of the circuit below. How should the source be changed in order that the voltage V be 1?

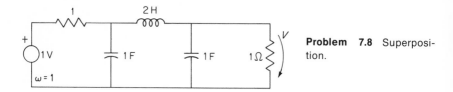

Problem 7.8 Superposition.

7.9   *Superposition*   Repeat Problem 7.8 for the dual of the circuit.

7.10   *Symmetry*   Find the current I_x and the voltage V_x for each of the three cases of excitation: **a.** $E_1 = E_2$; **b.** $E_1 = -E_2$; **c.** $E_1 = 1 + j$, $E_2 = 3 + j5$. ($\omega = 2$ rad/sec.)

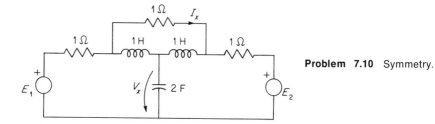

Problem 7.10 Symmetry.

7.11   *Symmetry*   Find the current I_x and the voltage V_x in the circuit shown below for the four cases of excitation: **a.** $E_1 = E_2$; **b.** $E_1 = -E_2$; **c.** $E_1 = 1 + j$, $E_2 = 1 - j$; **d.** $E_1 = 1 + j$, $E_2 = -1 + j$.

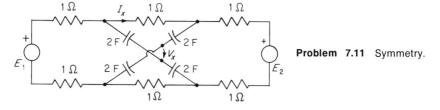

Problem 7.11 Symmetry.

7.12   *Constant-Resistance Lattice*   Show that if the condition $Z_A Z_B =$

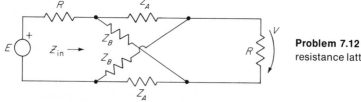

Problem 7.12 Constant-resistance lattice.

R^2 exists, then the input impedance $Z_{in} = R$, independent of frequency, and the output voltage V satisfies

$$\frac{V}{E} = \frac{1}{2}\frac{R - Z_A}{R + Z_A}$$

7.13 *Constant-Resistance Lattice* Show that when N constant-resistance lattices are connected in cascade as shown, the voltage V satisfies

$$\frac{V}{E} = \frac{1}{2}\prod_{k=1}^{N}\frac{R - Z_{Ak}}{R + Z_{Ak}}$$

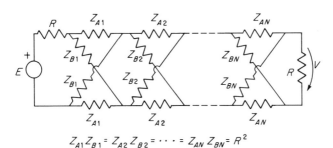

$$Z_{A1}Z_{B1} = Z_{A2}Z_{B2} = \cdots = Z_{AN}Z_{BN} = R^2$$

Problem 7.13   Constant-resistance lattices in cascade.

7.14 *Constant-Resistance Lattice* In Problem 7.12, calculate the steady-state response $v(t)$ for the excitation $e(t) = \cos t + \cos 10t + \cos 100t$, and **a.** Z_A is an inductor of 1 H; **b.** Z_A is a capacitor of 1 F; **c.** Z_A is a series combination of a 1-H inductor and a 1-F capacitor. Can you explain the fact that the amplitudes of all of the components of the voltage $v(t)$ are the same?

7.15 *Constant-Resistance Bridged-T* Show that the input impedance of the circuit shown is R, independent of frequency, and that $V/E = \frac{1}{2}/(1 + Z_A/R)$.

Problem 7.15   Bridged-*T*
network.

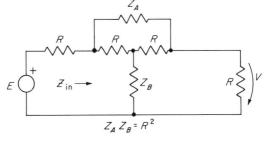

$$Z_A Z_B = R^2$$

7.16 *Constant-Resistance Bridged-T* Repeat Problem 7.14 for the constant-resistance bridged-*T* network. Omit the question.

7.17 *Constant-Resistance Bridged-T* Show that when N constant-resistance bridged-T networks are connected in cascade as shown, the output voltage satisfies

$$\frac{V}{E} = \frac{1}{2}\prod_{k=1}^{N}\frac{R}{Z_{Ak} + R}$$

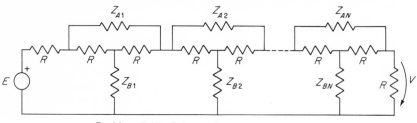

Problem 7.17 Bridged-T networks in cascade.

7.18 *Mesh Equations* Calculate the output voltage V for $\omega = 0, 0.1,$ 1, and 10 rad/sec. Suggest some use of the circuit.

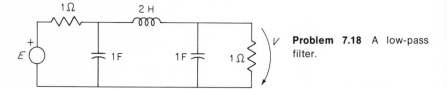

Problem 7.18 A low-pass filter.

7.19 *Mesh Equations* Repeat Problem 7.18 for the circuit shown below. Is there any difference between the two circuits as far as the output voltage V is concerned?

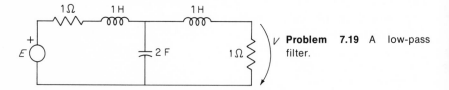

Problem 7.19 A low-pass filter.

7.20 *Mesh Equations* Calculate the output voltage V for the circuit shown at $\omega = 100, 1000,$ and 10,000 rad/sec.

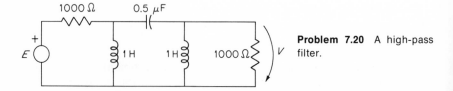

Problem 7.20 A high-pass filter.

7.21 *Band-Pass Filter* Calculate the voltage V for $\omega = 10^5$, 10^6, and 10^7 rad/sec.

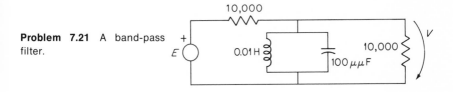

Problem 7.21 A band-pass filter.

7.22 *Maxwell Bridge* The following circuit is known as the Maxwell bridge, which can be used to measure the inductance and the un-avoidable resistance of a realistic inductor. Show that the bridge is balanced when $L_x = R_2R_3C$ and $R_x = R_2R_3/R_1$.

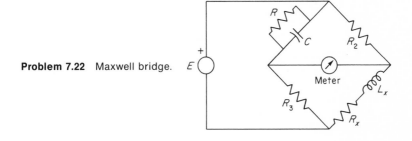

Problem 7.22 Maxwell bridge.

7.23 *Wien Bridge* The following circuit is known as the Wien bridge. It was used in the early days of electronics to measure the frequency of a source. Show that the bridge is balanced if the frequency $f = 1/(2\pi\sqrt{R_3R_4C_3C_4})$.

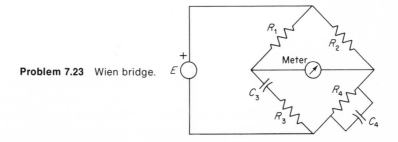

Problem 7.23 Wien bridge.

7.24 *Heaviside Bridge* The following circuit is known as the Heaviside bridge, which can be used to measure the mutual inductance of a coupled coil. Show that the bridge is balanced if $M = (R_2L_3 - R_1L_4)/(R_1 + R_2)$ and $R_1R_4 = R_2R_3$.

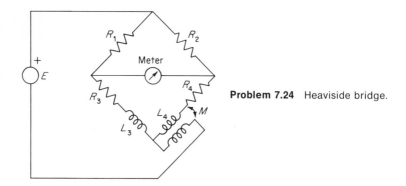

Problem 7.24 Heaviside bridge.

7.25 *Transistor Amplifier* The equivalent circuit of a transistor amplifier is shown below. Find the output voltage V at $\omega = 10^6$ rad/sec.

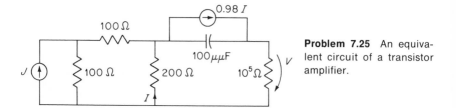

Problem 7.25 An equivalent circuit of a transistor amplifier.

7.26 *Power* Compute the average power delivered to the load in the following circuit. What is the power factor seen by the source? What is the power lost in the 100-Ω resistor?

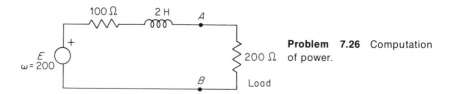

Problem 7.26 Computation of power.

7.27 *Power* In Problem 7.26, suppose a capacitor is placed across AB. What must be the capacitance if maximum power is to be delivered to the load? What is this power? What is the power lost in the 100-Ω resistor?

7.28 *Power* Find the total average power delivered to the load.

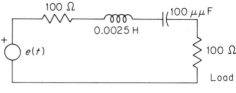

Problem 7.28 Average power when the excitation is a sum of sine waves.

$$e(t) = \cos 10^6 t + \cos 2 \times 10^6 t + \cos 3 \times 10^6 t$$

7.29 *Power* In Problem 7.28, calculate the average energy stored in the inductor and the capacitor.

7.30 *Power* The circuit below is a model of a microwave oscillator cavity which is coupled to a load R_2. The coupling mechanism is adjusted until maximum power is transferred to the load. Find the value of M such that this will take place.

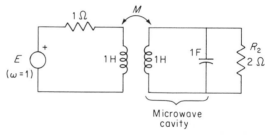

Problem 7.30 Power transmission by coupling.

7.31 *Three-Phase System* The following circuit (a) represents a three-phase power transmission system. The voltage sources have the same amplitude, but are 120° out of phase with respect to one another. Suppose the load calls for a voltage of V volts and an average power of P watts at a power factor of unity. Calculate the power lost in the cable. Now suppose we supply the same power to

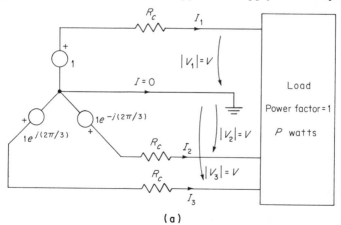

(a)

Problem 7.31 (a) A three-phase power transmission system

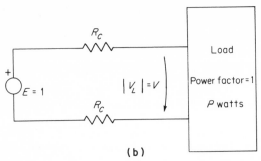

(b)

Problem 7.31   (b) A single-phase power transmission system

the load, using a single-phase system as shown in **(b)**. The load voltage is maintained at V. Show that the line current now is three times that in the three-phase system. Show that the cable loss is six times that in the three-phase system. This problem demonstrates why the power companies prefer to use a three-phase system to supply power to their customers.

7.32 *Three-Phase System* Compute the total instantaneous power of the three sources in a three-phase system. Show that it is proportional to the power factor of the load, but is constant for all times. This means that the torque on the generator is constant for all times rather than pulsing as would be the case if a single-phase system is used. See Figure 7.21.

7.33 *Oscillator* The circuit shown below is the equivalent circuit of an electronic oscillator. The dependent current source is realized by an amplifier. What must be the value of g_m in order that steady-state voltages and currents exist on the circuit? At what frequency can this occur?

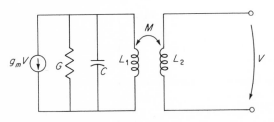

Problem 7.33 An oscillator. Sustained sinusoidal oscillation exists on the circuit in the absence of any excitation by virtue of the presence of the dependent source that is realized by an amplifier in practice.

7.34 *Star-Mesh* Derive the star-mesh transformation for admittances and impedance.

7.35 *Duality* Let the impedance of circuit N be Z. Let the admittance of circuit N' be Y. If it is found that Z and Y have the same numerical value for all frequencies, what can you say about N and N'? Explain. If Z is proportional to Y numerically, what can be said about N and N'? See Problems 7.12 and 7.15.

8

frequency-domain analysis

A signal may be regarded as composed of a sum of sine and cosine functions of different frequencies. In many applications, we wish to alter the signal by passing it through a circuit which may change the amplitude and phase of some of the frequency components while keeping the remainder relatively unaffected. All circuits except those composed entirely of resistors have this "frequency-selective" property. It is this property that is utilized in the extraction of a signal in telephone, telemetry, and data transmission systems; in the removal of unwanted interference in biomedical and geophysical observations; in the detection of signals in the presence of noise in radar and sonar systems; in the shaping of the dynamic response of an automatic control system, for example, by eliminating "overshoot" in an autopilot or tracking radar antenna.

In this chapter, we shall study the frequency-selective property of circuits in detail. Specifically, we wish to examine the steady-state response of a circuit to a sinusoidal excitation as the frequency of the excitation varies. The response is a sinusoid whose amplitude and phase will be functions of the frequency. These functions specify the "frequency characteristic," or "frequency response," of the circuit and are generically called "network functions." Network functions have many interesting properties. We shall derive a number of them and show that the frequency characteristics of a circuit are related to the unforced or natural response (homogeneous solution) of the circuit in the time domain.

The last point is worth noting and is not unexpected. It is the capacitors and inductors in a circuit that give rise to the dynamic response to an excitation. If the frequency of the current in a capacitor increases, we expect the voltage, which is proportional to the average value of the current, to decrease. If the same current is impressed on an inductor, we expect the voltage, which is proportional to the rate of change of the current, to increase. By connecting the elements in some suitable fashion, we can construct circuits which will smooth out the rapidly changing parts of a signal or suppress the slowly varying parts, depending upon the configuration of the circuits. We have seen examples of these in Chapter

7. We shall introduce additional ones which have interesting applications in practice.

Lastly, as we must have found by now, the computation of the steady-state response of circuits is a chore which is best relegated to a computer. For simple circuits, the frequency response can be computed by hand. For circuits with many nodes and loops, computer programs, which will take as input the configuration of a circuit and compute its response at the frequencies specified, have been written. We need not write any equations. At the end of the chapter, we shall present a general circuit analysis program DZNET. The program computes the frequency response of a circuit consisting of resistors, capacitors, inductors, mutual inductors, and dependent sources. For circuits that are in the form of a ladder network, a special computational scheme exists as we noted in Chapter 3. The computer program for computing the frequency response of ladder networks is also given.

8.1   Reactance Function

We start with a simple circuit shown in Figure 8.1. The response V to the excitation I is (in steady state)

$$V = j\omega L I \qquad (8.1)$$

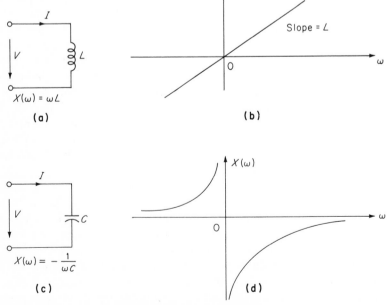

Figure 8.1   Simple reactances.

We define a function, $H(j\omega)$, as

$$H(j\omega) = \frac{\text{Response}}{\text{Excitation}} = \frac{V}{I} = j\omega L \qquad (8.2)$$

$H(j\omega)$ is called the network function, which in this case is the impedance $Z(j\omega) = j\omega L$, a function of the frequency variable $j\omega$. The absolute value of $H(j\omega)$ is ωL, and the phase is $\pi/2$ for $\omega \geq 0$. The significance is that the response is always 90° out of phase with respect to the excitation at all frequencies, and the amplitude is proportional to the frequency.

Now let

$$Z(j\omega) = jX(\omega) \qquad (8.3)$$

so that

$$X(\omega) = \omega L \qquad (8.4)$$

The function $X(\omega)$, which is a function of the real variable ω, is called the *reactance* of the inductor. It is a straight line with slope L, as shown in Figure 8.1(b). The function is extended for negative values of ω (negative frequency) for mathematical convenience.

In a similar fashion, the response V to an excitation I on a capacitor C shown in Figure 8.1(c) is related to I by

$$H(j\omega) = \frac{\text{Response}}{\text{Excitation}} = \frac{V}{I} = \frac{1}{j\omega C} \qquad (8.5)$$

The network function is again the impedance $Z(j\omega) = 1/j\omega C$. The reactance in this case is

$$X(\omega) = -\frac{1}{\omega C} \qquad (8.6)$$

which is sketched in Figure 8.1(d). From this plot and that of Figure 8.1(b), we see that the reactance of an inductor increases from zero (short circuit) at zero frequency to infinity (open circuit) at infinite frequency. The reactance of a capacitor is negative, and it is infinite at zero frequency and increases to zero at infinite frequency. It is reasonable to expect that by combining inductors and capacitors of various values, we would get a reactance that is zero or infinite at finite frequencies.

Consider a series combination of an inductor and capacitor (Figure 8.2). The reactance is

$$X(\omega) = \omega L - \frac{1}{\omega C} = \frac{\omega^2 LC - 1}{\omega C} \qquad (8.7)$$

which is plotted in Figure 8.2(b). It is zero at a finite frequency ω_0 given by

$$\omega_0 = \frac{1}{\sqrt{LC}} \qquad (8.8)$$

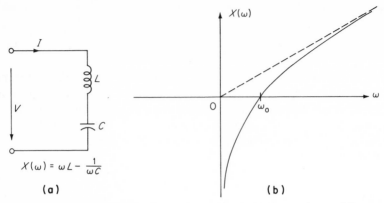

Figure 8.2 A series combination of an inductor and capacitor and its re-actance.

The reactance is negative for $\omega \leq \omega_0$ and positive for $\omega \geq \omega_0$. The significance of the frequency ω_0 can be seen as follows. Suppose the excitation to this simple circuit is a current I at frequency ω. Then the voltage across the inductor, V_L, and that across the capacitor, V_C, are respectively,

$$V_L = j\omega L I$$
$$V_C = \frac{1}{j\omega C I}$$

The respective time functions in steady state are

$$v_L(t) = I\omega L \cos\left(\omega t + \frac{\pi}{2}\right)$$
$$v_C(t) = \frac{I}{\omega C} \cos\left(\omega t - \frac{\pi}{2}\right)$$

The inductor voltage leads the current by 90°, and the capacitor voltage lags the current by 90°. The two voltages are exactly 180° out of phase with respect to each other. Their sum is identically zero at $\omega = \omega_0 (\omega_0 L =$

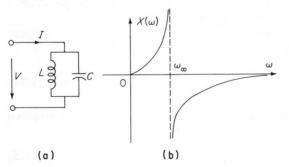

Figure 8.3 A parallel combination of an inductor and capacitor and its reactance.

$1/\omega_0 C$). The combination is equivalent to a short circuit at this frequency since the voltage is zero for all t, irrespective of the current.

Now consider a parallel combination of an inductor and a capacitor (Figure 8.3). The reactance is

$$X(\omega) = \frac{\omega L}{1 - \omega^2 LC} \tag{8.9}$$

which is plotted in Figure 8.3(b). It is seen that the reactance is infinite at a frequency ω_∞, given by

$$\omega_\infty = \frac{1}{\sqrt{LC}} \tag{8.10}$$

The significance of this frequency is as follows. Let the voltage across the combination be V. Then the current in the capacitor, I_C, and that in the inductor, I_L, are given respectively, by,

$$I_C = j\omega C V$$
$$I_L = \frac{1}{j\omega L} V$$

In the time domain in steady state, the respective time functions are

$$i_c(t) = \omega C V \cos\left(\omega t + \frac{\pi}{2}\right)$$
$$i_L(t) = \frac{V}{\omega L} \cos\left(\omega t - \frac{\pi}{2}\right)$$

The two currents are 180° out of phase at all frequencies. At $\omega = \omega_\infty$, the sum of the two currents is identically zero, irrespective of the voltage across the combination. The reactance at this frequency is, therefore, equivalent to an open circuit.

A more complicated combination is shown in Figure 8.4. The impedance is

$$Z(j\omega) = j\omega + \cfrac{1}{j\omega/6 + \cfrac{1}{j\omega(12/5) + \cfrac{1}{j\omega(5/18)}}} \tag{8.11}$$

After simplification, we find the reactance to be

$$X(\omega) = \frac{-(1 - \omega^2)(9 - \omega^2)}{\omega(4 - \omega^2)} \tag{8.12}$$

which is a *rational* function of ω. It is zero at $\omega = \pm 1$ and ± 3, and infinite at 0, ± 2, and ∞. Figure 8.4(b) shows $X(\omega)$ vs. ω. It is interesting to note that the frequencies at which the reactance becomes a short circuit and

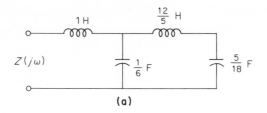

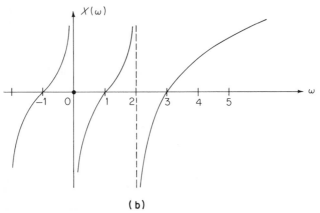

Figure 8.4 The reactance of a *LC* network.

the frequencies at which the reactance becomes an open circuit alternate. This is true for all reactance functions and for any combination of inductors and capacitors. The proof of this assertion is found in any text on advanced network theory.

8.2 Susceptance

What has been said about the impedance of a combination of inductors and capacitors can also be said in terms of the admittance of the combination. Thus the admittance of an inductor is

$$Y(j\omega) = \frac{1}{j\omega L}$$

Let

$$Y(j\omega) = jB(\omega)$$

so that

$$B(\omega) = -\frac{1}{\omega L} \qquad (8.13)$$

The function $B(\omega)$ is a function of the real variable ω and is called the *susceptance* of the inductor. The susceptance of a capacitor is

$$B(\omega) = \omega C$$

Proceeding as before, we find that for any combination of inductors and capacitors, the susceptance is the negative reciprocal of the reactance since

$$B(\omega) = -jY(j\omega) = -j\frac{1}{jX(\omega)} = -\frac{1}{X(\omega)} \qquad (8.14)$$

When the reactance becomes zero, the susceptance becomes infinite. Thus the susceptance of a short circuit is infinite, and the susceptance of an open circuit is zero.

How can we put the frequency characteristics of combinations of inductors and capacitors into good use? The following two sections give some clue.

8.3 *RLC* Series Resonant Circuit

Suppose we insert a series combination of an inductor and capacitor between a realistic voltage source and a "load" resistor as shown in Figure 8.5. Regarding the output voltage V as the response, we define the network function $H(j\omega)$ as

$$H(j\omega) = \frac{\text{Response}}{\text{Excitation}} = \frac{V}{E} = \frac{R_0}{R_g + R_0 + j(\omega L - 1/\omega C)}$$

$$= A(\omega)e^{j\theta(\omega)} \qquad (8.15)$$

where

$$A(\omega) = \frac{R_0}{[R^2 + (\omega L - 1/\omega C)^2]^{1/2}} \qquad (8.16)$$

$$\theta(\omega) = -\tan^{-1}\frac{(\omega L - 1/\omega C)}{R} \qquad (8.17)$$

and

$$R = R_g + R_0 \qquad (8.18)$$

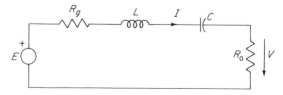

Figure 8.5 A series resonant circuit.

The network function is a ratio of two voltages and is, therefore, dimensionless. It can be interpreted as the output voltage per unit of input voltage and is commonly referred to as the *voltage transfer function*. The absolute value of $H(j\omega)$, $A(\omega)$, is sometimes called the amplitude function or gain of the circuit, and the function $\theta(\omega)$ is called the phase. These are plotted in Figure 8.6.

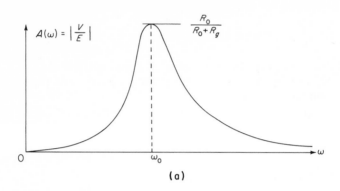

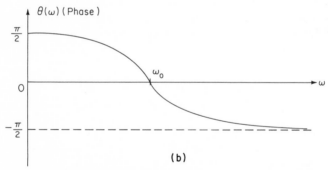

Figure 8.6 The (a) amplitude and (b) phase of the frequency response of a series resonant circuit.

The significance of these two quantities is that if the input voltage is $e(t) = \cos \omega t$, then the output voltage is $v(t) = A(\omega) \cos(\omega t + \theta(\omega))$. The amplitude of $v(t)$ is $A(\omega)$, and its phase is $\theta(\omega)$, both functions of the frequency ω.

The frequency at which the reactance of the circuit is zero is known as the *resonant* frequency, denoted as ω_0. It is given by

$$\omega_0 L - \frac{1}{\omega_0 C} = 0$$

or

$$\omega_0 = \frac{1}{\sqrt{LC}} \tag{8.19}$$

At this frequency the output is maximum and is in phase with the input voltage as can be seen from Figure 8.6. At frequencies above and below ω_0, the output voltage becomes smaller in amplitude and nearly 90° out of phase with the input voltage. This circuit can be used as a simple band-pass filter, since it attenuates frequency components of an input that are far below or above the frequency ω_0, while it leaves those components having frequencies near ω_0 relatively undisturbed. If the input voltage consists of a sum of sinusoids, the output voltage will consist of only those having frequencies near ω_0.

8.4 Energy, Power, and Q of a Resonant Circuit

To deduce further significance of the resonant frequency, we compute the energy stored and dissipated in the circuit and show that at resonance, the energy stored in the capacitor equals the energy stored in the inductor, and that the sharpness of the curve of Figure 8.6, which signifies how selective the circuit is, depends on the ratio of the energy stored to the energy dissipated in the circuit.

With reference to Figure 8.5, we have

$$\text{Total average power} = |I|^2 \frac{R}{2}$$

$$= \frac{1}{2} \frac{E^2 R}{R^2 + (\omega L - 1/\omega C)^2} \tag{8.20}$$

At resonance, $\omega = \omega_0$, and we have

$$\text{Total average power at resonance} = \frac{1}{2} \frac{E^2}{R}$$

Now the average energy stored in the inductor, W_L, is

$$W_L(\omega) = \frac{1}{4}|I|^2 L = \frac{1}{4} \frac{E^2 L}{R^2 + (\omega L - 1/\omega C)^2} \tag{8.21}$$

The average energy stored in the capacitor, W_C, is

$$W_C(\omega) = \frac{1}{4} \frac{|I|^2}{\omega^2 C} = \frac{1}{4} \frac{E^2}{R^2 + (\omega L - 1/\omega C)^2} \frac{1}{\omega^2 C} \tag{8.22}$$

At resonance, $W_L(\omega)$ and $W_C(\omega)$ become, respectively,

$$W_L(\omega_0) = \frac{1}{4} E^2 \frac{L}{R^2} \tag{8.23}$$

$$W_C(\omega_0) = \frac{1}{4} \frac{E^2}{R^2 \omega_0^2 C} = \frac{1}{4} E^2 \frac{L}{R^2} = W_L(\omega_0) \tag{8.24}$$

Thus the energy stored in the inductor equals that stored in the capacitor at resonance. The reactive power of the circuit is zero. (See Section 7.18.)

The total energy stored in the circuit at resonance is

$$W_L(\omega_0) + W_C(\omega_0) = \frac{1}{2} E^2 \frac{L}{R^2} \tag{8.25}$$

or twice that stored in the inductor or capacitor. We now define a very useful quantity known as the Q of the circuit as

$$Q = \frac{\omega_0 \text{ (Total average energy stored in circuit at resonance)}}{\text{(Total average power in circuit at resonance)}} \tag{8.26}$$

Note that Q is dimensionless. In our case, we have

$$Q = \frac{\omega_0 \ (1/2) \ E^2 L/R^2}{(1/2) \ E^2 R} = \frac{\omega_0 L}{R} \tag{8.27}$$

Since $\omega_0^2 = 1/LC$, we also have

$$Q = \frac{1}{\omega_0 C R} \tag{8.28}$$

The quantity Q then is a measure of the relative magnitudes of the energy stored in the circuit and the power dissipated in the circuit. The more power that is dissipated, the lower the Q. If the resistance is zero, the Q is infinite.

There is another point to note about the quantity Q. The gain function $A(\omega)$ [Equation (8.16)] can be written as follows:

$$A(\omega) = \frac{R_0}{\left[R^2 + \left(\omega L - \dfrac{1}{\omega C} \right)^2 \right]^{1/2}}$$

$$= \frac{1}{\left[1 + \left(\dfrac{\omega_0 L}{R} \right)^2 \left(\dfrac{\omega}{\omega_0} - \dfrac{\omega_0}{\omega} \right)^2 \right]^{1/2}} \frac{R_0}{R}$$

or

$$\frac{A(\omega)}{A(\omega_0)} = \frac{1}{\left[1 + Q^2 \left(\dfrac{\omega}{\omega_0} - \dfrac{\omega_0}{\omega} \right)^2 \right]^{1/2}} \tag{8.29}$$

with $A(\omega_0) = R_0/R$ being the maximum value of $A(\omega)$. This expression is plotted in Figure 8.7 for $Q = 10$. Let us calculate the frequencies ω_1 and ω_2 at which the attenuation drops to $1/\sqrt{2} = 0.707$ of its maximum value. Setting the denominator to $\sqrt{2}$, we find, after some algebra,

$$\omega_1 = \omega_0 \left(\sqrt{1 + \frac{1}{4Q^2}} - \frac{1}{2Q} \right) \tag{8.30}$$

$$\omega_2 = \omega_0 \left(\sqrt{1 + \frac{1}{4Q^2}} + \frac{1}{2Q} \right)$$

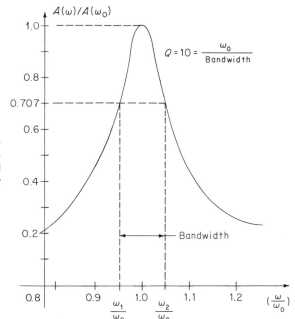

Figure 8.7 The amplitude of the frequency response in terms of Q and the resonant frequency ω_0.

We now note that

$$\omega_1\omega_2 = \omega_0^2 \tag{8.31}$$

so that ω_0 is the geometric mean of ω_1 and ω_2. Moreover, at these frequencies, the output voltage is $1/\sqrt{2}$ times its value at resonance, and the output power is half of its value at resonance. For this reason, ω_1 and ω_2 are also called the "half-power" frequencies.

We also note that

$$\omega_2 - \omega_1 = \frac{\omega_0}{Q} \tag{8.32}$$

The quantity $\omega_2 - \omega_1$ is known as the "bandwidth" of the resonant circuit. It is a measure of the sharpness of the frequency characteristic. It is seen that the higher the Q, the narrower the bandwidth and the more selective the circuit.

Example 8.1   Design a series *RLC* resonant circuit to have a resonance frequency of 10,000 Hz and a bandwidth of 100 Hz. The series and load resistances of the circuit are 10 and 15 Ω, respectively. See Figure 8.8.

Solution   The Q of the circuit is

$$Q = \frac{10,000}{100} = 100$$

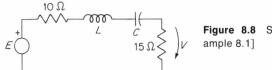

Figure 8.8 Series resonant circuit. [Example 8.1]

The inductance required is

$$L = \frac{QR}{\omega_0} = 39.8 \text{ mH}$$

The capacitance required is

$$C = \frac{1}{\omega_0 R Q} = 6360 \text{ pF}$$

The frequencies at which the output voltage drops to 0.707 times its maximum, which is $A(\omega_0) = 0.6$, are

$$f_1 = 9950 \text{ Hz}$$
$$f_2 = 10{,}050 \text{ Hz}$$

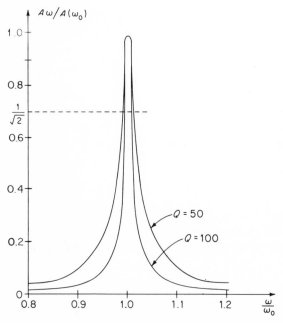

Figure 8.9 The response for two different values of Q. [Examples 8.1, 8.2]

The response is plotted in Figure 8.9. Note that at $f = 9900$ and 10,100 Hz, the output voltage drops to about 0.447 of its maximum. At 9000 and 11,000 Hz, the voltages are 0.0474 and 0.0524, respectively.

Example 8.2   Suppose the Q of the circuit of the last example is halved. Calculate the response of the circuit at the half-power frequencies of the last example.

Solution   The gain is found from

$$\frac{A(\omega)}{A(\omega_0)} = \frac{1}{\left[1 + 2500\left(\frac{\omega}{\omega_0} - \frac{\omega_0}{\omega}\right)^2\right]^{1/2}} \tag{8.33}$$

at $f = 9950$ and 10,050 Hz, we get

$$A(2\pi \times 9950) = 0.895 = A(2\pi \times 10,050)$$

The circuit is now much less selective. Figure 8.9 shows the response compared with that of the last example.   ∎

The frequency response of a circuit is best calculated from a computer program. A FORTRAN program that computes the gain and phase of the last two circuits is given below, together with some sample computer output.

```
C   FREQUENCY RESPONSE OF SERIES RLC CIRCUIT FOR Q=50 AND 100.
        COMPLEX H, J
        J=CMPLX(0., 1.)
        Q=0.
        DO 1 I=1,2
        Q=Q+50.
        W= .8
        WRITE(6,100)Q
 100 FORMAT(1H , 'Q=', F5.1)
        WRITE(6,101)
        DO 1 K=1,40
        W=W+ 0.01
        H= 1./(1. + J*Q*(W - 1./W))
        A=CABS(H)
        B=ATAN(AIMAG(H)/REAL(H))*57.2958
   1 WRITE(6,102)W,A,B
 102 FORMAT(3E15.6)
 101 FORMAT( ' FREQUENCY HZ', 6X, 'AMPLITUDE', 6X, 'PHASE')
        STOP
        END

     Q= 50.0
     FREQUENCY HZ       AMPLITUDE        PHASE
        0.810000E 00    0.470546E-01    0.873029E 02
        0.820000E CC    0.499985E-01    0.871340E 02
        0.830000E 00    0.532834E-01    0.869456E 02
        0.840000E 00    0.569727E-01    0.867339E 02
        0.850000E 00    0.611468E-01    0.864943E C2
        0.860000E 00    0.659087E-01    0.862209E 02
```

8.5   Parallel *RLC* Resonant Circuit

Suppose we insert a parallel combination of an inductor and capacitor between a source and load as shown in Figure 8.10. The voltage transfer function $H(j\omega) = V/E$ is given by

$$H(j\omega) = \frac{R_0 + (j\omega)^2 R_0 LC}{R + j\omega L + (j\omega)^2 RLC} = A(\omega)e^{j\theta(\omega)} \tag{8.34}$$

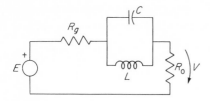

Figure 8.10 A parallel resonant circuit.

where $R = R_0 + R_g$. It is seen that the transfer function is a rational function of the variable $j\omega$. At the frequency ω_∞ given by

$$\omega_\infty = \frac{1}{\sqrt{LC}} \tag{8.35}$$

the output voltage is zero. This is the resonant frequency of the LC parallel combination. At this frequency, the combination is equivalent to an open circuit, so that the current in R_0 is zero, and hence the output voltage is zero.

The expression for $H(j\omega)$ can be put into the following form:

$$H(j\omega) = \frac{R_0}{R} \frac{\left(\dfrac{\omega_\infty}{\omega} - \dfrac{\omega}{\omega_\infty}\right)}{\dfrac{\omega_\infty L}{R}\left[\dfrac{R}{\omega_\infty L}\left(\dfrac{\omega_\infty}{\omega} - \dfrac{\omega}{\omega_\infty}\right) + j\right]} \tag{8.36}$$

Defining Q as before, we find, in this case,

$$Q = \frac{R}{\omega_\infty L} = \omega_\infty C R \tag{8.37}$$

The gain $A(\omega)$ and phase $\theta(\omega)$ are given, respectively, by

$$A(\omega) = \frac{R_0}{R} \frac{Q\left|\dfrac{\omega_\infty}{\omega} - \dfrac{\omega}{\omega_\infty}\right|}{\left[Q^2\left(\dfrac{\omega_\infty}{\omega} - \dfrac{\omega}{\omega_\infty}\right)^2 + 1\right]^{1/2}} \tag{8.38}$$

$$\theta(\omega) = -\tan^{-1}\frac{1}{Q\left(\dfrac{\omega_\infty}{\omega} - \dfrac{\omega}{\omega_\infty}\right)} \qquad \text{for } \omega < \omega_\infty \tag{8.39}$$

$$= -\tan^{-1}\frac{1}{Q\left(\dfrac{\omega_\infty}{\omega} - \dfrac{\omega}{\omega_\infty}\right)} + \pi \qquad \text{for } \omega > \omega_\infty$$

These functions of ω are plotted in Figure 8.11 for $Q = 22.6$, $\omega_\infty = 2\pi \times 60$ and $R_0/R = 1/1.1$. Note that $A(\omega_\infty) = 0$ as expected, and that the phase has a discontinuity of π at $\omega = \omega_\infty$.

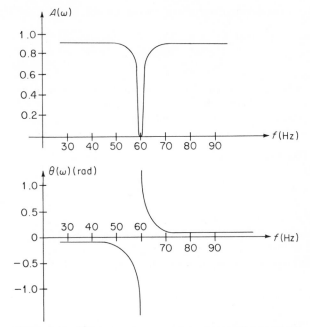

Figure 8.11 The frequency response of a parallel resonant circuit.

At frequencies ω_1 and ω_2 given by

$$\omega_1 = \omega_\infty \left(\sqrt{1 + \frac{1}{4Q^2}} - \frac{1}{2Q} \right)$$

$$\omega_2 = \omega_\infty \left(\sqrt{1 + \frac{1}{4Q^2}} + \frac{1}{2Q} \right) \qquad (8.40)$$

the gain is 0.707 times the maximum value, that is,

$$A(\omega_1) = A(\omega_2) = \frac{R_0}{R} \frac{1}{\sqrt{2}}$$

The bandwidth of the circuit is found to be

$$\text{Bandwidth} = \omega_2 - \omega_1 = \frac{\omega_\infty}{Q} \qquad (8.41)$$

Again, the higher the Q, the smaller the bandwidth and the more selective the circuit. The circuit can be used to remove a sinusoid of frequency ω_∞ from the input signal, and it is an example of a simple band-elimination or band-rejection filter.

Example 8.3   Let the input voltage of Figure 8.10 be

$$e(t) = 0.1(\cos 2\pi\ 10t + \tfrac{1}{9} \cos 2\pi\ 30t + \tfrac{1}{25} \cos 2\pi\ 50t$$
$$+ \tfrac{1}{49} \cos 2\pi\ 70t + \tfrac{1}{81} \cos 2\pi\ 90t)$$
$$+ \cos 2\pi\ 60t$$

which is seen to be composed of a weak signal superimposed on a strong 60-Hz interference. Let us choose values for L and C such that the output is free of the 60-Hz component and that the amplitudes of the other components are not less than 0.9 times their input values. Let $R_g = 1000$ and $R_0 = 10,000\ \Omega$.

Solution   The resonant frequency must be $2\pi\ 60$ so that

$$LC = \frac{1}{(2\pi\ 60)^2} \tag{8.42}$$

We must now select Q so that the amplitude requirements are satisfied. From Equation (8.38), we note that at low and high frequencies, the amplitude is about unity. Near the resonant frequency, the amplitude decreases rapidly. We should, therefore, find Q from the amplitude of the response at 70 Hz. We have

$$A(2\pi\ 70) = 0.9 = \frac{1}{1.1}\ \ \frac{Q(6/7 - 7/6)}{[Q^2(6/7 - 7/6)^2 + 1]^{1/2}} \tag{8.43}$$

Solving for Q, we get $Q = 22.6$. Using this value of Q, we find the response to be:

Frequency (Hz)	Amplitude
10	0.910
30	0.909
50	0.903
70	0.900
90	0.908

so that a Q of 22.6 will satisfy the design specifications. Using Equations (8.37) and (8.42), we find the element values to be

$$L = 1.29\ \text{H} \quad \text{and} \quad C = 5.45\ \mu\text{F}$$

The frequency response is plotted in Figure 8.11. To compute the steady-state output voltage, we make use of the following FORTRAN program, which includes an excerpt of computer output.

```
C  A SIMPLE BAND REJECTION FILTER TO REMOVE A 60 HZ
C  INTERFERENCE.
C
      DIMENSION A(10), B(10)
      COMPLEX H, J
      WRITE(6,99)
   99 FORMAT(1H , 12X, 'TIME', 11X, 'INPUT', 10X, 'OUTPUT')
      J=CMPLX(0., 1.)
C  ASSIGN ELEMENT VALUES.
      XL= 1.29
      C= 5.45E-06
      R=10000.
      RG=1000.
      W6= 60*6.283185
```

```
C   COMPUTE THE FREQUENCY RESPONSE AT 10, 30, 50, 70 AND 90 HZ.
        DO 1 I=1,5
        K=2*I -1
        W=62.83185*K
        H=R/(R +RG + J*W*XL/(1. - W*W*XL*C))
        A(K) =CABS(H)
    1 B(K) =ATAN(AIMAG(H)/REAL(H))
        H= R/(R +RG + J*W6*XL/(1. - W6*W6*XL*C))
        A(6)=CABS(H)
        B(6)=ATAN(AIMAG(H)/REAL(H))
C   WE NOW EVALUATE THE TIME RESPONSE.
        T=0.
        DO 2 N=1,100
        T=T+.001
        E =0.
        V =0.
        DO 3 M=1, 5
        ODD=2*M - 1
        E=E + COS(ODD*62.83185*T)/(ODD*ODD)
    3 V= V+A(2*M -1)*COS(ODD*62.83185*T + B(2*M-1))/(ODD*ODD)
        E = 0.1*E + COS(W6*T)
        V = 0.1*V + A(6)*COS(W6*T + B(6))
    2 WRITE(6, 100)T,E,V
  100 FORMAT(1H , 5X, 3E15.6)
        STOP
        END
```

TIME	INPUT	OUTPUT
0.100000E-02	0.104719E 01	0.113142E 00
0.200000E-02	0.843574E 00	0.115926E 00
0.300000E-02	0.536094E 00	0.115762E 00
0.400000E-02	0.167817E 00	0.112586E 00
0.500000E-02	-0.209751E 00	0.106658E 00
0.599999E-02	-0.543933E 00	0.984955E-01
0.699999E-02	-0.788292E 00	0.887896E-01
0.799999E-02	-0.909151E 00	0.783150E-01
0.899999E-02	-0.890291E 00	0.678505E-01
0.999999E-02	-0.735185E 00	0.581154E-01
0.110000E-01	-0.466451E 00	0.497166E-01
0.120000E-01	-0.122636E 00	0.431063E-01
0.130000E-01	0.247232E 00	0.385435E-01
0.140000E-01	0.590535E 00	0.360620E-01
0.150000E-01	0.858437E 00	0.354497E-01
0.160000E-01	0.101272E 01	0.362499E-01
0.170000E-01	0.103109E 01	0.377934E-01
0.180000E-01	0.910337E 00	0.392689E-01
0.190000E-01	0.666708E 00	0.398270E-01
0.200000E-01	0.333674E 00	0.387067E-01
0.210000E-01	-0.427608E-01	0.353630E-01
0.220000E-01	-0.410495E 00	0.295697E-01
0.230000E-01	-0.718618E 00	0.214747E-01
0.240000E-01	-0.924544E 00	0.115902E-01
0.250000E-01	-0.100000E 01	0.719177E-03
0.260000E-01	-0.935009E 00	-0.101745E-01
0.270000E-01	-0.739321E 00	-0.201254E-01
0.280000E-01	-0.441067E 00	-0.283262E-01
0.290000E-01	-0.828261E-01	-0.342550E-01
0.300000E-01	0.284352E 00	-0.377499E-01
0.310000E-01	0.608133E 00	-0.390177E-01
0.319999E-01	0.842271E 00	-0.385806E-01
0.329999E-01	0.953136E 00	-0.371776E-01
0.339999E-01	0.924455E 00	-0.356415E-01
0.349999E-01	0.759607E 00	-0.347771E-01
0.359999E-01	0.481134E 00	-0.352604E-01
0.369999E-01	0.127550E 00	-0.375683E-01
0.379999E-01	-0.252106E 00	-0.419433E-01
0.389999E-01	-0.605183E 00	-0.483861E-01
0.399999E-01	-0.882835E 00	-0.566678E-01
0.409999E-01	-0.104687E 01	-0.663558E-01
0.419999E-01	-0.107508E 01	-0.768467E-01
0.429999E-01	-0.964342E 00	-0.874089E-01

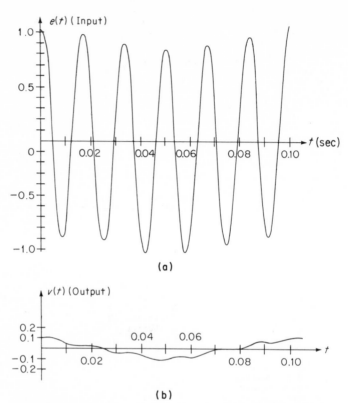

Figure 8.12 The input signal to the circuit of Figure 8.10 consists of a weak component buried in a strong 60-Hz interference. The circuit removes the interference, and the weak signal component is recovered at the output. [Example 8.3]

The input and output waveforms are plotted in Figure 8.12. Note that the weak signal is hardly noticeable in the input waveform. After the 60-Hz component is filtered out, the weak signal reveals itself at the output.

8.6 A Simple Broadband Circuit

As another illustration of what circuits can do, consider the circuit of Figure 8.13(a). The voltage transfer function is

$$\frac{V}{E} = H(j\omega) = \frac{R_2}{R_1 + R_2 + j\omega R_1 R_2 C} = A(\omega) e^{j\theta(\omega)} \tag{8.44}$$

The gain and phase are given by

$$A(\omega) = \frac{R_2}{R_1 + R_2} \frac{1}{[1 + \omega^2 R^2 C^2]^{1/2}} \tag{8.45}$$

$$\theta(\omega) = -\tan^{-1} \omega RC \tag{8.46}$$

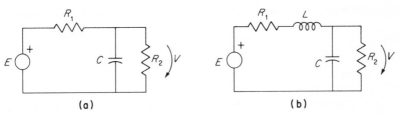

Figure 8.13 The addition of a series inductor as in (**b**) to the circuit of (**a**) improves its frequency response.

where $R = R_1 R_2/(R_1 + R_2)$. The gain is plotted in Figure 8.14 with label (a). It is seen that the output voltage drops to 0.707 times its maximum value at a frequency ω_0 given by

$$\omega_0 = \frac{1}{RC} \tag{8.47}$$

Figure 8.14 The frequency response of the circuit of Figure 8.13b for different values of Q or equivalently L.

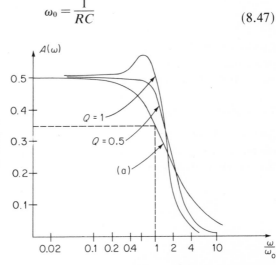

Applying the same terminology of the resonant circuit to this case, we call ω_0 the half-power frequency, and the bandwidth of the circuit is $\omega_0(0 \text{ to } \omega_0)$. In practice, the parallel RC combination may be the equivalent circuit of an antenna. The input signal is applied to the antenna for broadcasting, for example. Because of the frequency characteristics of the antenna, we see that the output voltage will be distorted. The high-frequency components are attenuated. In fact, if the frequency of the input is much higher than $\omega_0/2\pi$, there will be hardly any output. What can we do to broaden the bandwidth of the antenna? A simple scheme is to add a series inductor to the circuit as shown in Figure 8.13(b). The voltage transfer function is now

$$H(j\omega) = \frac{R_2}{R_1 + R_2 - \omega^2 LCR_2 + j\omega R_2 (CR_1 + L/R_2)} = A(\omega)e^{j\theta} \tag{8.48}$$

Solving for $A(\omega)$, we get

$$A(\omega) = \frac{R_2}{R_1 + R_2} \frac{1}{[(1 - Q\omega^2/\omega_0^2)^2 + (\omega/\omega_0)^2(1 + QR_1/(R_1 + R_2))^2]^{1/2}}$$
(8.49)

where $Q = \omega_0 L/R_1$.

Examination of Equation (8.49) shows that for values of ω such that

$$2 > Q(\omega/\omega_0)^2 + \frac{2R_1}{(R_1 + R_2)} + \frac{QR_1^2}{(R_1 + R_2)^2}$$
(8.50)

the output voltage will be greater than what it was before the addition of the inductor. For ω such that

$$2 < Q\frac{\omega^2}{\omega_0^2} + \frac{2R_1}{(R_1 + R_2)} + \frac{QR_1^2}{(R_1 + R_2)^2}$$
(8.51)

the output is less. Figure 8.14 shows the sketch of $A(\omega)$ for $Q = 0.5$ and 1.0. It is seen that the frequency characteristic has been broadened in the sense that the components of the input signal that have frequencies near ω_0 are not attenuated as much as before the insertion of the inductor.

8.7 Low-Pass Filter

An example of a circuit which is used to suppress high-frequency noise is shown in Figure 8.15. Writing the mesh equations and then solving for the output voltage V, we get

$$\begin{bmatrix} 1 + 1/(j\omega) & -1/(j\omega) & 0 \\ -1/(j\omega) & j2\omega + 2/(j\omega) & -1/(j\omega) \\ 0 & -1/(j\omega) & 1 + 1/(j\omega) \end{bmatrix} \begin{bmatrix} I_1 \\ I_2 \\ I_3 \end{bmatrix} = \begin{bmatrix} 1 \\ 0 \\ 0 \end{bmatrix}$$
(8.52)

$$\frac{V}{E} = H(j\omega) = \frac{I_3}{E}$$

$$H(j\omega) = \frac{0.5}{(j\omega)^3 + 2(j\omega)^2 + 2(j\omega) + 1} = A(\omega)e^{j\theta(\omega)}$$
(8.53)

The voltage transfer function is again a rational function of the variable $j\omega$. The gain $A(\omega)$ is given by

$$A(\omega) = \frac{0.5}{[1 + \omega^6]^{1/2}}$$
(8.54)

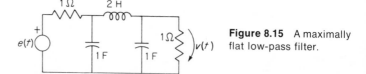

Figure 8.15 A maximally flat low-pass filter.

which has a Taylor series expansion about $\omega = 0$ given by

$$A(\omega) = 0.5(1 - \tfrac{1}{2}\omega^6 + \tfrac{3}{8}\omega^{12} \cdots) \tag{8.55}$$

Therefore, for small ω (low frequency), $A(\omega)$ is a constant up to the sixth derivative of $A(\omega)$. $H(j\omega)$ as given is the lowest-order rational function that has this property and is commonly called the *maximally flat function* of order 3. Much attention has been given to this class of rational functions and their corresponding circuits. Formulas that give the explicit expressions for the element values in terms of the complexity, or order, of the filter have been derived. (See **W.R. Bennett**, U.S. Patent No. 1,849,656.)

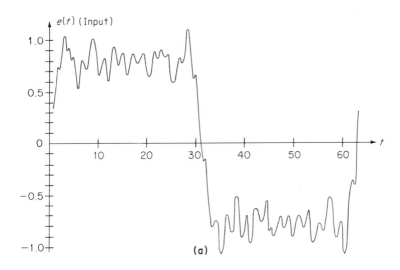

(a)

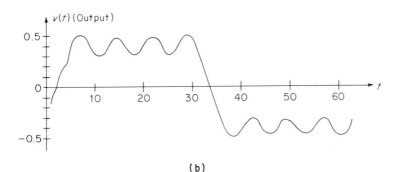

(b)

Figure 8.16 The input (a) to a low-pass filter consists of a signal and high-frequency noise. The filter removes the noise, and the output is essentially the original signal.

To show what we can do with this circuit, let us calculate the output voltage $v(t)$ when the input voltage $e(t)$ is given by

$$e(t) = \sin 0.1t + \tfrac{1}{3}\sin 0.3t + \tfrac{1}{5}\sin 0.5t + \tfrac{1}{7}\sin 0.7t$$
$$+ \tfrac{1}{9}\sin 0.9t + 0.1\sin 3t + 0.1\sin 5t$$

The first five terms constitute an approximation of a square wave. Superimposed on the square wave is noise, which is represented by the sum of the last two terms. Figure 8.16(a) shows a plot of $e(t)$. The output voltage is found from a computer program, and it is plotted in Figure 8.16(b). Notice how the low-pass filter has smoothed the signal by removing the high-frequency components.

A FORTRAN program to compute the input and output voltages is given below, together with some sample computer output.

```
C   A MAXIMALLY FLAT LOW PASS FILTER OF ORDER THREE USED TO
C   SUPPRESS HIGH FREQUENCY NOISE SUPERIMPOSED ON A SQUARE
C   WAVE.
C   DEFINE AMPLITUDE FUNCTION A(W) AND PHASE FUNCTION B(W).
        A(W) = 0.5/SQRT(1. + W**6)
        B(W) = -ATAN((2. - W*W)*W/(1. - 2.*W*W))
        WRITE(6,101)
  101 FORMAT(1H ,  '    TIME', 11X, 'INPUT', 10X, 'OUTPUT')
        T = 0.
        DO 1 I=1,315
        T = T + 0.2
        E = SIN(.1*T) + SIN(.3*T)/3. + SIN(.5*T)/5.
        E = E + SIN(.7*T)/7. + SIN(.9*T)/9.
        E = E + .1*SIN(3.*T) + .1*SIN(5.*T)
        V = A(.1)*SIN(.1*T + B(.1)) + A(.3)*SIN(.3*T + B(.3))/3.
        V = V + A(.5)*SIN(.5*T + B(.5))/5. + A(.7)*SIN(.7*T + B(.7))/7.
        V = V + A(.9)*SIN(.9*T + B(.9))/9. + .1*A(3.)*SIN(3.*T + B(3.))
        V = V + A(5.)*SIN(5.*T + B(5.)) * .1
    1 WRITE(6,100)T, E, V
  100 FORMAT(1H , 3E15.6)
        STOP
        END
```

TIME	INPUT	OUTPUT
0.200000E 00	0.240391E 00	-0.282617E 00
0.400000E 00	0.382382E 00	-0.252300E 00
0.600000E 00	0.405619E 00	-0.221560E 00
0.800000E 00	0.378047E 00	-0.190984E 00
0.100000E 01	0.391512E 00	-0.160639E 00

8.8 Lumped Delay Line

The examples of the last four sections were concerned mainly with the amplitude of the output voltage. There are occasions in practice in which the phase characteristic is more important.

Consider the problem of designing a circuit whose output voltage is the delayed replica of its input. If the input voltage is $e(t)$, we seek a circuit whose output is $v(t) = Ke(t - T)$, where K is a constant, and T is the delay in time. It is, of course, possible to use a length of transmission line for this purpose. However, the length of the line would be too long for practical use. For example, suppose the required delay is 1 μsec, the length would be $1 \times 10^{-6} \times 10^{10}$ cm or approximately 1000 ft.

Let us regard the input signal as a sum of sinusoids:

$$e(t) = A_1 \cos \omega_1 t + A_2 \cos \omega_2 t + \cdots + A_n \cos \omega_n t \qquad (8.56)$$

The desired output voltage $v(t)$ is

$$
\begin{aligned}
v(t) &= K\{A_1 \cos[\omega_1(t - T)] + A_2 \cos[\omega_2(t - T)] + \cdots \\
&\quad + A_n \cos[\omega_n(t - T)]\} \\
&= K\{A_1 \cos(\omega_1 t - \omega_1 T) + A_2 \cos(\omega_2 t - \omega_2 T) + \cdots \\
&\quad + A_n \cos(\omega_n t - \omega_n T)\}
\end{aligned}
\qquad (8.57)
$$

The phase of the first term is $-\omega_1 T$, the second $-\omega_2 T$, and so on; namely, the phase is proportional to frequency.

Let $H(j\omega)$ be the transfer function of the circuit that has this property. We are looking for a circuit whose transfer function is such that

$$H(j\omega) = A(\omega)e^{j\theta(\omega)}$$

with

$$A(\omega) = K \qquad \text{(independent of frequency)}$$

and

$$\theta(\omega) = -\omega T \qquad \text{(a linear function of frequency)} \qquad (8.58)$$

It turns out that such a frequency characteristic cannot be realized by any lumped circuit. It is possible, however, to approximate the linear phase characteristic with a circuit composed of lumped elements. One such circuit is shown in Figure 8.17.

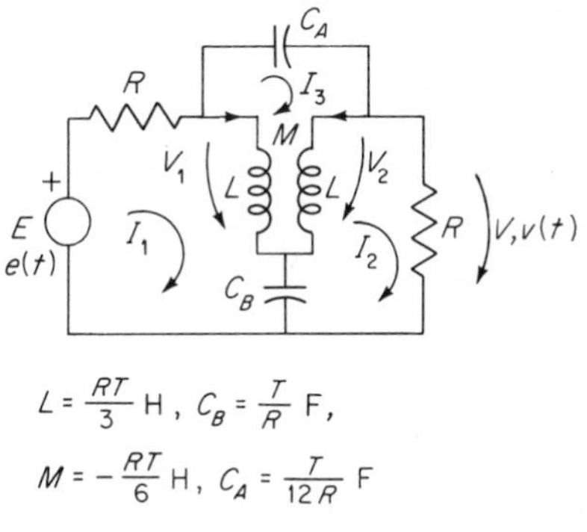

Figure 8.17 A circuit whose phase characteristic is approximately a linear function of frequency.

$$L = \frac{RT}{3} \text{ H}, \quad C_B = \frac{T}{R} \text{ F},$$

$$M = -\frac{RT}{6} \text{ H}, \quad C_A = \frac{T}{12R} \text{ F}$$

Writing loop equations, we get

$$
\begin{bmatrix}
R + 1/(j\omega C_B) & -1/(j\omega C_B) & 0 \\
-1/(j\omega C_B) & R + 1/(j\omega C_B) & 0 \\
0 & 0 & 1/(j\omega C_A)
\end{bmatrix}
\begin{bmatrix}
I_1 \\
I_2 \\
I_3
\end{bmatrix}
=
\begin{bmatrix}
E - v_1 \\
v_2 \\
v_1 - v_2
\end{bmatrix}
\qquad (8.59)
$$

with

$$v_1 = j\omega L(I_1 - I_3) + j\omega M(I_3 - I_2)$$
$$v_2 = j\omega M(I_1 - I_3) + j\omega L(I_3 - I_2)$$

Using the relations that exist among the element values, we get

$$R \begin{bmatrix} 1 + (1/3)j\omega T + 1/(j\omega T) & -1/(j\omega T) + (1/6)j\omega T & -(1/2)j\omega T \\ -1/(j\omega T) + (1/6)j\omega T & 1 + (1/3)j\omega T + 1/(j\omega T) & -(1/2)j\omega T \\ -(1/2)j\omega T & -(1/2)j\omega T & 12/(j\omega T) + j\omega T \end{bmatrix} \begin{bmatrix} I_1 \\ I_2 \\ I_3 \end{bmatrix} = \begin{bmatrix} E \\ 0 \\ 0 \end{bmatrix}$$

$$(8.60)$$

Solving for the voltage transfer function and canceling a common factor in the numerator and denominator, we get

$$H(j\omega) = \frac{1}{2} \frac{1 - (1/2)j\omega T + (1/12)(j\omega T)^2}{1 + (1/2)j\omega T + (1/12)(j\omega T)^2} = A(\omega)e^{j\theta(\omega)} \quad (8.61)$$

which is a rational function of the variable $j\omega$. The gain and phase are, respectively,

$$A(\omega) = 0.5 \qquad \text{(independent of frequency)} \qquad (8.62)$$

$$\theta(\omega) = -2\tan^{-1}\frac{(1/2)\omega T}{1 - (1/12)(\omega T)^2} \qquad (8.63)$$

Expanding $\theta(\omega)$ in Taylor series about $\omega = 0$, we obtain

$$\theta(\omega) = -\omega T + \frac{1}{720}(\omega T)^5 - \frac{1}{12,096}(\omega T)^7 + \cdots \qquad (8.64)$$

Clearly, $\theta(\omega)$ is approximately a linear function of ω. Moreover, the proportionality constant is T, the delay. The following table shows how good the approximation is.

ωT	$\theta(\omega)$ (Exact)
0.1	−0.10000
0.2	−0.19999
0.4	−0.39998
0.6	−0.59989
0.8	−0.79956
1.0	−0.99869
1.5	−1.49084
2.0	−1.96558

From this table, we see that if a signal is composed of frequency components not higher than $\omega = 1/T$, the output will be a delayed replica of the input. As an example, suppose the input signal is

$$e(t) = \sin 0.5t + 0.2\sin t \qquad (8.65)$$

The output voltage of the circuit is, for $T = 1$,

$$v(t) = 0.5\sin(0.5t - 0.49996) + 0.1\sin(t - 0.99869) \qquad (8.66)$$

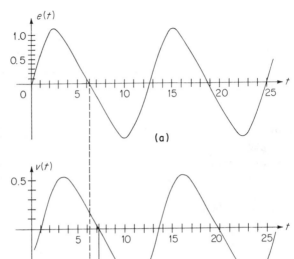

Figure 8.18 (a) The input to a lumped delay line; (b) the output, which is proportional to the input but is delayed by 1 sec.

The two voltages are plotted in Figure 8.18. It is seen that the output voltage is indeed delayed by 1 sec from the input. (The simplest way to see this is to note the zero crossings of the two waveforms.) The following FORTRAN program calculates the phase and the two voltages.

```
C   LUMPED DELAY LINE. PHASE CHARACTERISTIC AND CALCULATION
C   OF INPUT AND OUTPUT VOLTAGES OF CIRCUIT OF FIG. 8.17.
C   DEFINE PHASE FUNCTION.
      PHS(X) = -2.*ATAN(.5*X/(1. - X*X/12))
      X = 0.
      WRITE(6,101)
      DO 1 I=1,30
      X = X + 0.1
      P = PHS(X)
      ERROR = X + P
    1 WRITE(6,100)X, P, ERROR
  100 FORMAT(3E15.6)
  101 FORMAT(6X, 'WT', 11X, 'PHASE', 10X, 'ERROR')
      WRITE(6,102)
      T = 0.
      DO 2 I=1,60
      T = T + .5
      E = SIN(.5*T) + 0.2*SIN(T)
      V = .5*SIN(.5*T + PHS(.5)) + 0.1*SIN(T + PHS(1.))
    2 WRITE(6,103)T, E, V
  102 FORMAT(1H , 'TIME', 11X, 'INPUT', 10X, 'OUTPUT')
  103 FORMAT(3E15.6)
      STOP
      END
```

8.9 Transistor Amplifier

We saw examples of transistor amplifiers in Chapters 2 and 3 which were circuits consisting of resistors and dependent sources. We noted it was possible to obtain gain of value greater than unity in such circuits. When the input signal has high-frequency components, the circuit which was used to represent a transistor is no longer valid and must be modified to account for the diffusion and drift phenomena of the charge carriers in the material that makes up the transistor. Many equivalent circuits have been proposed. Figure 8.19 shows one such circuit. The transistor is inserted between a source and a load. Let us analyze the frequency characteristics of the amplifier.

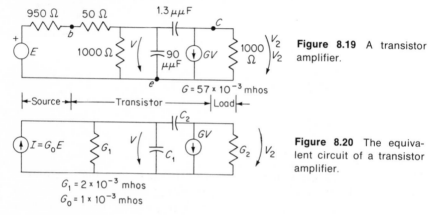

Figure 8.19 A transistor amplifier.

Figure 8.20 The equivalent circuit of a transistor amplifier.

$$G_1 = 2 \times 10^{-3}\ \text{mhos}$$
$$G_0 = 1 \times 10^{-3}\ \text{mhos}$$

Converting the source into a current source and writing node equations, we get, with reference to Figure 8.20

$$\begin{bmatrix} G_1 + j\omega(C_1 + C_2) & -j\omega C_2 \\ -j\omega C_2 + G & j\omega C_2 + G_2 \end{bmatrix} \begin{bmatrix} V_1 \\ V_2 \end{bmatrix} = \begin{bmatrix} I \\ 0 \end{bmatrix} \tag{8.67}$$

Solving for V_2, we obtain the voltage transfer function:

$$\frac{V_2}{E} = \frac{G_0(-G + j\omega C_2)}{G_1 G_2 + j\omega(C_2 G_1 + C_1 G_2 + C_2 G_2 + C_2 G) + (j\omega)^2 C_1 C_2} \tag{8.68}$$

Again, the transfer function is a rational function of the variable $j\omega$. At low frequency ($\omega \cong 0$), the voltage transfer function is

$$H(j\omega) = \frac{-G G_0}{G_1 G_2} = -28.5 \tag{8.69}$$

which is a gain of 28.5 with a phase of 180°. At high frequencies $\omega^2 C_1 C_2 \gg G_1 G_2$, the gain becomes

$$H(j\omega) = \frac{-jG_0}{\omega C_1} \tag{8.70}$$

The output voltage decreases as $1/\omega$ and the phase approaches $-90°$. For intermediate values of ω, the gain and phase are computed, and the results are plotted in Figure 8.21. Note that above 50 MHz, the gain is less than unity. Thus the presence of the capacitor C_1 causes the transistor to cease to be an amplifier at high frequencies, in spite of the presence of a dependent source.

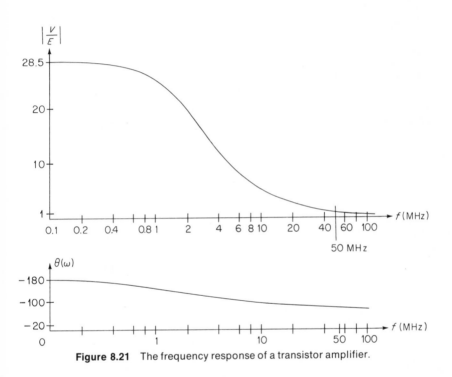

Figure 8.21   The frequency response of a transistor amplifier.

8.10   Network Functions of a Complex Variable*

By now, we should have noted that the network function (impedance, voltage transfer function, and so on) of a circuit is a rational function (ratio of two polynomials) of the variable $j\omega$. It is from the network function that we calculate the amplitude and phase of the various frequency components of the response when a signal is applied to the circuit.

Given an arbitrary circuit, we can always find the network function. Given a rational function of the variable $j\omega$, can we always find a circuit which has as its network function the given rational function? The answer is that we cannot always do so. The reason is that not every rational function can be realized as a network function of some physical circuit. Of all possible rational functions, only a small class are physically realizable. How do we find or define this class? To answer these and other questions, we must study the mathematical properties of network functions. This is done in an advanced course on network theory. In this book, we shall deduce a few obvious properties which are within our grasp.

To this end, we first generalize the frequency variable $j\omega$ to a complex frequency variable s:

$$s = \sigma + j\omega = \sqrt{\sigma^2 + \omega^2}\, e^{j\phi}$$

$$\phi = \tan^{-1}\frac{\omega}{\sigma}$$

where σ is the real part of s, and ω the imaginary part. The ordinary variable ω is seen to be simply the imaginary part of s. To each pair of values (σ, ω) there corresponds a complex value of s, which can be represented as a point on a complex plane as shown in Figure 8.22.

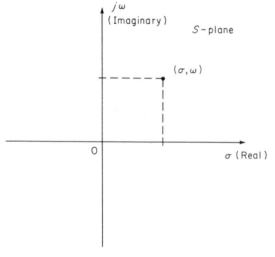

Figure 8.22 The complex frequency plane.

In terms of this new variable, our excitation takes the form Xe^{st}, and the response takes the form Ye^{st}. Wherever we had $j\omega$ before, we now have s. Thus the impedance of an inductor L is sL; the admittance of a capacitor C is sC, and so on. Continuing in this fashion, we find the network functions are rational functions of s.

As an example, consider the circuit of Figure 8.15. The loop equations are now

$$\begin{bmatrix} 1 + 1/s & -1/s & 0 \\ -1/s & 2s + 2/s & -1/s \\ 0 & -1/s & 1 + 1/s \end{bmatrix}\begin{bmatrix} I_1 \\ I_2 \\ I_3 \end{bmatrix} = \begin{bmatrix} 1 \\ 0 \\ 0 \end{bmatrix} \qquad (8.71)$$

and the voltage transfer function is

$$H(s) = \frac{1/2}{s^3 + 2s^2 + 2s + 1} \qquad (8.72)$$

The impedance of the LC circuit of Figure 8.4 is now

$$Z(s) = s + \cfrac{1}{(1/6)s + \cfrac{1}{(12/15)s + \cfrac{1}{(5/18)s}}}$$

$$= \frac{(s^2 + 1)(s^2 + 9)}{s(s^2 + 4)} = \frac{s^4 + 10s^2 + 9}{s^3 + 4s} \qquad (8.73)$$

The network function of the lumped delay line is now

$$H(s) = \frac{1}{2}\frac{1 - (1/2)s + (1/12)s^2}{1 + (1/2)s + (1/12)s^2} \qquad (8.74)$$

What do these network functions have in common?

8.11 Elementary Properties of Network Functions*

Consider an arbitrary circuit. Without loss of generality, we assume there is one excitation, and it is in the first loop. The loop equations are

$$\begin{bmatrix} Z_{11}(s) & Z_{12}(s) & \cdots & Z_{1M}(s) \\ Z_{21}(s) & Z_{22}(s) & \cdots & Z_{2M}(s) \\ \vdots & \vdots & \ddots & \vdots \\ Z_{M1}(s) & Z_{M2}(s) & \cdots & Z_{MM}(s) \end{bmatrix} \begin{bmatrix} I_1 \\ I_2 \\ \vdots \\ I_M \end{bmatrix} = \begin{bmatrix} E \\ 0 \\ \vdots \\ 0 \end{bmatrix}$$

where $Z_{ij}(s) = R_{ij} + sL_{ij} + 1/(sC_{ij})$, with R_{ij}, L_{ij}, and C_{ij} as defined in Chapter 7.

Solving for the current I_k, we get

$$I_k = \frac{\Delta_{1k}}{\Delta} E \qquad (8.76)$$

where

$$\Delta \quad = \text{determinant of the matrix of coefficients}$$
$$\Delta_{ik} = \text{the } (i,k) \text{ cofactor of } \Delta$$

Now suppose an impedance Z is in loop k. Let the voltage across it be V. Then the voltage transfer function $H(s)$ will be

$$H(s) = \frac{V}{E} = \frac{I_k Z(s)}{E} = \frac{Z\Delta_{1k}}{\Delta} \qquad (8.77)$$

The input impedance seen by the source is

$$Z_{\text{in}}(s) = \frac{E}{I} = \frac{\Delta}{\Delta_{11}} \tag{8.78}$$

In general, the network function is a ratio of two determinants. Now the expansion of a determinant of order M is the sum of $M!$ products of M terms, each product consisting of one element from each row such that no two elements from the same column appear in the same product. Each product will, therefore, be of the form

$$\left(R_{i_1 j_1} + sL_{i_1 j_1} + \frac{1}{sC_{i_1 j_1}}\right)\left(R_{i_2 j_2} + sL_{i_2 j_2} + \frac{1}{sC_{i_2 j_2}}\right)$$
$$\cdots (R_{i_M j_M} + sL_{i_M j_M} + 1/sC_{i_M j_M})$$

where $i_1 \neq i_2 \neq i_3 \cdots \neq i_M$, and $j_1 \neq j_2 \cdots \neq j_M$. When these factors are multiplied, the result is a polynomial in s divided by some power of s:

$$s^{-r}(a_M s^M + a_{M-1} s^{M-1} + \cdots + a_1 s + a_0)$$

in which the coefficients a_n's are sums of products of resistances, capacitances, and inductances and are therefore *real*. We deduce immediately the following property:

Property 1   The network function of a circuit is a ratio of two polynomials of s in which the coefficients are all real. It is of the form

$$H(s) = \frac{a_M s^M + \cdots + a_1 s + a_0}{b_N s^N + \cdots + b_N s + b_0} = \frac{P(s)}{Q(s)} \tag{8.79}$$

where $P(s)$ and $Q(s)$ denote, respectively, the numerator and denominator polynomials. We note the network functions enumerated above are all of this form.   ∎

A simple consequence of this property is that no matter how hard we try, we will never be able to find a circuit whose network function is any of the following: $\sqrt{s}$, $\sin s$, or Ke^{-sT}. Note that the last is the network function of a perfect delay line, since when $s = j\omega$, it becomes $Ke^{-j\omega T}$, which has an amplitude K and a phase $-\omega T$. [See Equation (8.58).]

8.12 Poles and Zeros*

A polynomial can be factored into a product of first-order polynomials. The network function of a circuit can, therefore, be written in the form

$$H(s) = \frac{P(s)}{Q(s)} = \frac{K(s - z_1)(s - z_2) \cdots (s - z_M)}{(s - p_1)(s - p_2) \cdots (s - p_N)} \tag{8.80}$$

where $z_1, z_2, \ldots, z_M$ are the M zeros of the polynomial $P(s)$, and $p_1, p_2, \ldots, p_N$ are the N zeros of the polynomial $Q(s)$. The z_i's are known as the *zeros* of $H(s)$ and the p_i's the *poles* of $H(s)$.

Because the coefficients of $P(s)$ and $Q(s)$ are all real, we have the following property.

Property 2   The poles and zeros of a network function are real or occur in complex conjugate pairs. ■

For example, the impedance $Z(s)$ of Equation (8.73)

$$Z(s) = \frac{(s^2 + 1)(s^2 + 9)}{s(s^2 + 4)}$$

has zeros at $\pm j$ and $\pm j3$, and poles at 0, $\pm j2$, and $\pm j\infty$. The poles and zeros are displayed in Figure 8.23(a). The voltage transfer function of Equation (8.72) is

$$H(s) = \frac{0.5}{s^3 + 2s^2 + 2s + 1} = \frac{0.5}{(s + 1)(s^2 + s + 1)}$$

The function has a pole at -1, and two complex poles at $-0.5 \pm j0.866$. It does not have any finite zero. The poles are displayed in Figure 8.23(b).

The transfer function of the lumped delay line is

$$H(s) = \frac{1}{2}\frac{1 - (1/2)s + (1/12)s^2}{1 + (1/2)s + (1/12)s^2}$$

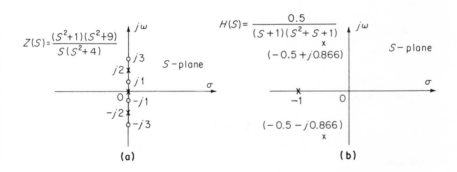

(a)   (b)

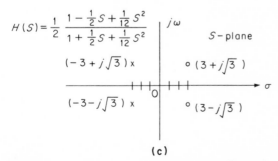

(c)

Figure 8.23 The pole-zero pattern of network functions. (The poles and zeroes at infinity, if any, are not shown. x = pole, o = zero.)

The zeros are $3 \pm j \sqrt{3}$, and the poles are $-3 \pm j \sqrt{3}$, as displayed in Figure 8.23(c). Note that the zeros are in the right half plane and are symmetrical with respect to the poles about the $j\omega$-axis.

Conceptually, one can think of the zeros as those values of the complex frequency variable s at which the function assumes its zero value. The poles are those values of s at which the function becomes undefined. Except for a multiplicative constant, a network function is completely specified by its poles and zeros.

8.13   Poles and Natural Time Response*

What is the physical significance of the poles? Consider the circuit of Figure 8.10. The transfer function is

$$H(s) = \frac{V}{E} = \frac{R_0^2 + s^2 \, LCR_0}{R_0 + R_g + sL + s^2LC(R_0 + R_g)} \tag{8.81}$$

The poles are

$$p_1 = \frac{-1 + \sqrt{1 - 4(R_0 + R_g)C/L}}{2C(R_0 + R_g)}$$

$$p_2 = \frac{-1 - \sqrt{1 - 4(R_0 + R_g)C/L}}{2C(R_0 + R_g)} \tag{8.82}$$

which are real and distinct if

$$1 > \frac{4(R_0 + R_g)C}{L}$$

real and equal if

$$1 = \frac{4(R_0 + R_g)C}{L}$$

and complex if

$$1 < \frac{4(R_0 + R_g)C}{L}$$

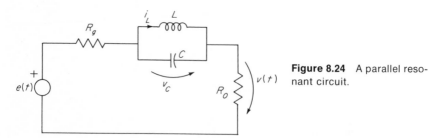

Figure 8.24   A parallel resonant circuit.

Now with reference to Figure 8.24 in which the circuit of Figure 8.10 is reproduced, the differential equations for the inductor current $i_L(t)$ and the capacitor voltage $v_C(t)$ are

$$L\frac{di_L}{dt} = v_C$$

$$C\frac{dv_C}{dt} = -i_L - \frac{v_C}{R_0 + R_g} + \frac{e(t)}{R_0 + R_g} \qquad (8.83)$$

The state equations are

$$\frac{d}{dt}\begin{bmatrix} i_L \\ \\ v_C \end{bmatrix} = \begin{bmatrix} 0 & \dfrac{1}{L} \\ \\ -\dfrac{1}{C} & -\dfrac{1}{(R_0 + R_g)C} \end{bmatrix}\begin{bmatrix} i_L \\ \\ v_C \end{bmatrix}\begin{bmatrix} 0 \\ \\ \dfrac{e}{(R_0 + R_g)C} \end{bmatrix} \qquad (8.84)$$

The eigenvalues of the system are found from

$$\begin{bmatrix} -\lambda & \dfrac{1}{L} \\ \\ -\dfrac{1}{C} & -\dfrac{1}{(R_0 + R_g)C} - \lambda \end{bmatrix} = 0 \qquad (8.85)$$

The characteristic equation is

$$\lambda^2 + \frac{1}{(R_0 + R_g)C}\lambda + \frac{1}{LC} = 0 \qquad (8.86)$$

The characteristic polynomial, except for a multiplicative constant, is precisely the denominator polynomial of the network function $H(s)$. It follows that the eigenvalues of the circuit, which are the roots of the characteristic equation (8.86), are the poles of the network function.

Recalling the meaning of the eigenvalues from Chapter 4, we now see that the poles of a network function specify the nature of the homogeneous solution, namely, the response of a circuit in the absence of any excitation. The latter is the same as the response resulting from the initial voltages on the capacitors and the initial currents in the inductors. This response is sometimes called the *natural* or *relaxed* response.

If the poles are real and distinct, the natural response (voltage or current) will be of the form

$$y(t) = K_1 e^{-\alpha_1 t} + K_2 e^{-\alpha_2 t} \qquad (8.87)$$

where $\alpha_1 > \alpha_2 > 0$, since the element values are all positive. [See Equations (8.82).]

If the poles are equal, we have

$$y(t) = K_1 t e^{-\alpha_1 t} + K_2 e^{-\alpha_2 t} \tag{8.88}$$

where

$$\alpha = \frac{-1}{2C(R_0 + R_g)} > 0$$

If the poles are complex, we have

$$y(t) = K_1 e^{-\alpha t} \cos(\beta t + K_2)$$

$$\tag{8.89}$$

where

$$\alpha = 1/[2C\ (R_g + R_0)]$$

and

$$\beta = \sqrt{-1 + 4(R_0 + R_g)C/L}/[2C(R_0 + R_g)]$$

In all of the above, K_1 and K_2 are arbitrary constants.

As another example, the voltage transfer function of the low-pass filter of Figure 8.15 is once more

$$H(s) = \frac{1/2}{s^3 + 2s^2 + 2s + 1} \tag{8.90}$$

Now suppose we write the differential equations for the circuit in terms of the state variables, which are the inductor current and capacitor voltages. With reference to Figure 8.25 and noting the element values, we have, by KVL,

$$v_3 = v_1 - v_2$$

or

$$2\frac{di_3}{dt} = v_1 - v_2 \tag{8.91}$$

By KCL,

$$i_1 = i_4 - i_3 = e - v_1 - i_3$$

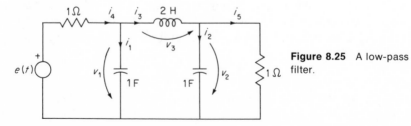

Figure 8.25　A low-pass filter.

or

$$\frac{dv_1}{dt} = -v_1 - i_3 + e \tag{8.92}$$

By KCL,

$$i_2 = i_3 - i_5 - i_3 - v_2$$

or

$$\frac{dv_2}{dt} = i_3 - v_2 \tag{8.93}$$

Combining the differential equations, we get the state equation:

$$\frac{d}{dt}\begin{bmatrix} v_1 \\ v_2 \\ i_3 \end{bmatrix} = \begin{bmatrix} -1 & 0 & -1 \\ 0 & -1 & 1 \\ 1/2 & -1/2 & 0 \end{bmatrix}\begin{bmatrix} v_1 \\ v_2 \\ i_3 \end{bmatrix} + \begin{bmatrix} e \\ 0 \\ 0 \end{bmatrix} \tag{8.94}$$

Assuming the homogeneous solution to be of the form

$$\begin{bmatrix} v_1 \\ v_2 \\ i_3 \end{bmatrix} = \begin{bmatrix} v_{10} \\ v_{20} \\ i_{30} \end{bmatrix} e^{\lambda t} \tag{8.95}$$

we substitute it into the state equation and obtain, in a manner similar to the case of second-order circuits, a homogeneous algebraic equation, which has nonzero solutions for a set of values of λ determined from

$$\begin{bmatrix} -1 - \lambda & 0 & -1 \\ 0 & -1 - \lambda & 1 \\ 1/2 & -1/2 & -\lambda \end{bmatrix} = 0 \tag{8.96}$$

Expanding the determinant, we get the characteristic equation

$$\lambda^3 + 2\lambda^2 + 2\lambda + 1 = 0 \tag{8.97}$$

whose roots are -1, $-0.5 \pm j0.866$, which are precisely the poles of the network function given by Equation (8.90). Once again, we see the poles and eigenvalues coincide and that the characteristic polynomial and the denominator polynomial of the network function are the same.

The natural response in this case will have the form

$$y(t) = K_1 e^{-t} + K_2 e^{-0.5t} \cos(0.866t + K_3) \tag{8.98}$$

where K_1, K_2, and K_3 are constants.

In the general case, the denominator polynomial of a network function is

$$Q(s) = b_0 + b_1 s + b_2 s^2 + \cdots + b_N s^N \tag{8.99}$$

which can always be factored into the form

$$Q(s) = Q_0(s + P_1)(s + P_2) \cdots (s + P_r)(s^2 + c_1 s + d_1)$$

$$(s^2 + c_2 s + d_2) \cdots (s^2 + c_q s + d_q) \qquad (8.100)$$

where the first r factors come from the real zeros of $Q(s)$ and the remainder from the complex zeros. The real zeros [poles of $H(s)$] give rise to a natural response of the form

$$e^{-\alpha t} \quad \text{and} \quad t^m e^{-\alpha t} \qquad (8.101)$$

where $m + 1$ denotes the multiplicity of the zero. The complex zeros give rise to a natural response of the form

$$e^{-\alpha t} \cos(\beta t + \theta) \quad \text{and} \quad t^m e^{-\alpha t} \cos(\beta t + \theta) \qquad (8.102)$$

The total natural response is simply a linear combination of these terms. We deduce, therefore, the following property:

Property 3 The natural response, namely, the response resulting from initial conditions, is of the form

$$y(t) = \Sigma \ K_i t^m \ e^{-\alpha_i t} \cos(\beta_i t + \theta_i)$$

8.14 Stability*

As a circuit without external excitation "relaxes" from its initial state, that is, the state specified by the initial capacitor voltages and inductor currents, one of three events could happen:

1. The natural response is in steady-state oscillation (Problem 4.17).
2. The natural response diminishes to zero eventually.
3. The response grows without limit.

We would all be quite uncomfortable if the third event were to take place, for then the power and energy in a circuit would grow to dangerous magnitude. We conclude, therefore, that only one of the first two events can take place. This is to say that the exponential terms in the natural response must go to zero as t increases to infinity. The consequence is that α in the expressions 8.101 and 102 must not be negative. Since α is, in fact, the real part of a pole, we now assert the following property:

Property 4 The poles of a network function of a circuit composed of positive elements must have nonpositive real parts; namely, they must all lie in the left half of the complex plane, or on the jω-axis. ∎

This is a very strong condition on network functions. Its significance cannot be overemphasized. All analytic properties of network functions are based on this property. It is outside the scope of this book to develop these theoretical considerations. Perhaps it is sufficient to note at this time that it is not possible to construct a physical circuit whose network function is

$$H(s) = \frac{1}{s^2 - s + 1}$$

or

$$H(s) = \frac{1}{s^4 + 1}$$

Both functions have poles in the right half plane.

8.15 Dynamic and Steady-State Response*

We noted in the last section that the denominator polynomial and the characteristic polynomial of the state equation are the same and that the poles specify the form of the natural response. What can we say about the numerator polynomial of a network function?

Consider the circuit of Figure 8.24 again. Let us obtain a differential equation for the output voltage $v(t)$. First, we have

$$v = Ri = \frac{R_0}{R_0 + R_g} (e - v_C) \tag{8.103}$$

Differentiating twice and making use of Equation (8.83), we get, after some algebra,

$$\frac{d^2v}{dt^2} + \frac{1}{(R_0 + R_g)C} \frac{dv}{dt} + \frac{1}{LC} v = \frac{R_0}{(R_0 + R_g)} \left(\frac{d^2e}{dt^2} + \frac{1}{LC} e \right) \tag{8.104}$$

Suppose the excitation is of the form $e(t) = E e^{st}$. Then from the theory of differential equations with constant coefficients, the particular solution of $v(t)$ is

$$v_p(t) = V e^{st} \tag{8.105}$$

Substituting the above into the differential equation, we get, after canceling the common factor e^{st}, an algebraic equation:

$$\left(s^2 + \frac{1}{(R_0 + R_g)C} s + \frac{1}{LC} \right) V = \frac{R_0}{R_0 + R_g} \left(s^2 + \frac{1}{LC} \right) E \tag{8.106}$$

Solving for V/E, we get

$$\frac{V}{E} = \frac{R_0}{R_0 + R_g} \frac{s^2 + \frac{1}{LC}}{s^2 + \frac{1}{(R_0 + R_g)C} s + \frac{1}{LC}} \tag{8.107}$$

which is the network function $H(s)$ given by Equation (8.81). Its numerator is precisely the right-hand side of the algebraic equation obtained from the differential equation for $v(t)$, and the denominator is the left-hand side of the algebraic equation.

In general, let $x(t) = Xe^{st}$ be the excitation to a circuit and $y(t) = Ye^{st}$ be a response. If the network function $H(s)$ is

$$H(s) = \frac{Y}{X} = \frac{a_0 + a_1 s + \cdots + a_M s^M}{b_0 + b_1 s + \cdots + b_N s^N} \qquad (8.108)$$

then the differential equation for $y(t)$ is

$$b_N \frac{d^N y}{dt^N} + b_{N-1} \frac{d^{N-1} y}{dt^{N-1}} + \cdots + b_1 \frac{dy}{dt} + b_0 y$$

$$= a_M \frac{d^M x}{dt^M} + a_{M-1} \frac{d^{M-1} x}{dt^{M-1}} + \cdots + a_1 \frac{dx}{dt} + a_0 x \qquad (8.109)$$

and conversely.

The network function of a circuit is obtained from sinusoidal steady-state considerations. Thus the observation that we just made has the significance that the time-domain description of the response can be found from the sinusoidal steady-state (frequency-domain) description. Conversely, if we had started with the differential equation for the response, the network function can be found from the differential equation by replacing each term $d^k(\cdot)/dt^k$ by s^k. The numerator polynomial is the right-hand side, and the denominator polynomial the left-hand side of the resultant algebraic equation. Thus the frequency-domain description of a circuit can be found from the time-domain description.

Actually, the relation between the frequency- and time-domain descriptions can be made more precise than what has been presented. For this purpose, certain transform theory of differential equations must be used. The details are found in texts on ordinary differential equations.

8.16 Impedance Scaling

In examples given throughout this text, we have used at times realistic element values (10,000 Ω, 1000 μF, 500 mH) and at times unrealistic values (1 Ω, 1 F, 100 H). Mathematically, it does not matter what numerical values are used. The analysis techniques remain the same. In practice, of course, one must deal with microfarads, megaohms, millihenrys. It will now be demonstrated that we can always change the element values of an unrealistic circuit to those of a realistic circuit by impedance and frequency scaling.

Consider a circuit N. Let it be described by a set of loop equations. Now construct another circuit N' obtained from N by the following scheme:

	Circuit N	Circuit N'
Resistor	$R = r$ ohms	$R' = Kr$ ohms
Inductor	$L = l$ henrys	$L' = Kl$ henrys
Capacitor	$C = c$ farads	$C' = c/K$ farads
Voltage-controlled current source	$J_D = g_D V$	$J'_D = (g_D/K)V'$
Current-controlled voltage source	$V_D = r_D I$	$V'_D = Kr_D I'$

To each resistor of r ohms in N, there corresponds a resistor of Kr ohms in N'. To each inductor of l henrys in N, there corresponds an inductor of Kl henrys in N', and so on as shown in the scheme above. Now we note that if Z is the impedance of some branch in N, then KZ is the impedance of the corresponding branch in N'. The loop equations of N' are the same as those of N, except that the coefficients are K times the coefficients of the loop equations of N. Let the determinant of the loop equations of N' be Δ' and the cofactors be Δ'_{ij}. Then it is clear that

$$\Delta' = K^M \Delta$$
$$\Delta'_{ij} = K^{M-1}\Delta_{ij} \qquad (8.110)$$

Let $V(j\omega)$ be the voltage across Z which is in loop 2 with loop current I_2, and let the voltage transfer function be $H(j\omega) = V(j\omega)/E$, where E is the excitation in loop 1. Then

$$H(j\omega) = \frac{I_2 Z}{E} = \frac{Z\Delta_{12}}{\Delta}$$

The corresponding transfer function in N' is

$$H'(j\omega) = \frac{Z'\Delta'_{12}}{\Delta'} = \frac{KZK^{M-1}\Delta_{12}}{K^M\Delta} = \frac{Z\Delta_{12}}{\Delta} = H(j\omega) \qquad (8.111)$$

Thus the voltage transfer function of N' is the same as that of N. On the other hand, the input impedance, or any other impedance function, of N' is now

$$Z'_{in} = \frac{\Delta'}{\Delta'_{11}} = \frac{K^M \Delta}{K^{M-1}\Delta_{11}} = K \frac{\Delta}{\Delta_{11}} = KZ_{in} \qquad (8.112)$$

The impedance of N' is K times that of N.

As examples, the circuits of Figure 8.26 all have the same voltage transfer function:

$$H(j\omega) = \frac{0.5}{(j\omega)^3 + 2(j\omega)^2 + 2(j\omega) + 1}$$

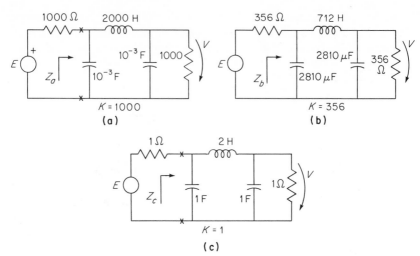

Figure 8.26 The three circuits have different impedance levels, but the same transfer function.

Circuit (a) is obtained from (c) by impedance scaling of factor $K = 1000$; circuit (b) is obtained from (c) by a factor of $K = 356$.

The input impedances, however, are given, respectively, by

$$Z_a(j\omega) = 1000\,\frac{2(j\omega)^2 + 2(j\omega) + 1}{2(j\omega)^3 + 2(j\omega)^2 + 2(j\omega) + 1}$$

$$Z_b(j\omega) = 356\,\frac{2(j\omega)^2 + 2(j\omega) + 1}{2(j\omega)^3 + 2(j\omega)^2 + 2(j\omega) + 1}$$

$$Z_c(j\omega) = \frac{2(j\omega)^2 + 2(j\omega) + 1}{2(j\omega)^3 + 2(j\omega)^2 + 2(j\omega) + 1}$$

8.17 Frequency Scaling

Suppose we construct a circuit N'' from N by the following scheme:

	Circuit N	Circuit N''
Resistor	$R = r$ ohms	$R'' = r$ ohms
Inductor	$L = l$ henrys	$L'' = Fl$ henrys
Capacitor	$C = c$ farads	$C'' = Fc$ farads

To each inductor in N of l henrys, there corresponds an inductor of Fl henrys in N'', c farads in N to Fc farads in N''. The other elements are unchanged. Now, let the frequency variable in N be ω and that in N'' be ω''. The impedance of an inductor L in N is $j\omega L$. The corresponding inductor L'' in N'' will have the same impedance value at $\omega'' = \omega/F$ since $j\omega''L'' = j\omega''FL = j\omega L$. Similarly, the impedance of a capacitor in N'' will have the same numerical value as that of its corresponding capacitor in

N if $\omega'' = \omega/F$. In general, the impedances of the corresponding branches in N and N'' will be equal numerically if $\omega'' = \omega/F$. Since the value of the determinant of the loop equations depends entirely on the impedances of the elements, we have

$$H''(j\omega'') = H(jF\omega'')$$
$$Z''(j\omega'') = Z(jF\omega'') \qquad (8.113)$$

In words, the network functions of N'' can be obtained from the corresponding functions of N by replacing the variable ω in the latter by $F\omega''$. As an example, the network function of the circuit of Figure 8.27 is

$$H''(j\omega'') = \frac{0.5}{(10^{-12})(j\omega'')^3 + 2(10^{-8})(j\omega'')^2 + 2(10^{-4})(j\omega'') + 1}$$

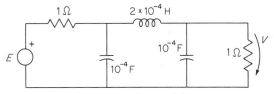

Figure 8.27 The circuit of Figure 8.26(c) after frequency scaling by a factor of 10^{-4}.

The circuit is obtained from that of Figure 8.28(c) by frequency scaling of factor $F = 10^{-4}$. In Figure 8.26(c), the voltage transfer function at 1 rad/sec is $0.354\, e^{-j3\pi/4}$, and in Figure 8.27, the function attains the same value at 10^4 rad/sec.

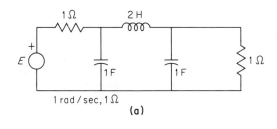

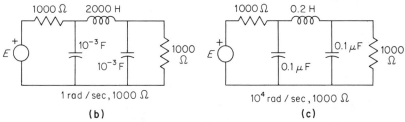

Figure 8.28 (a) The basic circuit; (b) the circuit after impedance scaling; (c) the circuit after impedance and frequency scaling.

Lastly, the unrealistic low-pass filter of Figure 8.28(a) is scaled up in impedance level by a factor of 1000 to give the circuit of Figure 8.28(b) as before. The latter is next scaled up in frequency by a factor of 10^{-4} to give the circuit of Figure 8.28(c), which has realistic element values. In (a), we deal with 1 Ω and 1 rad/sec. In (c), we deal with 1000 Ω and 10,000 rad/sec.

The point to note in the last two sections is that by impedance and frequency scaling, a circuit with unrealistic element values can be transformed into one with realistic values and vice versa.

8.18   Summary

By examining the response of a circuit in steady state under sinusoidal excitation as the frequency of the excitation varies, we obtain the frequency characteristics of the circuit. The characteristics can best be described in terms of a network function, which specifies the amplitude and phase of the various frequency components of the response. By putting together circuit elements of the right kinds and values, we can construct circuits to have interesting frequency characteristics. Examples of these are the low-pass filter and the lumped delay line. Other examples are found in the problems.

Given a circuit, we can always find the network function. Once we have the network function, the frequency characteristics can easily be computed from simple FORTRAN programs. A more interesting and difficult undertaking would be to obtain the frequency characteristics without having derived the network function. Many computer programs exist for this purpose. One such program is described at the end of the chapter.

By generalizing the frequency variable to take on complex values, a network function is seen to be a ratio of two polynomials of the complex variable with real coefficients. This description leads to the idea of poles and zeros of a network function. The poles specify the nature of the response resulting from the initial capacitor voltages and inductor currents.

Further examination of the relation between the differential equation that describes the circuit and the network function shows that the two are obtainable from each other. Thus in this sense, the time-domain description and the frequency-domain description of a circuit are equivalent.

SUGGESTED EXPERIMENTS

1. Construct the circuit of Example 8.3 and experimentally verify the input and output waveforms shown in Figure 8.12.
2. Construct the circuit of Figures 8.13(a) and (b). Measure the fre-

quency response. Use element values such that the frequencies of interest are in the audio range.

3. Construct an *RLC* series resonant circuit to resonate at 1000 Hz with $Q = 20$. Measure the frequency response. Do not forget the resistance associated with the coil in the calculation of Q.

4. Construct the low-pass filter of Section 8.7. Let $C = 0.1 \ \mu F$, $L = 0.2$ H, and $R = 1000 \ \Omega$. Measure the frequency response. Compare it with the theoretical values.

5. What must be the element values in Experiment 4 if the impedance level is raised by a factor of 100, and the frequency scale is raised by a factor of 1000? Construct the circuit and measure the frequency response.

6. Apply a square wave to the circuit of Experiment 4. Vary the frequency of the square wave. Observe the output. Explain what you see.

7. Apply a series of narrow pulses to the circuit of Experiment 4. Vary the width of the pulses and observe the output. Explain what you see.

8. Construct the high-pass filter of Problem 8.17. Measure the frequency response. Apply a square wave to the circuit. Explain what you see.

9. Construct the band-pass filter of Problem 8.19. Measure the frequency response. Compare it with the computed results.

10. Superimpose three sine waves of different frequencies and apply the composite wave to the circuit of the last experiment. Observe the output. Explain what you see.

11. Construct the band-elimination filter of Problem 8.21. Measure the frequency response.

12. Superimpose a high-frequency sine wave on a low-frequency sine wave. Apply the composite wave to the low-pass filter of Experiment 4. Explain what you see.

13. Superimpose a series of narrow pulses on a low-frequency sine wave. Apply the composite wave to the low-pass filter of Experiment 4. Explain what you see.

14. Construct the lumped delay line of Section 8.8. Scale the elements so that the resistances are of order 1000 Ω, and that the delay be 1 μsec at 1 MHz.

COMPUTER PROGRAMMING

Frequency Analysis of Circuits

Given a circuit, one can always solve for the network function by means of mesh, loop, or node equations. Once we have the network function, we can compute the frequency response by simply evaluating the function at the specified frequencies. For simple circuits, the network function is easy to obtain. For circuits with more than three loops or four nodes, the algebra becomes increasingly unpalatable. It is most desirable, therefore, to have computer programs which will take as input the configuration of the circuit, element types and values, and compute as output the frequency response. The programs should be sufficiently general so that circuits of arbitrary configuration and element types can be analyzed. Many such programs exist. We shall present one here. It is named DZNET.

Most of these general analysis programs are based on the method of tree-chord analysis that we presented in Chapter 6. If we note that an impedance can be regarded as "complex resistance" whose value depends on frequency, we see that frequency analysis of linear circuits is the same as analysis of linear resistor circuits. The program RNET described in Chapter 6 can be modified to include inductors, capacitors, mutual inductors, and dependent sources, as is done in DZNET.

There are situations, however, where the configurations of the circuits are so special that the analysis can be done in an iterative manner, or that the network function is expressible in canonical form in terms of the impedances of the circuits. In the former case, we have the ladder network. In the latter case, we have constant-resistance lattices and constant-resistance bridged-T networks. (See Problems 7.13 and 7.17.) Since these circuits are so widely used in practice, particularly in the design of communication systems and in signal processing, it is worthwhile to write computer programs especially for them. In this chapter, we shall present the ladder analysis program. The programs for the lattice and bridged-T are to be written as exercises. (See Problems 8.39 and 8.40).

Ladder Analysis

The general ladder network takes the form shown in Figure 8.29. With reference to the section "Ladder Network Analysis" of Chapter 3, if we replace the resistances by impedances, we obtain immediately the analysis scheme for ladder networks in the frequency domain. We must, how-

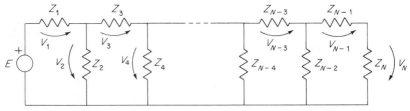

Figure 8.29 A general ladder network. The Z's are arbitrary impedances.

ever, identify or define the impedances and repeat the computation for each value of the frequency variable.

The FORTRAN program given below computes the frequency response of a ladder in which the impedances are defined in a FUNCTION subprogram.

```
C       FREQUENCY ANALYSIS OF LADDER NETWORK
C       INPUT DATA CONSIST OF THE NUMBER OF BRANCHES OF THE LADDER
C       AND THE FREQUENCIES AT WHICH COMPUTATION IS DESIRED.
C       THE BRANCH IMPEDANCES ARE DEFINED IN A FUNCTION SUBPROGRAM.
C       AS A CHECK,THE PROGRAM COMPUTES THE SUM OF THE VOLTAGES
C       AROUND THE PERIMETER OF THE LADDER.
        COMPLEX V(100),I(100),I1,I2,Z(100),VCHECK,BZ,V1,V2
        DIMENSION F(50)
C       SET NB TO THE NUMBER OF BRANCHES.
        DATA NB/4/
        F(1)=.01
        DO 1 J=2,50
        F(J)=F(J-1)+.02
      1 CONTINUE
      6 NFREQ=J-1
C       WRITE HEADING ON TOP
        WRITE(6,101)
    101 FORMAT(1H ,'FREQUENCY HZ',2X,'BRANCH')
C       FREQUENCY LOOP BEGINS HERE.
        DO 999 J=1,NFREQ
        WRITE(6,120)F(J)
    120 FORMAT(1PE13.3)
C       CALCULATION OF BRANCH IMPEDANCES.
        DO 99 JB=1,NB
     99 Z(JB)=BZ(JB,F(J))
C       CALCULATION OF V(N) BEGINS.
        V1=1
        I1=V1/Z(NB)
        I2=I1
        V2=Z(NB-1)*I2
        V1=V2+V1
        M=NB-1
        DO 2 K=3,M,2
        I1=V1/Z(NB+1-K)
        I2=I2+I1
        V2=I2*Z(NB-K)
      2 V1=V2+V1
        V(NB)=1./V1
C       WE NOW HAVE V(N).CALCULATION OF THE REMAINING BRANCH
C       VOLTAGES AND CURRENTS FOLLOWS.
        I(NB)=V(NB)/Z(NB)
        I(NB-1)=I(NB)
        M=NB-3
        DO 3 K=1,M,2
        V(NB-K)=I(NB-K)*Z(NB-K)
        V(NB-K-1)=V(NB-K)+V(NB-K+1)
        I(NB-K-1)=V(NB-K-1)/Z(NB-K-1)
      3 I(NB-K-2)=I(NB-K-1)+I(NB-K)
        V(1)=I(1)*Z(1)
C       CALCULATION OF AMPLITUDE AND PHASE OF BRANCH VOLTAGES AND
C       CURRENTS.
        WRITE(6,111)
    111 FORMAT(1H ,42X,'CURRENT'/26X,'REAL',8X,'IMAG',8X,'AMPL',8X,'PHSE')
        DO 4 L=1,NB
        AMPLI=CABS(I(L))
        PHSEI=ATAN(AIMAG(I(L))/REAL(I(L)))*57.2958
      4 WRITE(6,203)L,I(L),AMPLI,PHSEI
    203 FORMAT(17X,I2,2X,4(1PE12.3))
        WRITE(6,112)
    112 FORMAT(1H ,42X,'VOLTAGE'/26X,'REAL',8X,'IMAG',8X,'AMPL',8X,'PHSE')
        DO 20 L=1,NB
```

```
      PHSEV=ATAN(AIMAG(V(L))/REAL(V(L)))*57.2958
      AMPLV=CABS(V(L))
   20 WRITE(6,203)L,V(L),AMPLV,PHSEV
C     CALCULATION OF THE SUM OF THE VOLTAGES AROUND THE PERIMETER
C     OF THE LADDER.
      VCHECK=(0.,0.)
      NCK=N8-1
      DO 5 L=1,NCK,2
    5 VCHECK=VCHECK+V(L)
      VCHECK=VCHECK+V(NB)-1
      WRITE(6,104)VCHECK
  104 FORMAT(1H ,21X,'THE SUM OF THE VOLTAGES AROUND THE PERIMETER OF'/
     1 21X,'THE LADDER IS',2E15.6)
  999 CONTINUE
      STOP
      END
C     THE FOLLOWING FUNCTION SUBPROGRAM DEFINES THE BRANCH IMPEDANCES.
      COMPLEX FUNCTION BZ(I,F)
      COMPLEX J,CMPLX
      W=6.283185*F
      J=CMPLX(0.,1.0)
      GO TO (1,2,3,4),I
    1 BZ=1.0+1./(J*W*1.)
      RETURN
    2 BZ=J*W*0.5
      RETURN
    3 BZ=1.0/(J*W*1.0)
      RETURN
    4 BZ=1.0
      RETURN
      END
```

FREQUENCY HZ   B-RANCH
  1.000E-02

	\multicolumn{4}{c}{CURRENT}
	REAL	IMAG	AMPL	PHSE
1	3.948E-03	6.271E-02	6.283E-02	8.640E 01
2	3.563E-03	6.283E-02	6.296E-02	8.639E 01
3	-1.555E-05	-1.230E-04	1.240E-04	8.280E 01
4	-1.555E-05	-1.230E-04	1.240E-04	8.280E 01

	\multicolumn{4}{c}{VOLTAGE}
	REAL	IMAG	AMPL	PHSE
1	1.002E 00	-1.243E-04	1.002E 00	-7.106E-03
2	-1.974E-03	1.245E-04	1.978E-03	-3.609E 00
3	-1.958E-03	2.476E-04	1.974E-03	-7.205E 00
4	-1.555E-05	-1.230E-04	1.240E-04	8.280E 01

THE SUM OF THE VOLTAGES AROUND THE PERIMETER OF
THE LADDER IS -0.119209E-05   0.238535E-06

  3.000E-02

	\multicolumn{4}{c}{CURRENT}
	REAL	IMAG	AMPL	PHSE
1	3.555E-02	1.851E-01	1.885E-01	7.913E 01
2	3.679E-02	1.882E-01	1.918E-01	7.894E 01
3	-1.240E-03	-3.111E-03	3.349E-03	6.827E 01
4	-1.240E-03	-3.111E-03	3.349E-03	6.827E 01

	\multicolumn{4}{c}{VOLTAGE}
	REAL	IMAG	AMPL	PHSE
1	1.018E 00	-3.467E-03	1.018E 00	-1.952E-01
2	-1.774E-02	3.467E-03	1.808E-02	-1.106E 01
3	-1.650E-02	6.578E-03	1.776E-02	-2.173E 01
4	-1.240E-03	-3.111E-03	3.349E-03	6.827E 01

THE SUM OF THE VOLTAGES AROUND THE PERIMETER OF
THE LADDER IS -0.119209E-06   0.575092E-07

  5.000E-02

	\multicolumn{4}{c}{CURRENT}
	REAL	IMAG	AMPL	PHSE
1	9.508E-02	2.984E-01	3.144E-01	7.163E 01
2	1.083E-01	3.108E-01	3.291E-01	7.078E 01
3	-9.251E-03	-1.243E-02	1.550E-02	5.334E 01
4	-9.251E-03	-1.243E-02	1.550E-02	5.334E 01

	VOLTAGE			
	REAL	IMAG	AMPL	PHSE
1	1.049E 00	-1.702E-02	1.049E 00	-9.295E-01
2	-4.882E-02	1.702E-02	5.170E-02	-1.922E 01
3	-3.957E-02	2.945E-02	4.932E-02	-3.666E 01
4	-9.251E-03	-1.243E-02	1.550E-02	5.334E 01

THE SUM OF THE VOLTAGES AROUND THE PERIMETER OF
THE LADDER IS   -0.655651E-06   -0.294298E-06

7.000E-02

	CURRENT			
	REAL	IMAG	AMPL	PHSE
1	1.956E-01	3.944E-01	4.403E-01	6.362E 01
2	2.292E-01	4.203E-01	4.788E-01	6.140E 01
3	-3.356E-02	-2.589E-02	4.239E-02	3.765E 01
4	-3.356E-02	-2.589E-02	4.239E-02	3.765E 01

	VOLTAGE			
	REAL	IMAG	AMPL	PHSE
1	1.092E 00	-5.040E-02	1.094E 00	-2.642E 00
2	-9.243E-02	5.040E-02	1.053E-01	-2.860E 01
3	-5.888E-02	7.630E-02	9.637E-02	-5.235E 01
4	-3.356E-02	-2.589E-02	4.239E-02	3.765E 01

THE SUM OF THE VOLTAGES AROUND THE PERIMETER OF
THE LADDER IS   -0.137091E-05   0.305474E-06

Frequency Analysis of General Circuit – DZNET

DZNET is a subroutine for computing the frequency response of a linear circuit consisting of resistors, capacitors, inductors, mutual inductors, and dependent sources. The subroutine can handle a circuit with up to 80 elements and 40 nodes. All of the inductors are assumed to be lossy and to have the same Q, as are all the capacitors (see Problem 8.10). The subroutine will choose a reference node if one is not specified. As a check, the sum of the currents at the reference node is computed.

The frequency variable may be assigned complex values.

The program is based on the chord analysis of Chapter 6 with all resistances replaced by impedances. In addition, the set of terminal equations is enlarged to include mutual inductors and dependent sources. These elements are treated in a manner as explained in Sections 6.10 and 6.11. The program listing is given at the end of the chapter.

DZNET has two parts. The first part reads as input the configuration of the circuit. It then constructs the various matrices as described in Chapter 6. The second part computes the branch voltages and currents at the frequencies specified by the array $ZOMEGA$. The branch voltages and currents are contained in the two arrays $ZVOLT$ and $ZCURNT$, respectively. Their values are returned to the calling program as the values evaluated at the last frequency before exit.

The integer parameter NX determines to which part of the subroutine control is to be transferred. If $NX = 1$, both parts will be executed. If $NX > 1$, the first part is skipped, and only the second part will be executed. The matrices previously constructed remain unchanged. By this means, power and energy at any branches, network functions, input impedance, and so on, can be computed at each frequency by the calling program.

Printout is controlled by the integer parameter N. If $N = 0$, the branch voltages and currents will not be printed by the subroutine. If N is less than the total number of branches of the circuit, only the voltages and currents of those branches whose numbers are specified in the integer array *LISTP* will be printed. If N is equal to the total number of branches, all the branch voltages and currents will be printed, and the array *LISTP* need not be initialized in the calling program, though it must be dimensioned.

The subroutine may be called by the following statement. CALL DZNET (*ZOMEGA, NFREQZ, LISTP, N, ZVOLT, ZCURNT, QL, QC, QF, NX*) where

ZOMEGA	is a *complex array* of dimension at least 1, of complex frequencies $(\sigma + j\omega)$ in hertz at which the circuit is to be analyzed.
NFREQZ	is an *integer constant* and equal to the number of frequencies contained in *ZOMEGA*.
LISTP	is an *integer array* having a length of at least 1. It contains the branch numbers of the circuit whose voltages and currents are to be printed.
N	is a nonnegative *integer constant* and equal to the number of branches to be printed. If $N = 0$, no branch voltages and currents will be printed.
ZVOLT	is a *complex array* containing the voltages of the branches. The voltages are evaluated at the last frequency before exit. Its dimension must be equal to or greater than the total number of branches in the circuit.
ZCURNT	is a *complex array* containing the currents of the branches. The currents are evaluated at the last frequency before exit. Its dimension must be equal to or greater than the total number of branches in the circuit.
QL	is a *real variable* which is the Q of all of the inductors in the network.
QC	is a *real variable* which is the Q of all of the capacitors in the network.
QF	is a *real variable* which is the frequency in hertz at which *QL* and *QC* are measured.
NX	is a positive *integer constant*. It is set to 1 if both parts of the subroutine are to be executed. It is set to be greater than 1 if only the second part is to be executed.

Since *ZOMEGA, ZVOLT, ZCURNT* are complex, they must be declared as such in the calling program.

If the inductors and capacitors are ideal, put *QF* equal zero and assign arbitrary constants, but never zero, to *QL* and *QC*.

The input data deck is read by the first part of DZNET. It contains information on the structure of the circuit, its element types, and values. To prepare the input data deck, the following must be observed.

1. Number the nodes.
2. Number the branches in the following order: dependent voltage sources, independent voltage sources, capacitors, resistors, inductors, independent current sources, and dependent current sources.
3. Define current directions in each branch. Use conventional current directions for sources.
4. The mutual inductance is positive if the current flowing in one inductor produces a voltage whose direction agrees with the current direction in the coupled inductor.

The first card of the input data deck should be the following card:

cols.	2–3	7–8	12–13	17–18	22–23	27–28	32–33
	N	*B*	*DV*	*V*	*CAP*	*RES*	*IND*

37–38	42–43	47–48	52–53	61–80
MUT	*C*	*DC*	*GROUND*	NAME

where

$$N = \text{number of nodes}$$
$$B = \text{number of branches}$$
$$DV = \text{number of dependent voltage sources}$$
$$V = \text{number of independent voltage sources}$$
$$CAP = \text{number of capacitors}$$
$$RES = \text{number of resistors}$$
$$IND = \text{number of inductors}$$
$$MUT = \text{number of mutual inductors}$$
$$C = \text{number of independent current sources}$$
$$DC = \text{number of dependent current sources}$$
$$GROUND = \text{reference node; if 0, program will choose}$$
$$\text{a reference node}$$
$$NAME = \text{name of the circuit or user.}$$

All the above are two-digit integer constants, except NAME which is alphanumeric.

Following the first card are the element cards, each of which describes one element of the circuit. The element cards are arranged in accordance with their branch number. These cards have the following format:

$$\text{cols.} \quad \begin{array}{cccccc} 2\text{--}3 & 5\text{--}6 & 11\text{--}20 & 21\text{--}30 & 36 & 38\text{--}39 \\ L & K & Xr & Xi & A & J \end{array}$$

where

$L =$ the node from which the current in the element leaves
$K =$ the node into which the current in the element enters
$Xr =$ the real part of the element value
$Xi =$ the imaginary part of the element value

The two entries A and J are for dependent sources. A is a literal constant either V or C. If a V appears in column 36, it means the dependent source is controlled by the voltage of branch J; if a C appears in column 36, the dependent source is controlled by the current of branch J.

The quantities L, K, and J are two-digit integers. A is a literal constant, and Xr and Xi both have the format of $E10.3$. The units for Xr and Xi are the basic units, that is, volts, amperes, farads, henrys, and ohms.

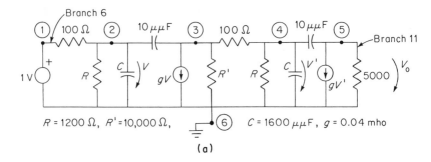

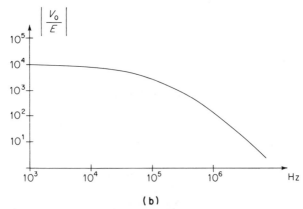

(b)

Figure 8.30 (a) The equivalent circuit of a transistor amplifier; (b) its frequency response as computed by DZNET.

Following the element cards are the mutual inductance cards, each of which describes one mutual inductance. The mutual inductance cards have the format as shown below:

$$\text{cols.} \quad \begin{array}{ccc} 2\text{-}3 & 5\text{-}6 & 11\text{-}20 \\ B1 & B2 & MUT \end{array}$$

where MUT is the mutual inductance in henrys that couples branches $B1$ and $B2$. $B1$ and $B2$ are two-digit integers, while MUT is a real number with format $E10.3$.

Example Suppose that we wish to find the frequency response of a transistor amplifier as shown in Figure 8.30 for frequencies from 1×10^3 Hz to 4×10^6 Hz. We also wish to print out the voltage and current of branches 6 and 11 and compute the input impedance at each frequency. The following program will accomplish this. The capacitors are assumed to be ideal. The computer output is given on page 433 at the end of the listing of DZNET.

```
C   DZNET EXAMPLE
        COMPLEX ZOMEGA(1), ZVOLT(13), ZCURNT(13), ZIN
        DIMENSION LISTP(2)
        NX=1
        F=500.
        LISTP(1)=6
        LISTP(2)=11
        DO 1 I=1,13
        F=F*2
        ZOMEGA(1)=CMPLX(0., F)
        CALL DZNET(ZOMEGA,1,LISTP,2,ZVOLT,ZCURNT,1.,1.,0.,NX)
        NX=NX+1
        ZIN=1./ZCURNT(6)
    1   WRITE(6,100)ZIN
  100   FORMAT(1H ,10X,'INPUT IMPEDANCE =',2E16.4//)
        STOP
        END
```

PROBLEMS

8.1 *Reactance* Obtain the reactance function of each of the following circuits. Sketch the function and note the resonant frequencies.

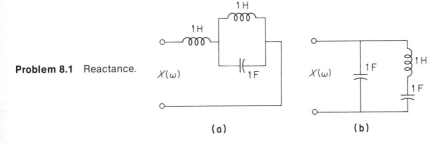

Problem 8.1 Reactance.

(a)   (b)

8.2 *Reactance* Show that the reactance function of the circuit below is $X(\omega) = \omega(\omega^4 - 6\omega^2 + 8)/(\omega^4 - 4\omega^2 + 3)$. Sketch the function and note all critical frequencies.

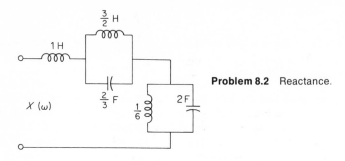

Problem 8.2 Reactance.

8.3 *Susceptance* Show that the susceptance function of the circuit shown below is the same as the reactance function of Problem 8.2.

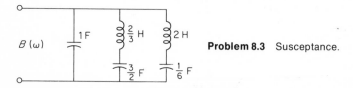

Problem 8.3 Susceptance.

8.4 *Reactance* Show that the reactance function of the circuit shown is $X(\omega) = \omega(120\omega^4 - 96\omega^2 + 9)/(120\omega^4 - 36\omega^2 + 1)$. Sketch the function and note all critical frequencies.

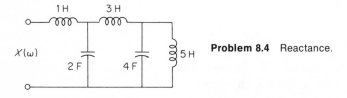

Problem 8.4 Reactance.

8.5 *Susceptance* Show that the susceptance function of the circuit shown below is the reactance function of Problem 8.4. Explain.

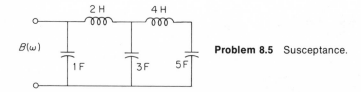

Problem 8.5 Susceptance.

8.6 *Resonance* Find the voltage transfer function of the following circuit. Sketch the gain and phase as functions of frequency. What is the bandwidth of the circuit?

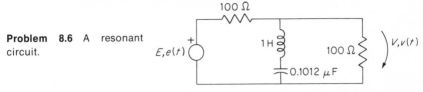

Problem 8.6 A resonant circuit.

8.7 *Resonance* Suggest some use for the circuit of Problem 8.6. Suppose the input voltage is $e(t) = 0.1(\sin 2\pi\ 200t + 0.5 \sin 2\pi\ 400t + 0.333 \sin 2\pi\ 600t) + \sin 2\pi\ 500t$. Find the steady-state output voltage. Sketch the waveforms. Write a computer program to calculate $e(t)$ and $v(t)$.

8.8 *Resonance* Find the voltage transfer function of the following circuit. Sketch the gain and phase functions. What is the bandwidth of the circuit?

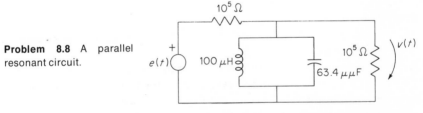

Problem 8.8 A parallel resonant circuit.

8.9 *Resonance* Let the input voltage of Problem 8.8 be an approximation of a square wave, that is, $e(t) = \sin 2\pi f_0 t + \frac{1}{3} \sin 6\pi f_0 t + \frac{1}{5} \sin 10\pi f_0 t + \frac{1}{7} \sin 14\pi f_0 t$, with $f_0 = 10^6$ Hz. Find the output voltage $v(t)$. Write a computer program to calculate $v(t)$ and $e(t)$. Sketch the waveforms. Explain your results.

8.10 *Q of an Element* A realistic inductor has resistance associated with it. A quantity known as the Q of an inductor is used to describe the amount of resistance in the inductor. It is given by $Q_L = \omega L/R$, where ω is the frequency at which the quantity is measured, L the inductance, and R the series resistance associated with the inductor. Similarly, the Q of a capacitor is defined as $Q_C = \omega C/G$, where C is the capacitance and G the conductance in parallel with the capacitor. Suppose in Problem 8.9 $Q_L = 10$ and $Q_C = \infty$. Find the voltage transfer function and sketch the waveforms. What is the effect of the Q on the selectivity of the circuit?

8.11 *Low-Pass Filter* Design a low-pass filter of the type shown in Figure 8.15 such that the 10,000-Hz component at the output is at least 100 times smaller than the 1000-Hz component. The input voltage has components at 1000 Hz, 3000 Hz, and 10,000 Hz. The source and load resistances are both 300 Ω.

8.12 *High-Pass Filter* Show that the gain of the following circuit is given by $A(\omega) = \frac{1}{2}\omega^3/\sqrt{1+\omega^6}$. Sketch the function and suggest some use for it.

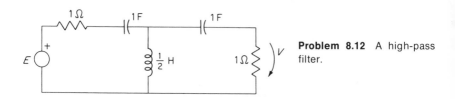

Problem 8.12 A high-pass filter.

8.13 *High-Pass Filter* Suppose the input voltage of Problem 8.12 is $e(t) = \cos 0.2t + \cos 20t$. Find the output voltage in steady state. Sketch the waveforms.

8.14 *High-Pass Filter* In Problem 8.13, suppose the Q of the inductor is 10 at $\omega = 1$. Find and sketch the gain functions. What is the effect of finite Q?

8.15 *Matching Network* In a communication system, one is interested in transmitting as much power as possible from a source to a load. No part of the signal power should be lost or dissipated. In practice, the load impedance, however, is never matched to the source impedance to effect maximum power transfer. Transformers are sometimes used, but they are not effective because the load is usually not a simple resistor and because a realistic transformer itself has frequency-selective property. A simple scheme to effect maximum power transfer, at least over a range of frequencies, is to add a network between the load and the source. The simplest network is a series inductor as shown in the figure. Find the value of the inductance such that the transfer function has the maximally flat characteristic. Sketch the power as a function of frequency.

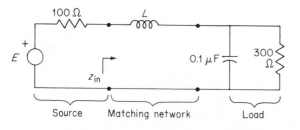

Problem 8.15 A simple matching network.

8.16 *Matching Network* A more complicated network than that of Problem 8.15 is as shown. Compute the power transferred to the load as a function of frequency. Find the input impedance and sketch its real and imaginary parts as functions of frequency. Use the ladder analysis program to do all the calculations if you wish.

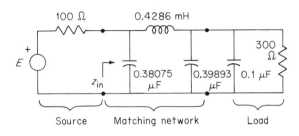

Problem 8.16 A more complicated matching network.

8.17 *High-Pass Filter* Design a high-pass filter of the type of Problem 8.12 such that the attenuation at 2×10^5 Hz is less than 0.01 times that at 4×10^6 Hz. The source and load resistors are both 1000 Ω.

8.18 *Band-Pass Filter* Find the gain function of the circuit shown below. Sketch the function. What can you do with the circuit? Scale the circuit so that the frequency of interest is 1000 Hz and the impedance level is 300 Ω instead of 1 Ω.

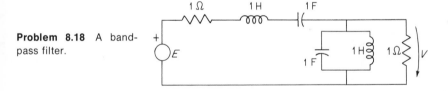

Problem 8.18 A band-pass filter.

8.19 *Band-Pass Filter* Use the ladder analysis program to compute the frequency response of the circuit shown. Repeat the computation for the same circuit except that the Q of the inductors is now finite and equal to 150. What is the effect of Q?

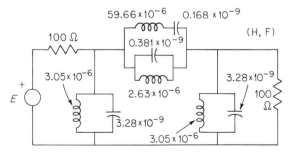

Problem 8.19 A realistic band-pass filter.

8.20 *Band-Elimination Filter* Use the ladder analysis program to compute the frequency response of the circuit shown below. Repeat the calculation for inductor $Q = 100$ and 300.

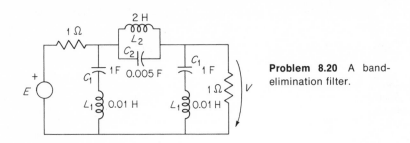

Problem 8.20 A band-elimination filter.

8.21 *Band-Elimination Filter* Scale the filter of Problem 8.20 so that the frequency of interest (resonant frequency of the parallel LC circuit) is 10^3 Hz and the impedance level is raised to 1000 Ω. Use the ladder analysis program to compute the frequency response.

8.22 *Constant-Resistance Lattice* Show that the voltage transfer function of the lattice is $H(s) = \frac{1}{2}(R - Z_A)/(R + Z_A)$. Show also that the input impedance is R, independent of frequency. Suppose the impedance Z_A is that of an LC network so that $Z_A(j\omega) = jX(\omega)$. Show that the gain is $\frac{1}{2}$, independent of frequency.

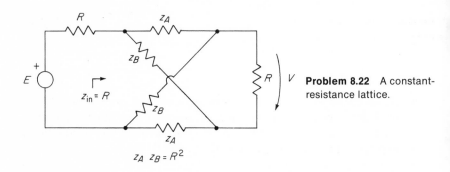

Problem 8.22 A constant-resistance lattice.

8.23 *Lumped Delay Line* Use the circuit of Problem 8.22 to realize the lumped delay line of Section 8.8. Let the delay be 100 nsec over a frequency range from dc to 10^6 Hz. The source and load resistances are both 1000 Ω.

8.24 *Lumped Delay Line* Connect N constant-resistance lattices of the type of Problem 8.23 in cascade. Show that the total delay is the sum of the delay of the N sections. Design a lumped delay line to give 1-μsec delay over a range of frequency from dc to 10^6 Hz.

8.25 *Transistor Amplifier* A transistor amplifier has an equivalent circuit shown below. Find the transfer function. Compute the frequency response. Beyond what frequency does the circuit cease to be an amplifier?

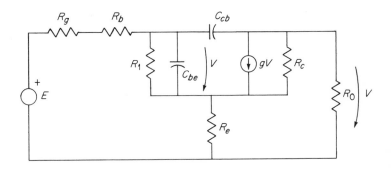

R_g = 150 Ω, R_b = 850 Ω, R_c = 1.67 x 10^6 Ω, R_e = 12.5 Ω
C_{be} = 2 x 10^{-9} F, C_{cb} = 10 x 10^{-12} F, g = 0.0782 mho
R_0 = 2 x 10^6 Ω, R_1 = 1000 Ω

Problem 8.25 A transistor amplifier.

8.26 *Transistor Amplifier* Use the computer program DZNET to compute the frequency response of the circuit of Problem 8.25.

8.27 *Transistor Amplifier* A realistic transistor amplifier is shown below. Use the computer program DZNET to compute the frequency response. Compute the input impedance also.

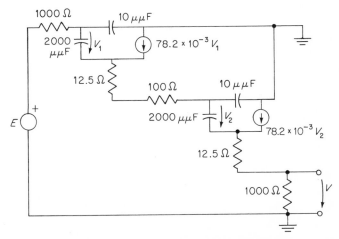

Problem 8.27 A realistic transistor amplifier.

8.28 *LC Impedance* Find the impedance function as a function of the complex variable s of each of the circuits of Problem 8.1. Find the poles and zeros of the impedance function.

8.29 *LC Impedance* Find the poles and zeros of the impedance function of the circuit shown below.

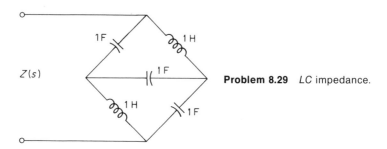

Problem 8.29 *LC* impedance.

8.30 *LC Impedance* Show that the impedance function of the circuit below is $Z(s) = (s^5 + 4s^3 + 3s)/(3s^4 + 4s^2 + 1)$. Find the poles and zeros.

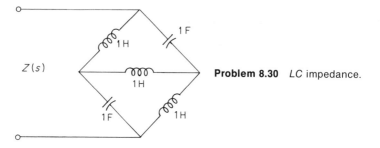

Problem 8.30 *LC* impedance.

8.31 *RC Impedance* Show that the impedance function of the following circuit is $Z(s) = (s^2 + 6s + 8)/(s^2 + 4s + 3)$. Find the poles and zeros. What can you say about the form of the response in the time domain resulting from initial conditions?

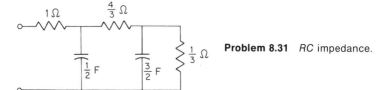

Problem 8.31 *RC* impedance.

8.32 *RL Impedance* Show that the impedance function of the following circuit is $Z(s) = (s^2 + 4s + 3)/(s^2 + 6s + 8)$. Find the poles
and zeros. What can you say about the response in the time domain resulting from the initial conditions? What can you say about
the circuit of this problem and that of Problem 8.31?

Problem 8.32 *RL* impedance.

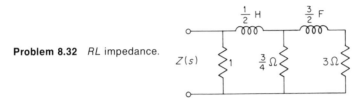

8.33 *RC Impedance* Show that the impedance functions of the following two circuits are the same as that of Problem 8.32. What can
you say about them?

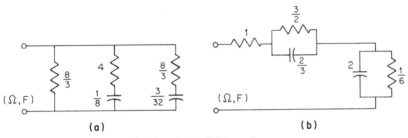

Problem 8.33 *RC* impedance.

8.34 *RL Impedance* Show that the impedance functions of the following circuits are the same as those of Problem 8.33. What can you
say about them?

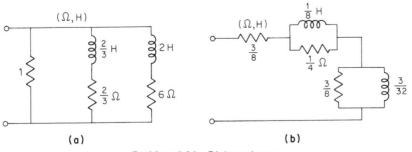

Problem 8.34 *RL* impedance.

8.35 *RC Transfer Function* Show that the transfer function of the following circuit is $H(s) = 1/(s^2 + 4s + 3)$. What is the form of the response in the time domain resulting from the initial conditions?

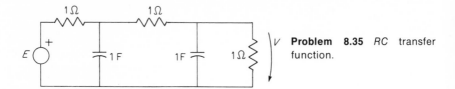

Problem 8.35 *RC* transfer function.

8.36 *RC Transfer Function* Show that the transfer function of the following circuit is $H(s) = \frac{1}{4}(s-1)/(s+1)$. What is the form of the homogeneous solution in the time domain?

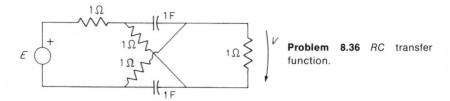

Problem 8.36 *RC* transfer function.

8.37 *RC Transfer Function* From the previous two problems, what conclusion can you draw as regards **a.** the location of the poles of the transfer function of an *RC* circuit? **b.** The form of the homogeneous solution in the time domain? **c.** The location of the zeros of the transfer function? (The zeros can be anywhere in the *s*-plane.)

8.38 *RL Transfer Function* Applying the principle of duality to Problem 8.37, what can you say about the transfer function of a circuit consisting of resistors and inductors only?

8.39 *Constant-Resistance Lattice* Write a computer program to compute the frequency response of *N* constant-resistance lattices connected in cascade. Define the impedances in a FUNCTION subprogram. Use the program to analyze the following circuit.

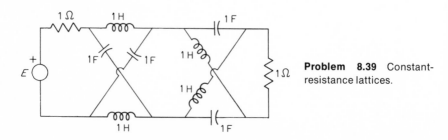

Problem 8.39 Constant-resistance lattices.

8.40 *Lattice* If the inductors in Problem 8.39 have finite Q, the circuit is no longer a constant-resistance network. Use DZNET to compute the frequency response and study the effect of finite Q.

8.41 *Constant-Resistance Bridged-T* Write a program to compute the frequency response of N constant-resistance bridged-T networks connected in cascade. Use the program to analyze the following circuit.

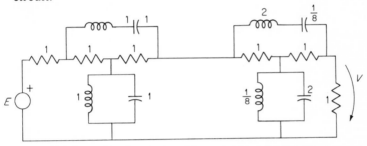

Problem 8.41 Constant-resistance bridged-T networks.

8.42 *Bridged-T* If the inductors have finite Q, the bridged-T networks of Problem 8.41 are no longer constant-resistance structures. Use DZNET to compute the frequency response and study the effect of finite Q.

8.43 *Fun Circuit* Show that the input impedance of circuit (a) is $1\ \Omega$, independent of frequency. Show that the input impedance of circuit (b) is also $1\ \Omega$. Show that the transfer function of circuit (b) is $H(s) = 1/(1+s)^N$.

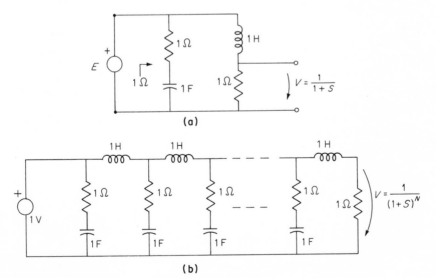

Problem 8.43 Fun circuit.

8.43 *Fun Circuit — continued*   Show that the transfer function of circuit
(c) is $H(s) = s^N/(1 + s)^N$.

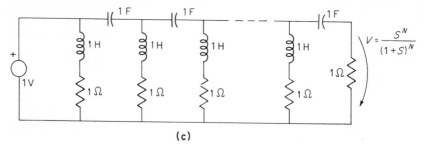

(c)
Problem 8.43 — continued

Program DZNET

```
      SUBROUTINE DZNET  (ZOMEGA,NFREQZ,LISTP,NN,ZVOLT,ZCURNT,QL,QC,QF,NX)  DZNET001
      IMPLICIT COMPLEX (Z)                                                 DZNET002
      INTEGER GROUND,DEP(80),BRNI(80),BRNJ(80),X,Y,U,V                     DZNET003
      REAL MUT(80),MAGE,MAGJ                                               DZNET004
      DIMENSION NODEI(80),NODEJ(80),ZVALUE(80),SOURCE(80)                  DZNET005
      DIMENSION H(2),INDEX(80),A(40,80),ATEMP(40,80)                       DZNET006
      DIMENSION ARRAY(1681),B(41,41),ZD1(41,80),ZD2(41,80)                 DZNET007
      DIMENSION B1(80,80),B2(80,80),ZG(80,80),ZRRAY(1681)                  DZNET008
      DIMENSION ZIMP(80,80),ZH(41,80),ZE(80),ZJ(80),ZA(41,41)              DZNET009
      DIMENSION ZB(41,41),ZC(41,41),ZR(80),ZS(80),LISTP(1)                 DZNET010
      DIMENSION LY(41),MY(41),ZVOLT(1),ZCURNT(1)                           DZNET011
      DIMENSION ZOMEGA(1)                                                  DZNET012
      EQUIVALENCE (ARRAY(1),ZRRAY(1)),(B2(1),ATEMP(1),ZC(1))               DZNET013
      EQUIVALENCE (A(1),B1(1)),(ZA(1),ZD1(1)),(ZB(1),ZD2(1))               DZNET014
      EQUIVALENCE (SOURCE(1),ZR(1)),(DEP(1),ZS(1))                         DZNET015
      DATA H/1HV,1HC/,X,Y,U,V/1,3,41,80/                                   DZNET016
      IF (NX .GT. 1) GO TO 1401                                           DZNET017
      M=0                                                                  DZNET017
      READ (5,1) NODE,NBRN,NDV,NIDV,NCAP,NRES,NIND,NMUT,NIDC,NDC,GROUND,   DZNET018
     1N1,N2,N3,N4,N5                                                       DZNET019
      WRITE (6,2) N1,N2,N3,N4,N5                                           DZNET020
      WRITE (6,3) NODE,NBRN,NDV,NIDV,NCAP,NRES,NIND,NMUT,NIDC,NDC,GROUND   DZNET021
      IF (NODE .GE. 2) GO TO 100                                           DZNET022
      WRITE (6,4)                                                          DZNET023
      STOP                                                                 DZNET024
100   IF (NBRN .GE. 2) GO TO 110                                           DZNET025
      WRITE (6,5)                                                          DZNET026
      STOP                                                                 DZNET027
110   IF (NIDV+NIDC .NE. 0) GO TO 120                                      DZNET028
      WRITE (6,6)                                                          DZNET029
      STOP                                                                 DZNET030
120   IF (NCAP+NRES+NIND .NE. 0) GO TO 130                                 DZNET031
      WRITE (6,7)                                                          DZNET032
      STOP                                                                 DZNET033
130   IF (NBRN .GE. NODE) GO TO 140                                        DZNET034
      WRITE (6,8)                                                          DZNET035
      STOP                                                                 DZNET036
140   NTREE=NODE-1                                                         DZNET037
      NLINK=NBRN-NTREE                                                     DZNET038
      NDEP=NDV+NDC                                                         DZNET039
      WRITE (6,9)                                                          DZNET040
      IF (NDV .EQ. 0) GO TO 200                                            DZNET041
      DO 210 I=1,NDV                                                       DZNET042
      READ (5,10) NODEI(I),NODEJ(I),ZVALUE(I),SOURCE(I),DEP(I)             DZNET043
      IF (SOURCE(I) .EQ. H(2)) GO TO 220                                   DZNET044
      IF (SOURCE(I) .EQ. H(1)) GO TO 240                                   DZNET045
      WRITE (6,12)                                                         DZNET046
      STOP                                                                 DZNET047
220   WRITE (6,11) I,NODEI(I),NODEJ(I),ZVALUE(I),DEP(I)                    DZNET048
      GO TO 210                                                            DZNET049
240   WRITE (6,13) I,NODEI(I),NODEJ(I),ZVALUE(I),DEP(I)                    DZNET050
210   CONTINUE                                                             DZNET051
200   IF (NIDV .EQ. 0) GO TO 300                                           DZNET052
      L=NDV+1                                                              DZNET053
      M=NDV+NIDV                                                           DZNET054
      DO 310 I=L,M                                                         DZNET055
      READ (5,10)  NODEI(I),NODEJ(I),ZVALUE(I),SOURCE(I),DEP(I)            DZNET056
      WRITE (6,14) I,NODEI(I),NODEJ(I),ZVALUE(I)                           DZNET057
310   CONTINUE                                                             DZNET058
300   IF (NCAP .EQ. 0) GO TO 320                                           DZNET059
      L=M+1                                                                DZNET060
      M=M+NCAP                                                             DZNET061
      DO 330 I=L,M                                                         DZNET062
      READ (5,10) NODEI(I),NODEJ(I),ZVALUE(I),SOURCE(I),DEP(I)             DZNET063
```

```
            WRITE (6,15) I,NODEI(I),NODEJ(I),ZVALUE(I)                  DZNET064
330         CONTINUE                                                    DZNET065
320         IF (NRES .EQ. 0) GO TO 340                                  DZNET066
            L=M+1                                                       DZNET067
            M=M+NRES                                                    DZNET068
            DO 350 I=L,M                                                DZNET069
            READ (5,10) NODEI(I),NODEJ(I),ZVALUE(I),SOURCE(I),DEP(I)    DZNET070
            WRITE (6,16) I,NODEI(I),NODEJ(I),ZVALUE(I)                  DZNET071
350         CONTINUE                                                    DZNET072
340         IF (NIND .EQ. 0) GO TO 360                                  DZNET073
            L=M+1                                                       DZNET074
            M=M+NIND                                                    DZNET075
            DO 370 I=L,M                                                DZNET076
            READ (5,10) NODEI(I),NODEJ(I),ZVALUE(I),SOURCE(I),DEP(I)    DZNET077
            WRITE (6,17) I,NODEI(I),NODEJ(I),ZVALUE(I)                  DZNET078
370         CONTINUE                                                    DZNET079
360         IF (NIDC .EQ. 0) GO TO 380                                  DZNET080
            L=M+1                                                       DZNET081
            M=M+NIDC                                                    DZNET082
            DO 390 I=L,M                                                DZNET083
            READ (5,10) NODEI(I),NODEJ(I),ZVALUE(I),SOURCE(I),DEP(I)    DZNET084
            WRITE (6,18) I,NODEI(I),NODEJ(I),ZVALUE(I)                  DZNET085
390         CONTINUE                                                    DZNET086
380         IF (NDC .EQ. 0) GO TO 500                                   DZNET087
            L=M+1                                                       DZNET088
            M=M+NDC                                                     DZNET089
            DO 400 I=L,M                                                DZNET090
            READ (5,10) NODEI(I),NODEJ(I),ZVALUE(I),SOURCE(I),DEP(I)    DZNET091
            IF (SOURCE(I) .EQ. H(2)) GO TO 410                          DZNET092
            IF (SOURCE(I) .EQ. H(1)) GO TO 430                          DZNET093
            WRITE (6,20)                                                DZNET094
            STOP                                                        DZNET095
410         WRITE (6,19) I,NODEI(I),NODEJ(I),ZVALUE(I),DEP(I)           DZNET096
            GO TO 400                                                   DZNET097
430         WRITE (6,21) I,NODEI(I),NODEJ(I),ZVALUE(I),DEP(I)           DZNET098
400         CONTINUE                                                    DZNET099
500         IF (NMLT .EQ. 0) GO TO 600                                  DZNET100
            WRITE (6,22)                                                DZNET101
            DO 510 I=1,NMUT                                             DZNET102
            READ (5,23) BRNI(I),BRNJ(I),MUT(I)                          DZNET103
            WRITE (6,24) MUT(I),BRNI(I),BRNJ(I)                         DZNET104
510         CONTINUE                                                    DZNET105
600         DO 610 K=1,NBRN                                             DZNET106
            INDEX(K)=K                                                  DZNET107
            DO 610 I=1,NODE                                             DZNET108
            IF (NODEI(K) .EQ. I) GO TO 620                              DZNET109
            IF (NODEJ(K) .EQ. I) GO TO 630                              DZNET110
            A(I,K)=0.0                                                  DZNET111
            GO TO 610                                                   DZNET112
620         A(I,K)=1.0                                                  DZNET113
            GO TO 610                                                   DZNET114
630         A(I,K)=-1.0                                                 DZNET115
610         CONTINUE                                                    DZNET116
            DO 640 I=1,NODE                                             DZNET117
            CHECK=0.0                                                   DZNET118
            DO 650 K=1,NBRN                                             DZNET119
            IF (A(I,K) .EQ. 0.0) GO TO 650                              DZNET120
            CHECK=CHECK+1.0                                             DZNET121
650         CONTINUE                                                    DZNET122
            IF (CHECK .GE. 2.0) GO TO 640                               DZNET123
            WRITE (6,25) I                                              DZNET124
            STOP                                                        DZNET125
640         CONTINUE                                                    DZNET126
            IF (GROUND .NE. 0) GO TO 700                                DZNET127
            CHECK=0.0                                                   DZNET128
            DO 660 I=1,NODE                                             DZNET129
            TEMP=0.0                                                    DZNET130
            DO 670 K=1,NBRN                                             DZNET131
670         TEMP=TEMP+ABS(A(I,K))                                       DZNET132
            IF (CHECK .GT. TEMP) GO TO 660                              DZNET133
            CHECK=TEMP                                                  DZNET134
            GROUND=I                                                    DZNET135
660         CONTINUE                                                    DZNET136
700         IF (GROUND .EQ. NODE) GO TO 710                             DZNET137
            DO 720 I=GROUND,NTREE                                       DZNET138
            J=I+1                                                       DZNET139
```

```
        DO 720 K=1,NBRN                          DZNET140
        A(I,K)=A(J,K)                            DZNET141
720     CONTINUE                                 DZNET142
710     DO 730 I=1,NTREE                         DZNET143
        DO 730 J=1,NBRN                          DZNET144
730     ATEMP(I,J)=A(I,J)                        DZNET145
        CALL TRIANG (A,NTREE,NBRN,INDEX)         DZNET146
        J=NDV+NIDV+NCAP+NRES+NIND                DZNET147
        DO 740 K=1,NTREE                         DZNET148
        IF (J .GE. INDEX(K)) GO TO 740           DZNET149
        WRITE (6,26)                             DZNET150
        STOP                                     DZNET151
740     CONTINUE                                 DZNET152
        J=NDV+NIDV                               DZNET153
        DO 750 K=NODE,NBRN                       DZNET154
        IF (J .LT. INDEX(K)) GO TO 750           DZNET155
        WRITE (6,27)                             DZNET156
        STOP                                     DZNET157
750     CONTINUE                                 DZNET158
        DO 760 K=1,NTREE                         DZNET159
        IF (A(K,K) .NE. 0.0) GO TO 760           DZNET160
        WRITE (6,28)                             DZNET161
        STOP                                     DZNET162
760     CONTINUE                                 DZNET163
        DO 800 M=1,2                             DZNET164
        J=NDV+(M-1)*NIDV                         DZNET165
        IF (J .EQ. 0) GO TO 800                  DZNET166
        DO 820 I=1,J                             DZNET167
        IF (INDEX(I) .LE. J) GO TO 820           DZNET168
        L=I+1                                    DZNET169
        DO 830 K=L,NTREE                         DZNET170
        IF (INDEX(K) .GT. J) GO TO 830           DZNET171
        TEMP=INDEX(K)                            DZNET172
        INDEX(K)=INDEX(I)                        DZNET173
        INDEX(I)=TEMP                            DZNET174
830     CONTINUE                                 DZNET175
820     CONTINUE                                 DZNET176
800     CONTINUE                                 DZNET177
        DO 810 M=1,2                             DZNET178
        J=NDC+NIDC*(M-1)                         DZNET179
        IF (J .EQ. 0) GO TO 810                  DZNET180
        L=NDV+NIDV+NCAP+NRES+NIND+NIDC*(2-M)     DZNET181
        DO 840 I=L,NBRN                          DZNET182
        IF (INDEX(I) .GE. L) GO TO 840           DZNET183
        DO 850 K=NODE,NBRN                       DZNET184
        IF (INDEX(K) .LT. L) GO TO 850           DZNET185
        TEMP=INDEX(K)                            DZNET186
        INDEX(K)=INDEX(I)                        DZNET187
        INDEX(I)=TEMP                            DZNET188
850     CONTINUE                                 DZNET189
840     CONTINUE                                 DZNET190
810     CONTINUE                                 DZNET191
900     DO 910 K=1,NBRN                          DZNET192
        DO 910 I=1,NTREE                         DZNET193
910     A(I,K)=ATEMP(I,INDEX(K))                 DZNET194
        DO 920 J=1,NTREE                         DZNET195
        DO 920 I=1,NTREE                         DZNET196
        K=I+NTREE*(J-1)                          DZNET197
        ARRAY(K)=A(I,J)                          DZNET198
920     CONTINUE                                 DZNET199
        CALL MINV (ARRAY,NTREE,DET,LY,MY)        DZNET200
        DO 930 J=1,NTREE                         DZNET201
        DO 930 I=1,NTREE                         DZNET202
        K=I+NTREE*(J-1)                          DZNET203
        A(I,J)=ARRAY(K)                          DZNET204
930     CONTINUE                                 DZNET205
        DO 940 I=1,NTREE                         DZNET206
        DO 940 J=1,NLINK                         DZNET207
        B(I,J)=0.0                               DZNET208
        DO 940 K=1,NTREE                         DZNET209
        B(I,J)=B(I,J)-A(I,K)*A(K,J+NTREE)        DZNET210
940     CONTINUE                                 DZNET211
        IF (NDEP .EQ. 0) GO TO 1250              DZNET212
        DO 950 I=1,NDEP                          DZNET213
        DO 950 J=1,NBRN                          DZNET214
        ZD1(I,J)=(0.0,0.0)                       DZNET215
```

```
          ZD2(I,J)=(0.0,0.0)                                    DZNET216
 950      CONTINUE                                              DZNET217
          IF (NDV .EQ. 0) GO TO 1000                            DZNET218
          M=1                                                   DZNET219
 1090     L=1+(M-1)*(NBRN-NDC)                                  DZNET220
          K=NDV+(M-1)*(NBRN-NDV)                                DZNET221
          DO 1010 I=L,K                                         DZNET222
          N=(I-L+1)+(M-1)*NDV                                   DZNET223
          DO 1020 J=1,NTREE                                     DZNET224
          IF (DEP(I) .EQ. INDEX(J)) GO TO 1030                  DZNET225
 1020     CONTINUE                                              DZNET226
          DO 1040 J=NODE,NBRN                                   DZNET227
          IF (DEP(I) .EQ. INDEX(J)) GO TO 1050                  DZNET228
 1040     CONTINUE                                              DZNET229
          WRITE (6,20)                                          DZNET230
          STOP                                                  DZNET231
 1030     IF (SOURCE(I) .EQ. H(1)) GO TO 1070                   DZNET232
          ZD2(N,J+NLINK-NDEP)=ZVALUE(I)                         DZNET233
          GO TO 1010                                            DZNET234
 1070     ZD1(N,J-NDV)=ZVALUE(I)                                DZNET235
          GO TO 1010                                            DZNET236
 1050     IF (SOURCE(I) .EQ. H(2)) GO TO 1070                   DZNET237
          ZD2(N,J-NTREE)=ZVALUE(I)                              DZNET238
 1010     CONTINUE                                              DZNET239
          IF (M .EQ. 2) GO TO 1100                              DZNET240
 1000     IF (NDC .EQ. 0) GO TO 1100                            DZNET241
          M=2                                                   DZNET242
          GO TO 1090                                            DZNET243
 1100     DO 1110 I=1,NBRN                                      DZNET244
          DO 1110 J=1,NBRN                                      DZNET245
 1110     B1(I,J)=0.0                                           DZNET246
          L=NLINK-NDC                                           DZNET247
          M=NTREE-NDV                                           DZNET248
          DO 1120 I=1,L                                         DZNET249
          DO 1120 J=1,M                                         DZNET250
          B1(I,J)=-B(NDV+J,I)                                   DZNET251
          B1(L+J,M+I)=B(NDV+J,I)                                DZNET252
 1120     CONTINUE                                              DZNET253
          DO 1130 I=1,NBRN                                      DZNET254
          DO 1130 J=1,NDEP                                      DZNET255
 1130     B2(I,J)=0.0                                           DZNET256
          IF (NDV .EQ. 0) GO TO 1111                            DZNET257
          DO 1140 I=1,L                                         DZNET258
          DO 1140 J=1,NDV                                       DZNET259
 1140     B2(I,J)=-B(J,I)                                       DZNET260
 1111     IF (NCC .EQ. 0) GO TO 1112                            DZNET261
          DO 1150 I=1,M                                         DZNET262
          DO 1150 J=1,NDC                                       DZNET263
 1150     B2(L+I,NDV+J)=B(NDV+I,L+J)                            DZNET264
 1112     L=NBRN-NDEP                                           DZNET265
          DO 1160 I=1,NDEP                                      DZNET266
          DO 1160 J=1,NDEP                                      DZNET267
          ZG(I,J)=(0.0,0.0)                                     DZNET268
          DO 1160 K=1,L                                         DZNET269
          ZG(I,J)=ZG(I,J)-ZD2(I,K)*B2(K,J)                      DZNET270
 1160     CONTINUE                                              DZNET271
          DO 1170 I=1,NDEP                                      DZNET272
 1170     ZG(I,I)=1.0+ZG(I,I)                                   DZNET273
          IF (NDEP .EQ. 1) GO TO 1300                           DZNET274
          DO 1180 J=1,NDEP                                      DZNET275
          DO 1180 I=1,NDEP                                      DZNET276
          K=I+NDEP*(J-1)                                        DZNET277
          ZRRAY(K)=ZG(I,J)                                      DZNET278
 1180     CONTINUE                                              DZNET279
          CALL CMINV (ZRRAY,NDEP,DET,LY,MY)                     DZNET280
          IF (DET) 1190,1200,1190                               DZNET281
 1200     WRITE (6,29)                                          DZNET282
          STOP                                                  DZNET283
 1190     DO 1210 J=1,NDEP                                      DZNET284
          DO 1210 I=1,NDEP                                      DZNET285
          K=I+NDEP*(J-1)                                        DZNET286
          ZG(I,J)=ZRRAY(K)                                      DZNET287
 1210     CONTINUE                                              DZNET288
          GO TO 1310                                            DZNET289
 1300     ZG(1,1)=1.0/ZG(1,1)                                   DZNET290
```

```
1310  DO 1320 I=1,NDEP                                        DZNET291
      DO 1320 J=1,L                                           DZNET292
      ZIMP(I,J)=(0.0,0.0)                                     DZNET293
      DO 1330 K=1,L                                           DZNET294
1330  ZIMP(I,J)=ZIMP(I,J)+ZD2(I,K)*B1(K,J)                    DZNET295
      ZIMP(I,J)=ZIMP(I,J)+ZD1(I,J)                            DZNET296
1320  CONTINUE                                                DZNET297
      DO 1340 I=1,NDEP                                        DZNET298
      DO 1340 J=1,L                                           DZNET299
      ZH(I,J)=(0.0,0.0)                                       DZNET300
      DO 1340 K=1,NDEP                                        DZNET301
      ZH(I,J)=ZH(I,J)+ZG(I,K)*ZIMP(K,J)                       DZNET302
1340  CONTINUE                                                DZENT303
      DO 1350 I=1,L                                           DZNET304
      DO 1350 J=1,L                                           DZNET305
      ZG(I,J)=(0.0,0.0)                                       DZNET306
      DO 1360 K=1,NDEP                                        DZNET307
1360  ZG(I,J)=ZG(I,J)+B2(I,K)*ZH(K,J)                         DZNET308
      ZG(I,J)=ZG(I,J)+B1(I,J)                                 DZNET309
1350  CONTINUE                                                DZNET310
      GO TO 1400                                              DZNET311
1250  DO 1260 I=1,NBRN                                        DZNET312
      DO 1260 J=1,NBRN                                        DZNET313
1260  ZG(I,J)=(0.0,0.0)                                       DZNET314
      L=NLINK-NDC                                             DZNET315
      M=NTREE-NDV                                             DZNET316
      DO 1270 I=1,L                                           DZNET317
      DO 1270 J=1,M                                           DZNET318
      ZG(I,J)=-B(NDV+J,I)                                     DZNET319
      ZG(L+J,M+I)=B(NDV+J,I)                                  DZNET320
1270  CONTINUE                                                DZNET321
1400  IF (NN .EQ. 0) GO TO 1401                               DZNET321
      WRITE (6,31)                                            DZNET322
      WRITE (6,32)                                            DZNET323
1401  DO 7000 IF=1,NFREQZ                                     DZNET324
      LA=NDV+NIDV+1                                           DZNET325
      LB=NBRN-NDC-NIDC                                        DZNET326
      LC=NDV+NIDV+NCAP                                        DZNET327
      LD=LC+NRES                                              DZNET328
      LE=NDV+1                                                DZNET329
      LF=NDV+NIDV                                             DZNET330
      LG=LB+1                                                 DZNET331
      LH=LB+NIDC                                              DZNET332
      DO 1410 I=1,NBRN                                        DZNET333
      ZE(I)=(0.0,0.0)                                         DZNET334
      ZJ(I)=(0.0,0.0)                                         DZNET335
      DO 1410 J=1,NBRN                                        DZNET336
      ZIMP(I,J)=(0.0,0.0)                                     DZNET337
1410  CONTINUE                                                DZNET338
      IF (NICV .EQ. 0) GO TO 1470                             DZNET339
      DO 1420 I=LE,LF                                         DZNET340
1420  ZE(I)=-ZVALUE(I)                                        DZNET341
1470  IF (NICC .EQ. 0) GO TO 1480                             DZNET342
      DO 1430 I=LG,LH                                         DZNET343
1430  ZJ(I)= ZVALUE(I)                                        DZNET344
1480  DO 1440 I=LA,LB                                         DZNET345
      J=I-LF                                                  DZNET346
      IF (INDEX(I) .LE. LC) GO TO 1450                        DZNET347
      IF (INDEX(I) .LE. LD) GO TO 1460                        DZNET348
      ZIMP(J,J)=(ZOMEGA(IF)+(QF/QL))*6.283185*ZVALUE(INDEX(I))DZNET349
      GO TO 1440                                              DZNET350
1460  ZIMP(J,J)=ZVALUE(INDEX(I))                             DZNET351
      GO TO 1440                                              DZNET352
1450  ZIMP(J,J)=1.0/(((ZOMEGA(IF)+(QF/QC))*6.283185*ZVALUE(INDEX(I)))  DZNET353
1440  CONTINUE                                                DZNET354
      IF (NMUT .EQ. 0) GO TO 1500                             DZNET355
      KA=LB-LF                                                DZNET356
      DO 1510 I=1,KA                                          DZNET357
      KB=LF+I                                                 DZNET358
      DO 1520 K=1,NMUT                                        DZNET359
      IF (INDEX(KB) .EQ. BRNI(K)) GO TO 1530                  DZNET360
      GO TO 1520                                              DZNET361
1530  DO 1540 J=1,KA                                          DZNET362
      KC=LF+J                                                 DZNET363
      IF (INDEX(KC) .EQ. BRNJ(K)) GO TO 1550                  DZNET364
1540  CONTINUE                                                DZNET365
```

```
1550   ZIMP(I,J)=ZOMEGA(IF)*MUT(K)*6.283185                        DZNET366
       ZIMP(J,I)=ZIMP(I,J)                                         DZNET367
1520   CONTINUE                                                    DZNET368
1510   CONTINUE                                                    DZNET369
1500   NT=NTREE-NDV-NIDV                                           DZNET370
       NC=NLINK-NDC-NIDC                                           DZNET371
       IF (NT .EQ. 0) GO TO 5000                                   DZNET372
       IF (NC .EQ. 0) GO TO 6000                                   DZNET373
       IF (NDEP .EQ. 0) GO TO 1600                                 DZNET374
       LA=NC+NIDC+NIDV                                             DZNET375
       CALL MATMUL(ZG,ZIMP,ZA,LA,NIDV,NT,NT,Y,Y,NT,X,V)           DZNET376
       DO 1610 I=1,NT                                              DZNET377
1610   ZA(I,I)=ZA(I,I)+1.0                                         DZNET378
       DO 1620 J=1,NT                                              DZNET379
       DO 1620 I=1,NT                                              DZNET380
       K=I+NT*(J-1)                                                DZNET381
       ZRRAY(K)=ZA(I,J)                                            DZNET382
1620   CONTINUE                                                    DZNET383
       CALL CMINV (ZRRAY,NT,DET,LY,MY)                             DZNET384
       IF (DET) 1630,1640,1630                                     DZNET385
1640   WRITE (6,30)                                                DZNET386
       STOP                                                        DZNET387
1630   DO 1650 J=1,NT                                              DZNET388
       DO 1650 I=1,NT                                              DZNET389
       K=I+NT*(J-1)                                                DZNET390
       ZA(I,J)=ZRRAY(K)                                            DZNET391
1650   CONTINUE                                                    DZNET392
       LB=NT+NIDV                                                  DZNET393
       DO 1660 I=1,NT                                              DZNET394
       DO 1660 J=1,NC                                              DZNET395
1660   ZC(I,J)=ZG(LA+I,LB+J)                                       DZNET396
       IF (NMUT .EQ. 0) GO TO 1670                                 DZNET397
       CALL MATMUL (ZG,ZIMP,ZB,LA,NIDV,NT,NT,Y,NT,NC,Y,V)         DZNET398
       DO 1680 I=1,NT                                              DZNET399
       DO 1680 J=1,NC                                              DZNET400
1680   ZC(I,J)=ZC(I,J)+ZB(I,J)                                     DZNET401
1670   CALL MATMUL (ZA,ZC,ZB,Y,Y,NT,NT,Y,Y,NC,Y,U)               DZNET402
       DO 1690 I=1,NT                                              DZNET403
       ZR(I)=(0.0,0.0)                                             DZNET404
1690   ZRRAY(I)=(0.0,0.0)                                          DZNET405
       IF (NIDV .EQ. 0) GO TO 1700                                 DZNET406
       DO 1710 I=1,NT                                              DZNET407
       DO 1710 J=1,NIDV                                            DZNET408
1710   ZR(I)=ZR(I)+ZG(LA+I,J)*ZE(NDV+J)                           DZNET409
1700   IF (NIDC .EQ. 0) GO TO 1720                                 DZNET410
       LC=NIDV+NT+NC                                               DZNET411
       DO 1730 I=1,NT                                              DZNET412
       DO 1730 J=1,NIDC                                            DZNET413
1730   ZRRAY(I)=ZRRAY(I)+ZG(LA+I,LC+J)*ZJ(NTREE+NC+J)             DZNET414
1720   DO 1740 I=1,NT                                              DZNET415
1740   ZRRAY(I)=ZRRAY(I)+ZR(I)                                     DZNET416
       DO 1750 I=1,NT                                              DZNET417
       ZR(I)=(0.0,0.0)                                             DZNET418
       DO 1750 J=1,NT                                              DZNET419
       ZR(I)=ZR(I)+ZA(I,J)*ZRRAY(J)                               DZNET420
1750   CONTINUE                                                    DZNET421
       GO TO 1800                                                  DZNET422
1600   LA=NC+NIDC+NIDV                                             DZNET423
       LB=NIDV+NT                                                  DZNET424
       DO 1760 I=1,NT                                              DZNET425
       DO 1760 J=1,NC                                              DZNET426
1760   ZB(I,J)=ZG(LA+I,LB+J)                                       DZNET427
       LC=LB+NC                                                    DZNET428
       DO 1770 I=1,NT                                              DZNET429
1770   ZR(I)=(0.0,0.0)                                             DZNET430
       IF (NIDC .EQ. 0) GO TO 1800                                 DZNET431
       DO 1780 I=1,NT                                              DZNET432
       DO 1780 J=1,NIDC                                            DZNET433
1780   ZR(I)=ZR(I)+ZG(LA+I,LC+J)*ZJ(NTREE+NC+J)                   DZNET434
1800   CALL MATMUL (ZG,ZIMP,ZA,Y,NIDV,NC,NT,Y,Y,NT,X,V)          DZNET435
       IF (NMUT .EQ. 0) GO TO 1810                                 DZNET436
       DO 1820 I=1,NC                                              DZNET437
       DO 1820 J=1,NT                                              DZNET438
1820   ZA(I,J)=ZA(I,J)+ZIMP(NT+I,J)                               DZNET439
1810   CALL MATMUL (ZA,ZB,ZC,Y,Y,NC,NT,Y,Y,NC,Y,U)               DZNET440
```

```
          DO 1830 I=1,NC                                          DZNET441
          ZS(I)=(0.0,0.0)                                         DZNET442
          DO 1830 J=1,NT                                          DZNET443
1830      ZS(I)=ZA(I,J)*ZR(J)+ZS(I)                               DZNET444
          DO 1840 I=1,NC                                          DZNET445
          DO 1840 J=1,NC                                          DZNET446
1840      ZC(I,J)=ZC(I,J)+ZIMP(NT+I,NT+J)                         DZNET447
          IF (NMUT .EQ. 0) GO TO 1900                             DZNET448
          CALL MATMUL (ZG,ZIMP,ZA,Y,NIDV,NC,NT,Y,NT,NC,X,V)       DZNET449
          DO 1850 I=1,NC                                          DZNET450
          DO 1850 J=1,NC                                          DZNET451
1850      ZC(I,J)=ZC(I,J)+ZA(I,J)                                 DZNET452
1900      IF (NDEP .EQ. 0) GO TO 1910                             DZNET453
          DO 1920 I=1,NC                                          DZNET454
          DO 1920 J=1,NC                                          DZNET455
1920      ZC(I,J)=ZC(I,J)-ZG(I,NIDV+NT+J)                         DZNET456
1910      DO 1930 J=1,NC                                          DZNET457
          DO 1930 I=1,NC                                          DZNET458
          K=I+NC*(J-1)                                            DZNET459
          ZRRAY(K)=ZC(I,J)                                        DZNET460
1930      CONTINUE                                                DZNET461
          CALL CMINV (ZRRAY,NC,DET,LY,MY)                         DZNET462
          IF (DET) 1940,1950,1940                                 DZNET463
1950      WRITE (6,30)                                            DZNET464
          STOP                                                    DZNET465
1940      DO 1991 J=1,NC                                          DZNET466
          DO 1991 I=1,NC                                          DZNET467
          K=I+NC*(J-1)                                            DZNET468
          ZC(I,J)=ZRRAY(K)                                        DZNET469
1991      CONTINUE                                                DZNET470
          DO 1960 I=1,NC                                          DZNET471
1960      ZJ(NTREE+I)=-ZS(I)                                      DZNET472
          IF (NIDV .EQ. 0) GO TO 1970                             DZNET473
          DO 1980 I=1,NC                                          DZNET474
          ZS(I)=(0.0,0.0)                                         DZNET475
          DO 1990 J=1,NIDV                                        DZNET476
1990      ZS(I)=ZS(I)+ZG(I,J)*ZE(NDV+J)                           DZNET477
          ZJ(NTREE+I)=ZJ(NTREE+I)+ZS(I)                           DZNET478
1980      CONTINUE                                                DZNET479
1970      IF (NDEP .EQ. 0) GO TO 2000                             DZNET480
          IF (NIDC .EQ. 0) GO TO 2000                             DZNET481
          DO 2010 I=1,NC                                          DZNET482
          ZS(I)=(0.0,0.0)                                         DZNET483
          DO 2020 J=1,NIDC                                        DZNET484
2020      ZS(I)=ZS(I)+ZG(I,NIDV+NT+NC+J)*ZJ(NTREE+NC+J)           DZNET485
          ZJ(NTREE+I)=ZJ(NTREE+I)+ZS(I)                           DZNET486
2010      CONTINUE                                                DZNET487
2000      DO 2030 I=1,NC                                          DZNET488
          ZS(I)=(0.0,0.0)                                         DZNET489
          DO 2030 J=1,NC                                          DZNET490
2030      ZS(I)=ZS(I)+ZC(I,J)*ZJ(NTREE+J)                         DZNET491
          DO 2040 I=1,NC                                          DZNET492
2040      ZJ(NTREE+I)=ZS(I)                                       DZNET493
          LA=NIDV+NDV                                             DZNET494
          DO 2050 I=1,NT                                          DZNET495
          ZS(I)=(0.0,0.0)                                         DZNET496
          DO 2060 J=1,NC                                          DZNET497
2060      ZS(I)=ZS(I)+ZB(I,J)*ZJ(NTREE+J)                         DZNET498
          ZJ(LA+I)=ZS(I)+ZR(I)                                    DZNET499
2050      CONTINUE                                                DZNET500
          IF (NMUT .EQ. 0) GO TO 2100                             DZNET501
          LB=NT+NC                                                DZNET502
          DO 2110 I=1,LB                                          DZNET503
          ZE(LA+I)=(0.0,0.0)                                      DZNET504
          DO 2110 J=1,LB                                          DZNET505
          ZE(LA+I)=ZE(LA+I)+ZIMP(I,J)*ZJ(LA+J)                    DZNET506
2110      CONTINUE                                                DZNET507
          GO TO 2120                                              DZNET508
2100      DO 2130 I=1,NT                                          DZNET509
          ZE(LA+I)=(0.0,0.0)                                      DZNET510
          DO 2130 J=1,NT                                          DZNET511
          ZE(LA+I)=ZE(LA+I)+ZIMP(I,J)*ZJ(LA+J)                    DZNET512
2130      CONTINUE                                                DZNET513
2190      DO 2140 I=1,NC                                          DZNET514
          ZE(NTREE+I)=(0.0,0.0)                                   DZNET515
          DO 2140 J=1,NC                                          DZNET516
```

```
          ZE(NTREE+I)=ZE(NTREE+I)+ZIMP(NT+I,NT+J)*ZJ(NTREE+J)      DZNET517
2140  CONTINUE                                                      DZNET518
2120  IF (NDEP .EQ. 0) GO TO 2500                                   DZNET519
      LA=NIDV+NT                                                    DZNET520
      IF (LA .EQ. 0) GO TO 2111                                     DZNET521
      DO 2150 I=1,LA                                                DZNET522
2150  ZR(NCV+I)=ZE(NDV+L)                                           DZNET523
2111  LB=NIDC+NC                                                    DZNET524
      IF (LB .EQ. 0) GO TO 2112                                     DZNET525
      DO 2160 I=1,LB                                                DZNET526
2160  ZR(NTREE+I)=ZJ(NTREE+I)                                       DZNET527
2112  LA=NBRN-NDEP                                                  DZNET528
      IF (NDV .EQ. 0) GO TO 2200                                    DZNET529
      DO 2210 I=1,NDV                                               DZNET530
      ZE(I)=(0.0,0.0)                                               DZNET531
      DO 2220 J=1,LA                                                DZNET532
2220  ZE(I)=ZE(I)+ZH(I,J)*ZR(NDV+J)                                 DZNET533
      ZR(I)=ZE(I)                                                   DZNET534
2210  CONTINUE                                                      DZNET535
2200  IF (NDC .EQ. 0) GO TO 2300                                    DZNET536
      LA=NBRN-NDC                                                   DZNET537
      LB=NBRN-NDEP                                                  DZNET538
      DO 2230 I=1,NDC                                               DZNET539
      ZJ(LA+I)=(0.0,0.0)                                            DZNET540
      DO 2240 J=1,LB                                                DZNET541
2240  ZJ(LA+I)=ZJ(LA+I)+ZH(NDV+I,J)*ZR(NEV+J)                       DZNET542
      ZR(LA+I)=ZJ(LA+I)                                             DZNET543
2230  CONTINUE                                                      DZNET544
2300  IF (NDV .EQ. 0) GO TO 2400                                    DZNET545
      DO 2310 I=1,NDV                                               DZNET546
      ZJ(I)=(0.0,0.0)                                               DZNET547
      DO 2310 J=1,NLINK                                             DZNET548
      ZJ(I)=ZJ(I)+B(I,J)*ZJ(NTREE+J)                                DZNET549
2310  CONTINUE                                                      DZNET550
2400  IF (NDC .EQ. 0) GO TO 2500                                    DZNET551
      LA=NBRN-NDC                                                   DZNET552
      DO 2410 I=1,NDC                                               DZNET553
      ZE(LA+I)=(0.0,0.0)                                            DZNET554
      DO 2410 J=1,NTREE                                             DZNET555
      ZE(LA+I)=ZE(LA+I)-B(J,NLINK-NDC+I)*ZE(J)                      DZNET556
2410  CONTINUE                                                      DZNET557
2500  IF (NIDV .EQ. 0) GO TO 2600                                   DZNET558
      DO 2510 I=1,NIDV                                              DZNET559
      ZJ(NCV+I)=(0.0,0.0)                                           DZNET560
      DO 2510 J=1,NLINK                                             DZNET561
      ZJ(NCV+I)=ZJ(NDV+I)+B(NDV+I,J)*ZJ(NTREE+J)                    DZNET562
2510  CONTINUE                                                      DZNET563
2600  IF (NIDC .EQ. 0) GO TO 2700                                   DZNET564
      LA=NBRN-NIDC-NDC                                              DZNET565
      DO 2610 I=1,NIDC                                              DZNET566
      ZE(LA+I)=(0.0,0.0)                                            DZNET567
      DO 2610 J=1,NTREE                                             DZNET568
      ZE(LA+I)=ZE(LA+I)+B(J,NC+I)*ZE(J)                             DZNET569
2610  CONTINUE                                                      DZNET570
2700  IF (LF .EQ. 0) GO TO 2702                                     DZNET571
      DO 2760 I=1,LF                                                DZNET572
2760  ZE(I)=-ZE(I)                                                  DZNET573
      DO 2780 I=LG,NBRN                                             DZNET574
2780  ZE(I)=-ZE(I)                                                  DZNET575
2702  DO 2701 I=1,NBRN                                              DZNEA575
      ZVOLT(INDEX(I))=ZE(I)                                         DZNEB575
2701  ZCURNT(INDEX(I))=ZJ(I)                                        DZNEC575
      IF (NN .GT. 0) GO TO 2770                                     DZNED575
      RETURN                                                        DZNEE575
2770  IF (NN .LT. NBRN) GO TO 2720                                  DZNET576
      DO 2710 K=1,NBRN                                              DZNET577
2710  LISTP(K)=K                                                    DZNET578
2720  DO 2730 J=1,NN                                                DZNET579
      K=LISTP(J)                                                    DZNET580
      DO 2740 I=1,NBRN                                              DZNET581
      IF (INDEX(I) .EQ. K) GO TO 2750                               DZNET582
2740  CONTINUE                                                      DZNET583
2750  SER=REAL(ZE(I))                                               DZNET584
      SEI=AIMAG(ZE(I))                                              DZNET585
      SJR=REAL(ZJ(I))                                               DZNET586
```

```
        SJI=AIMAG(ZJ(I))                                               DZNET587
        MAGE=(SER**2+SEI**2)**0.5                                      DZNET588
        MAGJ=(SJR**2+SJI**2)**0.5                                      DZNET589
        ANGLE=(ATAN2(SEI,SER)*180.0)/3.14159                           DZNET590
        ANGLJ=(ATAN2(SJI,SJR)*180.0)/3.14159                           DZNET591
        IF (K .LE. NDV) GO TO 2800                                     DZNET594
        IF (K .LE. NDV+NIDV) GO TO 2810                                DZNET595
        IF (K .LE. NDV+NIDV+NCAP) GO TO 2820                           DZNET596
        IF (K .LE. NDV+NIDV+NCAP+NRES) GO TO 2830                      DZNET597
        IF (K .LE. NDV+NIDV+NCAP+NRES+NIND) GO TO 2840                 DZNET598
        IF (K .LE. NDV+NIDV+NCAP+NRES+NIND+NIDC) GO TO 2850            DZNET599
        WRITE (6,33) K,ZOMEGA(IF),ZE(I),MAGE,ANGLE,ZJ(I),MAGJ,ANGLJ    DZNET600
        GO TO 2730                                                     DZNET601
 2850   WRITE (6,34) K,ZCMEGA(IF),ZE(I),MAGE,ANGLE,ZJ(I),MAGJ,ANGLJ    DZNET602
        GO TO 2730                                                     DZNET603
 2840   WRITE (6,35) K,ZCMEGA(IF),ZE(I),MAGE,ANGLE,ZJ(I),MAGJ,ANGLJ    DZNET604
        GO TO 2730                                                     DZNET605
 2830   WRITE (6,36) K,ZOMEGA(IF),ZE(I),MAGE,ANGLE,ZJ(I),MAGJ,ANGLJ    DZNET606
        GO TO 2730                                                     DZNET607
 2820   WRITE (6,37) K,ZCMEGA(IF),ZE(I),MAGE,ANGLE,ZJ(I),MAGJ,ANGLJ    DZNET608
        GO TO 2730                                                     DZNET609
 2810   WRITE (6,38) K,ZCMEGA(IF),ZE(I),MAGE,ANGLE,ZJ(I),MAGJ,ANGLJ    DZNET610
        GO TO 2730                                                     DZNET611
 2800   WRITE (6,39) K,ZOMEGA(IF),ZE(I),MAGE,ANGLE,ZJ(I),MAGJ,ANGLJ    DZNET612
 2730   CONTINUE                                                       DZNET613
        ZSUM=(0.0,0.0)                                                 DZNET614
        DC 2910 K=1,NBRN                                               DZNET615
        J=INDEX(K)                                                     DZNET616
        IF (NODEI(J) .EQ. GROUND) GO TO 2920                           DZNET617
        IF (NODEJ(J) .EQ. GROUND) GO TO 2930                           DZNET618
        GO TC 2910                                                     DZNET619
 2920   ZSUM=ZSUM+ZJ(K)                                                DZNET620
        GO TC 2910                                                     DZNET621
 2930   ZSUM=ZSUM-ZJ(K)                                                DZNET622
 2910   CCNTINUE                                                       DZNET623
        WRITE (6,41) GROUND,ZSUM                                       DZNET624
        GC TC 7000                                                     DZNET625
 5000   DO 5300 I=1,NC                                                 DZNET628
        DO 5300 J=1,NC                                                 DZNET629
 5300   ZA(I,J)=ZIMP(I,J)                                              DZNET630
        IF (NDEP .EQ. 0) GO TC 5010                                    DZNET631
        DO 5020 I=1,NC                                                 DZNET632
        DO 5020 J=1,NC                                                 DZNET633
 5020   ZA(I,J)=ZA(I,J)-ZG(I,NIDV+J)                                   DZNET634
        IF (NIDC .EQ. 0) GO TO 5010                                    DZNET635
        LA=NIDV+NC                                                     DZNET636
        LB=NBRN-NIDC-NDC                                               DZNET637
        DO 5030 I=1,NC                                                 DZNET638
        ZR(I)=(0.0,0.0)                                                DZNET639
        DC 5030 J=1,NIDC                                               DZNET640
        ZR(I)=ZR(I)+ZG(I,LA+J)*ZJ(LB+J)                               DZNET641
 5030   CCNTINUE                                                       DZNET642
        GC TC 5100                                                     DZNET643
 5010   DO 5040 I=1,NC                                                 DZNET644
 5040   ZR(I)=(0.0,0.0)                                                DZNET645
 5100   DC 5110 J=1,NC                                                 DZNET646
        DO 5110 I=1,NC                                                 DZNET647
        K=I+NC*(J-1)                                                   DZNET648
        ZRRAY(K)=ZA(I,J)                                               DZNET649
 5110   CONTINUE                                                       DZNET650
        CALL CMINV (ZRRAY,NC,DET,LY,MY)                                DZNET651
        IF (DET) 5120,5130,5120                                        DZNET652
 5130   WRITE (6,30)                                                   DZNET653
        STOP                                                           DZNET654
 5120   DO 5140 J=1,NC                                                 DZNET655
        DO 5140 I=1,NC                                                 DZNET656
        K=I+NC*(J-1)                                                   DZNET657
        ZA(I,J)=ZRRAY(K)                                               DZNET658
 5140   CONTINUE                                                       DZNET659
        IF (NIDV .EQ. 0) GO TO 5200                                    DZNET660
        DO 5210 I=1,NC                                                 DZNET661
        ZRRAY(I)=(0.0,0.0)                                            DZNET662
        DO 5220 J=1,NIDV                                               DZNET663
 5220   ZRRAY(I)=ZRRAY(I)+ZG(I,J)*ZE(NDV+J)                           DZNET664
        ZR(I)=ZRRAY(I)+ZR(I)                                           DZNET665
 5210   CONTINUE                                                       DZNET666
```

```
5200    DO 5230 I=1,NC                                              DZNET667
        ZJ(NTREE+I)=(0.0,0.0)                                       DZNET668
        DO 5230 J=1,NC                                              DZNET669
        ZJ(NTREE+I)=ZJ(NTREE+I)*ZA(I,J)*ZR(J)                       DZNET670
5230    CCNTINUE                                                    DZNET671
        GO TG 2190                                                  DZNET672
6000    DO 6500 I=1,NT                                              DZNET673
        DO 6500 J=1,NT                                              DZNET674
6500    ZA(I,J)=ZIMP(I,J)                                           DZNET675
        LA=NIDC+NIDV                                                DZNET676
        IF (NDEP .EQ. 0) GO TC 6400                                 DZNET677
        DO 6010 J=1,NT                                              DZNET678
        DC 6010 I=1,NT                                              DZNET679
        K=I+NT*(J-1)                                                DZNET680
        ZRRAY(K)=ZA(I,J)                                            DZNET681
6010    CCNTINUE                                                    DZNET682
        CALL CMINV (ZRRAY,NT,DET,LY,MY)                             DZNET683
        IF (DET) 6020,6030,6020                                     DZNET684
6030    WRITE (6,30)                                                DZNET685
        STOP                                                        DZNET686
6020    DO 6040 J=1,NT                                              DZNET687
        DO 6040 I=1,NT                                              DZNET688
        K=I+NT*(J-1)                                                DZNET689
        ZB(I,J)=ZRRAY(K)                                            DZNET690
        ZA(I,J)=ZRRAY(K)                                            DZNET691
6040    CCNTINUE                                                    DZNET692
        DO 6050 I=1,NT                                              DZNET693
        DO 6050 J=1,NT                                              DZNET694
6050    ZA(I,J)=ZA(I,J)-ZG(LA+I,NIDV+J)                             DZNET695
        DO 6C70 J=1,NT                                              DZNET696
        DO 6C7C I=1,NT                                              DZNET697
        K=I+NT*(J-1)                                                DZNET698
        ZRRAY(K)=ZA(I,J)                                            DZNET699
6070    CONTINUE                                                    DZNET700
        CALL CMINV (ZRRAY,NT,DET,LY,MY)                             DZNET701
        IF (DET) 6080,6090,6080                                     DZNET702
6090    WRITE (6,30)                                                DZNET703
        STOP                                                        DZNET704
6080    DO 6100 J=1,NT                                              DZNET705
        DC 6100 I=1,NT                                              DZNET706
        K=I+NT*(J-1)                                                DZNET707
        ZA(I,J)=ZRRAY(K)                                            DZNET708
6100    CCNTINUE                                                    DZNET709
6400    IF (NIDV .EQ. 0) GO TC 6200                                 DZNET710
        DO 6060 I=1,NT                                              DZNET711
        ZR(I)=(0.0,0.0)                                             DZNET712
        DO 6060 J=1,NIDV                                            DZNET713
6060    ZR(I)=ZR(I)+ZG(LA+I,J)*ZE(NDV+J)                            DZNET714
        GO TC 6250                                                  DZNET715
6200    DO 6110 I=1,NT                                              DZNET716
6110    ZR(I)=(0.0,0.0)                                             DZNET717
6250    IF (NIDC .EQ. 0) GO TO 6210                                 DZNET718
        LB=NIDV+NT                                                  DZNET719
        DO 6230 I=1,NT                                              DZNET720
        ZRRAY(I)=(0.0,0.0)                                          DZNET721
        DO 6220 J=1,NIDC                                            DZNET722
6220    ZRRAY(I)=ZRRAY(I)+ZG(LA+I,LB+J)*ZJ(NBRN-NIDC-NDC+J)         DZNET723
        ZR(I)=ZRRAY(I)+ZR(I)                                        DZNET724
6230    CCNTINUE                                                    DZNET725
6210    LA=NIDV+NDV                                                 DZNET726
        DO 6240 I=1,NT                                              DZNET727
        ZE(LA+I)=(0.0,0.0)                                          DZNET728
        DO 6240 J=1,NT                                              DZNET729
6240    ZE(LA+I)=ZE(LA+I)+ZA(I,J)*ZR(J)                             DZNET730
6240    CCNTINUE                                                    DZNET731
        DO 6300 I=1,NT                                              DZNET732
        ZJ(LA+I)=(0.0,0.0)                                          DZNET733
        DO 6300 J=1,NT                                              DZNET734
        ZJ(LA+I)=ZJ(LA+I)+ZB(I,J)*ZE(LA+J)                          DZNET735
6300    CONTINUE                                                    DZNET736
        GC TG 2120                                                  DZNET737
7000    CONTINUE                                                    DZNEA737
        RETURN                                                      DZNEB737
1       FORMAT (11(I3,2X),5X,5A4)                                   DZNET738
2       FORMAT (1H1,40X,5A4)                                        DZNET739
3       FORMAT (1H0,2X,'N =',I3,4X,'B =',I3,4X,'VD =',I3,4X,'V =',I3,4X,'CDZNET740
```

```
       1AP =',I3,4X,'RES =',I3,4X,'IND =',I3,4X,'MUT =',I3,4X,'C  =',I3,4XDZNET741
       2,'CD=',I3,4X,'GROUND =',I3)                                        DZNET742
    4  FORMAT (1H0,//5X,'**ERROR**   THERE MUST BE AT LEAST TWO NODES TO DZNET743
       1FORM A NETWORK. CHECK INPUT DATA CARDS.')                          DZNET744
    5  FORMAT (1H0,//5X,'**ERROR**   THERE MUST BE AT LEAST TWO BRANCHES DZNET745
       1TO FORM A NETWORK. CHECK INPUT DATA CARDS.')                       DZNET746
    6  FORMAT (1H0,//5X,'**ERROR**   NO INDEPENDENT SOURCES PRESENT.')     DZNET747
    7  FORMAT (1H0,//5X,'**ERROR**   NO PASSIVE ELEMENTS PRESENT IN THE NODZNET748
       1ETWORK.')                                                         DZNET749
    8  FORMAT (1H0,//5X,'**ERROR**   MORE NODES THAN BRANCHES.')          DZNET750
    9  FORMAT (////5X,'BRANCH  NO.',5X,'NODEI',3X,'NODEJ',5X,'TYPE',22X,'DZNET751
       1ELEMENT VALUE'/)                                                  DZNET752
   10  FORMAT (2I3,4X,2E10.3,5X,A1,I3)                                    DZNET753
   11  FORMAT (8X,I3,10X,I3,5X,I3,7X,'DEP. VOLTAGE SOURCE',5X,1PE10.3,3X,DZNET754
       11PE10.3,5X,'CONTROLLED BY CURRENT OF BRANCH NO. ',I3)             DZNET755
   12  FORMAT (1H0,//5X,'**ERROR**   CONTROLLING BRANCH FOR THE DEPENDENTDZNET756
       1VOLTAGE SOURCE IS MISSING')                                       DZNET757
   13  FORMAT (8X,I3,10X,I3,5X,I3,7X,'DEP. VOLTAGE SOURCE',6X,1PE10.3,3X,DZNET758
       11PE10.3,5X,'CONTROLLED BY VOLTAGE OF BRANCH NO. ',I3)             DZNET759
   14  FORMAT (8X,I3,10X,I3,5X,I3,7X,'VOLTAGE SOURCE',11X,1PE10.3,3X,1PE1DZNET760
       10.3)                                                              DZNET761
   15  FORMAT (8X,I3,10X,I3,5X,I3,7X,'CAPACITOR          ',11X,1PE10.3,3X,1PE1DZNET762
       10.3)                                                              DZNET763
   16  FORMAT (8X,I3,10X,I3,5X,I3,7X,'RESISTOR           ',11X,1PE10.3,3X,1PE1DZNET764
       10.3)                                                              DZNET765
   17  FORMAT (8X,I3,10X,I3,5X,I3,7X,'INDUCTOR           ',11X,1PE10.3,3X,1PE1DZNET766
       10.3)                                                              DZNET767
   18  FORMAT (8X,I3,10X,I3,5X,I3,7X,'CURRENT SOURCE',11X,1PE10.3,3X,1PE1DZNET768
       10.3)                                                              DZNET769
   19  FORMAT (8X,I3,10X,I3,5X,I3,7X,'DEP. CURRENT SOURCE',6X,1PE10.3,3X,DZNET770
       11PE10.3,5X,'CONTROLLED BY CURRENT OF BRANCH NO. ',I3)             DZNET771
   20  FORMAT (1H0,//5X,'**ERROR**   CONTROLLING BRANCH FOR THE DEPENDENTDZNET772
       1 CURRENT SOURCE IS MISSING.')                                     DZNET773
   21  FORMAT (8X,I3,10X,I3,5X,I3,7X,'DEP. CURRENT SOURCE',5X,1PE10.3,3X,DZNET774
       11PE10.3,5X,'CONTROLLED BY VOLTAGE OF BRANCH NO. ',I3)             DZNET775
   22  FORMAT (1H0,//5X,'MUTUAL INDUCTANCE',10X,'BETWEEN BRANCH',7X,'AND DZNET776
       1BRANCH'//)                                                        DZNET777
   23  FORMAT (2I3,4X,E10.3)                                              DZNET778
   24  FORMAT (5X,1PE10.3,20X,I3,16X,I3)                                  DZNET779
   25  FORMAT (1H0,//5X,'**ERROR**   NODE',I3,' IS DANGLING FROM CIRCUITDZNET780
       1.')                                                               DZNET781
   26  FORMAT (1H0,//5X,'**ERROR**   CURRENT SOURCES IN A CUTSET')        DZNET782
   27  FORMAT (1H0,//5X,'**ERROR**   VOLTAGE SOURCES IN A LOOP')          DZNET783
   28  FORMAT (1H0,//5X,'**ERROR**   PROGRAM IS UNABLE TO CHOOSE A TREE'DZNET784
   29  FORMAT (1H0,//5X,'**NOTE**   LINEAR RELATIONS EXIST BETWEEN THE DEDZNET785
       1PENDENT SOURCES.')                                                DZNET786
   30  FORMAT (1H0,//5X,'**NOTE**   FREQUENCY AT OR NEAR NATURAL FREQUENCDZNET787
       1Y OF NETWORK.')                                                   DZNET788
   31  FORMAT (1H1,2X,'BRANCH      FREQUENCY (C/S)      VOLTAGE ACROSS ELDZNET789
       1EMENT (VOLTS AND DEGREES)      CURRENT THROUGH BRANCH (AMPS. AND DZNET790
       2DEGREES)')                                                        DZNET791
   32  FORMAT (2X,'NO. TYPE    REAL        IMAGINARY       REAL      IMAGINADZNET792
       1RY   MAGNITUDE   PHASE        REAL        IMAGINARY    MAGNITUDE DZNET793
       2 PHASE'/)                                                         DZNET794
   33  FORMAT (2X,I2,'  D.CS  ',1P2E10.2,2(2X,1P2E13.4,1PE12.3,2X,0PF7.2))DZNET795
   34  FORMAT (2X,I2,'  CS   ',1P2E10.2,2(2X,1P2E13.4,1PE12.3,2X,0PF7.2))DZNET796
   35  FORMAT (2X,I2,'  IND  ',1P2E10.2,2(2X,1P2E13.4,1PE12.3,2X,0PF7.2))DZNET797
   36  FORMAT (2X,I2,'  RES  ',1P2E10.2,2(2X,1P2E13.4,1PE12.3,2X,0PF7.2))DZNET798
   37  FORMAT (2X,I2,'  CAP  ',1P2E10.2,2(2X,1P2E13.4,1PE12.3,2X,0PF7.2))DZNET799
   38  FORMAT (2X,I2,'  VS   ',1P2E10.2,2(2X,1P2E13.4,1PE12.3,2X,0PF7.2))DZNET800
   39  FORMAT (2X,I2,'  D.VS ',1P2E10.2,2(2X,1P2E13.4,1PE12.3,2X,0PF7.2))DZNET801
   41  FORMAT          (2X,'(THE SUM OF CURRENTS AT REFERENCE NODE (NODE # 'DZNET802
       1,I3,') = ',1PE20.8,1X,'J',1X,1PE20.8,')'/)                        DZNET803
   42  FORMAT (1H0,//////25X,'THIS IS THE END OF PROGRAM      ''',5A4,'''') DZNET804
       END                                                                DZNET805
       SUBROUTINE MATMUL (ZA,ZB,ZC,MA,NA,KA,LA,MB,NB,LB,MINJS,IA)         MATMUL01
       IMPLICIT COMPLEX (Z)                                               MATMUL02
       DIMENSION ZA(IA,IA),ZB(IA,IA),ZC(41,41)                           MATMUL03
       IF (MINUS .EQ. 1) GO TO 300                                        MATMUL04
       DO 100 I=1,KA                                                      MATMUL05
       DO 100 J=1,LB                                                      MATMUL06
       ZC(I,J)=(0.0,0.0)                                                  MATMUL07
       DO 100 K=1,LA                                                      MATMUL08
       ZC(I,J)=ZC(I,J)+ZA(MA+I,NA+K)*ZB(MB+K,NB+J)                        MATMUL09
  100  CONTINUE                                                           MATMUL10
       RETURN                                                             MATMUL11
```

```
300        DO 200 I=1,KA                                          MATMUL12
           DO 200 J=1,LB                                          MATMUL13
           ZC(I,J)=(0.0,0.0)                                      MATMUL14
           DO 200 K=1,LA                                          MATMUL15
           ZC(I,J)=ZC(I,J)-ZA(MA+I,NA+K)*ZB(MB+K,NB+J)            MATMUL16
200        CONTINUE                                               MATMUL17
           RETURN                                                 MATMUL18
           END                                                    MATMUL19
           SUBROUTINE CMINV (Z,N,D,L,M)                           CMINVOC1
           IMPLICIT COMPLEX (Z)                                   CMINVOO2
           DIMENSION Z(1),L(1),M(1)                               CMINVOC3
           D=1.0                                                  CMINVOO4
           NK=-N                                                  CMINVOC5
           DO 80 K=1,N                                            CMINVOO6
           NK=NK+N                                                CMINVOO7
           L(K)=K                                                 CMINVOO8
           M(K)=K                                                 CMINVOO9
           KK=NK+K                                                CMINVO10
           ZBIG=Z(KK)                                             CMINVO11
           DO 20 J=K,N                                            CMINVO12
           IZ=N*(J-1)                                             CMINVO13
           DO 20 I=K,N                                            CMINVO14
           IJ=IZ+I                                                CMINVO15
           TEMP1=REAL(ZBIG)**2+AIMAG(ZBIG)**2                     CMINVO16
           TEMP2=REAL(Z(IJ))**2+AIMAG(Z(IJ))**2                   CMINVO17
10         IF (TEMP1-TEMP2) 15,20,20                              CMINVO18
15         ZBIG=Z(IJ)                                             CMINVO19
           L(K)=I                                                 CMINVO20
           M(K)=J                                                 CMINVO21
20         CONTINUE                                               CMINVO22
           J=L(K)                                                 CMINVO23
           IF (J-K) 35,35,25                                      CMINVO24
25         KI=K-N                                                 CMINVO25
           DO 30 I=1,N                                            CMINVO26
           KI=KI+N                                                CMINVO27
           ZHOLD=-Z(KI)                                           CMINVO28
           JI=KI-K+J                                              CMINVO29
           Z(KI)=Z(JI)                                            CMINVO30
30         Z(JI)=ZHOLD                                            CMINVO31
35         I=M(K)                                                 CMINVO32
           IF (I-K) 45,45,38                                      CMINVO33
38         JP=N*(I-1)                                             CMINVO34
           DO 40 J=1,N                                            CMINVO35
           JK=NK+J                                                CMINVO36
           JI=JP+J                                                CMINVO37
           ZHOLD=-Z(JK)                                           CMINVO38
           Z(JK)=Z(JI)                                            CMINVO39
40         Z(JI)=ZHOLD                                            CMINVO40
45         BIG=REAL(ZBIG)**2+AIMAG(ZBIG)**2                       CMINVO41
           IF (BIG-0.000001) 46,46,48                             CMINVO42
46         D=0.0                                                  CMINVO43
           RETURN                                                 CMINVO44
48         DO 55 I=1,N                                            CMINVO45
           IF (I-K) 50,55,50                                      CMINVO46
50         IK=NK+I                                                CMINVO47
           Z(IK)=Z(IK)/(-ZBIG)                                    CMINVO48
55         CONTINUE                                               CMINVO49
           DO 65 I=1,N                                            CMINVO50
           IK=NK+I                                                CMINVO51
           ZHOLD=Z(IK)                                            CMINVO52
           IJ=I-N                                                 CMINVO53
           DO 65 J=1,N                                            CMINVO54
           IJ=IJ+N                                                CMINVO55
           IF (I-K) 60,65,60                                      CMINVO56
60         IF (J-K) 62,65,62                                      CMINVO57
62         KJ=IJ-I+K                                              CMINVO58
           Z(IJ)=ZHOLD*Z(KJ)+Z(IJ)                                CMINVO59
65         CONTINUE                                               CMINVO60
           KJ=K-N                                                 CMINVO61
           DO 75 J=1,N                                            CMINVO62
           KJ=KJ+N                                                CMINVO63
           IF (J-K) 70,75,70                                      CMINVO64
70         Z(KJ)=Z(KJ)/ZBIG                                       CMINVO65
75         CONTINUE                                               CMINVO66
           Z(KK)=1.0/ZBIG                                         CMINVO67
80         CONTINUE                                               CMINVO68
```

```
          K=N
100       K=(K-1)                                                    CMINV069
          IF (K) 150,150,105                                         CMINV070
105       I=L(K)                                                     CMINV071
          IF (I-K) 120,120,108                                       CMINV072
108       JQ=N*(K-1)                                                 CMINV073
          JR=N*(I-1)                                                 CMINV074
          DO 110 J=1,N                                               CMINV075
          JK=JQ+J                                                    CMINV076
          ZHOLD=Z(JK)                                                CMINV077
          JI=JR+J                                                    CMINV078
          Z(JK)=-Z(JI)                                               CMINV079
110       Z(JI)=ZHOLD                                                CMINV080
120       J=M(K)                                                     CMINV081
          IF (J-K) 100,100,125                                       CMINV082
125       KI=K-N                                                     CMINV083
          DO 130 I=1,N                                               CMINV084
          KI=KI+N                                                    CMINV085
          ZHOLD=Z(KI)                                                CMINV086
          JI=KI-K+J                                                  CMINV087
          Z(KI)=-Z(JI)                                               CMINV088
130       Z(JI)=ZHOLD                                                CMINV089
          GO TO 100                                                  CMINV090
150       RETURN                                                     CMINV091
          END                                                        CMINV092
                                                                     CMINV093
          SUBROUTINE TRIANG (A, N, M, INDEX)                         TRIANG01
          DIMENSION A(40,80),INDEX(80)                               TRIANG02
          DO 1000 LL=1,N                                             TRIANG03
          IMP=LL+1                                                   TRIANG04
100       IF (A(LL,LL) .EQ. 0.0) GO TO 400                           TRIANG05
          DO 300 I=IMP, N                                            TRIANG06
          IF (A(I,LL) .EQ. 0.0) GO TO 300                            TRIANG07
          DUMMY=A(LL,LL)/A(I,LL)                                     TRIANG08
          DO 200 K=LL, M                                             TRIANG09
200       A(I,K) = DUMMY*A(I,K)-A(LL,K)                              TRIANG10
300       CONTINUE                                                   TRIANG11
          GO TO 1000                                                 TRIANG12
400       IF (LL .EQ. N) GO TO 800                                   TRIANG13
          DO 700 I=IMP,N                                             TRIANG14
          IF (A(I,LL) .EQ. 0.0) GO TO 600                            TRIANG15
          DO 500 K=LL,M                                              TRIANG16
          TEMP = A(I,K)                                              TRIANG17
          A(I,K) = A(LL,K)                                           TRIANG18
          A(LL,K) = TEMP                                             TRIANG19
500       CONTINUE                                                   TRIANG20
          GO TO 100                                                  TRIANG21
600       IF (I .EQ. N ) GO TO 800                                   TRIANG22
700       CONTINUE                                                   TRIANG23
800       IF (LL .EQ. M) GO TO 1000                                 TRIANG24
          DO 900 J=IMP,M                                             TRIANG25
          IF (A(LL,J) .EQ. 0.0) GO TO 900                            TRIANG26
          DO 850 IT=1,N                                              TRIANG27
          TEMP=A(IT,J)                                               TRIANG28
          A(IT,J) = A(IT,LL)                                         TRIANG29
          A(IT,LL) = TEMP                                            TRIANG30
850       CONTINUE                                                   TRIANG31
          KEEP = INDEX(LL)                                           TRIANG32
          INDEX(LL) = INDEX(J)                                       TRIANG33
          INDEX(J) = KEEP                                            TRIANG34
          GO TO 100                                                  TRIANG35
900       CONTINUE                                                   TRIANG36
1000      CONTINUE                                                   TRIANG37
          RETURN                                                     TRIANG38
          END                                                        TRIANG39
          SUBROUTINE MINV (A,N,D,L,M)                                MINV0001
          DIMENSION A(1),L(1),M(1)                                   MINV0002
          D=1.0                                                      MINV0003
          NK=-N                                                      MINV0004
          DO 80 K=1,N                                                MINV0005
          NK=NK+N                                                    MINV0006
          L(K)=K                                                     MINV0007
          M(K)=K                                                     MINV0008
          KK=NK+K                                                    MINV0009
          BIGA=A(KK)                                                 MINV0010
          DO 20 J=K,N                                                MINV0011
          IZ=N*(J-1)                                                 MINV0012
```

```
            DO 20 I=K,N                                          MINV0013
            IJ=IZ+I                                              MINV0014
10          IF (ABS(BIGA)-ABS(A(IJ))) 15,20,20                   MINV0015
15          BIGA=A(IJ)                                           MINV0016
            L(K)=I                                               MINV0017
            M(K)=J                                               MINV0018
20          CONTINUE                                             MINV0019
            J=L(K)                                               MINV0020
            IF (J-K) 35,35,25                                    MINV0021
25          KI=K-N                                               MINV0022
            DO 30 I=1,N                                          MINV0023
            KI=KI+N                                              MINV0024
            HOLD=-A(KI)                                          MINV0025
            JI=KI-K+J                                            MINV0026
            A(KI)=A(JI)                                          MINV0027
30          A(JI)=HOLD                                           MINV0028
35          I=M(K)                                               MINV0029
            IF (I-K) 45,45,38                                    MINV0030
38          JP=N*(I-1)                                           MINV0031
            DO 40 J=1,N                                          MINV0032
            JK=NK+J                                              MINV0033
            JI=JP+J                                              MINV0034
            HOLD=-A(JK)                                          MINV0035
            A(JK)=A(JI)                                          MINV0036
40          A(JI)=HOLD                                           MINV0037
45          IF (BIGA) 48,46,48                                   MINV0038
46          D=0.0                                                MINV0039
            RETURN                                               MINV0040
48          DO 55 I=1,N                                          MINV0041
            IF (I-K) 50,55,50                                    MINV0042
50          IK=NK+I                                              MINV0043
            A(IK)=A(IK)/(-BIGA)                                  MINV0044
55          CONTINUE                                             MINV0045
            DO 65 I=1,N                                          MINV0046
            IK=NK+I                                              MINV0047
            HOLD=A(IK)                                           MINV0048
            IJ=I-N                                               MINV0049
            DO 65 J=1,N                                          MINV0050
            IJ=IJ+N                                              MINV0051
            IF (I-K) 60,65,60                                    MINV0052
60          IF (J-K) 62,65,62                                    MINV0053
62          KJ=IJ-I+K                                            MINV0054
            A(IJ)=HOLD*A(KJ)+A(IJ)                               MINV0055
65          CONTINUE                                             MINV0056
            KJ=K-N                                               MINV0057
            DO 75 J=1,N                                          MINV0058
            KJ=KJ+N                                              MINV0059
            IF (J-K) 70,75,70                                    MINV0060
70          A(KJ)=A(KJ)/BIGA                                     MINV0061
75          CONTINUE                                             MINV0062
            D=D*BIGA                                             MINV0063
            A(KK)=1.0/BIGA                                       MINV0064
80          CONTINUE                                             MINV0065
            K=N                                                  MINV0066
100         K=(K-1)                                              MINV0067
            IF (K) 150,150,105                                   MINV0068
105         I=L(K)                                               MINV0069
            IF (I-K) 120,120,108                                 MINV0070
108         JQ=N*(K-1)                                           MINV0071
            JR=N*(I-1)                                           MINV0072
            DO 110 J=1,N                                         MINV0073
            JK=JQ+J                                              MINV0074
            HOLD=A(JK)                                           MINV0075
            JI=JR+J                                              MINV0076
            A(JK)=-A(JI)                                         MINV0077
110         A(JI)=HOLD                                           MINV0078
120         J=M(K)                                               MINV0079
            IF (J-K) 100,100,125                                 MINV0080
125         KI=K-N                                               MINV0081
            DO 130 I=1,N                                         MINV0082
            KI=KI+N                                              MINV0083
            HOLD=A(KI)                                           MINV0084
            JI=KI-K+J                                            MINV0085
            A(KI)=-A(JI)                                         MINV0086
130         A(JI)=HOLD                                           MINV0087
            GO TO 100                                            MINV0088
150         RETURN                                               MINV0089
            END                                                  MINV0090
```

TRANSISTOR AMPLIFIER

N = 8   B = 16   VD = 0   V = 1   CAP = 5   RES = 8   IND = 0   MUT = 0   C = 0   CD= 2   GROUND = 8

BRANCH NO.	NODEI	NODEJ	TYPE	ELEMENT VALUE	
1	1	8	VOLTAGE SOURCE	1.000E 00	0.0
2	3	8	CAPACITOR	1.600E-09	0.0
3	3	4	CAPACITOR	1.000E-11	0.0
4	5	8	CAPACITOR	1.600E-09	0.0
5	5	6	CAPACITOR	1.000E-11	0.0
6	6	7	CAPACITOR	1.000E-09	0.0
7	1	2	RESISTOR	5.000E 01	0.0
8	2	3	RESISTOR	8.000E 01	0.0
9	3	8	RESISTOR	1.200E 03	0.0
10	4	8	RESISTOR	1.000E 05	0.0
11	5	8	RESISTOR	8.000E 01	0.0
12	6	8	RESISTOR	1.200E 03	0.0
13	6	8	RESISTOR	1.000E 04	0.0
14	7	2	RESISTOR	4.000E 03	0.0
15	4	8	DEP. CURRENT SOURCE	3.900E-02	0.0	CONTROLLED BY VOLTAGE OF BRANCH NO. 2
16	6	8	DEP. CURRENT SOURCE	3.900E-02	0.0	CONTROLLED BY VOLTAGE OF BRANCH NO. 4

BRANCH NO. TYPE	FREQUENCY (C/S) REAL IMAGINARY	VOLTAGE ACROSS ELEMENT REAL IMAGINARY	(VOLTS AND DEGREES) MAGNITUDE PHASE	CURRENT THROUGH BRANCH REAL IMAGINARY	(AMPS. AND DEGREES) MAGNITUDE PHASE
7 RES	0.0 1.00E 06	-7.6547E-01 -1.7513E-01	7.852E-01 -167.11	-1.5309E-02 -3.5026E-03	1.570E-02 -167.11
13 RES	0.0 1.00E 06	5.2703E 01 1.0488E 01	5.373E 01 11.26	5.2700E-03 1.0488E-03	5.373E-03 11.26
(THE SUM OF CURRENTS AT REFERENCE NODE (NODE # 8) = 1.4901162E-08 J 3.7252930E-09)					

INPUT IMPEDANCE = -0.6207E 02 0.1420E 02

7 RES	0.0 2.00E 06	-5.5939E-01 -1.1117E-01	5.806E-01 -168.96	-1.1398E-02 -2.2233E-03	1.161E-02 -168.96
13 RES	0.0 2.00E 06	2.7334E-01 3.3442E 00	2.754E-01 5.98	2.7334E-03 3.3442E-04	2.754E-03 6.98
(THE SUM OF CURRENTS AT REFERENCE NODE (NODE # 8) = 3.7252930E-09 J -3.7252930E-09)					

INPUT IMPEDANCE = -0.8452E 02 0.1649E 02

7 RES	0.0 3.00E 06	-4.8485E-01 -6.814E-02	4.896E-01 -172.00	-9.6971E-03 -1.3628E-03	9.792E-03 -172.00
13 RES	0.0 3.00E 06	-5.1564E-01 -5.2368E-01	1.517E 01 -1.98	1.5164E-03 -5.2368E-05	1.517E-03 -1.98
(THE SUM OF CURRENTS AT REFERENCE NODE (NODE # 8) = -1.3737008E-08 J 1.1175870E-08)					

INPUT IMPEDANCE = -0.1011E 03 0.1421E 02

7 RES	0.0 4.00E 06	-4.4358E-01 -4.7646E-02	4.462E-01 -173.87	-8.8735E-03 -9.5293E-04	8.925E-03 -173.87
13 RES	0.0 4.00E 06	8.9449E 00 -1.6746E 00	9.100E 00 -10.60	8.9449E-04 -1.6746E-04	9.100E-04 -10.60
(THE SUM OF CURRENTS AT REFERENCE NODE (NODE # 8) = -1.1641532E-09 J -1.1874362E-08)					

Sample Printout of DZNET

9

time-domain analysis

The dynamics of a circuit are determined by the rate of change of the capacitor voltages and the inductor currents in the circuit. To describe the behavior of the circuit for all times, we must have differential equations in terms of these variables. In this chapter we shall demonstrate how these equations can be formulated in a systematic manner for a general linear circuit, obtain the solution of these equations, and discuss the nature of the response. We shall see that the most important property in the time domain is that in a linear circuit, no matter how complicated it may be, the natural response, or that resulting from initial conditions, must always be a superposition of exponentially decaying, damped, or undamped sinusoidal time functions, and that the forced response, or that resulting from excitations, must always be a convolution of these time functions with the excitations. This property is really a consequence of the fact that the differential equations describing the circuit are linear and with constant coefficients.

We recall that the most basic mathematical description of a circuit is in terms of the KVL equations written for the chord voltages, the KCL or cut set equations written for the tree currents, and the set of terminal equations for the elements. What we must do now is to combine these equations into a set of first-order differential equations in the variables that are the capacitor voltages and inductor currents. We did this in Chapter 4 for first- and second-order circuits.

We now wish to extend the analysis to the general case, in which the circuit may be of any configuration and contain any number of resistors, capacitors, inductors, mutual inductors, dependent sources, and sources with arbitrary time functions.

In Chapter 6, using mesh currents, loop currents, or node voltages, we obtained a set of integral-differential equations in these variables that describes the dynamics of a circuit. However, we saw in Example 6.6 that these variables are not natural in that the solution of these equations must always depend on the knowledge of the capacitor voltages and inductor currents in the first place. The natural variables, as we have said on many occasions, are the state variables, which are the capacitor voltages and inductor currents, minimum in number, whose rate of change gives the dynamics of the circuit for all times. The minimum number is called the order of the circuit.

We shall begin with the formulation of state equations in simple circuits and proceed to the most general case by stages. The solution will be a generalization of the scheme developed in Chapter 4, using eigenvalues and eigenvectors. The complete solution will be once again a linear combination of the homogeneous and particular solutions.

For numerical computation, the set of first-order differential equations may be integrated by applying the Runge-Kutta algorithm or any other numerical methods. The computer program given in Chapter 4 can be extended to the general case in a straightforward manner.

Lastly, a numerical scheme for finding the solution of a circuit in the time domain without the use of state equations will be presented. The scheme is based on the use of node equations and implicit integration of the equations defining a capacitor and inductor. Each capacitor and inductor is replaced by a resistor in parallel with a current source whose value is "updated" at each time step. The resultant resistor circuit is solved by means of node equations at each time step. The advantage of this scheme over the state equation approach is that it does not require the complicated topological procedure for setting up the state equations and that the integration scheme leads to solutions that will never be unbounded, regardless of the step size used, if the exact solutions are bounded. This is not the case of "explicit" integration schemes, such as Runge-Kutta, in which the step size must be sufficiently small to keep the propagation error within limit.

9.1 Formulation of State Equations by Superposition

Consider a circuit consisting of resistors, capacitors, inductors, and sources. Suppose we pull out all the elements but the resistors and dependent sources as shown in Figure 9.1. For simplicity, we have shown only one element of each kind. We assume that there does not exist a

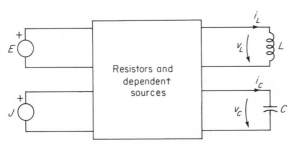

Figure 9.1 Formulation of state equations by superposition.

linear relation among the capacitor voltages or among the inductor currents in the circuit. The contrary case will be treated later. Now regarding the capacitor voltages and inductor currents as excitations and the capacitor currents and inductor voltages as responses, we can write, by superposition, a set of linear equations in these variables. For the circuit shown, we have

$$i_C = av_C + bi_L + cE + dJ$$
$$v_L = a'v_C + b'i_L + c'E + d'J \tag{9.1}$$

in which the coefficients of the variables are constants. Since the capacitor current $i_C = C\,dv_C/dt$ and the inductor voltage $v_L = L\,di_L/dt$, the left side of the equations consists of derivatives of the capacitor voltages and inductor currents. After dividing by C and L, we obtain a set of first-order differential equations in the state variables. In the case shown, we have

$$\frac{dv_C}{dt} = a_{11}v_C + a_{12}i_L + b_{11}E + b_{12}J$$

$$\frac{di_L}{dt} = a_{21}v_C + a_{22}i_L + b_{21}E + b_{22}J \tag{9.2}$$

In general, the state equation takes the form

$$\frac{d}{dt}\begin{bmatrix} x_1 \\ x_2 \\ \vdots \\ x_n \end{bmatrix} = \begin{bmatrix} a_{11} & a_{12} & \cdots & a_{n1} \\ a_{21} & a_{22} & \cdots & a_{n2} \\ \vdots & \vdots & \ddots & \vdots \\ a_{n1} & a_{n2} & \cdots & a_{nn} \end{bmatrix} \begin{bmatrix} x_1 \\ x_2 \\ \vdots \\ x_n \end{bmatrix} + \mathbf{u}(t) \tag{9.3}$$

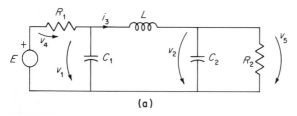

(a)

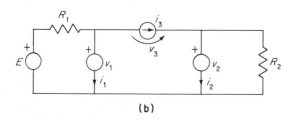

(b)

Figure 9.2 Formulation of state equation.

where the x's are the state variables; the a_{ij}'s are constants, and $\mathbf{u}(t)$ is a linear combination of the excitations. In an abbreviated form, we write

$$\dot{\mathbf{x}} = \mathbf{Ax} + \mathbf{u}(t) \tag{9.4}$$

where $\mathbf{x}$ and $\mathbf{u}$ are column vectors and $\mathbf{A}$ a square matrix.

Example 9.1 Formulate the state equation of the circuit shown in Figure 9.2.

Solution Replace each capacitor by a voltage source of value v_1 and v_2 and the inductor by a current source of value i_3. The resulting circuit is shown in Figure 9.2(b). By inspection,

$$i_1 = (E - v_1)G_1 - i_3$$
$$i_2 = i_3 - v_2G_2$$
$$v_3 = v_1 - v_2$$

Replacing the left-hand side by derivatives and dividing by C_1, C_2, and L, respectively, we get

$$\frac{d}{dt}\begin{bmatrix} v_1 \\ v_2 \\ i_3 \end{bmatrix} = \begin{bmatrix} -G_1C_1^{-1} & 0 & -C_1^{-1} \\ 0 & -G_2C_2^{-1} & C_2^{-1} \\ L^{-1} & -L^{-1} & 0 \end{bmatrix}\begin{bmatrix} v_1 \\ v_2 \\ i_3 \end{bmatrix} + \begin{bmatrix} G_1C_1^{-1} \\ 0 \\ 0 \end{bmatrix}E(t)$$

which has the form $\dot{\mathbf{x}} = \mathbf{Ax} + \mathbf{u}$. The other branch voltages and currents can be expressed in terms of the state variables. For example,

$$v_4 = E - v_1$$
$$v_5 = v_2$$

Example 9.2 Find the state equation of the circuit of Figure 9.3.

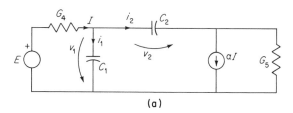

(a)

Figure 9.3 Formulation of state equation. [Example 9.2]

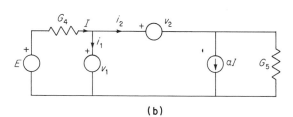

(b)

Solution   Replace each capacitor by a voltage source as in Figure 9.3(b). By inspection, we have

$$i_1 = I - i_2$$
$$i_2 = \alpha I + G_5(v_1 - v_2)$$
$$I = G_4(E - v_1) \tag{9.5}$$

Combining equations, we find

$$\frac{d}{dt}\begin{bmatrix} v_1 \\ v_2 \end{bmatrix} = \begin{bmatrix} -C_1^{-1}(G_4 + G_5 - \alpha G_4) & C_1^{-1}G_5 \\ C_2^{-1}(G_5 - \alpha G_4) & -C_2^{-1}G_5 \end{bmatrix}\begin{bmatrix} v_1 \\ v_2 \end{bmatrix} + \begin{bmatrix} C_1^{-1}G_4(1 - \alpha) \\ \alpha G_4 C_2^{-1} \end{bmatrix} E$$

Example 9.3   Formulate the state equation for the circuit of Figure 9.4.

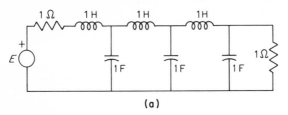

(a)

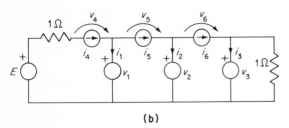

(b)

Figure 9.4 Formulation of state equation. [Example 9.3]

Solution   By inspection, the capacitor currents and inductor voltages are given by

$$i_1 = i_4 - i_5$$
$$i_2 = i_5 - i_6$$
$$i_3 = i_6 - v_3$$
$$v_4 = E - v_1 - i_4$$
$$v_5 = v_1 - v_2$$
$$v_6 = v_2 - v_3$$

Let $\mathbf{x}$ denote the state variables in the order $v_1, v_2,..., i_5, i_6$. We find

$$\frac{d\mathbf{x}}{dt} = \begin{bmatrix} 0 & 0 & 0 & 1 & -1 & 0 \\ 0 & 0 & 0 & 0 & 1 & -1 \\ 0 & 0 & -1 & 0 & 0 & 1 \\ -1 & 0 & 0 & -1 & 0 & 0 \\ 1 & -1 & 0 & 0 & 0 & 0 \\ 0 & 1 & -1 & 0 & 0 & 0 \end{bmatrix}\mathbf{x} + \begin{bmatrix} 0 \\ 0 \\ 0 \\ 1 \\ 0 \\ 0 \end{bmatrix} E \tag{9.6}$$

9.2 Formal Formulation of State Equations*

In order to treat the most general case, we must develop an algorithm by which the state equation of any linear circuit can be systematically formulated. The algorithm should lead to an explicit formula expressed in terms of the topology of the circuit.

To this end, we must go back to the use of tree and chords. It is best to do an example. Consider the circuit of Figure 9.5. Noting that tree voltages determine all of the branch voltages, and chord currents all of the branch currents, we select a tree such that all the voltage sources and as many capacitors as possible are branches of that tree, and that all the current sources and as many inductors as possible are chords. The remaining tree branches, if any, are made up of resistors. A tree constructed in this manner is known as a *proper tree*.

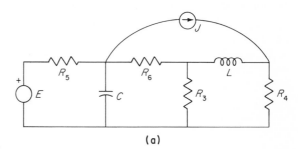

(a)

Figure 9.5 Formal formulation of state equations. (**a**) A circuit; (**b**) its proper tree.

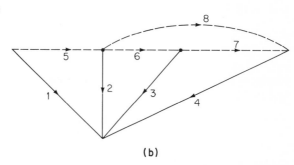

(b)

In our example, the tree selected is as shown in Figure 9.5(b). Writing KVL, KCL, and terminal equations, we have

KVL:

$$v_5 = v_1 - v_2 \qquad (9.7)$$
$$v_6 = v_2 - v_3 \qquad (9.8)$$
$$v_7 = v_3 - v_4 \qquad (9.9)$$
$$v_8 = v_2 - v_4 \qquad (9.10)$$

R chords

KCL:

$$i_1 = -i_5 \tag{9.11}$$
$$i_2 = i_5 - i_6 - i_8 \tag{9.12}$$
$$i_3 = i_6 - i_7 \tag{9.13}$$
$$i_4 = i_7 + i_8 \qquad R \text{ tree} \tag{9.14}$$

Terminal:

$$v_1 = E \tag{9.15}$$

$$i_2 = C \frac{dv_2}{dt} \tag{9.16}$$

$$i_3 = G_3 v_3 \tag{9.17}$$
$$i_4 = G_4 v_4 \qquad R \text{ tree} \tag{9.18}$$
$$v_5 = R_5 i_5 \tag{9.19}$$
$$v_6 = R_6 i_6 \qquad R \text{ chords} \tag{9.20}$$

$$v_7 = L \frac{di_7}{dt} \tag{9.21}$$
$$i_8 = J \tag{9.22}$$

Our strategy is to eliminate the resistor voltages and currents and to retain the tree capacitor voltages and inductor chord currents as unknowns.

First, take the KVL equations for the R chords [Equations (9.7) and (9.8)] and replace the R-chord voltages by R-chord currents [Equations (9.19) and (9.20)]:

$$R_5 i_5 = E - v_2$$
$$R_6 i_6 = v_2 - v_3$$

Replace the R-tree voltages by the R-tree currents [Equations (9.17) and (9.18)]:

$$R_5 i_5 = E - v_2$$
$$R_6 i_6 = v_2 - R_3 i_3$$

Replace the R-tree currents by chord currents, using Equations (9.13) and (9.14):

$$R_5 i_5 = E - v_2$$
$$R_6 i_6 = v_2 - R_3 i_6 + R_3 i_7$$

Solving for the R-chord currents in terms of the state variables, we get

$$i_5 = R_5^{-1} E - R_5^{-1} v_2 \tag{9.23}$$
$$i_6 = (R_6 + R_3)^{-1} v_2 + R_3 (R_3 + R_6)^{-1} i_7 \tag{9.24}$$

Take the terminal equation of the capacitor and replace all tree currents by chord currents [Equation (9.12)]:

$$C \frac{dv_2}{dt} = i_5 - i_6 - J$$

Replace the R-chord currents by those given by Equations (9.23) and (9.24) and multiply both sides of the equation by C^{-1}. We then have

$$\frac{dv_2}{dt} = -C^{-1}(R_5^{-1} + (R_3 + R_6)^{-1})v_2 - C^{-1}R_3(R_3 + R_6)^{-1}i_7$$
$$+ C^{-1}R_5^{-1}E - C^{-1}J \qquad (9.25)$$

which is the state equation for the capacitor.

Next, in a dual manner, we take the KCL for the R-tree currents [Equations (9.13) and (9.14)] and combine with the terminal equations and Equations (9.23) and (9.24). We then have

$$G_3v_3 = (R_6 + R_3)^{-1}v_2 + R_3(R_6 + R_3)^{-1}i_7 - i_7$$
$$G_4v_4 = i_7 + J$$

Solve for the R-tree voltages in terms of the state variables:

$$v_3 = R_3(R_6 + R_3)^{-1}v_2 - R_3R_6(R_3 + R_6)^{-1}i_7 \qquad (9.26)$$
$$v_4 = R_4i_7 + R_4J \qquad (9.27)$$

Take the terminal equation for the inductor [Equation (9.21)] and replace the chord voltage by tree voltages and the latter by Equation (9.26). Multiply both sides by L^{-1}. The state equation for the inductor becomes

$$\frac{di_7}{dt} = L^{-1}R_3(R_3 + R_6)^{-1}v_2 - L^{-1}(R_3R_6(R_3 + R_6)^{-1} + R_4)i_7$$
$$- L^{-1}R_4J \qquad (9.28)$$

This equation together with that for the capacitor constitute the system of state equations.

We see that this procedure is systematic and in fact is the algorithm on which computer formulation of state equations is based. We should note the similarity of the formulation of Chapter 6 for resistor circuits and that for the general case.

The solution of state equations will be given later. We note that once the state variables have been found, all the other voltages and currents in the circuit can be determined systematically as follows.

From Equations (9.23), (9.24), (9.26), and (9.27), we find the R-chord currents and R-tree voltages. We now have all the tree voltages, E, v_2, v_3, and v_4, and the chord currents, i_5, i_6, i_7, and J. By KVL and KCL, all the other voltages and currents are determined.

9.3 Circuits Containing Capacitor Loops*

In the preceding derivation of the state equation, we assume that there does not exist a linear relation among the capacitor voltages or among the inductor currents. We shall now treat the contrary case.

   Consider the circuit of Figure 9.6(a). Note the presence of a capacitor-voltage source loop. Select a proper tree as shown in Figure 9.6(b).

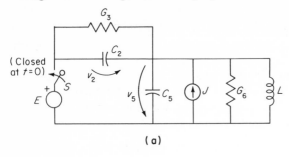

(a)

Figure 9.6 A circuit with a capacitor-voltage source loop.

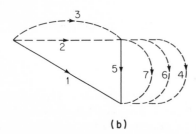

(b)

Writing KVL, KCL, and terminal equations and eliminating all resistor voltages and currents as in the preceding section, we find a set of differential equations in the unknowns that are the capacitor voltages and inductor currents given by

$$\frac{d}{dt}\begin{bmatrix} C_5 & -C_2 & 0 \\ 0 & 0 & 0 \\ 0 & 0 & L \end{bmatrix}\begin{bmatrix} v_5 \\ v_2 \\ i_4 \end{bmatrix} = \begin{bmatrix} -(G_3 + G_6) & 0 & -1 \\ -1 & -1 & 0 \\ 1 & 0 & 0 \end{bmatrix}\begin{bmatrix} v_5 \\ v_2 \\ i_4 \end{bmatrix}$$
$$+ \begin{bmatrix} G_3 & 1 \\ 1 & 0 \\ 0 & 0 \end{bmatrix}\begin{bmatrix} E \\ J \end{bmatrix} \tag{9.29}$$

It is seen that the matrix on the left is singular. The second row contains all zero elements. This means that not all of the variables of the set of equations are independent. In fact, the second equation reads

$$0 = -v_5 - v_2 + E \tag{9.30}$$

which is a statement of KVL for the capacitor-voltage source loop that we noted earlier. From this we see that the chord capacitor voltage v_2 cannot be an independent state variable and cannot be regarded as an excitation in the use of superposition to formulate state equations. Taking the derivative of the above, we get

$$\frac{dv_2}{dt} = -\frac{dv_5}{dt} + \frac{dE}{dt} \tag{9.31}$$

Substituting into the differential equations and simplifying, we get

$$\frac{d}{dt}\begin{bmatrix} C_5 + C_2 & 0 \\ 0 & L \end{bmatrix}\begin{bmatrix} v_5 \\ i_4 \end{bmatrix} = \begin{bmatrix} -(G_3 + G_6) & -1 \\ 1 & 0 \end{bmatrix}\begin{bmatrix} v_5 \\ i_4 \end{bmatrix}$$
$$+ \begin{bmatrix} G_3 & 1 \\ 0 & 0 \end{bmatrix}\begin{bmatrix} E \\ J \end{bmatrix} + \begin{bmatrix} C_2 \\ 0 \end{bmatrix}\frac{dE}{dt} \qquad (9.32)$$

The matrix on the left now is nonsingular. Multiplying both sides by its inverse, we find the state equation for the circuit to be

$$\frac{d}{dt}\begin{bmatrix} v_5 \\ i_4 \end{bmatrix} = \begin{bmatrix} -(G_3 + G_6)/(C_5 + C_2) & -1/(C_5 + C_2) \\ 1/L & 0 \end{bmatrix}\begin{bmatrix} v_5 \\ i_4 \end{bmatrix}$$
$$+ \begin{bmatrix} G_3/(C_5 + C_2) & 1/(C_5 + C_2) \\ 0 & 0 \end{bmatrix}\begin{bmatrix} E \\ J \end{bmatrix}$$
$$+ \begin{bmatrix} C_2/(C_5 + C_2) \\ 0 \end{bmatrix}\frac{dE}{dt} \qquad (9.33)$$

Note that the derivative of the source voltage appears. This comes about because of the presence of the capacitor-voltage loop. Note also that the number of equations is one less than the number of energy storage elements.

Lastly, the initial capacitor voltages immediately after the closure of switch S must adjust themselves instantaneously so that KVL is satisfied around the capacitor-voltage source loop. Let $v_2(0^-)$ and $v_5(0^-)$ be the capacitor voltages just before the closure of S, and $v_2(0^+)$ and $v_5(0^+)$ be their values immediately after the closure. By KVL, we must have

$$E = v_2(0^+) + v_5(0^+)$$

Since $v_2(0^-)$ and $v_5(0^-)$ are arbitrary, $v_2(0^+)$ and $v_5(0^+)$ must change instantaneously. This can happen only if charge is transferred from one capacitor to the other instantaneously. Because of conservation of charge, the charge lost by C_1 must be the same as that gained by C_2, and we have

$$C_2(v_2(0^+) - v_2(0^-)) = C_5(v_5(0^+) - v_5(0^-))$$

Combining with the KVL equation, we get

$$v_2(0^+) = \frac{C_5}{C_2 + C_5} E + \frac{C_2}{C_2 + C_5} v_2(0^-) - \frac{C_5}{C_2 + C_5} v_5(0^-)$$
$$v_5(0^+) = \frac{C_2}{C_2 + C_5} E - \frac{C_2}{C_2 + C_5} v_2(0^-) + \frac{C_5}{C_2 + C_5} v_5(0^-) \qquad (9.34)$$

It is these values of capacitor voltages that must be used as initial values in the solution of the circuit for $t \geq 0$.

9.4   Circuits Containing Inductor Cut Sets*

Consider the circuit of Figure 9.7. After switch S is opened at $t = 0$, its proper tree is as indicated. The circuit for $t \geq 0$ has an inductor cut set, $L_4 - L_7$. Writing the usual equations, we find a set of differential equations for the capacitor voltage and inductor currents given by

$$\frac{d}{dt} \begin{bmatrix} C_5 & 0 & 0 \\ 0 & L_4 & L_7 \\ 0 & 0 & 0 \end{bmatrix} \begin{bmatrix} v_5 \\ i_4 \\ i_7 \end{bmatrix} = \begin{bmatrix} -G_3 & -1 & 0 \\ 1 & -R_6 & 0 \\ 0 & 1 & -1 \end{bmatrix} \begin{bmatrix} v_5 \\ i_4 \\ i_7 \end{bmatrix} + \begin{bmatrix} G_3 \\ 0 \\ 0 \end{bmatrix} E \qquad (9.35)$$

which shows that the inductor tree current cannot be an independent state variable. Writing the KCL equation for the inductor cut set, we have

$$i_7 = i_4 \qquad (9.36)$$

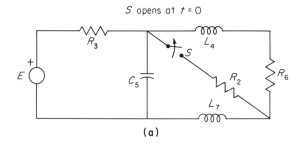

(a)

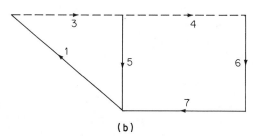

(b)

Figure 9.7 A circuit with an inductor cut set.

Eliminating i_7 from the differential equations, we find the state equations to be

$$\frac{d}{dt} \begin{bmatrix} v_5 \\ i_4 \end{bmatrix} = \begin{bmatrix} -G_3/C_5 & -1/C_5 \\ 1/(L_4 + L_7) & -R_6/(L_4 + L_7) \end{bmatrix} \begin{bmatrix} v_5 \\ i_4 \end{bmatrix} + \begin{bmatrix} G_3/C_5 \\ 0 \end{bmatrix} E \qquad (9.37)$$

The number of equations is one less than the total number of energy storage elements.

As in the case of a capacitor loop, the initial value of the currents in an inductor cut set must change instantaneously. Let $i_4(0^-)$ and $i_7(0^-)$ be the inductor currents before the opening of switch S, and $i_4(0^+)$ and $i_7(0^+)$ be those after opening. By KCL, we have

$$i_4(0^+) + i_7(0^+) = 0$$

Since flux is conserved, the net change of flux linkage around a loop containing the inductors must be zero, namely,

$$L_4[i_4(0^+) - i_4(0^-)] = L_7[i_7(0^+) - i_7(0^-)]$$

Combining with the KCL equation, we get

$$i_4(0^+) = \frac{L_4}{L_4 + L_7} i_4(0^-) - \frac{L_7}{L_4 + L_7} i_7(0^-)$$

$$i_7(0^+) = \frac{L_7}{L_4 + L_7} i_7(0^-) - \frac{L_4}{L_4 + L_7} i_4(0^-) \qquad (9.38)$$

These are the initial values of the inductor currents that must be used in the solution of the circuit for $t \geq 0$.

9.5 Circuits Containing Mutually Coupled Inductors*

Consider the circuit of Figure 9.8. Select a proper tree as shown in Figure 9.8(b). Writing KVL, KCL, and terminal equations, we find a set

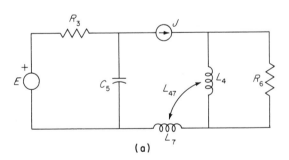

(a)

Figure 9.8 A circuit with a mutually coupled soil.

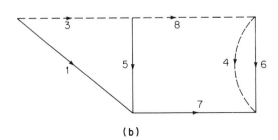

(b)

of differential equations in the unknowns which are the capacitor voltage and the two inductor currents:

$$\frac{d}{dt}\begin{bmatrix} C_5 & 0 & 0 \\ 0 & L_4 & L_{47} \\ 0 & 0 & 0 \end{bmatrix}\begin{bmatrix} v_5 \\ i_4 \\ i_7 \end{bmatrix} = \begin{bmatrix} -G_3 & 0 & 0 \\ 0 & -R_6 & 0 \\ 0 & 0 & 1 \end{bmatrix}\begin{bmatrix} v_5 \\ i_4 \\ i_7 \end{bmatrix} + \begin{bmatrix} G_3 & -1 \\ 0 & R_6 \\ 0 & -1 \end{bmatrix}\begin{bmatrix} E \\ J \end{bmatrix}$$

(9.39)

Once again, we see that the tree inductor current i_7 cannot be an independent state variable. After eliminating it from the differential equations, we find the state equation to be

$$\frac{d}{dt}\begin{bmatrix} v_5 \\ i_4 \end{bmatrix} = \begin{bmatrix} -G_3/C_5 & 0 \\ 0 & -R_6/L_4 \end{bmatrix}\begin{bmatrix} v_5 \\ i_4 \end{bmatrix} + \begin{bmatrix} G_3/C_5 & -1/C_5 \\ 0 & R_6/L_4 \end{bmatrix}\begin{bmatrix} E \\ J \end{bmatrix}$$
$$+ \begin{bmatrix} 0 \\ L_{47}/L_4 \end{bmatrix}\frac{dJ}{dt}$$

(9.40)

In this case the derivative of the current source appears. It comes about because of the presence of the mutual inductor.

9.6 Circuits Containing Dependent Sources*

In the preceding three sections, we saw examples of circuits in which a linear relation exists among the capacitor voltages and among the inductor currents by virtue of the presence of a capacitor loop and inductor cut set, respectively. In a circuit containing dependent sources, the relationship between the sources may be such as to cause linear relations to exist among the state variables.

Consider the circuit of Figure 9.9. The state equation is

$$\frac{d}{dt}\begin{bmatrix} C_2 & C_3 \\ \alpha C_2 & C_3 \end{bmatrix}\begin{bmatrix} v_2 \\ v_3 \end{bmatrix} = \begin{bmatrix} -G_4 & 0 \\ G_5 & -G_5 \end{bmatrix}\begin{bmatrix} v_2 \\ v_3 \end{bmatrix} + \begin{bmatrix} G_4 \\ 0 \end{bmatrix}E$$

(9.41)

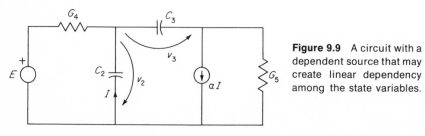

Figure 9.9 A circuit with a dependent source that may create linear dependency among the state variables.

If $\alpha = 1$, we see that the matrix on the left is singular. By elementary row operation, we transform the equations into

$$\frac{d}{dt} \begin{bmatrix} C_2 & C_3 \\ 0 & 0 \end{bmatrix} \begin{bmatrix} v_2 \\ v_3 \end{bmatrix} = \begin{bmatrix} -G_4 & 0 \\ (G_5 + G_4) & -G_5 \end{bmatrix} \begin{bmatrix} v_2 \\ v_3 \end{bmatrix} + \begin{bmatrix} G_4 \\ -G_4 \end{bmatrix} E \qquad (9.42)$$

The last equation states that

$$(G_4 + G_5)v_2 - G_5 v_3 = G_4 E$$

Thus a linear relation exists among the state variables. Eliminating v_3, we find the state equation to be

$$\frac{dv_2}{dt} = [C_2 + C_3(G_4 R_5 + 1)]^{-1} \left(-G_4 v_2 + G_4 E + C_3 R_5 G_4 \frac{dE}{dt} \right) \qquad (9.43)$$

The order of the circuit is reduced by one. Note the presence of the derivative of the source as a result of the linear relation among the capacitor voltages.

9.7 Solution of State Equations – Characteristic Matrix

We now have found that the equations describing the dynamics of a linear circuit are a set of linear first-order differential equations in terms of the independent capacitor voltages and inductor currents. A great deal is known about equations of this type, and there are numerous methods of solving these equations. Some are suitable for simple circuits, and others are efficient for hand calculation; still others are best suited for machine computation. We shall present a method in this text that is general and reveals the nature of the solution explicitly. The method will also lead to the identification of the class of time functions to which the response of a linear circuit belongs. In the process of obtaining the solution, we hope to deduce further properties about circuits in the time domain.

The method makes use of eigenvalues and eigenvectors.

Let us recall certain observations about first-order circuits and see how we can extend the analysis technique to the general case.

Consider the differential equation

$$\frac{dx}{dt} = \lambda_0 x + u(t) \qquad (9.44)$$

The homogeneous solution is

$$x_0(t) = e^{\lambda_0 t}$$

and the complete solution is

$$x(t) = e^{\lambda_0 t} x(0) + \int_0^t e^{\lambda_0 (t - \tau)} u(\tau) \, d\tau \qquad (9.45)$$

where $x(0)$ is the initial condition of $x(t)$. If we let

$$h(t) = e^{\lambda_0 t} \qquad (9.46)$$

the complete solution may be written as

$$x(t) = h(t)x(0) + \int_0^t h(t - \tau)u(\tau)\, d\tau$$

$$= h(t)x(0) + h(t)\int_0^t h^{-1}(\tau)u(\tau)\, d\tau \qquad (9.47)$$

The function $h(t)$ is called the characteristic function of the differential equation. It has two important properties:

1. It is a homogeneous solution of the differential equation.
2. Its initial value $h(0)$ is unity.

Once the characteristic function is found, the complete solution is given explicitly as a sum of $h(t)x(0)$ and the convolution of it and the excitation.

How can we extend the first-order case to the nth-order case? Suppose we construct a matrix $\mathbf{H}(t)$:

$$\mathbf{H}(t) = \begin{bmatrix} h_{11} & \cdots & h_{1n} \\ \vdots & \ddots & \vdots \\ h_{n1} & \cdots & h_{nn} \end{bmatrix} \qquad (9.48)$$

which has two properties

1. $\mathbf{H}(t)$ satisfies the homogeneous equation

$$\frac{d}{dt}\mathbf{H}(t) = \mathbf{A}\mathbf{H}(t) \qquad (9.49)$$

2. Its initial condition is

$$\mathbf{H}(0) = \begin{bmatrix} 1 & 0 & 0 & \cdots & 0 \\ 0 & 1 & 0 & \cdots & 0 \\ 0 & 0 & 1 & \cdots & 0 \\ \vdots & \vdots & \vdots & \ddots & \vdots \\ 0 & 0 & 0 & \cdots & 1 \end{bmatrix} = \mathbf{U} \qquad (9.50)$$

where $\mathbf{U}$ denotes the identity matrix. We shall call the matrix $\mathbf{H}(t)$ the characteristic matrix of the system $\dot{\mathbf{x}} = \mathbf{A}\mathbf{x} + \mathbf{u}$, where now $\mathbf{x}$ is a vector of state variables.

Now we assert that the complete solution of the system

$$\dot{\mathbf{x}} = \mathbf{Ax} + \mathbf{u}(t) \tag{9.51}$$

with initial conditions $\mathbf{x}(0)$ is

$$\mathbf{x}(t) = \mathbf{H}(t)\,\mathbf{x}(0) + \mathbf{H}(t)\int_0^t \mathbf{H}^{-1}(\tau)\mathbf{u}(\tau)\,d\tau \tag{9.52}$$

The solution is identical in form with that of the first-order case. To show that this is indeed the solution, we must show that it satisfies the inhomogeneous differential equation and the initial condition. The last part is obvious. For the first part, we have

$$
\begin{aligned}
\frac{d\mathbf{x}}{dt} &= \left[\frac{d}{dt}\mathbf{H}(t)\right]\mathbf{x}(0) + \left[\frac{d}{dt}\mathbf{H}(t)\right]\int_0^t \mathbf{H}^{-1}(\tau)\mathbf{u}(\tau)\,d\tau \\
&\quad + \mathbf{H}(t)\,\mathbf{H}^{-1}(t)\,\mathbf{u}(t) \\
&= \mathbf{A}\left[\mathbf{H}(t)\mathbf{x}(0) + \mathbf{H}(t)\int_0^t \mathbf{H}^{-1}(\tau)\mathbf{u}(\tau)\,d\tau\right] + \mathbf{u}(t) \\
&= \mathbf{Ax} + \mathbf{u}(t)
\end{aligned}
$$

which is what was required to be demonstrated.

The matrix $\mathbf{H}(t)$ has another important property, namely,

$$\mathbf{H}(t)\mathbf{H}^{-1}(\tau) = \mathbf{H}(t - \tau) \tag{9.53}$$

To see this, all we have to do is to show that the quantities on both sides of the equation separately satisfy the same differential equation and initial condition.

Let $\mathbf{Y}(t) = \mathbf{H}(t - \tau)$ for a fixed τ. Then

$$\frac{d}{dt}\mathbf{Y}(t) = \frac{d}{dt}\mathbf{H}(t - \tau) = \mathbf{AH}(t - \tau) = \mathbf{AY}(t)$$

and

$$\mathbf{Y}(\tau) = \mathbf{H}(0) = \mathbf{U}$$

Now for the left-hand side of Equation (9.53),

$$\frac{d}{dt}\left[\mathbf{H}(t)\mathbf{H}^{-1}(\tau)\right] = \mathbf{A}\left[\mathbf{H}(t)\mathbf{H}^{-1}(\tau)\right]$$

and

$$\mathbf{H}(\tau)\mathbf{H}^{-1}(\tau) = \mathbf{U}$$

It is seen that each side satisfies the same differential equation and initial condition at $t = \tau$. By uniqueness of solution of linear differential equations with constant coefficients, the two quantities must be the same.

Equation (9.52) may be written as

$$\mathbf{x}\,(t) = \mathbf{H}(t)\mathbf{x}(0) + \int_0^t \mathbf{H}(t-\tau)\mathbf{u}(\tau)\,d\tau \qquad (9.54)$$

In words, we say that the complete solution is the sum of the homogeneous solution that satisfies the initial condition and the convolution of the characteristic matrix and the excitation.

It remains to find the characteristic matrix $\mathbf{H}(t)$.

9.8 Eigenvalues and Eigenvectors

We recall in Chapter 4 that in a second-order circuit, the homogeneous solution is found by the use of eigenvalues and eigenvectors. Extension to the nth-order case is not difficult.

To review, suppose we have a homogeneous equation

$$\frac{d}{dt}\begin{bmatrix} x_1 \\ x_2 \end{bmatrix} = \begin{bmatrix} -1 & 1 \\ -1 & -1 \end{bmatrix}\begin{bmatrix} x_1 \\ x_2 \end{bmatrix} \qquad (9.55)$$

We assume the solution to be

$$\begin{bmatrix} x_1 \\ x_2 \end{bmatrix} = \begin{bmatrix} q_1 \\ q_2 \end{bmatrix} e^{\lambda t}$$

Substituting into the differential equation and dropping $\exp(\lambda t)$, we get

$$\begin{bmatrix} -1-\lambda & 1 \\ -1 & -1-\lambda \end{bmatrix}\begin{bmatrix} q_1 \\ q_2 \end{bmatrix} = 0 \qquad (9.56)$$

The equation has nonzero solution if, and only if,

$$\det \begin{bmatrix} -1-\lambda & 1 \\ -1 & -1-\lambda \end{bmatrix} = 0 \qquad (9.57)$$

and we obtain the characteristic equation

$$\lambda^2 + 2\lambda + 2 = 0$$

whose roots are $\lambda_1 = -1 + j$ and $\lambda_2 = -1 - j$. These values of λ for which nonzero solution exists are called the eigenvalues of the system of differential equations (9.55).

In the nth-order case, we would have

$$\frac{d\mathbf{x}}{dt} = \mathbf{A}\mathbf{x} \qquad (9.58)$$

Letting

$$\mathbf{x} = \mathbf{Q}e^{\lambda t}$$

we obtain from

$$\det (\mathbf{A} - \mathbf{U}\lambda) = 0 \tag{9.59}$$

the characteristic equation, after expanding the determinant,

$$a_n\lambda^n + a_{n-1}\lambda^{n-1} + \cdots + a_1\lambda + a_0 = 0 \tag{9.60}$$

in which all the coefficients are real. (Why?) The eigenvalues are, therefore, real or in complex conjugate pairs. This is one of the important properties of linear, time invariant circuits.

Returning to the second-order example, we now have two sets of homogeneous solutions, one for each of the two eigenvalues:

$$\begin{bmatrix} x_{11} \\ x_{21} \end{bmatrix} = \begin{bmatrix} Q_{11} \\ Q_{21} \end{bmatrix} e^{(-1+j)t} \qquad \begin{bmatrix} x_{12} \\ x_{22} \end{bmatrix} = \begin{bmatrix} Q_{12} \\ Q_{22} \end{bmatrix} e^{(-1-j)t}$$

The coefficients Q_{ij}'s are not all independent. Substituting each of the homogeneous solutions into the differential equations in turn, we get, after dropping $\exp(-1+j)t$ and $\exp(-1-j)t$, respectively,

$$\begin{bmatrix} -j & 1 \\ -1 & -j \end{bmatrix} \begin{bmatrix} Q_{11} \\ Q_{21} \end{bmatrix} = 0 \tag{9.61}$$

and

$$\begin{bmatrix} j & 1 \\ -1 & j \end{bmatrix} \begin{bmatrix} Q_{12} \\ Q_{22} \end{bmatrix} = 0$$

One of the possible sets of solutions of these two homogeneous equations is

$$\begin{bmatrix} Q_{11} \\ Q_{21} \end{bmatrix} = \begin{bmatrix} 1 \\ j \end{bmatrix} \qquad \begin{bmatrix} Q_{12} \\ Q_{22} \end{bmatrix} = \begin{bmatrix} 1 \\ -j \end{bmatrix}$$

The two sets of homogeneous solutions are linearly independent and may now be combined into a matrix form:

$$\mathbf{x}_0(t) = \begin{bmatrix} x_{11} & x_{12} \\ x_{21} & x_{22} \end{bmatrix} = \begin{bmatrix} Q_{11}e^{(-1+j)t} & Q_{12}e^{(-1-j)t} \\ Q_{21}e^{(-1+j)t} & Q_{22}e^{(-1-j)t} \end{bmatrix}$$

$$= \begin{bmatrix} Q_{11} & Q_{12} \\ Q_{21} & Q_{22} \end{bmatrix} \begin{bmatrix} e^{(-1+j)t} & 0 \\ 0 & e^{(-1-j)t} \end{bmatrix}$$

$$
= \begin{bmatrix} 1 & 1 \\ j & -j \end{bmatrix} \begin{bmatrix} e^{(-1+j)t} & 0 \\ 0 & e^{(-1-j)t} \end{bmatrix} \tag{9.62}
$$

which is of the form

$$
\mathbf{x}_0(t) = \mathbf{Q} \begin{bmatrix} e^{\lambda_1 t} & 0 \\ 0 & e^{\lambda_2 t} \end{bmatrix} \tag{9.63}
$$

where the matrix $\mathbf{Q}$ is given by

$$
\mathbf{Q} = \begin{bmatrix} Q_{11} & Q_{12} \\ Q_{21} & Q_{22} \end{bmatrix} = [\mathbf{Q}_1 \quad \mathbf{Q}_2]
$$

The column vectors $\mathbf{Q}_1$ and $\mathbf{Q}_2$ are known as the eigenvectors associated with the eigenvalues λ_1 and λ_2, respectively. The matrix $\mathbf{Q}$ is called the eigenvector matrix.

In the nth-order case, we would have that to each eigenvalue λ_i determined from Equation (9.60), there corresponds an eigenvector $\mathbf{Q}_i$ which satisfies the homogeneous equation [cf. Equation (9.61)]:

$$
[\mathbf{A} - \mathbf{U}\lambda_i]\mathbf{Q}_i = 0 \qquad i = 1, 2, \ldots, n
$$

The eigenvector matrix $\mathbf{Q}$ is an $n \times n$ matrix, whose columns are the eigenvectors

$$
\mathbf{Q} = [\mathbf{Q}_1 \ \mathbf{Q}_2 \ \cdots \ \mathbf{Q}_n]
$$

The homogeneous solution is then given by

$$
\mathbf{x}_0(t) = \mathbf{Q} \begin{bmatrix} e^{\lambda_1 t} & 0 & 0 & \cdots & 0 \\ 0 & e^{\lambda_2 t} & 0 & \cdots & 0 \\ \vdots & \vdots & \vdots & \ddots & \vdots \\ 0 & 0 & 0 & \cdots & e^{\lambda_n t} \end{bmatrix} \tag{9.64}
$$

provided that all of the eigenvalues are distinct. The case in which some or all of the eigenvalues are repeated will be taken up later.

We now have the homogeneous solution for the nth-order case. To find the characteristic matrix, all we have to do is to take linear combinations of the homogeneous solution given above and force the resultant matrix to satisfy the initial condition that it be an identity matrix.

Let $\mathbf{Z}$ be an unknown matrix of constants such that

$$
\mathbf{H}(t) = \mathbf{x}_0(t)\mathbf{Z} \tag{9.65}
$$

Imposing the initial condition on $\mathbf{H}(t)$, we have

$$
\mathbf{H}(0) = \mathbf{x}_0(0)\mathbf{Z} = \mathbf{Q}\mathbf{Z} = \mathbf{U}
$$

Therefore, the matrix Z must be

$$\mathbf{Z} = \mathbf{Q}^{-1} \tag{9.66}$$

It is shown in standard texts on matrix theory that the inverse of the eigenvector matrix always exists.

The characteristic matrix is, therefore, given by

$$\mathbf{H}(t) = \mathbf{Q} \begin{bmatrix} e^{\lambda_1 t} & 0 & 0 & \cdots & 0 \\ 0 & e^{\lambda_2 t} & 0 & \cdots & 0 \\ \vdots & \vdots & \vdots & \ddots & \vdots \\ 0 & 0 & 0 & \cdots & e^{\lambda_n t} \end{bmatrix} \mathbf{Q}^{-1} \tag{9.67}$$

9.9 Examples

Example 9.4 Find the complete solution of the circuit of Figure 9.10. The excitation $E(t)$ is an arbitrary function. The initial capacitor voltage is 1 V on one and 0 on the other as shown.

Solution The state equation is

$$\frac{d}{dt} \begin{bmatrix} v_1 \\ v_2 \end{bmatrix} = \begin{bmatrix} -\frac{3}{2} & \frac{1}{6} \\ \frac{3}{2} & -\frac{3}{2} \end{bmatrix} \begin{bmatrix} v_1 \\ v_2 \end{bmatrix} + \begin{bmatrix} \frac{1}{3} \\ 0 \end{bmatrix} E(t) \tag{9.68}$$

The characteristic equation is

$$\lambda^2 + 3\lambda + 2 = 0$$

and the eigenvalues are $\lambda_1 = -1$ and $\lambda_2 = -2$. The eigenvectors are found from

$$\begin{bmatrix} -\frac{3}{2} - \lambda_1 & \frac{1}{6} \\ \frac{3}{2} & -\frac{3}{2} - \lambda_1 \end{bmatrix} \begin{bmatrix} Q_{11} \\ Q_{21} \end{bmatrix} = 0$$

$$\begin{bmatrix} -\frac{3}{2} - \lambda_2 & \frac{1}{6} \\ \frac{3}{2} & -\frac{3}{2} - \lambda_2 \end{bmatrix} \begin{bmatrix} Q_{12} \\ Q_{22} \end{bmatrix} = 0$$

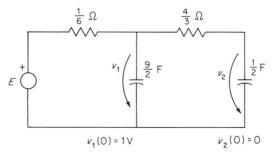

Figure 9.10 An *RC* circuit. [Example 9.4]

$\frac{1}{6}$ Ω $\frac{4}{3}$ Ω

E v_1 $\frac{9}{2}$ F v_2 $\frac{1}{2}$ F

$v_1(0) = 1\,\text{V}$ $v_2(0) = 0$

One possible eigenvector matrix and its inverse are

$$\mathbf{Q} = \begin{bmatrix} 1 & 1 \\ 3 & -3 \end{bmatrix} \qquad \mathbf{Q}^{-1} = \begin{bmatrix} \frac{1}{2} & \frac{1}{6} \\ \frac{1}{2} & -\frac{1}{6} \end{bmatrix}$$

The characteristic matrix is found to be

$$\mathbf{H}(t) = \mathbf{Q} \begin{bmatrix} e^{-t} & 0 \\ 0 & e^{-2t} \end{bmatrix} \mathbf{Q}^{-1}$$

$$= \begin{bmatrix} \frac{1}{2}(e^{-t} + e^{-2t}) & \frac{1}{6}(e^{-t} - e^{-2t}) \\ \frac{3}{2}(e^{-t} - e^{-2t}) & \frac{1}{2}(e^{-t} + e^{-2t}) \end{bmatrix}$$

Note that

$$\mathbf{H}(0) = \begin{bmatrix} 1 & 0 \\ 0 & 1 \end{bmatrix}$$

The complete solution is given by

$$\begin{bmatrix} v_1 \\ v_2 \end{bmatrix} = \mathbf{H}(t) \begin{bmatrix} 1 \\ 0 \end{bmatrix} + \int_0^t \mathbf{H}(t - \tau) \begin{bmatrix} \frac{4}{3} \\ 0 \end{bmatrix} E(\tau)\, d\tau$$

or

$$v_1(t) = \frac{1}{2}(e^{-t} + e^{-2t}) + \frac{2}{3} \int_0^t \left[e^{-(t-\tau)} + e^{-3(t-\tau)} \right] E(\tau)\, d\tau$$

$$v_2(t) = \frac{3}{2}(e^{-t} - e^{-2t}) + 2 \int_0^t \left[e^{-(t-\tau)} - e^{-2(t-\tau)} \right] E(\tau)\, d\tau$$

We note that the method of variational parameters of Chapter 4 is not used here. The matrix method simplifies the algebra a great deal.

Example 9.5 The initial conditions on the inductors of the circuit shown in Figure 9.11 are zero. Find the complete solution for the output voltage $v(t)$, $t \geq 0$.

Solution The state equation is

$$\frac{d}{dt} \begin{bmatrix} i_1 \\ i_2 \end{bmatrix} = \begin{bmatrix} -12/5 & 9/35 \\ 14/15 & -13/5 \end{bmatrix} \begin{bmatrix} i_1 \\ i_2 \end{bmatrix} + \begin{bmatrix} 15/98 \\ 0 \end{bmatrix} E(t) \qquad (9.70)$$

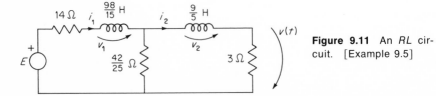

Figure 9.11 An *RL* circuit. [Example 9.5]

The characteristic equation is $\lambda^2 + 5\lambda + 6 = 0$, and the eigenvalues are $\lambda_1 = -2$ and $\lambda_2 = -3$. The eigenvector matrix and its inverse are found to be

$$\mathbf{Q} = \begin{bmatrix} 1 & 1 \\ 14/9 & -7/3 \end{bmatrix} \qquad \mathbf{Q}^{-1} = \begin{bmatrix} 3/5 & 9/35 \\ 2/5 & -9/35 \end{bmatrix}$$

The characteristic matrix is

$$\mathbf{H}(t) = \mathbf{Q} \begin{bmatrix} e^{-2t} & 0 \\ 0 & e^{-3t} \end{bmatrix} \mathbf{Q}^{-1}$$

$$= \begin{bmatrix} (3/5)e^{-2t} + (2/5)e^{-3t} & (9/35)e^{-2t} - (9/35)e^{-3t} \\ (14/15)e^{-2t} - (14/15)e^{-3t} & (2/5)e^{-2t} + (3/5)e^{-3t} \end{bmatrix}$$

Since the initial conditions are all zero, the complete solution has contribution only from the convolution term:

$$\begin{bmatrix} i_1 \\ i_2 \end{bmatrix} = \int_0^t \mathbf{H}(t - \tau) \begin{bmatrix} 15/98 \\ 0 \end{bmatrix} E(\tau) \, d\tau$$

The output voltage $v(t)$ is then

$$v(t) = 3i_2 = \frac{3}{7} \int_0^t e^{-2(t-\tau)} E(\tau) \, d\tau - \frac{3}{7} \int_0^t e^{-3(t-\tau)} E(\tau) \, d\tau$$

As a check, let us find v_1 and v_2.

$$v_1(t) = \frac{98}{15} \frac{di_1}{dt} = E(t) - \frac{6}{5} \int_0^t e^{-2(t-\tau)} E(\tau) \, d\tau$$

$$- \frac{6}{5} \int_0^t e^{-3(t-\tau)} E(\tau) \, d\tau$$

$$v_2(t) = -\frac{18}{35} \int_0^t e^{-2(t-\tau)} E(\tau) \, d\tau + \frac{27}{35} \int_0^t e^{-3(t-\tau)} E(\tau) \, d\tau \tag{9.71}$$

Taking the sum of the voltages around the perimeter of the circuit, we find, after substitution,

$$14i_1 + v_1 + v_2 + 3i_2 - E = 0$$

Example 9.6 Find the output voltage resulting from a unit step input in the circuit of Figure 9.12. Assume all initial conditions to be zero.

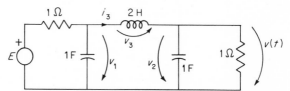

Figure 9.12 A third-order circuit. [Example 9.6]

Solution   Assign state variables as shown. The state equation is

$$\frac{d}{dt}\begin{bmatrix} v_1 \\ v_2 \\ i_3 \end{bmatrix} = \begin{bmatrix} -1 & 0 & -1 \\ 0 & -1 & 1 \\ 1/2 & -1/2 & 0 \end{bmatrix}\begin{bmatrix} v_1 \\ v_2 \\ i_3 \end{bmatrix} + \begin{bmatrix} 1 \\ 0 \\ 0 \end{bmatrix} \qquad (9.72)$$

The characteristic equation is $\lambda^3 + 2\lambda^2 + 2\lambda + 1 = 0$ and the eigenvalues are $\lambda_1 = -1$, $\lambda_2 = (-1 + j\sqrt{3})/2$, and $\lambda_3 = (-1 - j\sqrt{3})/2$. One possible eigenvector matrix is

$$\mathbf{Q} = \begin{bmatrix} 1 & 1 & 1 \\ 1 & -1 & -1 \\ 0 & (-1-j\sqrt{3})/2 & (-1+j\sqrt{3})/2 \end{bmatrix}$$

Its inverse is

$$\mathbf{Q}^{-1} = \begin{bmatrix} 1/2 & 1/2 & 0 \\ (\sqrt{3}+j)/4\sqrt{3} & (-\sqrt{3}-j)/4\sqrt{3} & j/\sqrt{3} \\ (\sqrt{3}-j)/4\sqrt{3} & (-\sqrt{3}+j)/4\sqrt{3} & -j/\sqrt{3} \end{bmatrix}$$

The characteristic matrix is found in the usual way and will not be displayed. The complete solution is

$$v_1(t) = \frac{1}{2}\left(1 - e^{-t} + \frac{2}{\sqrt{3}} e^{-0.5t} \sin\frac{\sqrt{3}}{2}t\right)$$

$$v_2(t) = \frac{1}{2}\left(1 - e^{-t} - \frac{2}{\sqrt{3}} e^{-0.5t} \sin\frac{\sqrt{3}}{2}t\right)$$

$$i_3(t) = \frac{1}{\sqrt{3}}\left(-\frac{1}{2} e^{-0.5t} \sin\frac{\sqrt{3}}{2}t - \frac{\sqrt{3}}{2} e^{-0.5t} \cos\frac{\sqrt{3}}{2}t + \frac{\sqrt{3}}{2}\right)$$

As a check, the inductor voltage is found to be

$$v_3(t) = 2\frac{di_3}{dt} = \frac{2}{\sqrt{3}} e^{-0.5t} \sin\frac{\sqrt{3}}{2}t$$

which is the same as $v_1 - v_2$. The output voltage $v(t)$ is simply the state variable $v_2(t)$.

We see that in this third-order circuit, the response is a sum of exponentially decaying and damped sinusoidal time functions. In the nth-order case, we would have similar results since the eigenvalues, in general, are negative real or complex with negative real parts. See Section 8.14.

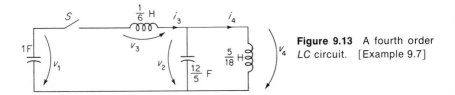

Figure 9.13 A fourth order *LC* circuit. [Example 9.7]

Example 9.7 In the lossless circuit of Figure 9.13, the switch was open for $t < 0$. At $t = 0$, it is closed. Find the response of the circuit. The initial voltage on the 1-F capacitor is 1 V.

Solution The state equation is

$$\frac{d}{dt}\begin{bmatrix} v_1 \\ v_2 \\ i_3 \\ i_4 \end{bmatrix} = \begin{bmatrix} 0 & 0 & -1 & 0 \\ 0 & 0 & 5/12 & -5/12 \\ 6 & -6 & 0 & 0 \\ 0 & 18/5 & 0 & 0 \end{bmatrix} \begin{bmatrix} v_1 \\ v_2 \\ i_3 \\ i_4 \end{bmatrix} \tag{9.74}$$

The characteristic equation is

$$\lambda^4 + 10\lambda^2 + 9 = 0$$

which has imaginary roots:

$$\lambda_1 = j \quad \lambda_2 = -j \quad \lambda_3 = j3 \quad \lambda_4 = -j3$$

This is not surprising. (Why?) We expect the response will consist of terms $\sin t$, $\cos t$, $\sin 3t$, and $\cos 3t$.

One possible eigenvector matrix is

$$\mathbf{Q} = \begin{bmatrix} 1 & 1 & 1 & 1 \\ 5/6 & 5/6 & -1/2 & -1/2 \\ -j & j & -j3 & j3 \\ -j3 & j3 & j3/5 & -j3/5 \end{bmatrix}$$

Its inverse is found to be

$$\mathbf{Q}^{-1} = \begin{bmatrix} 3/16 & 3/8 & j/32 & j5/32 \\ 3/16 & 3/8 & -j/32 & -j5/32 \\ 5/16 & -3/8 & j5/32 & -j5/96 \\ 5/16 & -3/8 & -j5/32 & j5/96 \end{bmatrix}$$

Since there is no excitation in the circuit, the only response is that resulting from the initial conditions. The response is found from

$$\begin{bmatrix} v_1 \\ v_2 \\ i_3 \\ i_4 \end{bmatrix} = \mathbf{H}(t)\begin{bmatrix} 1 \\ 0 \\ 0 \\ 0 \end{bmatrix}$$

It is noted that only the first column of $\mathbf{H}(t)$ need be found. The solution is then given by

$$v_1(t) = \frac{3}{8}\cos t + \frac{5}{8}\cos 3t$$

$$v_2(t) = \frac{5}{16}\cos t - \frac{5}{16}\cos 3t$$

$$i_3(t) = \frac{3}{8}\sin t + \frac{15}{8}\sin 3t$$

$$i_4(t) = \frac{9}{8}\sin t - \frac{3}{8}\sin 3t \tag{9.75}$$

As a check, the voltage across the $\frac{1}{6}$-H inductor is

$$\frac{1}{6}\frac{di_3}{dt} = \frac{1}{16}\cos t + \frac{15}{16}\cos 3t$$

which is the same as $v_1(t) - v_2(t)$.

We see once again that in a lossless circuit, the response due to the initial conditions is a superposition of sine or cosine functions. Once an initial voltage or current is established in the circuit, steady, undamped oscillations will take place.

9.10 A Pause and Some Observations

The problem of finding the response of a circuit in the time domain amounts to solving a set of first-order differential equations. The method of solution given here is general. A great deal of algebra is required in analyzing a high-order circuit. This is unavoidable. Numerical techniques which would aid in finding the eigenvalues and eigenvectors of a matrix exist, and one should make use of these. Most computer systems, in fact, have these numerical algorithms implemented as subroutines. To emphasize the systematic nature of the solution scheme, we summarize it below:

1. From the given state equation $\dot{\mathbf{x}} = \mathbf{A}\mathbf{x} + \mathbf{u}$, we find the characteristic equation from

$$\det |\mathbf{A} - \lambda\mathbf{U}| = 0$$

   Upon expansion of the determinant, we obtain

$$a_n\lambda^n + a_{n-1}\lambda^{n-1} + \cdots + a_1\lambda + a_0 = 0$$

2. Find the roots of the characteristic equation, $\lambda_1, \lambda_2,..., \lambda_n$. Suppose they are all distinct. If not, go to Section 9.11.
3. For each eigenvalue λ_i, find the eigenvector from

$$[\mathbf{A} - \lambda_i\mathbf{U}]\mathbf{Q}_i = 0$$

4. Construct the eigenvector matrix $\mathbf{Q}$ by placing the eigenvectors together as columns in the same order as the eigenvalues.

$$\mathbf{Q} = [\mathbf{Q}_1 \quad \mathbf{Q}_2 \quad \cdots \quad \mathbf{Q}_n]$$

5. Find the inverse of $\mathbf{Q}$.
6. The characteristic matrix is given by

$$\mathbf{H}(t) = \mathbf{Q}\begin{bmatrix} e^{\lambda_1 t} & 0 & 0 & \cdots & 0 \\ 0 & e^{\lambda_2 t} & 0 & \cdots & 0 \\ \vdots & & & \ddots & \vdots \\ 0 & 0 & 0 & \cdots & e^{\lambda_n t} \end{bmatrix}\mathbf{Q}^{-1} \tag{9.76}$$

7. The complete solution is

$$\mathbf{x}(t) = \mathbf{H}(t)\mathbf{x}(0) + \int_0^t \mathbf{H}(t - \tau)\mathbf{u}(\tau) \, d\tau \tag{9.77}$$

The examples of the preceding section are solved in the manner outlined above. The examples also illustrate the nature of the response of circuits of certain types. In fact, we can make the following observations:

1. In an RC or RL circuit, the eigenvalues are negative real and distinct. The natural response is a sum of exponentially decaying functions. See Examples 9.4 and 9.5.

2. In an LC circuit, the eigenvalues are imaginary and distinct. The natural response is a superposition of sine and cosine functions. The frequencies of oscillation are precisely the eigenvalues. See Example 9.7.

3. In an RLC circuit, the eigenvalues are real or complex. The natural response consists of exponentially decaying functions and damped oscillations. See Example 9.6.

It turns out that these observations are true in general. The proof is given in advanced texts on network theory.

9.11 Repeated Eigenvalues

If in a circuit, the eigenvalues are not distinct, then the characteristic matrix cannot be written in the form of Equation (9.76). We saw in Chapter 4 that in the second-order case, repeated eigenvalues gave rise to terms of the form $te^{-\alpha t}$ in addition to $e^{-\alpha t}$. We shall now demonstrate by examples what we must do to handle repeated eigenvalues in the general case.

We first consider some special forms of state equations that have repeated eigenvalues.

Suppose we have the following state equation:

$$\frac{d}{dt}\begin{bmatrix} x_1 \\ x_2 \end{bmatrix} = \begin{bmatrix} -2 & 0 \\ 0 & -2 \end{bmatrix}\begin{bmatrix} x_1 \\ x_2 \end{bmatrix} \tag{9.78}$$

The root of the characteristic equation is -2 and is repeated twice, or of multiplicity two. Now note that the equations are decoupled. One does not have anything to do with the other. The solution can, therefore, be obtained separately:

$$\begin{aligned} x_1(t) &= x_1(0)e^{-2t} \\ x_2(t) &= x_2(0)e^{-2t} \end{aligned} \tag{9.79}$$

Now suppose we have

$$\frac{d}{dt}\begin{bmatrix} x_1 \\ x_2 \end{bmatrix} = \begin{bmatrix} -2 & 0 \\ 1 & -2 \end{bmatrix}\begin{bmatrix} x_1 \\ x_2 \end{bmatrix} \tag{9.80}$$

The eigenvalue is repeated and is the same as before. The first equation is decoupled from the second, and the solution is

$$x_1(t) = x_1(0)e^{-2t} \tag{9.81}$$

The second equation is

$$\frac{dx_2}{dt} = x_1 - 2x_2 \tag{9.82}$$

We may regard the term x_1 as an excitation. Using Equation (9.81), we obtain the solution for x_2 given by

$$\begin{aligned} x_2(t) &= x_2(0)e^{-2t} + \int_0^t e^{-2(t-\tau)} x_1(0)e^{-2\tau} \, d\tau \\ &= x_2(0)e^{-2t} + x_1(0)te^{-2t} \end{aligned} \tag{9.83}$$

Thus the homogeneous solution to the state equation (9.80) consists of e^{-2t} and te^{-2t}. Imposing the initial condition on the homogeneous solution, we find the characteristic matrix to be

$$\mathbf{H}(t) = \begin{bmatrix} e^{-2t} & 0 \\ te^{-2t} & e^{-2t} \end{bmatrix} \tag{9.84}$$

In general, if the state equation is given in the form

$$\frac{d\mathbf{x}}{dt} = \begin{bmatrix} \lambda_0 & 0 & 0 & \cdots & 0 & 0 \\ 1 & \lambda_0 & 0 & \cdots & 0 & 0 \\ 0 & 1 & \lambda_0 & \cdots & 0 & 0 \\ 0 & 0 & 1 & \cdots & 0 & 0 \\ \vdots & \vdots & \vdots & \ddots & \vdots & \vdots \\ 0 & 0 & 0 & \cdots & 1 & \lambda_0 \end{bmatrix} \mathbf{x} \tag{9.85}$$

The characteristic matrix is given by

$$\mathbf{H}(t) = \begin{bmatrix} e^{\lambda_0 t} & 0 & 0 & \cdots & 0 \\ te^{\lambda_0 t} & e^{\lambda_0 t} & 0 & \cdots & 0 \\ \frac{t^2}{2!} e^{\lambda_0} & te^{\lambda_0 t} & e^{\lambda_0 t} & \cdots & 0 \\ \frac{t^3}{3!} e^{\lambda_0 t} & \frac{t^2}{2!} e^{\lambda_0 t} & te^{\lambda_0 t} & \cdots & 0 \\ \vdots & \vdots & \vdots & \vdots & \vdots \\ \frac{t^{m-1}}{(m-1)!} e^{\lambda_0 t} & \cdots & & & e^{\lambda_0 t} \end{bmatrix} \tag{9.86}$$

with m being the multiplicity of the eigenvalue λ_0.

To obtain $\mathbf{H}(t)$ from Equation (9.85), we use the solution of the first equation in the second equation, the solution of the second in the third, and so on. $\mathbf{H}(t)$ follows when we impose the initial condition that $\mathbf{H}(0) = \mathbf{U}$.

Now we must note the special form of the state equation of (9.85). Not every state equation that has repeated eigenvalues has the form shown. However, by elementary row and column operations of the $\mathbf{A}$ matrix, we can always transform it into the special form. The transformation amounts to taking linear combinations of the state variables.

Consider the following example:

$$\frac{d}{dt}\begin{bmatrix} x_1 \\ x_2 \end{bmatrix} = \begin{bmatrix} -2 & 0 \\ 4 & -2 \end{bmatrix}\begin{bmatrix} x_1 \\ x_2 \end{bmatrix} \tag{9.87}$$

Define a new set of variables:

$$\begin{matrix} y_1 = 4x_1 \\ \text{or} \\ y_2 = x_2 \end{matrix} \qquad \begin{bmatrix} y_1 \\ y_2 \end{bmatrix} = \begin{bmatrix} 4 & 0 \\ 0 & 1 \end{bmatrix}\begin{bmatrix} x_1 \\ x_2 \end{bmatrix}$$

and

$$\begin{bmatrix} x_1 \\ x_2 \end{bmatrix} = \begin{bmatrix} 4 & 0 \\ 0 & 1 \end{bmatrix}^{-1}\begin{bmatrix} y_1 \\ y_2 \end{bmatrix} = \begin{bmatrix} \frac{1}{4} & 0 \\ 0 & 1 \end{bmatrix}\begin{bmatrix} y_1 \\ y_2 \end{bmatrix}$$

Then

$$\frac{d}{dt}\begin{bmatrix} y_1 \\ y_2 \end{bmatrix} = \begin{bmatrix} 4 & 0 \\ 0 & 1 \end{bmatrix}\begin{bmatrix} -2 & 0 \\ 4 & -2 \end{bmatrix}\begin{bmatrix} \frac{1}{4} & 0 \\ 0 & 1 \end{bmatrix}\begin{bmatrix} y_1 \\ y_2 \end{bmatrix}$$

$$= \begin{bmatrix} -2 & 0 \\ 1 & -2 \end{bmatrix}\begin{bmatrix} y_1 \\ y_2 \end{bmatrix} \tag{9.88}$$

which is in the desired form.

In general, let $\mathbf{T}$ be a matrix such that

$$\mathbf{TAT}^{-1} = \mathbf{J}_0 \tag{9.89}$$

Then the transformation

$$\mathbf{y} = \mathbf{Tx} \tag{9.90}$$

will transform the state equation

$$\dot{\mathbf{x}} = \mathbf{Ax}$$

into the form

$$\dot{\mathbf{y}} = \mathbf{J}_0\mathbf{y} \tag{9.91}$$

with the matrix J_0 given by

$$
J_0 = \begin{bmatrix}
\lambda_0 & 0 & 0 & 0 & \cdots & 0 & 0 \\
1 & \lambda_0 & 0 & 0 & \cdots & 0 & 0 \\
0 & 1 & \lambda_0 & 0 & \cdots & 0 & 0 \\
0 & 0 & 1 & \lambda_0 & \cdots & 0 & 0 \\
\vdots & \vdots & \vdots & \vdots & \ddots & \vdots & \vdots \\
0 & 0 & 0 & \cdots & & 1 & \lambda_0
\end{bmatrix} \tag{9.92}
$$

In matrix theory J_0 is known as the elementary Jordan matrix.

Example 9.8   Find the complete solution of the circuit of Figure 9.14.

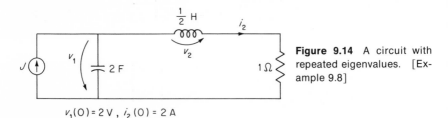

Figure 9.14 A circuit with repeated eigenvalues. [Example 9.8]

$v_1(0) = 2\,V$, $i_2(0) = 2\,A$

Solution   The state equation is

$$
\frac{d}{dt}\begin{bmatrix} v_1 \\ i_2 \end{bmatrix} = \begin{bmatrix} 0 & -1/2 \\ 2 & -2 \end{bmatrix}\begin{bmatrix} v_1 \\ i_2 \end{bmatrix} + \begin{bmatrix} 1/2 \\ 0 \end{bmatrix} J(t) \tag{9.93}
$$

The eigenvalue is repeated and is equal to -1. Now we seek a transformation T such that

$$
T\begin{bmatrix} 0 & -1/2 \\ 2 & -2 \end{bmatrix}T^{-1} = \begin{bmatrix} -1 & 0 \\ 1 & -1 \end{bmatrix}
$$

$$
T\begin{bmatrix} 0 & -1/2 \\ 2 & -2 \end{bmatrix} = \begin{bmatrix} -1 & 0 \\ 1 & -1 \end{bmatrix}T
$$

Expanding the equation and equating coefficients, we find

$$
T = \begin{bmatrix} 1 & -1/2 \\ -1 & 1 \end{bmatrix} \qquad T^{-1} = \begin{bmatrix} 2 & 1 \\ 2 & 2 \end{bmatrix}
$$

The new state variables are

$$
\begin{bmatrix} y_1 \\ y_2 \end{bmatrix} = \begin{bmatrix} 1 & -1/2 \\ -1 & 1 \end{bmatrix}\begin{bmatrix} v_1 \\ i_2 \end{bmatrix} = \begin{bmatrix} v_1 - (1/2)i_2 \\ -v_1 + i_2 \end{bmatrix}
$$

The new state equation is

$$\frac{d}{dt}\begin{bmatrix} y_1 \\ y_2 \end{bmatrix} = \begin{bmatrix} -1 & 0 \\ 1 & -1 \end{bmatrix}\begin{bmatrix} y_1 \\ y_2 \end{bmatrix} + \begin{bmatrix} 1 & -1/2 \\ -1 & 1 \end{bmatrix}\begin{bmatrix} 1/2 \\ 0 \end{bmatrix} J(t) \tag{9.94}$$

with initial condition given by

$$\begin{bmatrix} y_1(0) \\ y_2(0) \end{bmatrix} = \begin{bmatrix} 1 & -1/2 \\ -1 & 1 \end{bmatrix}\begin{bmatrix} 2 \\ 2 \end{bmatrix} = \begin{bmatrix} 1 \\ 0 \end{bmatrix}$$

The characteristic matrix is

$$\mathbf{H}(t) = \begin{bmatrix} e^{-t} & 0 \\ te^{-t} & e^{-t} \end{bmatrix}$$

The complete solution for **y** is

$$\begin{bmatrix} y_1(t) \\ y_2(t) \end{bmatrix} = \mathbf{H}(t)\begin{bmatrix} 1 \\ 0 \end{bmatrix} + \int_0^t \mathbf{H}(t-\tau)\begin{bmatrix} 1/2 \\ -1/2 \end{bmatrix} J(\tau)\ d\tau$$

or

$$y_1(t) = e^{-t} + \frac{1}{2}\int_0^t e^{-(t-\tau)}J(\tau)\ d\tau$$

$$y_2(t) = te^{-t} + \frac{1}{2}\int_0^t (t-\tau)e^{-(t-\tau)}J(\tau)\ d\tau$$

$$-\frac{1}{2}\int_0^t e^{-(t-\tau)}J(\tau)\ d\tau \tag{9.95}$$

Transforming back to the original state variables, we find

$$\begin{bmatrix} v_1 \\ i_2 \end{bmatrix} = \begin{bmatrix} 2 & 1 \\ 2 & 2 \end{bmatrix}\begin{bmatrix} y_1 \\ y_2 \end{bmatrix}$$

and

$$v_1(t) = 2e^{-t} + te^{-t} + \frac{1}{2}\int_0^t e^{-(t-\tau)}J(\tau)\ d\tau$$

$$+ \frac{1}{2}\int_0^t (t-\tau)\ e^{-(t-\tau)}J(\tau)\ d\tau$$

$$i_2(t) = 2e^{-t} + 2te^{-t} - \int_0^t (t-\tau)\ e^{-(t-\tau)}J(\tau)\ d\tau \tag{9.96}$$

Example 9.9 In the circuit of Figure 9.15, switch S was open for $t < 0$. At $t = 0$, it is closed. Find the response of the circuit for $t \geq 0$. The initial voltage on the $\frac{8}{3}$-F capacitor is 1 V.

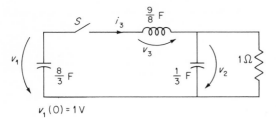

Figure 9.15 A circuit with repeated eigenvalue of multiplicity three. [Example 9.9]

$v_1(0) = 1\,\mathrm{V}$

Solution   The state equation is

$$\frac{d}{dt}\begin{bmatrix} v_1 \\ v_2 \\ i_3 \end{bmatrix} = \begin{bmatrix} 0 & 0 & -3/8 \\ 0 & -3 & 3 \\ 8/9 & -8/9 & 0 \end{bmatrix}\begin{bmatrix} v_1 \\ v_2 \\ i_3 \end{bmatrix} \qquad (9.97)$$

The characteristic equation is

$$\lambda^3 + 3\lambda^2 + 3\lambda + 1 = 0$$

and the eigenvalue is -1 with multiplicity 3.

We look for a transformation **T** such that

$$\mathbf{T}\begin{bmatrix} 0 & 0 & -3/8 \\ 0 & -3 & 3 \\ 8/9 & -8/9 & 0 \end{bmatrix} = \begin{bmatrix} -1 & 0 & 0 \\ 1 & -1 & 0 \\ 0 & 1 & -1 \end{bmatrix}\mathbf{T}$$

It is found to be

$$\mathbf{T} = \begin{bmatrix} 1 & 1/2 & -9/8 \\ 1 & -1/4 & 0 \\ 1 & 1/8 & 0 \end{bmatrix} \quad \text{and} \quad \mathbf{T}^{-1} = \begin{bmatrix} 0 & 1/3 & 2/3 \\ 0 & -8/3 & 8/3 \\ -8/9 & -8/9 & 16/9 \end{bmatrix}$$

Defining a new set of state variables

$$\mathbf{y} = \mathbf{T}\begin{bmatrix} v_1 \\ v_2 \\ i_3 \end{bmatrix}$$

We have a new state equation

$$\dot{\mathbf{y}} = \begin{bmatrix} -1 & 0 & 0 \\ 1 & -1 & 0 \\ 0 & 1 & -1 \end{bmatrix}\mathbf{y}$$

with initial conditions given by

$$\mathbf{y}(0) = \mathbf{T}\begin{bmatrix} 1 \\ 0 \\ 0 \end{bmatrix} = \begin{bmatrix} 1 \\ 1 \\ 1 \end{bmatrix}$$

The complete solution for **y** is

$$y_1(t) = e^{-t}$$
$$y_2(t) = (t+1)e^{-t}$$
$$y_3(t) = \left(\frac{t^2}{2} + t + 1\right)e^{-t} \qquad (9.98)$$

and those for the voltages and current are:

$$v_1(t) = \left(\frac{t^2}{3} + t + 1 \right) e^{-t}$$

$$v_2(t) = \frac{4}{3} t^2 e^{-t}$$

$$i_3(t) = \frac{8}{9} (t^2 + t) e^{-t} \tag{9.99}$$

As a check, the inductor voltage is found to be

$$v_L(t) = \frac{9}{8} \frac{di_3}{dt} = (-t^2 + t + 1) e^{-t}$$

which is the same as $v_1 - v_2$.

9.12 Repeated Eigenvalues—General Case

In the previous section, we considered state equations in which all of the eigenvalues are the same. In the most general case, we would have eigenvalues $\lambda_1, \lambda_2, \ldots, \lambda_r$ with multiplicity $m_1, m_2, \ldots, m_r$, respectively. As expected, the response will be a superposition of responses, each being of the form of Equation (9.99); for example, consider the following state equation:

$$\frac{d}{dt} \begin{bmatrix} x_1 \\ x_2 \\ x_3 \\ x_4 \end{bmatrix} = \begin{bmatrix} -1 & 0 & 0 & 0 \\ 1 & -1 & 0 & 0 \\ 0 & 0 & -2 & 0 \\ 0 & 0 & 1 & -2 \end{bmatrix} \begin{bmatrix} x_1 \\ x_2 \\ x_3 \\ x_4 \end{bmatrix} \tag{9.100}$$

The eigenvalues are -1 of multiplicity 2 and -2 of multiplicity 2. We note that the system of equation can be decomposed into two uncoupled systems:

$$\frac{d}{dt} \begin{bmatrix} x_1 \\ x_2 \end{bmatrix} = \begin{bmatrix} -1 & 0 \\ 1 & -1 \end{bmatrix} \begin{bmatrix} x_1 \\ x_2 \end{bmatrix}$$

$$\frac{d}{dt} \begin{bmatrix} x_3 \\ x_4 \end{bmatrix} = \begin{bmatrix} -2 & 0 \\ 1 & -2 \end{bmatrix} \begin{bmatrix} x_3 \\ x_4 \end{bmatrix}$$

Each system may now be solved as in the previous section.

The **A** matrix of Equation (9.100) is of the form

$$\mathbf{J} = \begin{bmatrix} \mathbf{J}_1 & 0 \\ 0 & \mathbf{J}_2 \end{bmatrix} \tag{9.101}$$

in which $\mathbf{J}_1$ and $\mathbf{J}_2$ are of the form of Equation (9.92). Not every state equation, which has more than one repeated eigenvalue, has this form.

By linear transformation of the state variables, the **A** matrix can be transformed into it. The matrix **J** is known as the Jordan matrix.

Example 9.10   Solve the state equation

$$\frac{d}{dt}\begin{bmatrix} x_1 \\ x_2 \\ x_3 \\ x_4 \end{bmatrix} = \begin{bmatrix} -1 & 0 & 0 & 0 \\ 1 & -1 & 0 & 0 \\ 0 & 0 & -2 & 0 \\ 1 & 0 & 1 & -2 \end{bmatrix}\begin{bmatrix} x_1 \\ x_2 \\ x_3 \\ x_4 \end{bmatrix} \tag{9.102}$$

The transformation matrix is found to be

$$\mathbf{T} = \begin{bmatrix} 1 & 0 & 0 & 0 \\ 0 & 1 & 0 & 0 \\ 0 & 0 & 1 & 0 \\ -1 & 0 & 0 & 1 \end{bmatrix} \qquad \mathbf{T}^{-1} = \begin{bmatrix} 1 & 0 & 0 & 0 \\ 0 & 1 & 0 & 0 \\ 0 & 0 & 1 & 0 \\ 1 & 0 & 0 & 1 \end{bmatrix}$$

The new state variables satisfy the decoupled systems of differential equations:

$$\frac{d}{dt}\begin{bmatrix} y_1 \\ y_2 \end{bmatrix} = \begin{bmatrix} -1 & 0 \\ 1 & -1 \end{bmatrix}\begin{bmatrix} y_1 \\ y_2 \end{bmatrix} \tag{9.103}$$

$$\frac{d}{dt}\begin{bmatrix} y_3 \\ y_4 \end{bmatrix} = \begin{bmatrix} -2 & 0 \\ 1 & -2 \end{bmatrix}\begin{bmatrix} y_3 \\ y_4 \end{bmatrix} \tag{9.104}$$

The original state variables are related to the new ones by

$$\begin{aligned} x_1 &= y_1 \\ x_2 &= y_2 \\ x_3 &= y_3 \\ x_4 &= y_1 + y_4 \end{aligned} \tag{9.105}$$

Thus in the general case, our strategy is to transform the state equation into a system of decoupled state equations, each of which has a single repeated eigenvalue. Each system is solved separately. The solution to the original system is a linear combination of the solutions of the separate systems.

9.13   Other Methods of Solution

It must be pointed out that the state equation of a circuit is but a system of linear, first-order differential equations with constant coefficients. Many solution schemes have been developed over the years. Notable among them is the method of Laplace transform. It will require considerable space to present the method. It is usually given in standard texts on theory of differential equations and will be omitted here. Another method is the use of exponential matrix, which has certain computational advantages. However, these and other schemes are more an aid in computation than an analytic tool. To gain an insight into the nature of the response of a circuit, the use of eigenvalues and eigenvectors is preferred.

9.14 Computer Solution – Runge-Kutta

The computation of eigenvalues, eigenvectors, the transformation matrix, and its inverse is a tedious chore for a circuit of order greater than four. If we are interested only in the values of the response over a given period of time, then numerical solution of the state equation should be used. Again, the Runge-Kutta algorithm that we studied in Chapter 4 can be extended to the nth-order case.

Let the state equation be written in matrix form:

$$\dot{\mathbf{x}} = \mathbf{A}\mathbf{x} + \mathbf{u}(t) \qquad (9.106)$$

with initial values $\mathbf{x}(0)$. Let the step size be h and denote the vector $\mathbf{x}(kh)$ as $\mathbf{x}_k$. In the case under consideration, matrix $\mathbf{A}$ is a constant matrix. We generate four vectors:

$$\mathbf{K}_1 = h\mathbf{A}\mathbf{x}_k + h\mathbf{u}(t_k)$$
$$\mathbf{K}_2 = h\mathbf{A}\left[\mathbf{x}_k + \frac{\mathbf{K}_1}{2}\right] + h\mathbf{u}\left(t_k + \frac{h}{2}\right)$$
$$\mathbf{K}_3 = h\mathbf{A}\left[\mathbf{x}_k + \frac{\mathbf{K}_2}{2}\right] + h\mathbf{u}\left(t_k + \frac{h}{2}\right)$$
$$\mathbf{K}_4 = h\mathbf{A}\left[\mathbf{x}_k + \mathbf{K}_3\right] + h\mathbf{u}(t_k + h) \qquad (9.107)$$

The next value of $\mathbf{x}$ is given by

$$\mathbf{x}_{k+1} = \mathbf{x}_k + \frac{(\mathbf{K}_1 + 2\mathbf{K}_2 + 2\mathbf{K}_3 + \mathbf{K}_4)}{6} \qquad (9.108)$$

The accuracy of the numerical results depends critically on the step size. It must be chosen to be much smaller than the reciprocal of the absolute value of the largest eigenvalue. Since the latter is generally not known, we select a "reasonable" value for h first and then do the integration twice, the second time with a step size half of that of the first time. The two solutions should agree up to a predetermined number of places; otherwise, the step size must be reduced, say halved, and the process is repeated.

9.15 Solution Without State Equation – Implicit Integration

A moment's reflection would reveal that our objective in this chapter is to develop an analytic scheme to calculate the response of a circuit in the time domain. So far, we have presented a scheme which requires setting up the state equation and solving it by analytic or numerical means. A great deal of manipulation is necessary to obtain the state equation in the form of a set of first-order differential equations. A question naturally arises as to whether or not it is possible to obtain the solution of the circuit in the time domain without the state equation. After all, mesh and node equations describe the circuit equally well. It is true that these have certain disadvantages as noted in Chapter 6, but perhaps for numerical purposes, they offer advantages over the state equation approach. For example, node equations can be set up trivially from the topology of the circuit. If we could combine this with a numerical integration scheme which is less cumbersome than Runge-Kutta, perhaps we would achieve our objective without using the complicated topological procedure of setting up the state equation.

Many such schemes exist. We shall present one in which node equations and implicit integration of the equations defining a capacitor and inductor are used.

With reference to Figure 9.16(a), the terminal equation for the capacitor is

$$v(t) = \frac{1}{C} \int_a^t i(y) \, dy + v(a) \qquad (9.109)$$

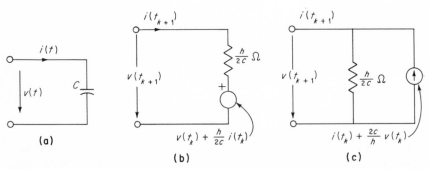

Figure 9.16 At integration step $k + 1$, a capacitor **(a)** is equivalent to a resistor in series with a voltage source **(b)**; **(c)** the Norton equivalent of **(b)**

In the numerical solution of a circuit, we predict the values of the circuit variables at time t_{k+1} from those at time t_k $(k = 0, 1, 2,...)$. For a capacitor, we have

$$v(t_{k+1}) = \frac{1}{C} \int_{t_k}^{t_{k+1}} i(y) \, dy + v(t_k) \qquad (9.110)$$

Suppose we approximate the integral by the trapezoidal rule, namely,

$$v(t_{k+1}) = \frac{h}{2C} \left[i(t_{k+1}) + i(t_k) \right] + v(t_k)$$

$$= \frac{h}{2C} i(t_{k+1}) + \left[\frac{h}{2C} i(t_k) + v(t_k) \right] \qquad (9.111)$$

The equation states that the capacitor voltage at time t_{k+1} is the sum of two voltages, the first being proportional to the capacitor current at t_{k+1}, the second being a "memory" term which is a constant dependent on the capacitor current and voltage at the previous time step. It follows that at time t_{k+1}, the capacitor is equivalent to a resistor in series with a voltage source. The value of the resistance is $h/2C$, and the value of the voltage source is $v(t_k) + (h/2C)i(t_k)$. The equivalent circuit is shown in Figure 9.16(b). It can be converted into a Norton equivalent shown in Figure 9.16(c).

In a similar manner, with reference to Figure 9.17(a), we have

$$i(t_{k+1}) = \frac{1}{L} \int_{t_k}^{t_{k+1}} v(y)\ dy + i(t_k)$$

$$= \frac{h}{2L}\, v(t_{k+1}) + \left[\frac{h}{2L}\, v(t_k) + i(t_k) \right] \qquad (9.112)$$

and the equivalent circuit consists of a resistor in parallel with a current source as shown in Figure 9.17(b).

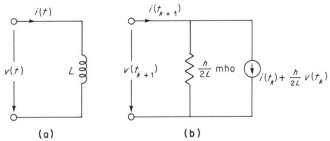

(a)   **(b)**

Figure 9.17 At integration step $k + 1$, an inductor (**a**) is equivalent to a resistor in parallel with a current source (**b**).

The two observations just noted form the basis of a simple scheme for calculating the time response of a circuit without the use of a state equation. Starting at time $t = 0$ $(k = 0)$, we replace each capacitor and inductor by their respective equivalent circuit. The resultant circuit is a *resistor* circuit, which can be solved algebraically by means of node equations. The solution will yield the values of the circuit variables at time $t = h$ $(k = 1)$. We then update the circuit by changing the values of the sources. The node equations are solved again with the new set of sources to give the circuit variables at time $t = 2h$ $(k = 2)$, and so on.

As an example, consider the circuit of Figure 9.18(a). Replacing the capacitor and inductor by their equivalent circuits, we obtain the circuit of Figure 9.18(b). Writing node equations, we get

$$\begin{bmatrix} 1 + \dfrac{2}{h} + \dfrac{h}{4} & -\dfrac{h}{4} \\[2ex] -\dfrac{h}{4} & 1 + \dfrac{2}{h} + \dfrac{h}{4} \end{bmatrix} \begin{bmatrix} v_1(t_{k+1}) \\[2ex] v_2(t_{k+1}) \end{bmatrix} = \begin{bmatrix} j_1(t_{k+1}) \\[2ex] j_2(t_{k+1}) \end{bmatrix} \qquad (9.113)$$

with

$$j_1(t_{k+1}) = J(t_{k+1}) + i_1(t_k) + \frac{2}{h} v_1(t_k) - i_3(t_k) - \frac{h}{4} v_3(t_k)$$

$$j_2(t_{k+1}) = i_3(t_k) + \frac{h}{4} v_3(t_k) + i_2(t_k) + \frac{2}{h} v_2(t_k)$$

To start the solution, we must have $v_1(0)$, $i_1(0)$, $v_2(0)$, $i_2(0)$, $v_3(0)$, and $i_3(0)$. The initial capacitor voltages and inductor currents are assumed

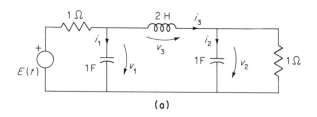

(a)

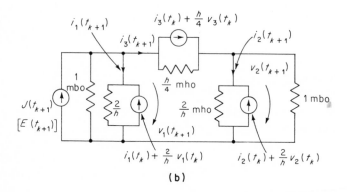

(b)

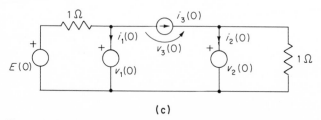

(c)

Figure 9.18 Computation of responses of a low-pass filter **(a)** by implicit integration; **(b)** its equivalent circuit at integration step $k + 1$; **(c)** the circuit from which the initial values of the circuit variables are determined.

given. To find the others, we note that at time $t = 0$, the circuit is that shown in Figure 9.18(c) from which we obtain

$$i_1(0) = E(0) - v_1(0) - i_3(0)$$
$$i_2(0) = i_3(0) - v_2(0)$$
$$v_3(0) = v_1(0) - v_2(0) \qquad (9.114)$$

This scheme of computing the initial values is applicable if there are no capacitor loops or inductor cut sets. If there are, then the initial values must be determined as explained in Sections 9.3 and 9.4.

The computer results of the example are shown below. Zero initial conditions are assumed on the capacitors and inductors. The step size used was 0.01 sec. It is noted that the results agree well with the exact values, which are obtained from the analytic expression for $v_2(t)$ given by that of Equation (9.73). As a check on the effect of round-off errors, i_3 is also computed by taking KCL at the output node and is indicated as I-3-CHECK on the computer print-out.

```
C IMPLICIT INTEGRATION- EXAMPLE
      REAL I1Z,I2Z,I3Z,I3CK,J1,J2
      WRITE(6,101)
      IWRT=0
C     SET INITIAL VALUES OF CURRENTS AND VOLTAGES
      V1Z=0.
      I1Z=1.
      V2Z=0.
      I2Z=0.
      V3Z=0.
      I3Z=0.
      I3CK=0.
C     STEP SIZE IS H.
      H=.01
      TMAX=10.
      DEL=2.+4./(H*H) + 4./H +H/2.
      DEL11 =1.+ 2./H + H/4.
      DEL12=H/4.
      KMAX=TMAX/H
      T = H
      DO 1 K=1,KMAX
C     V2EXCT IS THE EXACT VALUE OF V2.
      V2EXCT=(1. -EXP(-T) - 2.*EXP(-.5*T)*SIN(SQRT(3.)*T/2.)/SQRT(3.))
     C/2.
      J1=1.+I1Z+2.*V1Z/H-I3Z-H*V3Z/4.
      J2= I3Z +H*V3Z/4. + I2Z +2.*V2Z/H
      V1NXT=(DEL11*J1 +DEL12*J2)/DEL
      V2NXT=(DEL11*J2 +DEL12*J1)/DEL
      I1Z=-I1Z + 2.*(V1NXT -V1Z)/H
      I2Z=-I2Z +2.*(V2NXT-V2Z)/H
      V3NXT=V1NXT -V2NXT
      I3Z=I3Z+H*(V3NXT+V3Z)/4.
C     CHECK I3 BY KCL.
      I3CK=I2Z +V2NXT
      V1Z=V1NXT
      V2Z=V2NXT
      V3Z=V3NXT
C     PRINT OUT EVERY 50TH SET OF VALUES OF CIRCUIT VARIABLES.
      IWRT=IWRT+1
      IF(IWRT.NE.50)GO TO 99
      WRITE(6,100)T,V1Z,I1Z,V2Z,V2EXCT,I2Z,V3Z,I3Z,I3CK
      IWRT=0
   99 T=T+H
    1 CONTINUE
```

```
100 FORMAT(1H ,F6.2,8F15.6)
101 FORMAT(1H ,1X,'TIME',7X,'V-1',12X,'I-1',12X,'V-2',12X,
C 'V-2-EXACT',6X,'I-2',12X,'V-3',12X,'I-3',12X,'I-3-CHECK')
STOP
END
```

TIME	V-1	I-1	V-2	V-2-EXACT
0.50	0.385389	0.562346	0.008062	0.008062
1.00	0.582773	0.246977	0.049301	0.049307
1.50	0.651087	0.043582	0.125700	0.125716
2.00	0.641904	-0.066663	0.222648	0.222678
2.50	0.595942	-0.107647	0.321841	0.321882
3.00	0.541660	-0.103868	0.408410	0.408463
3.50	0.495899	-0.076736	0.473753	0.473818
4.00	0.466025	-0.042630	0.515511	0.515594
4.50	0.452673	-0.011958	0.536058	0.536164
5.00	0.452608	0.010025	0.540489	0.540603
5.50	0.461037	0.022077	0.534708	0.534823
6.00	0.473247	0.025534	0.524105	0.524215
6.50	0.485554	0.022923	0.512774	0.512876
7.00	0.495638	0.017082	0.503277	0.503373
7.50	0.502485	0.010311	0.496792	0.496871
8.00	0.506082	0.004326	0.493413	0.493477
8.50	0.507040	-0.000108	0.492586	0.492642
9.00	0.506233	-0.002790	0.493476	0.493534
9.50	0.504514	-0.003839	0.495242	0.495309
10.00	0.502579	-0.003720	0.497204	0.497282

I-2	V-3	I-3	I-3-CHECK
0.044133	0.377327	0.052199	0.052195
0.120827	0.533471	0.170136	0.170128
0.179453	0.525386	0.305220	0.305153
0.201983	0.419256	0.424677	0.424632
0.189740	0.274101	0.511638	0.511581
0.153668	0.133250	0.562134	0.562077
0.106973	0.022145	0.580783	0.580727
0.060911	-0.049486	0.576526	0.576422
0.023026	-0.083385	0.559201	0.559084
-0.003271	-0.087881	0.537288	0.537218
-0.017970	-0.073671	0.516816	0.516738
-0.023072	-0.050858	0.501161	0.501033
-0.021427	-0.027122	0.491444	0.491347
-0.016194	-0.007639	0.487202	0.487084
-0.009744	0.005693	0.487096	0.487047
-0.003363	0.012669	0.489513	0.489450
0.000353	0.014454	0.492993	0.492939
0.002928	0.012757	0.496447	0.496404
0.003929	0.009273	0.499220	0.499170
0.003786	0.005374	0.501047	0.500990

SUGGESTED EXPERIMENTS

1. Construct the circuit of Problem 9.1(a). Scale the elements so that the time response is in the millisecond range. Apply a square wave to the circuit and observe the response. Apply a series of narrow pulses to the circuit and observe the response. Explain what you see.

2. Experimentally verify Problem 9.14.

3. Construct the circuit of Problem 9.12. Scale the elements so that the time response is in the microsecond range. Apply in turn a square wave, triangular wave, and a series of pulses to the circuit. Compare with your calculation.

4. Construct the circuit of Problem 9.23. Repeat the last experiment with this circuit. Explain your observations.
5. Construct the lumped delay line of Problem 9.22. Scale the elements so that the delay is about 1 μsec. Compare the experimental results with the computer output.

COMPUTER PROGRAMMING (PROJECTS)

1. Write a general program to integrate a system of differential equations. The computation should be done twice, once with a step size h and once with $h/2$. The difference of the two outputs should be printed. Include the plotting routine *EEPLOT* to facilitate display of the outputs.
2. Write a program to find the eigenvalues and eigenvectors of a square matrix. Consult texts on numerical analysis for algorithms. (See, for example, Carnahan, B., Luther, H.A., and Wilkes, J.O., *Applied Numerical Methods*, New York: Wiley, 1969.)
3. Write a computer program to implement the Runge-Kutta algorithm with variable step size. (See, for example, Pennington, R.H., *Introductory Computer Methods and Numerical Analysis*, New York: Macmillan, 1965.)

PROBLEMS

9.1   *Exercise*   Write the state equation for each of the following circuits.

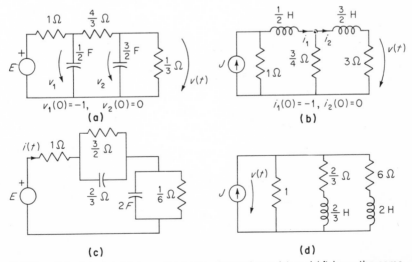

Problem 9.1   (a) and (b) have the same eigenvalues; (c) and (d) have the same eigenvalues.

9.2   *Exercise* Write the state equation for each of the following circuits. Express the output quantity in each in terms of the state variables.

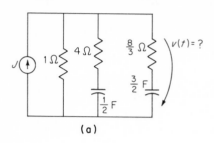

(a)

Problem 9.2 State equations.

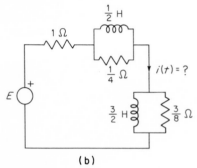

(b)

9.3   *Exercise* Find the frequencies of oscillation of each of the following circuits.

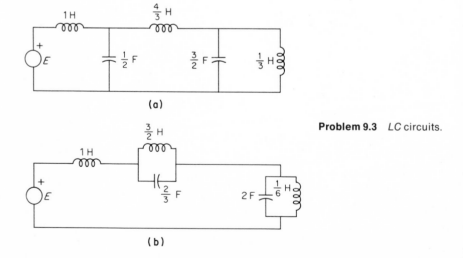

(a)

Problem 9.3 *LC* circuits.

(b)

9.4 Write the state equation for each of the following circuits. Note the presence of a capacitor loop and inductor cut set.

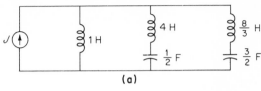

(a)

Problem 9.4 (a) Inductor cut set; (b) capacitor loop.

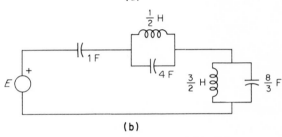

(b)

9.5 *Capacitor Loop* Write the state equation for the following circuit.

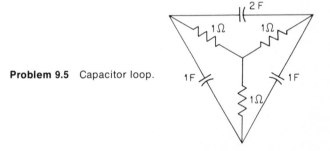

Problem 9.5 Capacitor loop.

9.6 *Inductor Cut Set* Write the state equation of the following circuit.

Problem 9.6 Inductor cut set.

9.7 *Dependent Source* Write the state equation of the circuit shown.

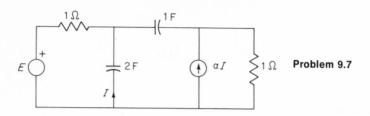

Problem 9.7

9.8 *Mutual Inductor* Write the state equation of the circuit shown.

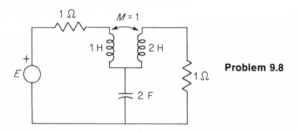

Problem 9.8

9.9 *Step Response* Find the step response of the circuit of Problem 9.1(a). Repeat the problem for the circuit of Problem 9.1(b). Are you surprised?

9.10 *Step Response* Find the current $i(t)$ in the circuit of Problem 9.1(c). Assume all initial conditions are zero. What is $v(t)$ in the circuit of Problem 9.1(d)?

9.11 *Pulse Response* Write a computer program based on the Runge-Kutta algorithm to calculate the output voltage $v(t)$. The excitation is specified as follows: $e(t) = t$, $0 \le t \le 1$; $e(t) = 1$, $1 < t \le 2$; $e(t) = 3 - t$, $2 < t \le 3$; $e(t) = 0$, $t > 3$.

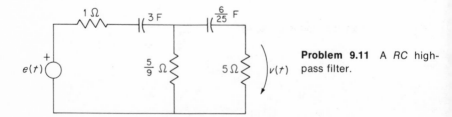

Problem 9.11 A *RC* high-pass filter.

9.12 *Low-Pass Filter* Find the step response of the circuit shown. Assume all initial conditions to be zero.

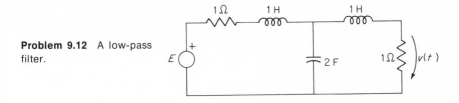

Problem 9.12 A low-pass filter.

9.13 *High-Pass Filter* Find the step response of the circuit shown. Assume all initial conditions to be zero. Compare the results with those of Problem 9.12.

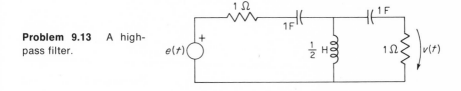

Problem 9.13 A high-pass filter.

9.14 *Smoothing Filter* Let the input to the circuit of Problem 9.12 be specified as follows: $e(t) = \sin 0.5t + p(t)$, $0 \leq t \leq 2\pi$; $e(t) = 0$, $t > 2\pi$; $p(t)$ is a narrow pulse of height 0.3, width 0.1 sec, and is centered at $t = \pi$. Write a computer program based on the implicit integration method to calculate the output voltage $v(t)$.

9.15 Repeat Problem 9.14 for the circuit of Problem 9.13. Explain the results. Are you surprised?

9.16 *Repeated Eigenvalue* Find the output voltage of the circuit shown. The input is a step.

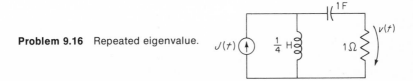

Problem 9.16 Repeated eigenvalue.

9.17 *Repeated Eigenvalue* Write a computer program to calculate the response of the circuit of Problem 9.16 for an excitation specified as follows: $J(t) = t$, $0 \le t \le 1$; $J(t) = 2 - t$, $1 \le t \le 2$; $J(t) = 0$, $t > 2$.

9.18 *Repeated Eigenvalue* Find the step response of the circuit shown. Compare the results with those of Problem 9.16. Are you surprised?

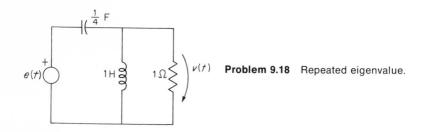

Problem 9.18 Repeated eigenvalue.

9.19 Repeat Problem 9.18 for the circuit shown.

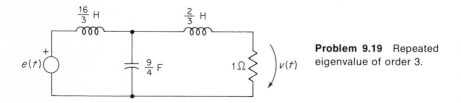

Problem 9.19 Repeated eigenvalue of order 3.

9.20 *Repeated Eigenvalue* Find the step response of the circuit shown. Assume all initial conditions to be zero.

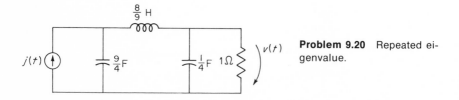

Problem 9.20 Repeated eigenvalue.

9.21 *Pulse Response* Apply a series of narrow pulses to the circuit of Problem 9.20 and compute $v(t)$. Let the pulses be of unit height, width, 0.1 sec, and period 1 sec. Repeat the computation for a period of 0.2 sec. Use the Runge-Kutta algorithm.

9.22 *Lumped Delay Line* The circuit shown is an approximation of a delay line. Compute the step response of the circuit. Plot the output voltage.

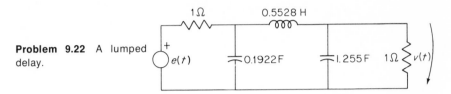

Problem 9.22 A lumped delay.

9.23 *Low-Pass Filter* Use implicit integration to compute the response $v(t)$ of the circuit shown.

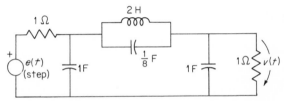

Problem 9.23 A low-pass filter with a capacitor loop.

10

multiterminal circuits

In this chapter we consider the characterization of multiterminal circuits in the frequency domain and develop analysis techniques for circuits consisting of these as component parts. We recall that in Chapter 2 we had to treat such circuits in a special way. We shall see in this chapter that circuit equations can be systematically set up, provided that we have the proper terminal characterization of the multiterminal circuits.

Examples of multiterminal circuits are transistors, operational amplifiers, differential amplifiers, and vacuum tubes. These are the components of large circuits, such as voltage amplifiers used in the detection of weak signals, power amplifiers to convert a sensing signal (displacement, velocity, acceleration, or pressure) into a voltage or current to drive a motor or switch, radio receivers, active filters (circuits composed of operational amplifiers, resistors, and capacitors), analog computers, in fact, any electronic systems. Figure 10.1 shows a simplified version of a power amplifier used for guidance control in an aircraft. Figure 10.2 shows an oscillator that generates a sinusoidal voltage. Note the 3-terminal circuits that are the transistors.

Given a circuit such as that of the oscillator, how does one calculate the response of the circuit? There are two distinct approaches. One is to replace every multiterminal circuit by an equivalent circuit that consists of two-terminal elements only. The resulting circuit now has only two-terminal elements, and we can proceed with the analysis using loop, mesh, or node equations. The configuration and the element values of the equivalent circuit are obtained from physical considerations or from external measurement at the terminals. It may have many branches and nodes. This approach is most suitable for circuits containing one or two multiterminal elements.

Another approach is that we first obtain a terminal characterization of each multiterminal circuit in the form of a set of equations, or parameters, that relate the terminal voltages and currents in some explicit manner, and then apply the current and voltage equations at the nodes to which the terminals of the multiterminal circuits are connected. The internal structure of each is not revealed; in fact, it is unimportant. The choice

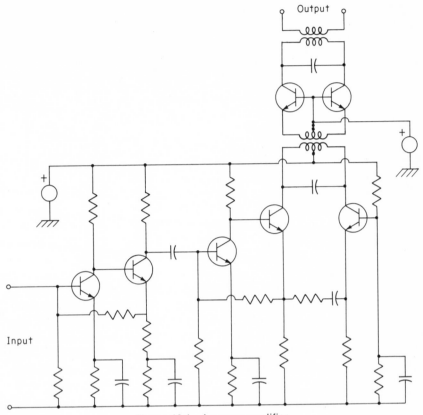

Figure 10.1 A power amplifier.

of terminal characterization is important, however, as there are many different modes by which the terminal voltages and currents are related. One mode may be more convenient than another, depending upon how the circuits are connected.

The first approach leads to a situation with which we are familiar and will not be discussed further. We shall consider the second approach in this chapter. As we shall see later, it offers a distinctive advantage over the first from the computational point of view.

10.1 Motivation

Circuits are used as signal processors, as we have noted on many an occasion. The processing may be one of amplification, extraction in the presence of noise, or smoothing. In all cases, we are interested in the response or "output" at a pair of terminals when an excitation or "input"

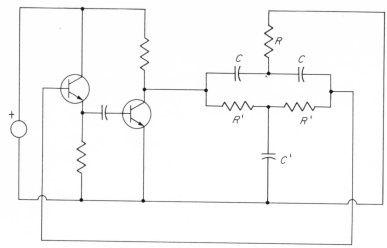

Figure 10.2 A bridged-T oscillator.

is applied at another. The voltages and currents elsewhere are of no interest as long as the input-output relation is what is desired. Schematically, the situation is depicted in Figure 10.3, in which N_S is a circuit containing the source or excitation, N_L a circuit containing the "load" at which the output is taken, and N a 3-terminal circuit. Many of the circuits studied in Chapter 8, filters, equalizers, amplifiers, are of this form.

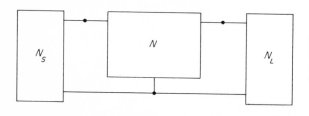

Figure 10.3 A 3-terminal circuit excited by a source network N_s and terminated in a load N_L.

Replacing the circuits N_S and N_L by their Thévenin equivalents, we obtain a circuit as shown in Figure 10.4. In this configuration, the 3-terminal circuit N is said to be "terminated" in impedance Z_L at the output and Z_S at the input. The impedances are also referred to as the load

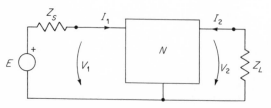

Figure 10.4 The source network and the load of the circuit of Figure 10.3 are each replaced by their respective Thévenin equivalent circuit.

and source impedances, respectively. Our objective is to obtain a terminal characterization of N, namely, a set of parameters relating the terminal voltages V_1 and V_2 and currents I_1 and I_2, so that the input-output relations, or network functions, of the overall circuit can be expressed in terms of them, and not in terms of the individual elements in the interior of N, which may not even be identifiable, as in the case of a transistor.

The circuit N may consist of a collection of 2-, 3-, or N-terminal circuits. (See Figure 10.1.) We must, therefore, develop schemes by which these circuits can be combined into an overall 3-terminal circuit N.

10.2 Indefinite Admittance Parameters

Consider the N-terminal circuit of Figure 10.5. Assume it does not contain any internal independent sources. Assign currents and voltages as shown. Note that the voltages are defined with respect to a reference node, which may or may not be connected to the circuit. Note also that in the voltage graph of the circuit, the terminal voltages form a tree. Regarding the currents as responses and the voltages as excitations, we can write, by superposition, the following equations:

$$I_1 = y_{11}V_1 + y_{12}V_2 + \cdots + y_{1N}V_N$$
$$I_2 = y_{21}V_1 + y_{22}V_2 + \cdots + y_{2N}V_N$$
$$\vdots$$
$$I_N = y_{N1}V_1 + y_{N2}V_2 + \cdots + y_{NN}V_N \qquad (10.1)$$

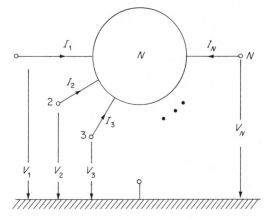

Figure 10.5 An N-terminal circuit with its terminal voltages and currents defined.

The admittances y_{ij} are known as the *indefinite admittance parameters* and can be found from the following:

$$y_{ij} = \frac{I_i}{V_j}\bigg|_{V_k = 0} \qquad \text{all } k \neq j \qquad (10.2)$$

In words, the parameter y_{ij} is a measure of the current I_i when all but the jth terminal are shorted to the reference node, and when a voltage of 1 V is applied at terminal j. If we denote the set of currents and voltages in matrix form as **I** and **V**, respectively, then the circuit is characterized by

$$\mathbf{I} = \mathbf{YV}$$

where the matrix **Y** is known as the indefinite admittance matrix. As an example, the indefinite admittance matrix of the circuit of Figure 10.6 is found to be, after application of Equation (10.2),

$$\begin{bmatrix} 4 & -1 & -3 \\ -1 & 3 & -2 \\ -3 & -2 & 5 \end{bmatrix} \qquad (10.3)$$

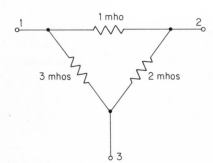

Figure 10.6 An example of a 3-terminal circuit.

That of the circuit of Figure 10.7 is

$$\mathbf{Y} = \begin{bmatrix} 2(1+\alpha) & 0 & -2(1+\alpha) \\ -2\alpha & 3 & -3+2\alpha \\ -2 & -3 & 5 \end{bmatrix} \qquad (10.4)$$

It is not difficult to show that by writing loop equations for the circuit with the terminal voltages as excitations and solving for the currents in

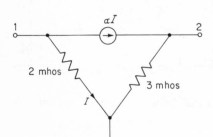

Figure 10.7 An example of a 3-terminal circuit containing a dependent source.

question, the indefinite admittance matrix of a circuit containing resistors, inductors, capacitors, and mutual inductances is symmetric. If the circuit contains dependent sources, the matrix may or may not be symmetric. (See Sections 10.6 and 10.7.)

Example 10.1 Suppose the 3-terminal circuit of Figure 10.7 is terminated as shown in Figure 10.8. Find the minimum value of α for which the circuit is a voltage amplifier.

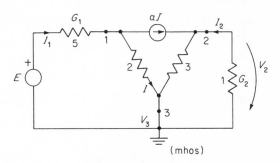

Figure 10.8 The circuit of Figure 10.7 terminated in G_1 and G_2. [Example 10.1]

(mhos)

Solution The terminal conditions on the 3-terminal circuit are

$$V_3 = 0$$

$$V_1 = E - \frac{I_1}{G_1}$$

$$I_2 = -G_2 V_2$$

The terminal characterization of N is (with $V_3 = 0$)

$$I_1 = y_{11}V_1 + y_{12}V_2$$
$$I_2 = y_{21}V_1 + y_{22}V_2$$

Eliminating V_1, I_1, and I_2, we get

$$\frac{V_2}{E} = \frac{-y_{21}G_1}{(y_{11} + G_1)(y_{22} + G_2) - y_{12}y_{21}}$$

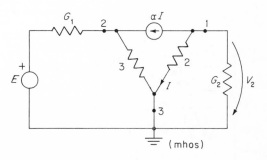

Figure 10.9 The circuit of Figure 10.8 with the 3-terminal circuits reversed. [Example 10.2]

(mhos)

Substituting the numerical values, we find

$$\frac{V_2}{E} = \frac{5\alpha}{2(7+2\alpha)}$$

The gain V_2/E will be greater than one if

$$\alpha > 14$$

Example 10.2   In Figure 10.8, suppose the circuit N is reversed, so that it is terminated as shown in Figure 10.9. Find the gain V_2/E.

Solution   Interchanging the subscripts 1 and 2 in the expression for V_2/E of the last example, we get in this case

$$\frac{V_2}{E} = \frac{-y_{12}G_1}{(y_{11}+G_2)(y_{22}+G_1) - y_{12}y_{21}}$$

Since $y_{12} = 0$, we have

$$\frac{V_2}{E} = 0$$

No matter what the terminations are, the output voltage will be zero. No power is transferred from the source to the load.

10.3   Properties of Indefinite Admittance Matrix

We now make a number of observations on the indefinite admittance matrix. With reference to Figure 10.10, by KCL, we have that

$$I_1 + I_2 + \cdots + I_N = 0 \tag{10.5}$$

Adding the equations in (10.1), we have

$$0 = \sum_{k=1}^{N} y_{k1}V_1 + \sum_{k=1}^{N} y_{k2}V_2 + \cdots + \sum_{k=1}^{N} y_{kN}V_N \tag{10.6}$$

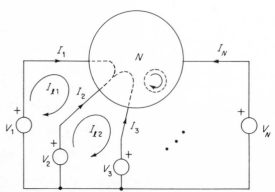

Figure 10.10 The circuit used in the proof of Property 1.

independent of V_1, V_2,..., V_N. Therefore,

$$\sum_{k=1}^{N} y_{ki} = 0 \qquad \text{for all } i \qquad (10.7)$$

and we have the following property:

Property 1   The sum of the admittances of any column of an indefinite admittance matrix is zero.   ■

Second, suppose we write loop equations for the circuit with the terminal voltages as excitations as shown in Figure 10.10. Since the voltage sources have a common node and form a cut set, the set of fundamental loops chosen for the equations will have the property that in each loop, there will be either none or exactly two voltage sources. [See Problem 1.20(e).] A typical loop equation would be

$$Z_{p1}I_{l1} + Z_{p2}I_{l2} + \cdots + Z_{pm}I_{lm} = 0$$

or

$$Z_{p1}I_{l1} + Z_{p2}I_{l2} + \cdots + Z_{pm}I_{lm} = V_s - V_t \qquad (10.8)$$

where $I_{lk}(k = 1, 2,..., m)$ are the loop currents, and V_s and V_t are the voltage sources in loop p.

Now suppose we add an arbitrary voltage ΔV to all of the voltage sources so that the value of V_1 is now $V_1 + \Delta V$, the value of V_2 is now $V_2 + \Delta V$, and so on. Clearly, the values of the loop currents are unchanged, since in the set of loop equations, the right-hand side consists of either 0 or terms of the form

$$(V_s + \Delta V) + (V_t + \Delta V) = V_s - V_t \qquad (10.9)$$

Every current, in particular I_1, I_2,..., I_N, can be expressed uniquely in terms of the loop currents. We, therefore, conclude that if all of the terminal voltages are increased by ΔV, the currents I_1, I_2,..., I_N are unchanged, and we have, from Equations (10.1),

$$I_1 = y_{11}(V_1 + \Delta V) + y_{12}(V_2 + \Delta V) + \cdots + y_{1N}(V_N + \Delta V)$$
$$I_2 = y_{21}(V_1 + \Delta V) + y_{22}(V_2 + \Delta V) + \cdots + y_{2N}(V_N + \Delta V)$$
$$\vdots$$
$$I_N = y_{N1}(V_1 + \Delta V) + y_{N2}(V_2 + \Delta V) + \cdots + y_{NN}(V_N + \Delta V)$$

or

$$(y_{11} + y_{12} + \cdots + y_{1N})\Delta V = 0$$
$$(y_{21} + y_{22} + \cdots + y_{2N})\Delta V = 0$$
$$\vdots$$
$$(y_{N1} + y_{N2} + \cdots + y_{NN})\Delta V = 0 \qquad (10.10)$$

for arbitrary ΔV. It follows that

$$\sum_{i=1}^{N} y_{ki} = 0 \qquad \text{for all } k \qquad (10.11)$$

and we have the following properties:

Property 2   The sum of the admittances of any row of an indefinite admittance matrix is zero.   ■

The two observations are verified in Equations (10.3) and (10.4). The significance is that any one row and its corresponding column of an indefinite admittance matrix are redundant. Thus, we need at most one parameter to characterize a 2-terminal circuit, four parameters for a 3-terminal circuit, and $(N - 1)(N - 1)$ parameters for an N-terminal circuit.

Next, suppose that we take a 3-terminal circuit and add to it an isolated node as shown in Figure 10.11. Since this node is not connected to any other, we can write the following equations:

$$I_1 = y_{11}V_1 + y_{12}V_2 + y_{13}V_3 + 0V_4$$
$$I_2 = y_{21}V_1 + y_{22}V_2 + y_{23}V_3 + 0V_4$$
$$I_3 = y_{31}V_1 + y_{32}V_2 + y_{33}V_3 + 0V_4$$
$$I_4 = \quad 0V_1 + \quad 0V_2 + \quad 0V_3 + 0V_4 \qquad (10.12)$$

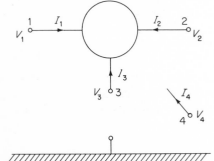

Figure 10.11 Addition of an isolated node to a 3-terminal circuit.

The indefinite admittance matrix of the 4-terminal circuit is simply that of the 3-terminal circuit appended by a row and column of zeros. In general, we have the following property:

Property 3   If an isolated node is added to an N-terminal circuit, the resultant indefinite admittance matrix is $(N + 1)(N + 1)$ and is obtained by adding a row and column of zeros to the $N \times N$ matrix of the original circuit. The position of the new row and column corresponds to the number assigned to the new node.   ■

This seemingly trivial observation is the basis on which we may regard all interconnections of multiterminal circuits as parallel connections, as will be seen in the next section.

10.4 Parallel Connection of *N*-Terminal Circuits

Consider two *N*-terminal circuits. The terminals of one are connected to the corresponding terminals of the other as shown in Figure 10.12. The two circuits are said to be in parallel. Let the voltages of circuit *A* be $\mathbf{V}_A$ and those of circuit *B* be $\mathbf{V}_B$, similarly for the currents $\mathbf{I}_A$ and $\mathbf{I}_B$. The parallel connection leads to a set of terminal conditions:

$$\mathbf{V}_A = \mathbf{V}_B = \mathbf{V}$$
$$\mathbf{I}_A + \mathbf{I}_B = \mathbf{I} \qquad (10.13)$$

Let

$$\mathbf{I}_A = \mathbf{Y}_A \mathbf{V}_A$$
$$\mathbf{I}_B = \mathbf{Y}_B \mathbf{V}_A$$

and

$$\mathbf{I} = \mathbf{Y}\mathbf{V}$$

We, therefore, have

$$\mathbf{Y} = \mathbf{Y}_A + \mathbf{Y}_B \qquad (10.14)$$

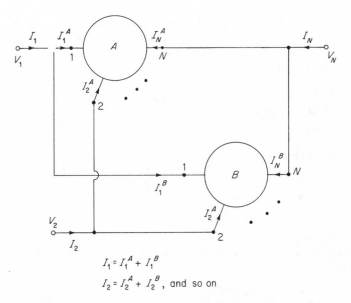

$$I_1 = I_1^A + I_1^B$$
$$I_2 = I_2^A + I_2^B, \text{ and so on}$$

Figure 10.12 Two *N*-terminal circuits connected in parallel.

The indefinite admittance matrix of the overall circuit is the sum of the admittance matrices of the two circuits.

Example 10.3   The circuit shown in Figure 10.13 is known as a notch filter. Find the condition satisfied by the element values under which, and the frequency at which, the output is zero.

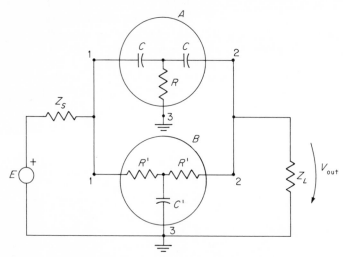

Figure 10.13   A notch filter.   [Example 10.3]

Solution   The 3-terminal circuit inserted between the source and the load is clearly a parallel connection of two 3-terminal circuits. With reference to the figure, it is found that the admittance matrix of circuit A is (with row 3 and column 3 omitted for brevity)

$$\mathbf{Y}_A = \frac{1}{1 + j2\omega RC} \begin{bmatrix} j\omega C(1 + j\omega RC) & \omega^2 RC^2 \\ \omega^2 RC^2 & j\omega C(1 + j\omega RC) \end{bmatrix} \quad (10.15)$$

and that of circuit B is

$$\mathbf{Y}_B = \frac{1}{2R' + j\omega C'R'R'} \begin{bmatrix} 1 + j\omega C'R' & -1 \\ -1 & 1 + j\omega C'R' \end{bmatrix} \quad (10.16)$$

The admittance matrix of the overall 3-terminal circuit is the sum of $\mathbf{Y}_A$ and $\mathbf{Y}_B$. The output voltage will be zero if the 2-1-term is zero. (See Example 10.2.) In our case,

$$\frac{\omega^2 RC^2}{1 + j2\omega RC} + \frac{-1}{2R' + j\omega C'R'R'} = 0$$

Equating the real and imaginary parts to zero, we find the frequency at which the output is zero to be

$$\omega_0 = \frac{1}{\sqrt{2RC^2R'}}$$

and the condition on the element values $4RC = R'C'$. This example shows that it is possible to construct a band-elimination filter using resistors and capacitors only.

Example 10.4 Obtain the terminal characterization of the 4-terminal circuit shown in Figure 10.14.

Solution The circuit may be regarded as a parallel connection of two 4-terminal circuits shown in Figures 10.14(b) and (c). Adding the admittance matrices of the two component circuits, we get

$$
\begin{bmatrix}
4 & -1 & 0 & -3 \\
-1 & 3 & 0 & -2 \\
0 & 0 & 0 & 0 \\
-3 & -2 & 0 & 5
\end{bmatrix}
+
\begin{bmatrix}
0 & 0 & 0 & 0 \\
0 & 4 & -1 & -3 \\
0 & -1 & 3 & -2 \\
0 & -3 & -2 & 5
\end{bmatrix}
=
\begin{bmatrix}
4 & -1 & 0 & -3 \\
-1 & 7 & -1 & -5 \\
0 & -1 & 3 & -2 \\
-3 & -5 & -2 & 10
\end{bmatrix}
$$

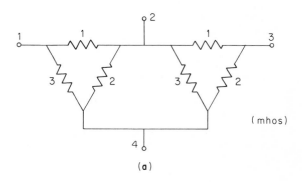

(a)

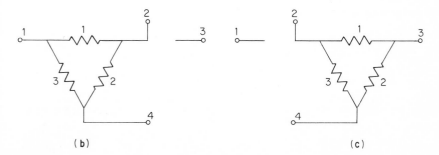

(b)   (c)

Figure 10.14 (a) A 4-terminal circuit; (b) and (c) its two components. [Example 10.4.]

10.5 Pivotal Condensation

Suppose in an N-terminal circuit, we no longer regard one of the terminals as an external terminal, but rather as an internal node. The resultant $(N-1)$-terminal circuit has an admittance matrix which can be obtained from the original $N \times N$ matrix in a systematic way.

With reference to Figure 10.15, let the admittance matrix be

$$\mathbf{Y} = \begin{bmatrix} y_{11} & \cdots & y_{1k} & \cdots & y_{1N} \\ \vdots & & \vdots & & \vdots \\ y_{k1} & & y_{kk} & & y_{kN} \\ \vdots & & \vdots & & \vdots \\ y_{N1} & \cdots & y_{Nk} & \cdots & y_{NN} \end{bmatrix} \qquad (10.17)$$

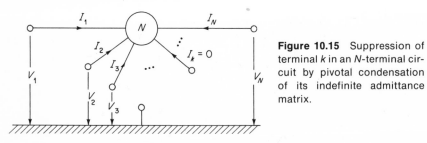

Figure 10.15 Suppression of terminal k in an N-terminal circuit by pivotal condensation of its indefinite admittance matrix.

Let us create an $(N-1)$-terminal circuit by abandoning terminal k. This is to say that we force I_k to be zero. The kth equation becomes

$$I_k = 0 = y_{k1}V_1 + \cdots + y_{kk}V_k + \cdots + y_{kN}V_N \qquad (10.18)$$

A linear relation exists among the voltages. Solving for V_k, we have

$$V_k = \frac{-1}{y_{kk}} (y_{k1}V_1 + \cdots + y_{k(k-1)}V_{k-1}$$
$$+ y_{k(k+1)}V_{k+1} + \cdots + y_{kN}V_N) \qquad (10.19)$$

Substituting it into all but the kth equations, we have

$$I_j = \left(y_{j1} - \frac{y_{jk}y_{k1}}{y_{kk}}\right) V_1 + \left(y_{j2} - \frac{y_{jk}y_{k2}}{y_{kk}}\right) V_2 + \cdots$$
$$+ \left(y_{jN} - \frac{y_{jk}y_{kN}}{y_{kk}}\right) V_N \qquad j = 1, 2, \ldots, N, \text{ and } j \neq k. \qquad (10.20)$$

Let the elements of the new $(N-1)(N-1)$ matrix be y'_{ij}. We, therefore, have

$$y'_{ij} = y_{ij} - \frac{y_{ik}y_{kj}}{y_{kk}} \qquad (10.21)$$

for $i, j = 1, 2, \ldots, N$, and $i \neq k$ and $j \neq k$.

Example 10.5 Eliminate terminal 2 of the 4-terminal circuit of Example 10.4 and obtain a new admittance matrix for the resultant 3-terminal circuit.

Solution Applying the formula of Equation (10.21), we find the new matrix to be

$$Y' = \begin{bmatrix} 27 & -1 & -26 \\ -1 & 20 & -19 \\ -26 & -19 & 45 \end{bmatrix} \left(\frac{1}{7}\right)$$ ∎

It is by the systematic application of the technique of pivotal condensation, combined with that of parallel connection, that the characterization of a circuit composed of a collection of multiterminal circuits is obtained. Examples follow.

Example 10.6 *(Operational Amplifier)* An operational amplifier is a 3-terminal device which has the following properties:

1. The gain, or the ratio of the output voltage to input voltage, is very high.
2. The input current is very low.
3. The input voltage and current are not affected by the output voltage or current.
4. The Thévenin equivalent resistance of the output circuit is very low.

An idealized equivalent circuit of such a device is shown in Figure 10.16. The resistance R_{in} is very high and R_{out} very low. The gain K is of the order of 500 to several thousand. A simplified realization of the operational amplifier is shown in Figure 10.17. It consists of two transistors, each having an equivalent circuit as shown in the adjoining figure. Find a 3-terminal characterization of the equivalent circuit and that of the simplified realization. Obtain an expression for the gain K and resistances R_{in} and R_{out}.

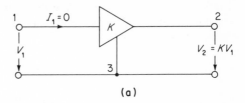

(a)

Figure 10.16 (a) An operational amplifier; (b) its equivalent circuit. [Example 10.6]

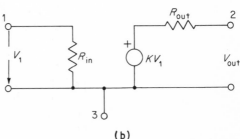

(b)

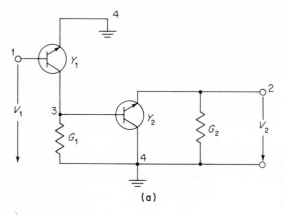

Figure 10.17 (a) A practical realization of an operational amplifier; (b) an equivalent circuit of the transistor. [Example 10.6]

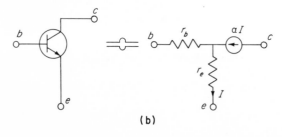

Solution Let the terminals of the transistor be labeled b, e, and c, for base, emitter, and collector, respectively. The admittance matrix is found to be

$$\mathbf{Y}_t = \begin{bmatrix} y_{bb} & y_{be} & y_{bc} \\ y_{eb} & y_{ee} & y_{ec} \\ y_{cb} & y_{ce} & y_{cc} \end{bmatrix} = \frac{1}{(1-\alpha)r_b + r_e} \begin{bmatrix} 1-\alpha & \alpha-1 & 0 \\ -1 & 1 & 0 \\ \alpha & -\alpha & 0 \end{bmatrix} \quad (10.22)$$

The circuit of Figure 10.17 may be regarded as a parallel connection of three 4-terminal circuits, one for each of the transistors and one consisting of the two resistors G_1 and G_2. Denoting the admittance matrices of these as $\mathbf{Y}_1$, $\mathbf{Y}_2$, and $\mathbf{Y}_G$, respectively, we have

$$\mathbf{Y}_1 = \begin{bmatrix} y_{bb} & 0 & 0 & y_{be} \\ 0 & 0 & 0 & 0 \\ y_{cb} & 0 & 0 & y_{ce} \\ y_{eb} & 0 & 0 & y_{ee} \end{bmatrix} \quad (10.23)$$

$$\mathbf{Y}_2 = \begin{bmatrix} 0 & 0 & 0 & 0 \\ 0 & y_{ee} & y_{eb} & 0 \\ 0 & y_{be} & y_{bb} & 0 \\ 0 & y_{ce} & y_{cb} & 0 \end{bmatrix} \quad (10.24)$$

$$\mathbf{Y}_G = \begin{bmatrix} 0 & 0 & 0 & 0 \\ 0 & G_2 & 0 & -G_2 \\ 0 & 0 & G_1 & -G_1 \\ 0 & -G_2 & -G_1 & G_1 + G_2 \end{bmatrix} \quad (10.25)$$

The overall admittance matrix is

$$\mathbf{Y} = \mathbf{Y}_1 + \mathbf{Y}_2 + \mathbf{Y}_G = \begin{bmatrix} y_{bb} & 0 & 0 & y_{be} \\ 0 & y_{ee} + G_2 & y_{eb} & -G_2 \\ y_{cb} & y_{be} & y_{bb} + G_1 & -G_1 + y_{ce} \\ y_{eb} & y_{ee} - G_2 & y_{cb} - G_1 & y_{ee} + G_1 + G_2 \end{bmatrix} \quad (10.26)$$

Eliminating terminal 3 by pivotal condensation, we find the admittance matrix of the resultant 3-terminal circuit to be

$$\mathbf{Y}' = \begin{bmatrix} y_{bb} & 0 & y_{be} \\ -y_{cb}y_{eb}R & G_2 + y_{ee} - y_{be}y_{eb}R & -G_2 - y_{eb}(y_{ce} - G_1)R \\ y_{eb} - y_{cb}(y_{cb} - G_1)R & y_{ce} - G_2 - y_{be}(y_{cb} - G_1)R & y_{44} - (y_{ce} - G_1)(y_{cb} - G_1)R \end{bmatrix}$$

$$(10.27)$$

where $R = (G_1 + y_{bb})^{-1}$ and $y_{44} = y_{ee} + G_1 + G_2$.

The admittance matrix of the equivalent circuit of an operational amplifier [Figure 10.16(b)] is

$$\mathbf{Y}_A = \begin{bmatrix} G_{\text{in}} & 0 & -G_{\text{in}} \\ -KG_{\text{out}} & G_{\text{out}} & (K-1)G_{\text{out}} \\ -G_{\text{in}} + KG_{\text{out}} & -G_{\text{out}} & -(K-1)G_{\text{out}} + G_{\text{in}} \end{bmatrix} \quad (10.28)$$

Comparing this with $\mathbf{Y}'$, we find the gain K, input resistance R_{in}, and output resistance R_{out} to be

$$K = \frac{y_{cb}y_{eb}}{(G_2 + y_{ee})(G_1 + y_{bb}) - y_{eb}y_{be}}$$

$$= \frac{-\alpha}{(G_2(r_e + r_b - \alpha r_b) + 1)(G_1(r_e + r_b - \alpha r_b) + 1 - \alpha) + \alpha - 1} = -490$$

$$(10.29)$$

$$R_{\text{in}} = \frac{1}{G_{\text{in}}} = \frac{1}{y_{bb}} = \frac{r_e - \alpha r_b + r_b}{1 - \alpha} = 1010 \ \Omega \quad (10.30)$$

$$R_{\text{out}} = \frac{1}{G_{\text{out}}} = \left[G_2 + \frac{G_1}{G_1(r_e - \alpha r_b + r_b) + 1 - \alpha} \right]^{-1} = 59 \ \Omega \quad (10.31)$$

In the above, typical element values of the transistor parameters are used: $\alpha = 0.99$, $r_e = 10 \ \Omega$, $r_b = 10 \ \Omega$, $G_1 = 0.2 \times 10^{-3}$ mhos, and $G_2 = 0.3 \times 10^{-3}$ mhos. It is seen that the circuit approximates an ideal operational amplifier well.

Example 10.7 (Bridged-T Oscillator) The circuit of Figure 10.18 is a parallel connection of an ideal operational amplifier and the circuit of Example 10.3. Sinusoidal voltage and current exist on the circuit in the absence of any excitation, provided that the gain of the operational amplifier is sufficiently large. The circuit is known as a bridged-*T* oscillator. Find the condition that must be satisfied by the element values for oscillation to occur and the frequency of oscillation.

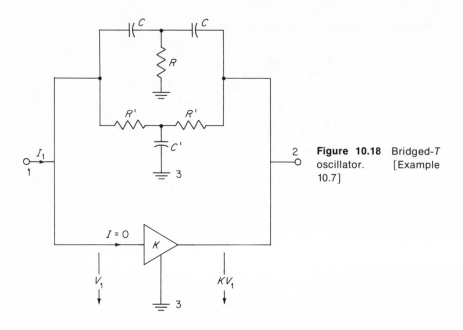

2

Figure 10.18 Bridged-*T* oscillator. [Example 10.7]

Solution   Let the admittance matrix of the *RC* circuit be

$$\mathbf{Y}_{RC} = \begin{bmatrix} y_{11} & y_{12} & y_{13} \\ y_{21} & y_{22} & y_{23} \\ y_{31} & y_{32} & y_{33} \end{bmatrix}$$ (10.32)

and that of the operational amplifier be given as in Equation (10.28). The overall admittance matrix is the sum of the two. Noting that the terminal conditions are

$$I_1 = 0$$
$$V_2 = KV_1$$
$$V_3 = 0$$

and also that in practice, G_{in} is much smaller than $|y_{11}|$ at the frequency of oscillation, we have

$$y_{11} + Ky_{12} = 0$$ (10.33)

Using the expressions for y_{11} and y_{12} found in Example 10.3 and equating the real and imaginary parts separately to zero, we find that when $K \gg 1$, the frequency of oscillation is

$$\omega_0 = \frac{1}{\sqrt{2RC^2R'}}$$ (10.34)

and the condition satisfied by the elements is

$$4RC = R'C'$$ (10.35)

■

The foregoing examples show how we can analyze a large circuit by parts. The circuit is decomposed into a small number of 3-terminal circuits. The admittance matrix of each is obtained. Each is expanded by adding a row and column of zeros. The overall matrix is found by adding the component matrices and by eliminating the unwanted terminals by pivotal condensation. In each step, the number of admittances that must be handled is small. If we had obtained the overall admittance matrix directly from the large circuit (Figure 10.2), we would have to deal with a large matrix whose elements are difficult to find. Computer programs that are based on analysis by decomposition, addition, and pivotal condensation have been written. In some programs, the admittance parameters of standard circuits such as transistors are stored so that the user need only specify the types of transistors and the terminals to which each is connected.

10.6 Impedance Parameters

Characterization of an N-terminal circuit in terms of its indefinite admittance parameters is seen, from the foregoing, to be a convenient way of analyzing a circuit composed of multiterminal circuits. However, it is not the only way. Moreover, there are situations in which the parameters do not exist. An example is the circuit shown in Figure 10.19. For this and other circuits whose admittance parameters are not defined or are difficult to find, we must seek a different characterization. In what follows, we give three other modes of characterization. We consider 3-terminal circuits only. The generalization to N-terminal is straightforward and will be omitted.

With reference to Figure 10.20, suppose we regard the terminal voltages V_1 and V_2 as responses, and currents I_1 and I_2 as excitations. By superposition, we can write

$$V_1 = z_{11}I_1 + z_{12}I_2$$
$$V_2 = z_{21}I_1 + z_{22}I_2 \qquad (10.36)$$

It is not necessary to use V_3 and I_3 in the characterization, since by KVL, $V_3 = V_1 - V_2$, and by KCL $I_3 = I_1 + I_2$. The four parameters, z_{11}, z_{12}, z_{21}, and z_{22}, are known as the *open-circuit impedance parameters* of N. They

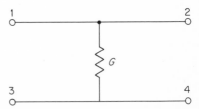

Figure 10.19 A circuit whose indefinite admittance parameters do not exist.

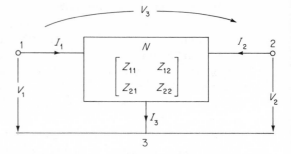

may be found as follows. By noting that by setting currents I_1 and I_2 to
zero in turn (opening terminal pairs 1-3 and 2-3), we obtain

$$z_{11} = \frac{V_1}{I_1}\bigg]_{I_2=0}$$

$$z_{12} = \frac{V_1}{I_2}\bigg]_{I_1=0}$$

$$z_{21} = \frac{V_2}{I_1}\bigg]_{I_2=0}$$

$$z_{22} = \frac{V_2}{I_2}\bigg]_{I_1=0} \tag{10.37}$$

Impedance z_{11} is seen to be the input impedance across 1-3 with 2-3
open, and z_{22} is the input impedance across 2-3 with 1-3 open. The
parameter z_{12} is a measure of the open-circuit voltage across 1-3 when a
current I_2 is applied at 2-3, and z_{21} is a measure of the open-circuit
voltage across 2-3 when a current I_1 is applied at 1-3.

Example 10.8 Obtain the open-circuit impedance parameters of the circuit
shown in Figure 10.21.

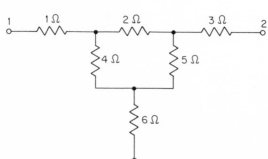

Figure 10.21 Open-circuit impedance parameters of a
3-terminal circuit. [Example
10.8]

Solution Apply a current source across the terminal pair 1-3 and leave 2-3 open. By inspection, we get

$$z_{11} = \frac{V_1}{I_1}\bigg]_{I_2=0} = \frac{105}{11}\ \Omega$$

$$z_{21} = \frac{V_2}{I_1}\bigg]_{I_2=0} = \frac{86}{11}\ \Omega$$

Applying a current source across 2-3 and leaving 1-3 open, we find

$$z_{12} = \frac{V_1}{I_2}\bigg]_{I_1=0} = \frac{86}{11}\ \Omega$$

$$z_{22} = \frac{V_2}{I_2}\bigg]_{I_1=0} = \frac{129}{11}\ \Omega$$

The 3-terminal circuit is, therefore, characterized by the equations

$$V_1 = \frac{105}{11}\ I_1 + \frac{86}{11}\ I_2$$

$$V_2 = \frac{86}{11}\ I_1 + \frac{129}{11}\ I_2$$

Note that $z_{12} = z_{21}$. We shall show later that under certain conditions, this is always true.

Example 10.9 A 3-terminal circuit is embedded in a circuit as shown in Figure 10.22. Find V_0/E and the input impedance Z_{in}.

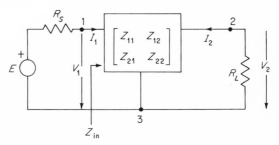

Figure 10.22 A 3-terminal circuit terminated in R_S and R_L. [Example 10.9]

Solution The external circuit imposes a set of conditions at the terminal pairs 1-3 and 2-3, namely,

$$V_1 = E - R_s I_1$$
$$V_2 = -R_L I_2$$

The 3-terminal circuit is characterized by

$$V_1 = z_{11}I_1 + z_{12}I_2$$
$$V_2 = z_{21}I_1 + z_{22}I_2$$

Making substitution and simplifying, we get

$$\frac{V_2}{E} = \frac{z_{21}R_L}{(R_s + z_{11})(R_L + z_{22}) - z_{12}z_{21}}$$

and

$$Z_{in} = \frac{V_1}{I_1} = \frac{z_{11}R_L + z_{11}z_{22} - z_{12}z_{21}}{z_{22} + R_L}$$

Thus, once we have the characterization of the 3-terminal circuit, the network functions of the overall network can be expressed in terms of the impedance parameters. In a way, the parameters z_{11}, z_{12}, z_{21}, and z_{22} are the generalization of the Thévenin equivalent impedance of a 2-terminal circuit.

Example 10.10 Obtain the open-circuit impedance of the circuit of Figure 10.23.

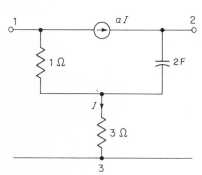

Figure 10.23 Open-circuit impedance parameters of a circuit containing a capacitor. [Example 10.10]

Solution With reference to Figure 10.23, we find, by inspection,

$$z_{11} = 4 - \alpha$$
$$z_{21} = 3 + \frac{\alpha}{j\omega 2}$$
$$z_{12} = 3 - \alpha$$
$$z_{22} = 3 + \frac{1 + \alpha}{j\omega 2}$$

In this case, $z_{21} \neq z_{12}$.

10.7 Reciprocity

We noted in Examples 10.8 and 10.10 that in one case $z_{12} = z_{21}$ and in the other $z_{12} \neq z_{21}$. These results are not accidental. We now wish to derive the condition under which $z_{12} = z_{21}$. A 3-terminal circuit in which this is true is called a *reciprocal* circuit (with respect to terminal pairs 1-3 and 2-3). Its significance will be pointed out later.

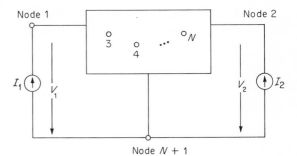

Figure 10.24 A circuit used in the proof of reciprocity.

Node 1   Node 2

Node $N + 1$

Consider the 3-terminal circuit of Figure 10.24. Let there be $n + 1$ nodes, and number the nodes as shown. The node equations of the circuit are

$$Y_{11}V_1 + Y_{12}V_2 + \cdots + Y_{1n}V_n = I_1$$
$$Y_{21}V_1 + Y_{22}V_2 + \cdots + Y_{2n}V_n = I_2$$
$$Y_{31}V_1 + Y_{32}V_2 + \cdots + Y_{3n}V_n = 0$$
$$\vdots$$
$$Y_{n1}V_1 + Y_{n2}V_2 + \cdots + Y_{nn}V_n = 0 \qquad (10.38)$$

Solving for V_1 and V_2, we have

$$V_1 = \frac{\Delta_{11}}{\Delta} I_1 + \frac{\Delta_{21}}{\Delta} I_2$$

$$V_2 = \frac{\Delta_{12}}{\Delta} I_1 + \frac{\Delta_{22}}{\Delta} I_2 \qquad (10.39)$$

where Δ is the determinant of the coefficients of the node equations, and Δ_{ij} is the i, j cofactor of Δ. Comparing the above with Equation (10.36), we get

$$z_{11} = \frac{\Delta_{11}}{\Delta}$$

$$z_{12} = \frac{\Delta_{21}}{\Delta}$$

$$z_{21} = \frac{\Delta_{12}}{\Delta}$$

$$z_{22} = \frac{\Delta_{22}}{\Delta} \qquad (10.40)$$

and the following observation:

The necessary and sufficient condition that a 3-terminal circuit be reciprocal with respect to its terminal pairs 1-3 and 2-3 is that $\Delta_{12} = \Delta_{21}$.

Now we recall that in a circuit containing resistors, capacitors, inductors, and mutual inductors, the coefficients of the node equations are symmetric, namely, $Y_{ij} = Y_{ji}$. It follows that $\Delta_{12} = \Delta_{21}$. Thus we conclude that every $RCLM$ circuit is reciprocal.

The converse is not true, however. Not every reciprocal circuit is an $RCLM$ circuit. For example, z_{21} and z_{12} of the circuit of Figure 10.25 are given, respectively, by

$$z_{21} = 3 + (1 + \mu)\alpha$$
$$z_{12} = 3 - \alpha + \mu(1 + \alpha) \tag{10.41}$$

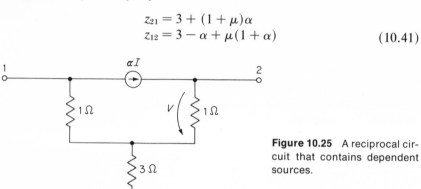

Figure 10.25 A reciprocal circuit that contains dependent sources.

The circuit is reciprocal if $\mu = 2\alpha$.

The significance of reciprocity can be deduced from the definition of the parameters z_{12} and z_{21} as given in Equation (10.37). In words, z_{12} is the voltage one would measure across 1-3 when a current source of 1 A is connected across 2-3, and z_{21} is the voltage across 2-3 when a current source is connected across 1-3. Invoking homogeneity, we may, therefore, make the following statement:

> Let N be a 3-terminal reciprocal circuit with respect to terminal pairs 1-3 and 2-3. Let a current source be connected across 1-3 and a voltmeter across 2-3. The reading of the voltmeter is unchanged when the current source and the voltmeter are exchanged in place.

10.8 Short-Circuit Admittance Parameters

If in Section 10.6 we replace every quantity (voltage, current, impedance) by its dual quantity, we obtain a characterization in terms of the *short-circuit admittance parameters*. Specifically, we now have, with reference to Figure 10.26,

$$I_1 = y_{11}V_1 + y_{12}V_2$$
$$I_2 = y_{21}V_1 + y_{22}V_2 \tag{10.42}$$

and we note immediately that these parameters are the same as the indefinite admittance parameters, but the characterization now is without the superfluous variables I_3 and V_3 $(= 0)$.

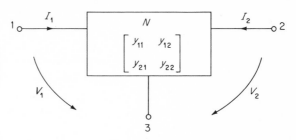

Figure 10.26 A 3-terminal circuit characterized by its short-circuit admittance parameters.

Equations (10.42) and (10.36) give two different characterizations of the same circuit. The impedance and admittance parameters must, therefore, be related. In Equation (10.42), solving for V_1 and V_2 in terms of the currents and comparing with Equation (10.36), we have

$$z_{11} = \frac{y_{22}}{y_{11}y_{22} - y_{12}y_{21}}$$

$$z_{12} = \frac{-y_{12}}{y_{11}y_{22} - y_{12}y_{21}}$$

$$z_{21} = \frac{-y_{21}}{y_{11}y_{22} - y_{12}y_{21}}$$

$$z_{22} = \frac{y_{11}}{y_{11}y_{22} - y_{12}y_{21}} \tag{10.43}$$

From this we see that $z_{12} = z_{21}$ implies $y_{12} = y_{21}$ and conversely. The physical significance of the statement $y_{12} = y_{21}$ is that in a 3-terminal circuit that is reciprocal with respect to terminal pairs 1-3 and 2-3, the reading of an ammeter connected across 2-3, when a voltage source is connected across 1-3, is unchanged if the ammeter and voltage source are exchanged in place.

Example 10.11 Find the network function $V_2/E = H(j\omega)$ of the circuit of Figure 10.27(a).

Solution Let us obtain a characterization of the 3-terminal LC circuit first. By inspection, we have

$$y_{11} = y_{22} = \frac{1 - 18\omega^2}{j2\omega}$$

$$y_{12} = y_{21} = \frac{16\omega^2 - 1}{j2\omega}$$

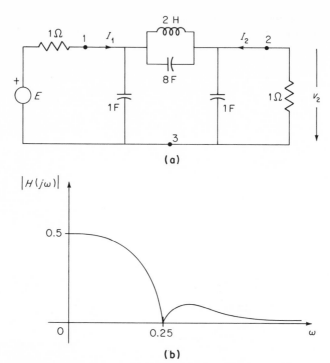

Figure 10.27   **(a)** A low-pass filter; **(b)** its frequency response.
[Example 10.11]

The terminal conditions at the input and output terminal pairs are

$$E = I_1 + V_1$$
$$V_2 = -I_2$$
$$\frac{V_2}{E} = \frac{-y_{21}G_s}{(G_s + y_{11})(G_L + y_{22}) - y_{12}y_{21}}$$
$$= \frac{1 - 16\omega^2}{(2 - 36\omega^2) + j(4\omega - 34\omega^3)}$$

Note that the output voltage is zero at $\omega = \frac{1}{4}$ rad/sec, which is a zero of y_{21} and the resonant frequency of the parallel LC circuit. The frequency characteristic of the circuit is sketched in Figure 10.27(b), which shows that the circuit is a low-pass filter with a sharp cutoff feature around $\omega = \frac{1}{4}$ rad/sec.

10.9   Equivalent Circuits

Thévenin and Norton equivalent circuits, which have been found to be so useful in analysis, can now be extended to the case of a 3-terminal circuit. We begin with the terminal equations in terms of the open-circuit impedance parameters:

$$V_1 = z_{11}I_1 + z_{12}I_2$$
$$V_2 = z_{21}I_1 + z_{22}I_2 \qquad (10.44)$$

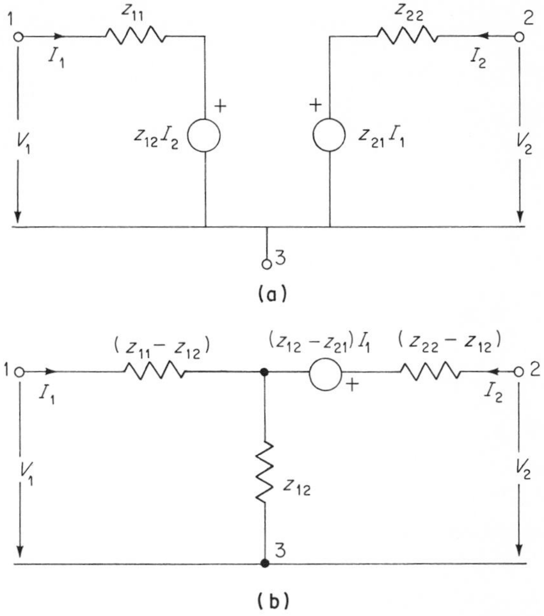

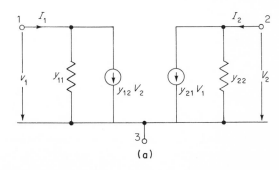

Figure 10.28 Two possible equivalent circuits of a 3-terminal circuit characterized by its impedance parameters.

(a)

(b)

and deduce immediately an equivalent circuit shown in Figure 10.28(a). It has two dependent sources. Another equivalent circuit is that shown in Figure 10.28(b). That these are equivalent circuits can be seen if one writes voltage equations around the two loops and notes that they are the same as Equation (10.44).

The equivalent circuit consists of two-terminal elements. Thus we have succeeded in replacing a 3-terminal circuit by a circuit of impedances and dependent sources. If in a large circuit composed of a collection of 3-terminal circuits we replace each by its equivalent, we get a circuit composed of only two-terminal elements. The analysis of the overall circuit can now be carried out by means of node, loop, or mesh equations. In a dual manner, if we use the characterization in terms of the short-circuit admittance parameters, we obtain the equivalent circuits shown in Figures 10.29(a) and (b).

Figure 10.29 Two possible equivalent circuits of a 3-terminal circuit characterized by its admittance parameters.

(a)

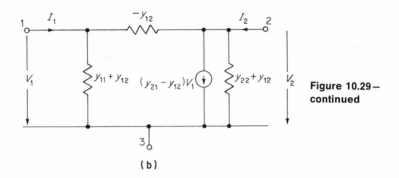

Figure 10.29—
continued

(b)

10.10   Hybrid Parameters

A set of parameters that is widely used in the characterization of transistors is that of *hybrid* parameters. The terminal voltages and currents are related as follows:

$$V_1 = h_{11}I_1 + h_{12}V_2$$
$$I_2 = h_{21}I_1 + h_{22}V_2 \qquad (10.45)$$

As before, the parameters are given physical interpretation by the way in which each is determined. Thus

$$h_{11} = \frac{V_1}{I_1}\bigg]_{V_2=0} = \text{(input impedance at 1-3 with 2-3 short-circuited)}$$

$$h_{12} = \frac{V_1}{V_2}\bigg]_{I_1=0} = \text{(ratio of voltage at 1-3 to voltage at 2-3 with 1-3 open)}$$

$$h_{21} = \frac{I_2}{I_1}\bigg]_{V_2=0} = \text{(ratio of current in short-circuit at 2-3 to current applied at 1-3)}$$

$$h_{22} = \frac{I_2}{V_2}\bigg]_{I_1=0} = \text{(input admittance at 2-3 with 1-3 open)} \quad (10.46)$$

Example 10.12   A transistor is found to have a terminal characterization given by

$$V_1 = 1400I_1 + 3.37 \times 10^{-4}V_2$$
$$I_2 = \quad 44I_1 + \quad 27 \times 10^{-6}V_2$$

It is embedded in another circuit as shown in Figure 10.30. Find a value of G such that the gain V_2/E is -1000.

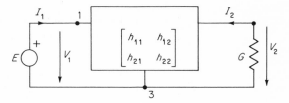

Figure 10.30 A circuit characterized by its hybrid parameters and terminated in a source E and load G.

Solution The terminal equations are

$$V_1 = E = h_{11}I_1 + h_{12}V_2$$
$$I_2 = -GV_2 = h_{21}I_1 + h_{22}V_2$$

Solving for V_2, we find

$$\frac{V_2}{E} = \frac{h_{21}}{h_{21}h_{12} - h_{11}(h_{22} + G)}$$

Making substitutions and setting V_2/E to -1000, we get

$$G = 15 \times 10^{-6} \text{ mhos}$$

10.11 *ABCD* Parameters

Many filter circuits consist of a cascade connection of 3-terminal circuits. The characterization of the latter is best done in terms of the set of *ABCD* parameters, also known as the *line* parameters. They relate the output voltage and current to the input voltage and current. Thus, with reference to Figure 10.31, we write

$$V_1 = AV_2 + BI_2$$
$$I_1 = CV_2 + DI_2 \qquad (10.47)$$

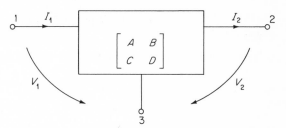

Figure 10.31 A circuit characterized by its *ABCD* parameters.

Note that the direction of I_2 is reversed from what it was in the previous characterizations. This change is deliberate, and the reason for it will be apparent shortly.

The determination of the *ABCD* parameters is done in the usual way, except that it is the reciprocal of each that is found, as follows.

$$\frac{1}{A} = \frac{V_2}{V_1}\bigg]_{I_2=0} = \begin{matrix} \text{(ratio of voltage at output to that at} \\ \text{input with output open)} \end{matrix}$$

$$\frac{1}{B} = \frac{I_2}{V_1}\bigg]_{V_2=0} = \begin{matrix} \text{(ratio of current at output to voltage} \\ \text{applied at input with output shorted)} \end{matrix}$$

$$\frac{1}{C} = \frac{V_2}{I_1}\bigg]_{I_2=0} = \begin{matrix} \text{(ratio of voltage at output to current} \\ \text{applied at input with output open)} \end{matrix}$$

$$\frac{1}{D} = \frac{I_2}{I_1}\bigg]_{V_2=0} = \begin{matrix} \text{(ratio of current at output to current} \\ \text{applied at input with output shorted)} \end{matrix} \qquad (10.48)$$

It is seen that the excitations are the current and voltage at the input terminals, and the responses are taken at the output under open- or short-circuit condition. The parameters A and D are dimensionless; B has a dimension of ohms, and C mhos.

The $ABCD$ parameters are related to the open-circuit impedances and short-circuit admittances. The relations are given below.

$$A = \frac{z_{11}}{z_{21}} = \frac{-y_{22}}{y_{21}}$$

$$B = \frac{z_{11}z_{22} - z_{12}z_{21}}{z_{21}} = \frac{-1}{y_{21}}$$

$$C = \frac{1}{z_{21}} = \frac{-(y_{11}y_{22} - y_{12}y_{21})}{y_{21}}$$

$$D = \frac{z_{22}}{z_{21}} = \frac{-y_{11}}{y_{21}} \qquad (10.49)$$

From the above, we deduce that in a reciprocal circuit, $AB - CD = 1$.

Example 10.13 Find the $ABCD$ parameters of each of the circuits shown in Figure 10.32.

Solution Applying the definitions as given in Equations (10.48), we find for (a)

$$\begin{bmatrix} V_1 \\ I_1 \end{bmatrix} = \begin{bmatrix} 1 & Z \\ 0 & 1 \end{bmatrix} \begin{bmatrix} V_2 \\ I_2 \end{bmatrix} \qquad (10.50)$$

and for (b)

$$\begin{bmatrix} V_1 \\ I_1 \end{bmatrix} = \begin{bmatrix} 1 & 0 \\ Y & 1 \end{bmatrix} \begin{bmatrix} V_2 \\ I_2 \end{bmatrix} \qquad (10.51)$$

Note that the open-circuit impedance parameters do not exist for circuit (a), and the short-circuit admittance parameters do not exist for circuit (b).

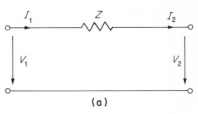

Figure 10.32 The *ABCD* parameters of two simple circuits. [Example 10.13]

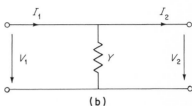

10.12 Circuits in Cascade

ABCD parameters are most useful in the characterization of a 3-terminal circuit composed of other 3-terminal circuits connected in cascade as shown in Figure 10.33. The *ABCD* parameters of the overall circuit can be obtained simply from those of the individual circuits.

Let each component circuit be characterized by

$$
\begin{bmatrix} V_{1(k)} \\ I_{1(k)} \end{bmatrix} = \begin{bmatrix} A_k & B_k \\ C_k & D_k \end{bmatrix} \begin{bmatrix} V_{2(k)} \\ I_{2(k)} \end{bmatrix} \qquad k = 1,\ldots, N
$$

(10.52)

Matching the voltages and currents at the terminals, we get

$$
\begin{bmatrix} V_1 \\ I_1 \end{bmatrix} = \begin{bmatrix} A_1 & B_1 \\ C_1 & D_1 \end{bmatrix} \begin{bmatrix} V_{2(1)} \\ I_{2(1)} \end{bmatrix} = \begin{bmatrix} A_1 & B_1 \\ C_1 & D_1 \end{bmatrix} \begin{bmatrix} A_2 & B_2 \\ C_2 & D_2 \end{bmatrix} \begin{bmatrix} V_{2(2)} \\ I_{2(2)} \end{bmatrix} = \cdots
$$

$$
= \prod_{k=1}^{N} \begin{bmatrix} A_k & B_k \\ C_k & D_k \end{bmatrix} \begin{bmatrix} V_2 \\ I_2 \end{bmatrix}
$$

(10.53)

Figure 10.33 *N* 3-terminal circuits connected in cascade.

The overall $ABCD$ matrix is the product of the $ABCD$ matrices of the component circuits. This observation leads to a simple scheme to find the network functions of circuits of the type shown in Figure 10.33.

Applying the terminal conditions at the input and output, we find the input impedance Z_{in} to be

$$Z_{in} = \frac{V_1}{I_1} = \frac{AZ_L + B}{CZ_L + D} \qquad (10.54)$$

and voltage gain to be

$$H = \frac{V_2}{E} = \frac{1}{Z_s(C + D/Z_L) + A + B/Z_L}$$

In the special case of a ladder network, the $ABCD$ matrix of the overall circuit is, with reference to Figure 10.34 and Equations (10.50) and (10.51),

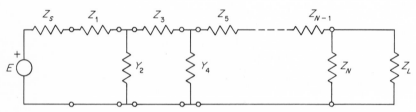

Figure 10.34 A ladder network.

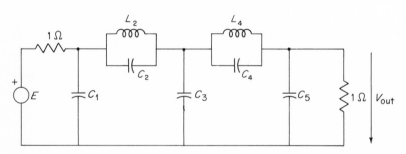

$C_1 = 0.89318\ F$

$C_2 = 0.1022\ F$ $\qquad\qquad L_2 = 1.26033\ H$

$C_3 = 1.57677\ F$

$C_4 = 0.29139\ F$ $\qquad\qquad L_4 = 1.03950\ H$

$C_5 = 0.74177\ F$

Figure 10.35 An elliptic low-pass filter.

$$
\begin{bmatrix} A & B \\ C & D \end{bmatrix} = \begin{bmatrix} 1 & Z_1 \\ 0 & 1 \end{bmatrix} \begin{bmatrix} 1 & 0 \\ Y_2 & 1 \end{bmatrix} \begin{bmatrix} 1 & Z_3 \\ 0 & 1 \end{bmatrix} \begin{bmatrix} 1 & 0 \\ Y_4 & 1 \end{bmatrix} \cdots \qquad (10.55)
$$

The ladder network is regarded as a cascade connection of N 3-terminal circuits, each consisting of either a series impedance or a shunt admittance.

A FORTRAN program to compute the frequency response of a low-pass filter shown in Figure 10.35 is given below. The $ABCD$ parameters of each section are defined in the subroutine $ABCD(...)$. In the program N is the number of sections, NF the number of frequency points at which calculation is desired, $FONE$ the starting frequency, FM a frequency factor by which the current frequency value is multiplied or divided to obtain the next frequency value, ZL the load impedance, ZS the source impedance, ZIN the input impedance to be found, and $VOUT$ the output voltage.

```
C FREQUENCY RESPONSE OF ELLIPTIC FILTER
C NOTE: F IN RADIANS PER SECOND.
      COMPLEX A,B,C,D,ATEM,BTEM,CTEM,DTEM
      COMPLEX ASEC,BSEC,CSEC,DSEC,ZL,ZS,ZIN,VOUT
      DATA N,NF,FONE,FM,ZL,ZS/5,100,1.,1.05,(1.,0.),(1.,0.)/
      WRITE(6,99)
      NFF=NF/2
      DO 1 K=1,2
      F=FONE
      DO 1 I=1,NFF
      GO TO (11,12),K
   11 F=F*FM
      GO TO 13
   12 F=F/FM
   13 CONTINUE
      ATEM=1.
      BTEM=0.
      CTEM=0.
      DTEM=1.
      DO 2 J=1,N
      CALL ABCD(J,F,ASEC,BSEC,CSEC,DSEC)
      A=ATEM*ASEC + BTEM*CSEC
      B=ATEM*BSEC + BTEM*DSEC
      C=CTEM*ASEC + DTEM*CSEC
      D=CTEM*BSEC + DTEM*DSEC
      ATEM=A
      BTEM=B
      CTEM=C
      DTEM=D
    2 CONTINUE
      ZIN=(A*ZL+B)/(C*ZL+D)
      VOUT=1./(ZS*(C+D/ZL)+A+B/ZL)
      VABS=CABS(VOUT)
      VPHASE=ATAN(AIMAG(VOUT)/REAL(VOUT))*57.29577
    1 WRITE(6,100)F,VABS,VPHASE,ZIN
  100 FORMAT(1H ,5E16.6)
   99 FORMAT(1H ,'FREQUENCY RAD/SEC',6X,'VOUT AMP',8X,'VOUT PHASE',
     C   6X,'ZIN REAL',8X,'ZIN IMAG')
      STOP
      END
      SUBROUTINE ABCD(I,W,A,B,C,D)
      COMPLEX A,B,C,D,WJ
      WJ=CMPLX(0.,W)
      GO TO (1,2,3,4,5),I
```

The computer output is plotted in Figure 10.36. It is seen that the circuit is indeed a low-pass filter with an interesting characteristic: It approximates a constant 0.5 for frequencies less than 1.0 and a constant 0.0 for frequencies greater than 1.0. The approximation is optimum in the sense that all extremal values in the pass band are the same, and all extremal values in the stop band are the same. The filter belongs to the class of filters known as elliptic function filters. The frequencies at which the

```
1 A=1.
  B=0.
  C=WJ*0.89318
  D=1.
  RETURN
2 A=1.
  B=1./(WJ*0.1022+1./(WJ*1.26033))
  C=0.
  D=1.
  RETURN
3 A=1.
  B=0.
  C=WJ*1.57677
  D=1.
  RETURN
4 A=1.
  B=1./(WJ*0.29139+1./(WJ*1.0395))
  C=0.
  D=1.
  RETURN
5 A=1.
  B=0.
  C=WJ*0.74177
  D=1.
  RETURN
  END
```

FREQUENCY RAD/SEC	VOUT AMP	VOUT PHASE	ZIN REAL	ZIN IMAG
0.105000E 01	0.478858E 00	-0.423244E 02	0.131916E 01	0.611843E 0●
0.110250E 01	0.421958E 00	-0.678745E 02	0.242364E 01	0.137511E 0:
0.115762E 01	0.322448E 00	0.853348E 02	0.685892E 01	-0.205088E 0:
0.121550E 01	0.216821E 00	0.619877E 02	0.851159E 00	-0.383125E 0●
0.127627E 01	0.136146E 00	0.437458E 02	0.124566E 00	-0.233573E 0:
0.134009E 01	0.829711E-01	0.297140E 02	0.276786E-01	-0.172175E 0●
0.140709E 01	0.495479E-01	0.186227E 02	0.724914E-02	-0.139218E 0.
0.147744E 01	0.287739E-01	0.954660E 01	0.198868E-02	-0.118221E 0:
0.155131E 01	0.159086E-01	0.189310E 01	0.523955E-03	-0.103359E 0:
0.162888E 01	0.800253E-02	-0.471376E 01	0.118445E-03	-0.920927E 00
0.171032E 01	0.323193E-02	-0.105198E 02	0.175861E-04	-0.831392E 00
0.179583E 01	0.456107E-03	-0.156919E 02	0.302901E-06	-0.757771E 00
0.188562E 01	0.104920E-02	-0.203482E 02	0.160660E-05	-0.695682E 00
0.197990E 01	0.174957E-02	-0.245751E 02	0.434623E-05	-0.642291E 00
0.207890E 01	0.194902E-02	-0.284380E 02	0.519131E-05	-0.595665E 00
0.218284E 01	0.184429E-02	-0.319875E 02	0.442217E-05	-0.554448E 00
0.229198E 01	0.156528E-02	-0.352637E 02	0.310254E-05	-0.517641E 00
0.240657E 01	0.119641E-02	-0.382995E 02	0.159300E-05	-0.464498E 0●
0.252690E 01	0.792092E-03	-0.411214E 02	0.642225E-06	-0.454444E 00
0.265324E 01	0.386686E-03	-0.437520E 02	0.151235E-06	-0.427028E 00
0.278590E 01	0.116729E-05	-0.462100E 02	-0.513160E-07	-0.401893E 00
0.292520E 01	0.352451E-03	-0.485116E 02	0.104254E-06	-0.378747E 00
0.307145E 01	0.668081E-03	-0.506708E 02	0.422839E-06	-0.357352E 00
0.322502E 01	0.943519E-03	-0.526995E 02	0.909654E-06	-0.337511E 00
0.338627E 01	0.117904E-02	-0.546085E 02	0.156883E-05	-0.319057E 00
0.355558E 01	0.137646E-02	-0.564069E 02	0.213046E-05	-0.301850E 00
0.373336E 01	0.153847E-02	-0.581031E 02	0.258991E-05	-0.285770E 00
0.392002E 01	0.166824E-02	-0.597045E 02	0.305441E-05	-0.270713E 00
0.411602E 01	0.176907E-02	-0.612177E 02	0.335536E-05	-0.256589E 00
0.432182E 01	0.184429E-02	-0.626487E 02	0.360147E-05	-0.243321E 00
0.453790E 01	0.189707E-02	-0.640029E 02	0.380916E-05	-0.230839E 0●

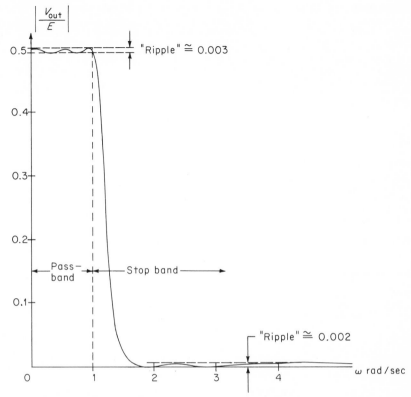

Figure 10.36 The frequency response of the filter of Figure 10.35.

extremal values occur are given in terms of elliptic functions of a parameter which is related to the sharpness of the cutoff characteristic.

10.13 Summary

The four modes of characterization of a multiterminal circuit that have been presented so far are not all of the possible ones that one can use. For practical purposes, however, they are quite adequate. As to which mode is preferred, it would depend on how the multiterminal circuit is connected to the rest of the circuit and on the ease with which the parameters can be determined. The hybrid parameters are suitable for transistors because they can be measured directly. If a circuit is a cascade connection of 3-terminal circuits, the characterization should be in terms of the *ABCD* parameters. In the general case in which no discernible pattern of connection exists, then the indefinite admittance matrix should be used. In all cases, our strategy is to analyze a large circuit by parts, thereby avoiding the necessity of writing and solving a large system of equations.

PROBLEMS

10.1 *Exercise* Find the indefinite admittance parameter of each of the following circuits.

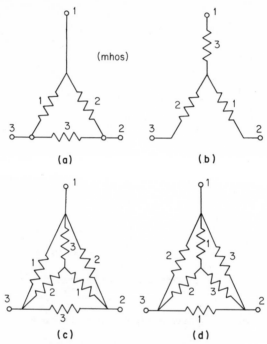

(a)   (b)

(c)   (d)

Problem 10.1   Indefinite admittance matrix.

10.2 *Exercise* Find the indefinite admittance matrix of the following circuit.,

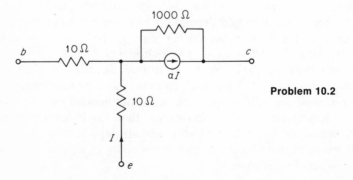

Problem 10.2

10.3 *Exercise* Each of the 3-terminal circuits shown in the figure is that of Problem 10.2. Find the gain V_0/E.

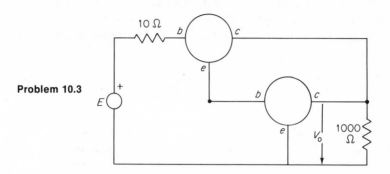

Problem 10.3

10.4 *Exercise* Find the input resistance and gain of the circuit.

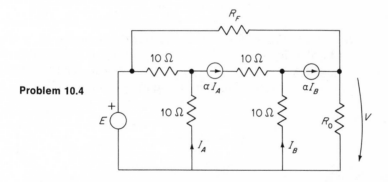

Problem 10.4

10.5 *Active RC Filter* Show that the transfer function of the following circuit is that of a low-pass filter given by $H(s) = K/(As^2 + Bs + 1)$ with $A = R_1 C_1 R_2 C_2$ and $B = R_2 C_2 + R_1 C_2 + R_1 C_1(1 - K)$. Select a suitable set of element values so that the transfer function is $1/(s^2 + \sqrt{2}s + 1)$. Sketch the frequency characteristics.

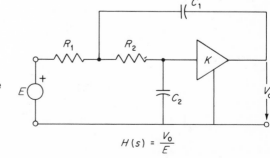

Problem 10.5 An *RC* active filter.

$$H(s) = \frac{V_0}{E}$$

10.6   *Active RC Filter*   Show that the transfer function of the following circuit is that of a high-pass filter given by $H(s) = Cs^2/(As^2 + Bs + 1)$, with $C = KR_1C_1R_2C_2$, $A = R_1C_1R_2C_2$, and $B = R_1C_1 + R_1C_2 + R_2C_2(1 - K)$. Select a set of element values to realize $H(s) = s^2/(s^2 + \sqrt{2}s + 1)$. Sketch the frequency characteristics.

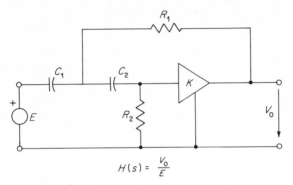

$$H(s) = \frac{V_0}{E}$$

Problem 10.6   An active RC filter.

10.7   *Active RC Filter*   Show that the transfer function is that of a third-order low-pass filter given by $H(s) = \lambda_1\lambda_2\lambda_3/(s^3 + \lambda_1s^2 + \lambda_1\lambda_2s + \lambda_1\lambda_2\lambda_3)$ with $\lambda_1 = 1/R_1C_1$, $\lambda_2 = 1/R_2C_2$, and $\lambda_3 = 1/R_3C_3$. Realize the transfer function of the low-pass filter of Section 8.7.

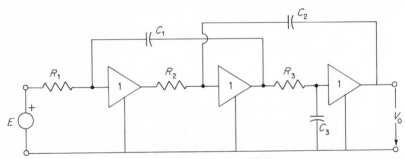

Problem 10.7   An active RC filter.

10.8   *Transistor Oscillator*   A transistor has an equivalent circuit as

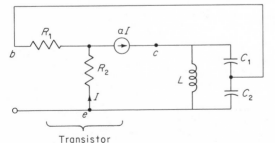

Problem 10.8   A transistor oscillator.

shown. It is connected to an LC circuit to form an oscillator.
Find the frequency of oscillation.

10.9 *Equivalent Circuit* A 3-terminal circuit N is characterized by
its hybrid parameters. $V_1 = h_{11}I_1 + h_{12}V_2$ and $I_2 = h_{21}I_1 + h_{22}V_2$.
Show that the circuit shown in the figure is an equivalent circuit
of N.

Problem 10.9 An equivalent circuit based on the
hybrid parameters.

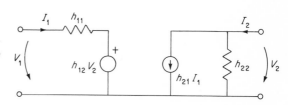

10.10 *Exercise* A 3-terminal circuit N_1 is characterized by $V_1 = 2I_1$
$+ 3I_2$ and $V_2 = 3I_1 + 5I_2$. Another one N_2 is characterized by
$V_1 = 2V_2 - 3I_2$ and $I_1 = 3V_2 - 5I_2$. The two circuits are connected
in parallel. Find the $ABCD$ parameters of the resultant circuit.

10.11 *Reciprocity* Let N_1 and N_2 be two reciprocal 3-terminal circuits.
Show that the 3-terminal circuit, which is the cascade connec-
tion of N_1 and N_2, is reciprocal.

10.12 *Reciprocity* In circuit (a) of the figure, $I = 2$ A. Find the voltage
V_1 in circuit (b).

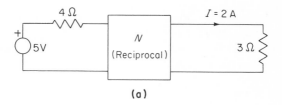

(a)

Problem 10.12

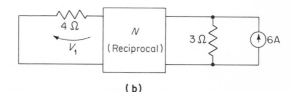

(b)

10.13 *Reciprocity* Determine whether or not the circuit shown in the figure is reciprocal with respect to terminal pairs 1-3 and 2-3 and with respect to 1-3 and 2-1.

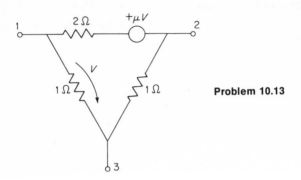

Problem 10.13

10.14 *Filter* Write a computer program to compute the frequency response of the maximally flat low-pass filter. Sketch $H(j\omega)$ and compare it with the exact expression $0.5/(1 + \omega^{10})^{0.5}$.

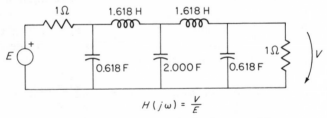

$$H(j\omega) = \frac{V}{E}$$

Problem 10.14   A maximally flat low-pass filter.

10.15 *Filter* The circuit shown is a low-pass filter whose amplitude response $H(j\omega)$ has the property that its extremal values are the same in the pass band. Write a computer program to calculate $H(j\omega)$ and compare with that of Problem 10.14.

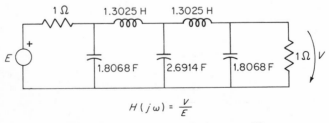

$$H(j\omega) = \frac{V}{E}$$

Problem 10.15   An equiripple low-pass filter.

10.16 *Filter* The circuit shown is an elliptic low pass filter. Compute the frequency response.

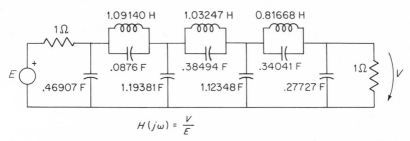

$$H(j\omega) = \frac{V}{E}$$

Problem 10.16 An elliptic low-pass filter.

10.17 *Fun Circuit* A 3-terminal circuit consists of infinite number of identical 3-terminal circuits in cascade, each being characterized by its *ABCD* parameters. Find the input impedance and the voltage at the output terminal pair of the second circuit.

10.18 *Fun Circuit* One hundred identical 3-terminal circuits are connected as shown. Find V_0/E. Each circuit is characterized by $V_1 = (12/7)I_1 + (11/7)I_2$ and $V_2 = (-1/7)I_1 + (2/7)I_2$.

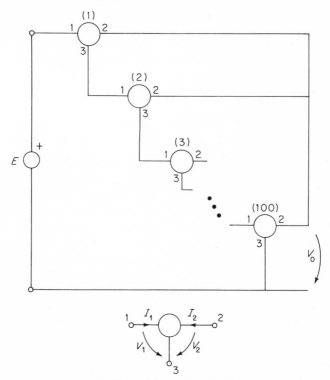

Problem 10.18 A fun circuit.

answers to selected problems

Chapter 1

1.3 $v_1 = -1$ V, $v_2 = 2$ V.
1.4 $v_4 = 2$ V, $v_5 = 3$ V.
1.6 No. The edges associated with these voltages do not form a tree.
1.7 $i_1 = -1$, $i_2 = -1$, $i_3 = -1$, $i_4 = 3$.
1.8 $i_1 = -A - B$, $i_4 = B - C$, $i_5 = A + B$, $i_7 = B - C$.
1.9 $i_1 = -7$, $i_2 = -3$, $i_3 = -4$, $i_4 = -5$.
1.11 Yes. v_6, v_2, v_4, and v_7 each appears in one, and only one, equation. $b - n + 1 = 4$. The set is a maximally independent set.
1.13 Equations are not independent. $Q_4 = Q_2 - Q_1$.
1.14 Yes. These are $n - 1$ KCL equations.
1.15 No. $Q_3 = Q_1 - Q_2$.
1.18 (a) F, (b) F, (c) F, (d) F, (e) T, (f) T.
1.19 (a) T, (b) F, (c) F, (d) F, (e) T.

Chapter 2

2.1 On R_1, $v_1 = R_1/(R_1 + R_2)e(t)$, $i_1 = e(t)/(R_1 + R_2)$. On R_2, $v_2 = R_2/(R_1 + R_2)e(t)$, $i_2 = i_1$.
2.2 On the 1-Ω resistor, from l to r, $v_1 = 4$ V, $i_1 = 4$ A. On the 2-Ω resistor, from t to b, $v_2 = -2$ V, $i_2 = -1$ A. On the 3-Ω resistor, from t to b, $v_3 = 5$ V, $i_3 = \frac{5}{3}$ A.
2.3 On the 1-Ω resistor, from l to r, $v_1 = 4$ V, $i_1 = 4$ A. On the $\frac{1}{2}$-Ω resistor, from t to b, $v_2 = -1$ V, $i_2 = -2$ A.

On the $\frac{1}{3}$-Ω resistor, from t to b, $v_3 = \frac{5}{3}$ V, $i_3 = 5$ A.
2.4 $e(t) = 4dv/dt + v$.
2.5 $j(t) = 4di/dt + i$.
2.6 The high-frequency component has an amplitude about 100 times larger than that of the low-frequency component.
2.7 The amplitude of the high-frequency component is about 1×10^6 times that of the low-frequency component.
2.8 1000, for pulse width = 1 μsec and height = 0.05 V.
2.9 $v = (RC)^{-1}\alpha/(\alpha - 1) \int_0^t e(u)\, du + v(0)$.
2.10 $v = \alpha L/[(\alpha - 1)R]\, de/dt$.
2.11 $e(t)$ is a square wave.
2.12 $e(t) \simeq -10^3 \cos 10^3 t - 10^3 \cos 10t$; high-pass filter.
2.13 Voltage across left inductor $= -0.425 \cos t$. Current in right inductor $= -10.5 \sin t$.
2.14 $v/e = -R_2/(-R_2 + (\alpha - 1)^3 R_1)$.
2.15 $v/e = \alpha^2 R_2/R_1(\alpha - 1)^2$.
2.16 Current in $\frac{1}{3}$-F capacitor $= \cos 3t$. Current in 2-Ω resistor $= \cos 3t$. Current in 2-H inductor $= \frac{1}{3} \sin 3t - \frac{1}{6} \cos 3t + \frac{1}{6}$. Current in 1-$\Omega$ resistor $= \frac{1}{3} \sin 3t + \frac{5}{6} \cos 3t + \frac{1}{6}$.
2.17 Voltage across 3000-Ω resistor $= 60 \, \alpha e/(\alpha - 2)$; $i = e/50(\alpha - 2)$.
2.18 $v_1 = 1/91$, $v_2 = -90/91$, $i_1 = 11/910$, $i_2 = -1/455$, $v_3 = 90/91$.

2.19 $v/E = -1$.

2.21 $i = 19/8 \sin \omega t - 9/16 \sin 3\omega t + 1/16 \sin 5\omega t$.

2.22 $i = \frac{3}{2}, 0 \leq t \leq \frac{2}{3}$; $i = 3, \frac{2}{3} < t \leq 2$; $i = -3, 2 < t \leq \frac{10}{3}$; $i = -\frac{3}{2}, \frac{10}{3} < t \leq 4$; $i = \frac{3}{4}, 4 < t \leq 6$; $i = -\frac{3}{4}, 6 < t \leq 8$. The waveform repeats itself with period $t = 8$ sec.

2.25 $\phi = \phi_1, 0 \leq t \leq 6$;
$\phi = \phi_1 + (\phi_2 - \phi_1)(t - 6)/2$, $6 < t \leq 8$;
$\phi = \phi_2, 8 < t \leq 16$;
$\phi = \phi_1 - (\phi_2 - \phi_1)(t - 18)/2$, $16 < t \leq 18$;
$\phi = \phi_1, 18 < t \leq 66$.
Waveform for $66 < t \leq 80$ is the same as that for $6 < t \leq 20$. Voltage is a positive rectangular pulse for $6 \leq t \leq 8$ and $66 \leq t \leq 68$, and a negative rectangular pulse for $16 \leq t \leq 18$ and $76 \leq t \leq 78$. It is zero elsewhere.

2.26 $i_{EA} = -\frac{1}{2}$ A, $i_{AD} = \frac{1}{2}$ A, $i_{BF} = -\frac{3}{2}$ A, $i_{BC} = \frac{1}{2}$ A, $i_{EC} = \frac{1}{2}$ A, $i_{DC} = -1$ A, $i_{FD} = -\frac{3}{2}$ A, $i_{AB} = -1$ A.

2.27 R_s; $e^2/4R$.

2.28 $n_1/n_2 = (R_s/R)^{1/2}$.

Chapter 3

3.1

	v_x (volts)	i_x (amperes)
(a)	$(-1/2)E$	$-(1/6)E$
(b)	$(3/7)E$	$(2/7)E$
(c)	$(8/23)E$	$(7/23)E$
(d)	$(1/5)E$	$(-2/5)E$

3.2

	v_x (volts)	i_x (amperes)
(a)	I	$(1/2)I$
(b)	$(8/7)I$	$(6/7)I$
(c)	$(1/2)I$	$(5/6)I$
(d)	0	0

3.3

	v_x (volts)	i_x (amperes)
(a)	$(3/11)E_1$ $- (9/11)E_2$	$(3/11)E_1$ $+ (2/11)E_2$
(b)	$(2/3)E$	$(1/3)E - I$

3.4

	v_x (volts)	i_x (amperes)
(a)	$-3/4$	$-3/2$
(b)	13	-2
(c)	$-10/3$	0
(d)	$-\frac{1}{3}(E + 10J)$	$\frac{1}{3}(E - 2J)$

3.5 $E = 250/37$ V.

3.6 $E = 9.25$ or 4.65 V.

3.7 $i_1 = (2e_1 - e_2 - e_3)/(3R_2 + R_1)$,
$i_2 = (-e_1 + 2e_2 - e_3)/(3R_2 + R_1)$,
$i_3 = (-e_1 - e_2 + 2e_3)/(3R_2 + R_1)$.

3.8 **(a)** Cable loss $= 2R \times 10^6$ W. **(b)** Cable loss $= 2R \times 10^2$ W. **(c)** System B. **(d)** $E = [2000(R + R_g) + 100]$ $\cos \omega t$. $E' = [20(R + R_g) + 10,000]$ $\cos \omega t$.

3.9 $v/E = R_2/(R_2 + R_1(1 - \alpha))$.

3.10 $v_{AB} = 43/128$ E, $v_{CD} = 11/64$ E, $v_{EF} = 3/32$ E, $v = 1/16$ E.

3.12 $R_{eq} = R_2(1 - \alpha)$, $v_{oc} = -\alpha R_2 E / R_1$.

3.13 $R_x = (R + \sqrt{R^2 + 4Rr})/2$.

3.21 Seven.

3.22 $v_{oc} = R_2/(R_2 + R'_g)E'$, $R_{eq} = R_2 R'_g$ $(1 - \alpha)/(R_2 + R'_g) + R_1$, with $R'_g = R_1$ $+ R_2 R_g(1 - \alpha)/(R_2 + R_g)$, $E' = R_2 E/$ $(R_2 + R_g)$.

3.23 **(a)** $V_1 = (4/7)E_s$, $V_2 = (2/7)E_s$, $I_1 = 0$, $I_2 = 0$. **(b)** $V_1 = 0$, $V_2 = 0$, I_1 $= (2/11)E_s$, $I_2 = (1/11)E_s$. **(c)** $V_1 = 16/7$, $V_2 = 8/7$, $I_1 = 2/11$, $I_2 = 1/11$.

3.24 **(a)** $V_1 = 4/7$, $V_2 = 0$, $I_1 = 0$, I_2 $= 1/7$. **(b)** $V_1 = 0$, $V_2 = 2/5$, $I_1 = 1/5$, I_2 $= 0$. **(c)** $V_1 = 16/7$, $V_2 = 2/5$, $I_1 = 1/5$, $I_2 = 4/7$.

3.27 $V = -1/3$ V, $R = 38/11 \ \Omega$.

Chapter 4

4.1 **(a)** $v_x(0) = E - V$, $v_x(\infty)$ $= \dfrac{R_1}{R_1 + R_2} E$; $i_x(0) = \dfrac{E}{R_1} - \dfrac{R_1 + R_2}{R_1 R_2} V$, $i_x(\infty) = 0$; $T = C \cdot \dfrac{R_1 R_2}{R_1 + R_2}$.

(b) $v_x(0) = \dfrac{R_2}{R_1 + R_2} E$, $v_x(\infty) = 0$; $i_x(0) = \dfrac{E}{R_1 + R_2}$, $i_x(\infty) = 0$; $T = C(R_1 + R_2)$.

(c) $v_x(0) = \dfrac{R_2}{R_1 + R_2} E - \dfrac{R_1 R_2}{R_1 + R_2} I$, $v_x(\infty) = 0$; $i_x(0) = \dfrac{E}{R_1 + R_2} - \dfrac{R_1}{R_1 + R_2} I$, $i_x(\infty) = 0$; $T = L(G_1 + G_2)$.

(d) $v_x(0) = JR_2$, $v_x(\infty) = \dfrac{R_2}{R_1 + R_2}$

$E + \dfrac{R_1 R_2}{R_1 + R_2} \, J$; $\ i_x(0) = 0$, $\ i_x(\infty)$

$= \dfrac{E}{R_1 + R_2} - \dfrac{R_2}{R_1 + R_2} J$; $T = \dfrac{L}{R_1 + R_2}$.

4.2 **(a)** $v_x(0) = (11/12)E$, $v_x(\infty) = E$; $i_x(0) = E/12$, $i_x(\infty) = 0$; $T = 16$.

(b) $v_x(0) = 0$, $v_x(\infty) = 0$; $i_x(0) = J$, $i_x(\infty) = J$; $T = 3$.

4.3 For $0 \le t \le 1$, $v_x = 0$. $1 \le t \le 2$, $v_x(t) = 5(1 - e^{-(t-1)})$. $2 \le t$, $v_x(x) = 3.16 e^{-(t-2)}$.

4.4 $v_x(t) = 3(1 - e^{-t/9})$, $0 \le t \le 20$; $v_x(t) = -3.99 e^{-(t-20)/9}$, $t > 20$.

4.5 $L = 0.92$ H.

4.6 $C = 0.11$ F.

4.7 $T_0 = 10.4 \times 10^{-5}$ sec.

4.8 $v_x(t)$ is periodic with period 20 sec. $v_x(0^+) = 15$ V, $v_x(10^-) = 5$ V, $v_x(10^+) = -15$ V, $v_x(20) = -5$ V.

4.9 R_1, R_2, and C in series. Voltage taken across R_2. $R_1 = 4R_2$, $C \ll 2 \times 10^{-9}$ F.

4.10 30 kΩ, C and 20 kΩ in series. Voltage taken across C and 20 kΩ combination. $C \ll 0.2 \times 10^{-9}$ F.

4.11 Choose R_1, R_2, R_3 such that $0.98 R_3 = 100(R_2 + 0.02 R_1)$ and C such that $R_3 C = 3.12 \, \mu$sec.

4.12 Choose R_1, R_2 such that $\dfrac{0.98E}{R_2 + 0.02 R_1} = 100 \times 10^{-3}$ A. Choose R_3, L such that time constant $T = 4.35 \times 10^{-6}$ sec.

4.13 $C = 7 \times 10^{-9}$ F.

4.14 $v(t) = \frac{5}{2} e^{-t}$; $v_c(t) = \frac{5}{2}(2 + t)e^{-t}$; $i_L(t) = -\frac{5}{2} t e^{-t}$.

4.15 $v_c(t) = 0.842 e^{-0.6t} \cos 3.155t$ $+ 0.138 e^{-0.6t} \sin 3.155t$, $i_L(t) = 0.098 e^{-0.6t}$ $\cos 3.155t - 2.71 e^{-0.6t} \sin 3.155t$, $v(t) = 0.1 i_L(t)$.

4.16 $v_C(t) = 1.9 \sin t - 1.745 \cos t$ $+ 2.8 e^{-2t} - 1.055 e^{-7/2t}$, $i_L(t) = 9.35$ $\sin t - 5.07 \cos t + 5.6 e^{-2t} - 0.529 e^{-7/2t}$.

4.17 $v_c(t) = -2 \cos(0.25 \times 10^7 t) + 1$, $i_L = 4 \times 10^{-4} \sin(0.25 \times 10^7 t)$.

4.18 *Case 1:* $R_2 = 1000 \ \Omega$, overdamped. $\lambda_1 = -8.95 \times 10^3$, $\lambda_2 = -55.55 \times 10^3$, $v_c(0) = 6.8$ V, $i_L(0) = 6.8 \times 10^{-3}$ A. *Case 2:* $R_2 = 1800 \ \Omega$, critically damped. $\lambda_1 = \lambda_2 = -22.9 \times 10^3$, $v_c(0) = 7.93$ V, $i_L(0) = 4.41 \times 10^{-3}$ A. *Case 3:* $R_2 = 100{,}000 \ \Omega$, underdamped. λ_1, $\lambda_2 = (-3.5 \pm j23.8)10^3$. $v_c(0) \cong 10$ V, $i_L(0) \simeq 0.1 \times 10^{-3}$ A.

4.19 $\lambda_1 = -2$, $\lambda_2 = -\frac{1}{2}$.

  (a) $v_1(t) = \frac{5}{3} e^{-2t} - \frac{2}{3} e^{-1/2t}$

        $v_2(t) = -\frac{5}{3} e^{-2t} - \frac{1}{3} e^{-1/2t}$

  (b) $v_1(t) = -\frac{1}{3} e^{-2t} + \frac{4}{3} e^{-1/2t}$

        $v_2(t) = \frac{1}{3} e^{-2t} + \frac{2}{3} e^{-1/2t}$

  (c) $v_1(t) = \frac{1}{3} e^{-2t} + \frac{2}{3} e^{-1/2t}$

        $v_2(t) = -\frac{1}{3} e^{-2t} + \frac{1}{3} e^{-1/2t}$.

4.20 $\mu = 10$, λ_1, $\lambda_2 = -1 \pm j20$; underdamped. $\mu = 12$, λ_1, $\lambda_2 = \pm j19.48$; voltages and currents are sinusoidal. Sustained oscillation takes place.

4.21 $C \cong 0.0087$ F.

4.22 $R = 0.5 \ \Omega$, $t_0 = 3.89$ sec.

4.23 $C = 0.01565$ F.

4.24 $R_1 = 10$ kΩ, $R_2 = 238 \ \Omega$, $C = 100 \times 10^{-9}$ F, $E = 11.5$ V.

Chapter 5

5.7 AND gate.

5.10 Third harmonic. $q = -\frac{3}{4}v + v^3$.

5.11 $v = 0.654367$ V, $i = 0.934564$ A.

5.12 $t = 0.1$ sec, $v = 0.049299$ V, $v_0 = 0.050536$ V. $t = 1.0$ sec, $v = 0.379669$ V, $v_0 = 0.461802$ V.

5.13 $t = 0.1$ sec, $v = 0.936930$ V, $i = 0.552135$ A. $t = 1.0$ sec, $v = 0.729562$ V, $i = 0.0741730$ A.

Chapter 6

6.1 $v_1 = \frac{26}{21}$ V, $v_2 = -\frac{124}{21}$ V.

6.2 12 Ω.

6.3 $v_{AB} = \frac{15}{4}$ V, $v_{AC} = -\frac{45}{4}$ V, $v_{CB} = 15$ V.

6.4 $E = 12.5$ V.

6.5 $R_2 = \frac{2}{3} \ \Omega$, $R_3 = 1 \ \Omega$.

6.6 $v_{AB} = -1$ V, $v_{AC} = 0$, $v_{CB} = -1$ V.

6.7 $i_1 = -0.342343$ A, $i_2 = 0.765765$ A.
6.10 Four constant-R bridged-T sections. $R_{A1} = R$, $R_{A2} = \frac{1}{2} R$, $R_{A3} = \frac{1}{3} R$, $R_{A4} = \frac{1}{4} R$.
6.13 $R_A = \frac{1}{3} R$.
6.14 $\frac{9}{13} J^2$ W.
6.15 12 mhos.
6.16 $i_1 = \frac{7}{37} J$, $i_2 = -\frac{18}{37} J$, $i_3 = -\frac{12}{37} J$.
6.17 $v(t) = \frac{3}{11} + \frac{156}{275} \exp(-\frac{11}{100} t)$.
6.18 $v(t) \equiv 0$.
6.19 $0 \le R < 3$, underdamped; $R = 3$, critically damped; $R > 3$, overdamped.
6.20 $E^2 R_2 \left[\dfrac{R_1}{N_1} + \dfrac{(N_1 - N_2)^2 R_3}{(N_1 - 1)^2 N_2} + N_2 R_2 \right]^{-2}$.
6.21 $R_1 = 10\ \Omega$, $R_2 = 84{,}100\ \Omega$.
6.22 2.5 mi, 35 Ω.
6.27 Power gain $= 10^5$.

Chapter 7

7.1 Impedances are: **(a)** $2 - j$; **(b)** $1/(2 - j)$.
7.2 Impedances are: **(a)** $(13 - 3j)/(9 + 2j)$; **(b)** $(-13 + 3j)/(-2 + 9j)$; **(c)** zero.
7.3 $14.75 \cos(2t + \theta)$, $\theta = -116.6°$; $0.149 \sin(2t + \theta')$, $\theta' = -71.6°$.
7.4 $149 \cos(2t + \theta)$, $\theta = -116.6°$; $0.149 \cos(2t + \theta')$, $\theta' = -56.6°$.
7.5 Zero.
7.6 $v_1(t) = 0.638 \sin(t + \theta_1) + 0.64 \cos(2t + \theta_2)$, $\theta_1 = 28.6°$, $\theta_2 = -0.61°$. $i_c(t) = 0.658 \sin(t + \theta_3) + 0.659 \cos(2t + \theta_4)$, $\theta_3 = 14.56°$, $\theta_4 = 14.31°$.
7.7 No.
7.8 $E = -2 + j2$.
7.10 **(a)** $I_x = 0$, $V_x = \dfrac{1}{-3 + j2} E$; **(b)** $I_x = \dfrac{j4}{j6 + 1} E$, $V_x = 0$; **(c)** $I_x = \dfrac{8 - j4}{1 + j6}$, $V_x = \dfrac{2 + j3}{-3 + j2}$.
7.11 **(c)** $I_x = \frac{1}{3} j$, $V_x = -\frac{1}{3} j$; **(d)** $I_x = \frac{1}{3}$, $V_x = -\frac{1}{3}$.
7.14 $v(t) = 0.5[\cos(t + \theta_1) + \cos(10t + \theta_2) + \cos(100t + \theta_3)]$, **(a)** $\theta_1 = -90°$, $\theta_2 = -168.6°$, $\theta_3 = -178.8°$. **(b)** $\theta_1 = 90°$, $\theta_2 = 11.4°$, $\theta_3 = 1.16°$.
7.16 **(a)** $v(t) = 0.354 \cos(t + \theta_1) + 0.0498 \cos(10t + \theta_2) + 0.005 \cos(100t + \theta_3)$; $\theta_1 = -45°$, $\theta_2 = -84.3°$, $\theta_3 = -89.4°$. **(b)** $v(t) = 0.354 \cos(t + \theta_1) + 0.498 \cos(10t + \theta_2) + 0.5 \cos(100t + \theta_3)$; $\theta_1 = 45°$, $\theta_2 = 5.7°$, $\theta_3 = 0.58°$.
7.18 $V(j0) = \frac{1}{2} E$, $V(j) = \dfrac{\sqrt{2}}{2} E e^{j\theta}$, $\theta = -135°$, $V(j10) \cong j\frac{1}{2} \times 10^{-3} E$.
7.19 No.
7.20 $V(j100) \cong -j5 \times 10^{-4} E$; $V(j1000) = \dfrac{\sqrt{2}}{2} E e^{j\theta}$, $\theta = 135°$; $V(j10^4) \simeq 0.5$.
7.21 $V(j10^5) = 0.099 e^{j\theta}$, $\theta = 78.6°$; $V(j10^6) = 0.5E$; $V(j10^7) = 0.099 e^{j\theta}$, $\theta = -78.6°$.
7.25 $V(j10^6) \simeq j2455J$.
7.26 Power to load $= 4 \times 10^{-4}$ W, power factor $= 0.6$.
7.27 18.75 μF.
7.28 12.5×10^{-4} W.
7.29 $W_c = W_L = 1.56 \times 10^{-8}$ J.
7.30 $M = 1$ H.
7.33 $g_m = -\dfrac{L_1 G}{M}$, $\omega = \dfrac{1}{\sqrt{L_1 C}}$.

Chapter 8

8.1 **(a)** $X(\omega) = \omega(2 - \omega^2)(1 - \omega^2)^{-1}$, **(b)** $X(\omega) = -(1 - \omega^2)\omega^{-1}(2 - \omega^2)^{-1}$.
8.6 $\omega_0 = 1000\pi$. Bandwidth $= 50$ rad/sec.
8.7 $\sin 1000\pi t$ component eliminated.
8.8 Bandwidth $= 0.1\pi \times 10^6$ rad/sec.
8.11 $L = 44.1$ mH, $C = 0.245\ \mu$F.
8.13 $v(t) \simeq \frac{1}{2} \cos 20t + 4 \times 10^{-3} \cos(0.2t + \theta)$, $\theta = -113.1°$.
8.15 $L = 0.515$ mH.
8.17 $L = 86\ \mu$H, $C = 172\ \mu\mu$F.
8.21 $C_1 = 0.159\ \mu$F, $L_1 = 1.59$ mH, $C_2 = 795\ \mu\mu$F, $L_2 = 0.318$ H.
8.23 $L = 50\ \mu$H, $C = 16.7\ \mu\mu$F.

8.24   7 sections, $L = 79.5\ \mu$H, $C = 26.5$ $\mu\mu$F.

8.29   Poles at $0, \pm j\sqrt{3}$; zeros at $\pm j\dfrac{1}{\sqrt{3}}$.

Chapter 9

9.1 $\lambda_1 = -2$, $\lambda_2 = -4$.

9.2 $v(t) = \frac{2}{13}\, v_1 + \frac{3}{13}\, v_2 + \frac{8}{13}\, J$; $\lambda_1 = -0.441$, $\lambda_2 = -0.1745$.

9.3 $\lambda_1 = 0$, $\lambda_{2,3} = \pm j\sqrt{2}$, $\lambda_{4,5} = \pm j2$.

9.4 $\lambda_{1,2} = \pm j0.664$, $\lambda_{3,4} = \pm j0.418$.

9.5 $\dfrac{dv_1}{dt} = -\frac{4}{15}\, v_1 - \frac{1}{15}\, v_2$, $\dfrac{dv_2}{dt} = -\frac{1}{15}$ $v_1 - \frac{4}{15}\, v_2$.

9.6 $\dfrac{di_1}{dt} = -\frac{1}{3}\, i_1$, $\dfrac{di_2}{dt} = -\frac{1}{3}\, i_2$.

9.7 $\dfrac{dv_1}{dt} = -\dfrac{1}{1+\alpha}\, v_1 + \dfrac{1-\alpha}{1+\alpha}\, v_2 + \dfrac{\alpha}{1+\alpha}\, E$, $\dfrac{dv_2}{dt} = \dfrac{0.5}{1+\alpha}\, v_1 - \dfrac{1}{1+\alpha}\, v_2 + \dfrac{0.5}{1+\alpha}\, E$

9.8 $\dfrac{dv_1}{dt} = 0.5 i_2 + 0.5 i_3$, $\dfrac{di_2}{dt} = -v_1 - 2 i_2$ $+ i_3 + 2E$, $\dfrac{di_3}{dt} = i_2 - i_3 - E$.

9.9 $v(t) = \frac{1}{8} - \frac{1}{2}\, e^{-2t} + \frac{3}{8}\, e^{-4t}$.

9.10 $i(t) = \frac{3}{8} + \frac{1}{4}\, e^{-2t} + \frac{3}{8}\, e^{-4t}$.

9.11   Characteristic equation $\lambda^2 + \lambda + \frac{1}{6} = 0$.

9.13 $v(t) = \dfrac{1}{2}\, e^{-t} - \dfrac{1}{\sqrt{3}}\, e^{-(1/2)t} \sin \dfrac{\sqrt{3}}{2}\, t$.

9.16 $v(t) = e^{-2t} - 2t e^{-2t}$.

9.19 $v(t) = 1 - e^{-(1/2)t}\, \left(1 + \frac{1}{2}t + \frac{1}{8}t^2\right)$.

9.20 $v(t) = 1 - e^{-2t} - 2t e^{-t}$.

CHAPTER 10

10.1   (a) $y_{11} = 3$, $y_{12} = -2$, $y_{13} = -1$,

$y_{22} = 5$, $y_{23} = -3$. (b) $y_{11} = \dfrac{3}{2}$, $y_{12} = -\dfrac{1}{2}$, $y_{13} = -1$, $y_{22} = \dfrac{5}{6}$, $y_{23} = -\dfrac{1}{3}$. (c) $y_{11} = \dfrac{9}{2}$, $y_{12} = -\dfrac{5}{2}$, $y_{13} = -2$, $y_{22} = \dfrac{35}{6}$, $y_{23} = -\dfrac{10}{3}$.

(d) $y_{11} = \dfrac{35}{6}$, $y_{12} = -\dfrac{7}{2}$, $y_{13} = -\dfrac{7}{3}$, $y_{22} = \dfrac{11}{2}$, $y_{23} = -2$.

10.2 $y_{bb} = \dfrac{(1.01 - \alpha)}{d}$, $y_{cb} = \dfrac{(-0.01 + \alpha)}{d}$, $y_{eb} = \dfrac{-1}{d}$, $d = 10(2.01 - \alpha)$.

10.3 -16.25 for $\alpha = 1$.

10.4   Gain $= \dfrac{(1 - \alpha)\, G_F d + \alpha^2}{(1 - \alpha)(G_F + G_0)d}$, $d = (20 - 10\alpha)$.

10.8 $\omega^2 = \dfrac{(1 - \alpha)}{LC_1}$.

10.10 $A = \frac{4}{5}$, $B = \frac{3}{10}\ \Omega$, $C = 2$ mhos, $D = 2$.

10.12 $V_1 = \frac{144}{5}$ V.

10.13   No. No.

10.17 $Z_{\text{in}} = -\dfrac{D - A}{2C} + \left[\dfrac{(D - A)^2}{4C^2} + \dfrac{B}{C}\right]^{1/2}$, $\dfrac{V_2}{V_1} = \dfrac{1}{A + B/Z_{\text{in}}}$.

10.18 $\dfrac{V_0}{E} = \dfrac{1}{1 - A_0 - B_0 C_0 / (1 - D_0)}$. $A_0 = \frac{5}{12}\,(1 + \frac{7}{5}3^{100})$. $B_0 = \frac{5}{12}\,(-1 + 3^{100})$, $C_0 = \frac{5}{12}\,(-\frac{1}{5} + \frac{7}{5}3^{100})$, $D_0 = \frac{5}{12}\,(\frac{7}{5} + 3^{100})$.

INDEX

FALL 77

INVENTORY 1983